Acoustical Imaging

Volume 13

Acoustical Imaging

A Continuation Order Plan is available for this series. A continuation order will bring
delivery of each new volume immediately upon publication. Volumes are billed only upon
actual shipment. For further information please contact the publisher.

Acoustical Imaging

Volume 13

Edited by

M. Kaveh
R. K. Mueller

University of Minnesota
Minneapolis, Minnesota

and

J. F. Greenleaf

Mayo Clinic
Rochester, Minnesota

PLENUM PRESS · NEW YORK AND LONDON

The Library of Congress cataloged the first volume of this series as follows:

International Symposium on Acoustical Holography.

  Acoustical holography; proceedings. v. 1–
New York, Plenum Press, 1967–

  v. illus. (part col.), ports. 24 cm.

  Editors: 1967–   A. F. Metherell and L. Larmore (1967 with H. M. A. el-Sum)
  Symposium for 1967–   held at the Douglas Advanced Research Laboratories,
Huntington Beach, Calif.

  1. Acoustic holography—Congresses—Collected works. I. Metherell. Alexander A.,
ed. II. Larmore, Lewis, ed. III. el-Sum, Hussein Mohammed Amin, ed. IV. Douglas
Advanced Research Laboratories, v. Title.
QC244.5.I.5   69-12533

ISBN-13: 978-1-4612-9715-4   e-ISBN-13: 978-1-4613-2779-0
DOI: 10.1007/978-1-4613-2779-0

Proceedings of the Thirteenth International Symposium on
Acoustical Imaging, held October 26–28, 1983, in
Minneapolis, Minnesota

© 1984 Plenum Press, New York
Softcover reprint of the hardcover 1st edition 1984
A Division of Plenum Publishing Corporation
233 Spring Street, New York, N.Y. 10013

PREFACE

This volume constitutes the proceedings of the Thirteenth International Symposium on Acoustical Imaging which was held in Minneapolis, Minnesota during October 26-28, 1983. Forty-eight research papers were presented during the meeting by researchers from twelve countries, again demonstrating the true international character of these meetings. Of these presentations this volume contains forty-two complete manuscripts. The abstracts for additional papers that were not available at publication time are also included.

According to the recent tradition of these symposia an interdisciplinary program under the general theme of acoustical imaging was organized. This can clearly be observed from the wide range of topics and approaches contained in the following manuscripts. There are papers of mathematical nature dealing with the basis of image formation and algorithms for digitally carrying out specific imaging tasks. One finds manuscripts dealing with the design and construction of imaging transducers as well as complete imaging systems. Applications include medical imaging and nondestructive testing, seismic and underwater imaging. This volume, therefore, should be of interest to active researchers in acoustical imaging as a report on current research and to workers in signal processing, sonics and ultrasonics who are interested in exploring the diverse areas of application for their fields of interest.

These proceedings are organized in seven topical sections, paralleling the sessions of the conference. These are: Inversion and Tomography, Microscopy, Scattering and Propagation, Tissue and Material Characterization, Signal Processing, Transducers and Arrays, Imaging Systems and Special Techniques.

The editors would like to express their gratitude to many individuals and several organizations. Foremost, our thanks to the authors for their excellent contributions and cooperation in providing us with camera-ready drafts of their manuscripts in a prompt fashion. Medical ultrasound pioneer, Dr. J. Wild, delivered the opening lecture of the symposium with an historical overview of ultrasonic visualization techniques and their impact on biomedical investigation. The session chairmen were instrumental in the smooth running of the conference. These gentlemen were Drs. J. Ridder, G. Wade, R. Waag, J. Jones, O. Tretiak, L. Hargrove, J. Powers, W. Robbins and P. Alais. Members of the International Committee Drs. Ridder, Waag, Alais, Powers, van der Wal, Ermert, Ash, Hill, Chubachi and Burckhardt aided us with the review of abstracts and the planning of future symposia. Professor M. Soumekh helped with the preparation of the abstract booklet, and Ms. Betsy Rhame provided expert editorial consulting in the review of the completed manuscripts. The symposium was generously supported by the Physics Program of the Office of Naval Research, directed by Dr. Logan Hargrove, under Grant #N00014-83-C-0019. The logistical organization of the meeting was facilitated by the Department of Conferences, University of Minnesota, under the coordination of Ms. Leslie Denny. The symposium was co-sponsored by the Institute of Electrical and Electronic Engineers, Twin Cities Section.

Finally, the editors would like to apologize for any uncorrected typographical errors and nonuniformities in the format and style of the manuscripts in this volume. It has been decided by consensus that rapid publication of the proceedings of these symposia is preferable to attempted editorial perfection. The Fourteenth International Symposium on Acoustical Imaging will be held during April 22-25, 1985 at the Hague, the Netherlands, under the Chairmanship of Drs. Berkhout, Ridder and van der Wal.

M. Kaveh and R. K. Mueller   J. F. Greenleaf
Minneapolis, Minnesota   Rochester, Minnesota

CONTENTS

INVERSION AND TOMOGRAPHY

MICROSCOPY

SCATTERING AND PROPAGATION

TISSUE AND MATERIAL CHARACTERIZATION

SIGNAL PROCESSING

SPECIAL TECHNIQUES

INVERSE SOLUTIONS FOR MULTIPLE SCATTERING

IN INHOMOGENEOUS ACOUSTIC MEDIA

J. M. Blackledge and L. Zapalowski

Physics Department, Queen Elizabeth College
University of London, Campden Hill Road
Kensington, London W8 7AH UK

INTRODUCTION

Recent advances in the theory of quantitative acoustic scatter imaging have been made under the assumption that: (i) the acoustic continuum has uniform, frequency dependent absorption characteristics and (ii) the scattering of acoustic radiation from material inhomogeneities is weak enough for the first Born or Rytov approximations to hold. These scatter imaging techniques all depend on extensive computation with aquired data bases in order to generate the final image. Jones et al. (1982) have demonstrated that the final result and hence, image fidelity, is highly dependant on the physical model assumed for the propogation and scattering of acoustic waves in the material under investigation. The possibility that even high resolution data aquisition and computational methods will lead to manifestly inaccurate and semi-quantitative images has been called image 'fuzziness'.

In this paper, we demonstrate that a solution to the inverse problem exists for an acoustic medium with spatial variations in absorption arising from viscous relative motion. The solution is also valid for multiple scattering under the assumption that the wavelength of radiation has a magnitude that is small compared with the inter-scattering distance. Finally, we look at seismic methods and discuss the one dimensional inverse scattering problem for a stratified continuum model.

ACOUSTIC FIELD EQUATIONS

The field equations for an acoustic continuum can be specified by three conversation equasions for mass, momentum and energy which can be written in the following form

1

$$\frac{\partial \rho}{\partial t} + \nabla \cdot (\rho \underline{v}) = 0 \tag{1}$$

$$\rho \hat{D} \underline{v} + \nabla \cdot \hat{\underline{\underline{T}}} = 0 \tag{2}$$

$$\frac{\partial}{\partial t} \left(\tfrac{1}{2}\rho v^2 + U \right) + \nabla \cdot \left(\tfrac{1}{2}\rho v^2 \underline{v} + U\underline{v} + \hat{\underline{\underline{T}}} \cdot \underline{v} + \underline{q} \right) = 0 \tag{3}$$

where $\hat{\underline{\underline{T}}}$ is the total material stress tensor, ρ is the material density, $\underline{v}$ is the acoustic velocity vector field, U is the internal energy which we write as $U = \tfrac{1}{2}\mathrm{trace}\underline{\underline{P}} = 3p/2$ where p is the scalar pressure, $\hat{D}$ is the convective derivative and $\underline{q}$ is the heat flux. In many applications we can assume that there is no heat flux. The energy conservation equation is then specified by the adiabatic equation of state

$$c^2 \hat{D}\rho = \hat{D}p \tag{4}$$

where c is the acoustic wave speed given by $c^2 = 1/\rho\kappa$ and κ is the compressibility. However, we shall investigate the case where an adiabatic equation of state is not applicable and generate a fourth dynamical equation which is coupled to Eq.(3) via the heat flux and describes the irreversible thermodynamic nature of an acoustic continuum.

IRREVERSIBLE THERMODYNAMICS OF AN ACOUSTIC CONTINUUM

Let us consider the Gibbs' equation which is given by

$$\frac{TdS}{dt} = \frac{dU}{dt} - \frac{pdV}{dt} \tag{5}$$

where S, V and T are the entropy, volume and temperature per unit mass respectively. Changing to quantities per unit volume, we can write

$$\frac{TdS}{dt} = \frac{dU}{dt} - \frac{\mu d\rho}{dt} \tag{6}$$

where μ is the chemical potential defined by

$$\text{Gibbs' potential} = \rho\mu = U + p - TS \tag{7}$$

If we now take the energy equation (3) $- v$. momentum equation (2) $+ \tfrac{1}{2}v^2$ mass conservation equation (1) we can write

$$\frac{\partial U}{\partial t} = -\nabla \cdot (U\underline{v} + \underline{q}) - (\hat{\underline{\underline{T}}} \cdot \nabla) \cdot \underline{v} \tag{8}$$

and if we now employ the mass conservation equation $d\rho/dt = -\rho\nabla \cdot \underline{v}$, then from Eq.(6) we have

$$T \frac{dS}{dt} + T \underline{v} \cdot \nabla S = - U \nabla \cdot \underline{v} - \nabla \cdot \underline{q} - (\hat{\underline{\underline{T}}} \cdot \nabla) \cdot \underline{v} + \mu \rho \nabla \cdot \underline{v} \tag{9}$$

We now split the total material stress tensor $\hat{\underline{\underline{T}}}$ into $\hat{\underline{\underline{T}}} = \underline{\underline{I}} p + \underline{\underline{\tau}}$ where $\underline{\underline{I}}$ is the unit tensor and $\underline{\underline{\tau}}$ is the material stress tensor. The term $(\hat{\underline{\underline{T}}} \cdot \nabla) \cdot \underline{v}$ is then given by

$$(\underline{\underline{T}} \cdot \nabla) \cdot \underline{v} = p \nabla \cdot \underline{v} + (\underline{\underline{\tau}} \cdot \nabla) \cdot \underline{v} \tag{10}$$

and represents the reversible compressional heating described by the term $p \nabla \cdot \underline{v}$ and the irreversible viscous heating described by the term $(\underline{\underline{\tau}} \cdot \nabla) \cdot \underline{v}$. We note, that the local internal energy changes as a result of convection, heat flow, reversible compressional heating or cooling by expansion and through viscous and frictional heating. Using the definition of μ we can now manipulate Eq.(9) into the form

$$\frac{\partial S}{\partial t} + \nabla \cdot \left(\frac{U \underline{v}}{T} + P \frac{\underline{v}}{T} + \frac{\underline{q}}{T} - \rho \mu \frac{\underline{v}}{T} \right) = \underline{q} \cdot \nabla (1/T) - \underline{\underline{\tau}} : \frac{\nabla \underline{v}}{T} \tag{11}$$

and using Eq.(7) we can identify the entropy flux as

$$\underline{S}_f = S \underline{v} + \underline{q}/T \tag{12}$$

and write

$$\frac{\partial S}{\partial t} + \nabla \cdot \underline{S}_f = - \sigma \tag{13}$$

where σ is the entropy source term defined by

$$\sigma = (\underline{\underline{\tau}} : \nabla \underline{v} + \underline{q} \cdot \nabla \ln T) / T \tag{14}$$

This equasion shows, by contrast to the conversation equations for the mass, momentum and energy, that the entropy is not necesserily conserved, i.e. it is possible for σ to be non zero. Furthermore, the equation allows a mathematical statement for the second law of thermodynamics, namely

$$\sigma \geqslant 0 \tag{15}$$

Finally, we can define the material stress tensor $\underline{\underline{\tau}}$ in terms of the Navier-Stokes velocity stress tensor given by

$$\underline{\underline{\tau}} = \underline{\underline{I}} \lambda \nabla \cdot \underline{v} + \mu (\nabla \underline{v} + \underline{v} \nabla) \tag{16}$$

where λ and μ are the elastic Lamé parameters. The complete set of acoustic field equations can then be written as

$$\frac{\partial \rho}{\partial t} + \nabla \cdot (\rho \underline{v}) = 0 \tag{17}$$

$$\rho \hat{D} \underline{v} = - \nabla p + \nabla \cdot [\underline{\underline{I}} \lambda \nabla \cdot \underline{v} + \mu (\nabla \underline{v} + \underline{v} \nabla)] \tag{18}$$

$$\frac{\partial}{\partial t} \left(\tfrac{1}{2} \rho v^2 + \tfrac{3}{2} p \right) + \nabla \cdot [\tfrac{1}{2} \rho v^2 \underline{v} + \tfrac{3}{2} p \underline{v} + \underline{v} \lambda \nabla \cdot \underline{v} + \mu \underline{v} \cdot (\nabla \underline{v} + \underline{v} \nabla)$$
$$+ \underline{q}] = 0 \tag{19}$$

$$\frac{\partial S}{\partial t} + \nabla \cdot (S\underline{v} + \underline{q}/T) = -\frac{1}{T}\left[\underline{\underline{I}}\,\lambda\nabla\cdot\underline{v}:\nabla\underline{v} + \mu(\nabla\underline{v}+\nabla\underline{v}):\nabla\underline{v} + \underline{q}\cdot\nabla\ell nT\right] \quad (20)$$

These field equations represent the complete dynamical behaviour of an acoustic continuum supporting acoustic waves. In this form, the field parameters are strongly coupled which necessitates numerical procedures. Preliminary results on numerical solutions to these equations will be published elsewhere.

ACOUSTIC SCATTERING EQUATION

We shall consider an inhomogeneous acoustic continuum model based on the following equations of motion (Blackledge et al. 1982)

$$\frac{\partial p}{\partial t} = \kappa^{-1}\nabla\cdot\underline{v} \quad (21)$$

$$\rho\frac{\partial \underline{v}}{\partial t} = -\nabla p + \nabla\cdot[\lambda\underline{\underline{I}}\,\nabla\cdot\underline{v} + \mu(\nabla\underline{v} + \underline{v}\nabla)] \quad (22)$$

Using equations (21) and (22) we can generate the longitudinal acoustic wave equation which, for a spherical wave source, is given by (Blackledge et al. 1983)

$$\nabla[\nabla^2\phi(\underline{r}|\underline{r}_0,\omega) + \xi^2\phi(\underline{r}|\underline{r}_0,\omega)] = \hat{\mathcal{L}}\phi(\underline{r}|\underline{r}_0,\omega) - \underline{\hat{n}}\,\delta^3(\underline{r}-\underline{r}_0) \quad (23)$$

where ϕ is the velocity potential

$$\hat{\mathcal{L}} = -\xi^2\gamma_\rho\nabla - \xi^2 c_0^2\omega^{-2}\nabla\gamma_\kappa\nabla^2 - \frac{i\xi^2\lambda_0}{\rho_0\omega}\nabla\gamma_\lambda\nabla^2 + 2i\frac{\xi^2\mu_0}{\rho_0\omega}\nabla\cdot\gamma_\mu\nabla\nabla \quad (24)$$

$$\frac{\omega}{c_0}(1+\omega^2\tau_0^2)^{-\frac{1}{4}}exp(\frac{i}{2}arctan\,\omega\tau_0) \quad (25)$$

and

$$\tau_0 = (\lambda_0 + 2\mu_0)/\rho_0 c_0^2 \quad (26)$$

The solution to Eq.(23) for the condition

$$\langle \gamma_f \rangle = f_0 \quad (27)$$

at a position vector $\underline{r}_s$ is then given by

$$\nabla\phi(\underline{r}_s|\underline{r}_0,\omega) = \underline{\hat{n}}\cdot\underline{\underline{G}}(\underline{r}_s|\underline{r}_0,\omega) + \int_{\mathcal{D}^3} d^3\underline{r}\,\hat{\mathcal{L}}\phi(\underline{r}|\underline{r}_0,\omega)\cdot\underline{\underline{G}}(\underline{r}|\underline{r}_s,\omega) \quad (28)$$

where $\mathcal{D}$ is the scattering domain and $\underline{\underline{G}}$ is the Green's dyadic which is the solution of

$$(\nabla^2 + \xi^2)\, \underline{G}(\underline{r}|\underline{r}_j, \omega) = \hat{\underline{n}}_j \hat{\underline{n}}_j\, \delta^3(\underline{r} - \underline{r}_j)\,;\, j = 0, s \quad (29)$$

We can now proceed to generate the Born series for Eq.(28) by substituting for the incident field

$$\nabla\phi(\underline{r}|\underline{r}_0, \omega) = \hat{\underline{n}} \cdot \underline{G}(\underline{r}|\underline{r}_0, \omega) \quad (30)$$

which yields an iterative solution of the form (Blackledge, 1983)

$$d(\underline{r}_s|\underline{r}_0, \omega) = \sum_{n=1}^{\infty} \prod_{j=1}^{n} \lambda^{(n)} \int_{\mathcal{D}_j^3} d^3\underline{r}_j\, f(\underline{r}_j, \theta_j, \omega)\, g(\underline{r}_j|\underline{r}_{j-1})\, g(\underline{r}_n|\underline{r}_s)$$

for the condition \hspace{6cm} (31)

$$|\underline{\xi}|\,|\underline{r}_j - \underline{r}_{j-1}| \gg 1$$

where

$$\lambda^{(n)} = (-1)^n i\, \xi^{2n+1}(1 - i\omega\tau_0)^{-n} \quad (32)$$

and

$$d(\underline{r}_s|\underline{r}_0, \omega) = \nabla^2\phi(\underline{r}_s|\underline{r}_0, \omega) \quad (33)$$

and $f(\underline{r}_j, \theta_j, \omega)$ is the complex scatter generating function given by

$$f(\underline{r}_j, \theta_j, \omega) = c(\underline{r}_j, \theta_j) + i\omega\tau_0 \alpha(\underline{r}_j, \theta_j) \quad (34)$$

where

$$c(\underline{r}_j, \theta_j) = \gamma_\kappa(\underline{r}_j) + \cos\theta_j\, \gamma_\rho(\underline{r}_j) \quad (35)$$

and

$$\alpha(\underline{r}_j, \theta_j) = \gamma_\rho(\underline{r}_j)\cos\theta_j + [\lambda_0\gamma_\lambda(\underline{r}_j) + 2\mu_0\gamma_\mu(\underline{r}_j)\cos\theta_j]/(\lambda_0 + 2\mu_0) \quad (36)$$

The field d is the acoustic dilatation which represents the imaging data base and is defined in terms of the acoustic pressure by

$$d(\underline{r}_s|\underline{r}_0)/d_0(\underline{r}_s|\underline{r}_0) = p(\underline{r}_s|\underline{r}_0)/p_0(\underline{r}_s|\underline{r}_0) \quad (37)$$

where d_0 and p_0 are the dilatation and pressure fields respectively in the absence of any acoustic scatter generating parameters.

INVERSE SOLUTION FOR MULTIPLE SCATTERING

We now solve Eq.(31) for $f(\underline{r}, \theta, \omega)$ by applying the procedure first adopted by Jost and Kohn (1952). Thus, if we substitute

$$d = \epsilon d_\epsilon \qquad (38)$$

and

$$f = \sum_{k=1}^{\infty} \epsilon^k f_k \qquad (39)$$

into Eq.(31) and equate powers of ϵ, we obtain

$$k = 1 \quad d_\epsilon(r_s|r_o) = \lambda^{(1)} \int_{\mathcal{D}_1^3} f_1(r_1, \theta_1) g(r_1|r_o) g(r_s|r_1) d^3 r_1 \qquad (40)$$

$$k = 2 \quad \lambda^{(1)} \int_{\mathcal{D}_1^3} f_2(r_1, \theta_1) g(r_1|r_o) g(r_1|r_s) d^3 r_1$$

$$= \lambda^{(2)} \int_{\mathcal{D}_1^3} \int_{\mathcal{D}_2^3} d^3 r_1 d^3 r_2 f_1(r_1, \theta_1) g(r_1|r_o) f_1(r_2, \theta_2) g(r_1|r_2) g(r_2|r_s)$$

$$(41)$$

and so on for k=3 to ∞. The solution is then given by

$$f(r, \theta, \omega) = \sum_{k=1}^{\infty} \epsilon^k f_k(r, \theta, \omega) \qquad (42)$$

Equation (42) then allows us to calculate $f(r,\theta,\omega)$ to the desired accuracy by iteration. Clearly, to obtain $f_k(r,\theta,\omega)$ we require the exact solution to an equation of the form

$$\int_{\mathcal{D}^3} f_k(r, \theta, \omega) g(r|r_o) g(r|r_s) d^3 r = I_k(r_s|r_o, \omega) \qquad (43)$$

The fundamental difference between this equation and earlier equations that have been considered by other authors is that the scatter generating function is now both spatially and frequency dependent. This means that inversion techniques which have been developed by Ball et al. (1979) and Norton and Linzer (1981) are not directly applicable. Details of exact inverse solutions to Eq.(43) that generate a frequency dependent scatter generating function will be published elsewhere.

3D PROFILE EXTRACTION

The angular and frequency dependence of the scatter generating function given in Eq.(34) allows us to extract the individual profiles γ_ρ, γ_k, γ_λ and γ_μ by constructing a data base of the form

$$d_{nm} = d(r_s | r_o, \theta_n, \omega_m); \quad n=1,2; \quad m=1,2$$

which, via a suitable inversion algorithm will map the scatter generating function

$$f_{nm} = f(r, \theta_n, \omega_m); \quad n=1,2; \quad m=1,2$$

Using a data base of this form, we can then construct a set of linear simultaneous equations given by

$$
\begin{bmatrix} f_{11} \\ f_{12} \\ f_{21} \\ f_{22} \end{bmatrix}
=
\begin{bmatrix}
1 & (1-i\omega_1\tau_0)\cos\theta_1 & c_1\omega_1 & c_2\omega_1\cos^2\theta_1 \\
1 & (1-i\omega_2\tau_0)\cos\theta_1 & c_1\omega_2 & c_2\omega_2\cos^2\theta_1 \\
1 & (1-i\omega_1\tau_0)\cos\theta_2 & c_1\omega_1 & c_2\omega_1\cos^2\theta_2 \\
1 & (1-i\omega_2\tau_0)\cos\theta_2 & c_1\omega_2 & c_2\omega_2\cos^2\theta_2
\end{bmatrix}
\begin{bmatrix} \gamma_k \\ -\gamma_\rho \\ \gamma_\lambda \\ \gamma_\mu \end{bmatrix}
\quad (44)
$$

where

$$c_1 = i\tau_0\lambda_0 / (\lambda_0 + 2\mu_0) \quad (45)$$

and

$$c_2 = 2c_1\mu_0/\lambda_0 \quad (46)$$

The solution to these equations then yields the four parameters γ_k, γ_ρ, γ_λ and γ_μ. Thus, by recording the scattered acoustic field at two different angles of incidence and two different incident frequencies, we can in principle generate a quantitative image of the four intrinsic scatter generating parameters.

SEISMIC METHODS: 1D ANALYSIS

The analysis presented so far, has been dependent on considering a three dimensional continuum model. We shall now demonstrate how inverse solutions may be developed for a continuum which is modelled as a stratified or layered medium. This approach is strongly related to seismic methods applied to data obtained from backscattered fields. We shall consider the propagation of a longitudinal wave in a stratified medium with spatial variations in density and compressibility. For the time harmonic dependence $\exp(i\omega t)$, the hydrodynamic equations of motion show that the pressure field $p(y,z,k)$ satisfies the equation

$$\left(\frac{\partial^2}{\partial y^2} + \frac{\partial^2}{\partial z^2} + k^2\rho(z)\kappa(z)\right)p(y,z,k) = \frac{d}{dz}\ln\rho(z)\frac{\partial}{\partial z}p(y,z,k) \quad (47)$$

where $k = \omega/c_o$ is the free space wave number, $\rho(z)$ is the density profile and $\kappa(z)$ is the compressibility profile. Equation (47) can now be transformed into the one dimensional Schrödinger equation by writing

$$p(y, z, k) = \phi(z, k) \exp(-iky \sin\theta); \quad z \geqslant 0 \quad (48)$$

and

$$ds/dz = \rho \qquad (49)$$

where s is the 'apparent depth', which gives

$$\left(\frac{\partial^2}{\partial s^2} + k^2 \chi^2(s, \theta) \right) \phi(s, k) = 0 \qquad (50)$$

where

$$\chi^2(s, \theta) = [\rho(s) K(s) - \sin^2\theta]/\rho^2(s) \qquad (51)$$

The usual method of generating the Schrödinger equation is to now write $f(s,\theta) = 1 - \chi^2(s,\theta)$ so that Eq.(50) becomes

$$\left(\frac{d^2}{ds^2} + k^2 \right) \phi(s, k) = k^2 f(s, \theta) \phi(s, k) \qquad (52)$$

This equation resembles the one dimensional Schrödinger equation with the exception that the scatter generating function depends quadratically on the wave number k. For fixed k, the behaviour of the inverse solutions to the Schrödinger equation and Eq.(52) are identical but for k dependence, the normalization conditions and also the spectral function of the Sturm-Liouville equation are changed. This difference renders the Gel'fand-Levitan algorithm inapplicable. A method of overcoming this difficulty is to use the Liouville transform

$$\begin{rcases} \gamma(x, k) = g(s, \theta) \phi(s, k) \\ ds/dx = g^{-2}(s, \theta) \\ g(s, \theta) = \chi^{\frac{1}{2}}(s, \theta) \end{rcases} \qquad (53)$$

which now gives the one dimensional Schrödinger equation

$$\left(\frac{d^2}{dx^2} + k^2 \right) \gamma(x, k) = f(x) \gamma(x, k) \qquad (54)$$

where the acoustic scatter generating function $f(x,\theta)$ is given by

$$f(x, \theta) = g^{-1}(x, \theta) \frac{\partial^2 g(x, \theta)}{\partial x^2} \qquad (55)$$

The problem has thus been reduced to that of finding $f(x,\theta)$ in terms of the wave field $\Psi(x,k)$ which, for an incident plane wave $\exp(ikx)$ satisfies the integral equation

$$\Psi(x',k) = e^{ikx'} - \frac{i}{2|k|} \int_{-\infty}^{\infty} e^{i|k||x-x'|} f(x)\Psi(x,k)dx$$

(56)

This solution represents the 'outgoing wave solution' and has the asymptotic form

$$\Psi(x',k) = e^{ikx'} + r(k)e^{-ikx'}; \quad x' \to \infty \quad (57)$$

where $r(k)$ is the reflection coefficient given by

$$r(k) = \frac{-i}{2|k|} \int_{-\infty}^{\infty} e^{ikx} f(x)\Psi(x,k)dx \quad (58)$$

Following Razavy (1975), we can now write the basic equations of the inverse problem as

$$T(p,k) = f(p,k) - \int_{-\infty}^{\infty} \frac{dq}{q^2-k^2-i\epsilon} f(p,q)T(q,k)dq \quad (59)$$

where

$$f(p,k) = \int_{-\infty}^{\infty} e^{-ipx} f(x)e^{ikx} dx \quad (60)$$

and

$$T(p,k) = \int_{-\infty}^{\infty} e^{-ipx} f(x)\Psi(x,k) dx \quad (61)$$

$$f(p,k) = T(p,k) + \int_{-\infty}^{\infty} \frac{dq}{q^2-k^2-i\epsilon} T^*(k,q)T(p,q) \quad (62)$$

$$W(k) = \frac{ir(k)}{2k} + \int_{-\infty}^{\infty} dq\, T^*(k,q)T(-k,q)$$

$$\cdot [H(k)(q^2-k^2-i\epsilon)^{-1} + H(-k)(q^2+k^2+i\epsilon)^{-1}]$$

(63)

where

$$W(k) = f(-k, k) = \frac{1}{2\pi} \int_{-\infty}^{\infty} f(x) e^{2ikx} dx \qquad (64)$$

and $H(k)$ is the unit step function and

$$f(k, k') = W\left(\frac{k'-k}{2}\right) \qquad (65)$$

The iterative method of solution is then obtained using the Jost and Kohn procedure by replacing $r(k)$ by $\varepsilon r(k)$ and writing

$$
\left.
\begin{aligned}
W(k) &= \sum_{n=1}^{\infty} \varepsilon^n W_n(k) \\
T(k, k') &= \sum_{n=1}^{\infty} \varepsilon^n T_n(k, k') \\
f(k, k') &= \sum_{n=1}^{\infty} \varepsilon^n f_n(k, k')
\end{aligned}
\right\} \qquad (66)
$$

Substituting these equations into equations (59), (62) and (65) and equating powers of ε gives the desired solution where the function $f(x)$ is given by

$$f(x) = \sum_{n=1}^{\infty} \varepsilon^n \int_{-\infty}^{\infty} W_n(k) e^{-2ikx} dk \qquad (67)$$

Equations (63) and (67) enable us to calculate $f(x)$ to the desired accuracy by iteration. Alternatively, we can calculate $f(x)$ from $r(k)$ by using the relationship

$$\frac{1}{2\pi} \int_{-\infty}^{\infty} k^2 e^{-ikx} \left[\frac{H(k)}{q^2-k^2-i\varepsilon} + \frac{H(-k)}{q^2-k^2-i\varepsilon} \right] = \frac{1}{q} \frac{d^2}{dx^2} [H(-x)\sin(qx)] \qquad (68)$$

Multiplying Eq. (63) by $2k^2 \exp(-2ikx)$, integrating over k from $-\infty$ to $+\infty$ and simplifying the terms, we can write

$$\frac{1}{2} \frac{d^2}{dx^2} f(x) = -\frac{d}{dx} \int_{-\infty}^{\infty} r(k) e^{-2ikx} dk$$

$$+ 2 \int_{-\infty}^{\infty} \frac{dq}{q} \int_{-\infty}^{\infty} dx'' \int_{-\infty}^{2x-x''} dx' f(x'') \Upsilon(x'', q) \sin q(x'+x''-2x) \frac{d^2}{dx'^2} [f(x') \Upsilon(x', q)] \qquad (69)$$

and

$$\gamma(x', k) = e^{ikx'} - \frac{i}{2|k|} \int_{-\infty}^{\infty} e^{i|k||x-x'|} f(x) \gamma(x, k) \, dx \tag{70}$$

Equations (69) and (70) together form a pair of coupled integral equations for f(x) and $\gamma(x,k)$ which can be solved numerically. Another method of approach to solving the one dimensional inverse scattering problem is to write the reflection coefficient in terms of f(x) via the transformation

$$R(x+\tau) + K(x,\tau) + \int_{-\infty}^{x} K(x,y) R(y+\tau) \, dy = 0 \tag{71}$$

where

$$R(x) = \frac{1}{2\pi} \int_{-\infty}^{\infty} r(k) e^{-ikx} \, dk \tag{72}$$

and

$$\frac{dK(x,x)}{dx} = \frac{1}{2} f(x) \tag{73}$$

If Eq.(71) can be solved for $K(x,\tau)$, then Eq.(73) gives the solution to the inverse scattering problem. Equation (71) is the Gel'fand-Levitan equation and an exact solution to this equation has recently been developed by Pechenick and Cohen (1981) given by

$$1 + \sum_{\alpha=1}^{n} (F_{j\alpha} e^{a_\alpha x} + G_{j\alpha} e^{-a_\alpha x}) b_\alpha(x) = 0 \tag{74}$$

j = 1,2,......,n where

$$F_{j\alpha} = (a_\alpha - ik_j)^{-1} \tag{75}$$

$$G_{j\alpha} = r(-ia_\alpha)/(a_\alpha - ik_j) \tag{76}$$

and

$$K(x,x) = \sum_{\alpha=1}^{n} \{ [e^{a_\alpha x} - r(-ia_\alpha) e^{-a_\alpha x}] b_\alpha(x) \} H(x) \tag{77}$$

and where r(k) is a rational function of k and is assumed to satisfy the following requirements:

    (i) r(0) = -1

(ii) $[r(k)]^* = r(-k)$ $\forall$ Re k

(iii) $|r(k)| \leqslant 1$ $\forall$ Re k

(iv) r(k) is analytic for all k in the upper half plane

Equation (74) is a set of n linear simultaneous equations which are solved numerically for a set of values of x. The acoustic scatter generating function f(x) can then be found from equations (77) and (73). The function $\chi^2(s)$ can be determined from the Liouville transformation via the following equations

$$g(x) = g(0)\left[1 + \tfrac{1}{2}\int_0^x f(x')dx'\right]$$
$$s = \int_0^x g^{-2}(x')dx' \qquad\qquad\qquad (78)$$
$$\chi(s) = g^2(x(s))$$

We note that x = x(s) has a <u>unique inverse</u> s = s(x) provided that $\chi(s)$ is positive for all $0 \leqslant s < \infty$. Having obtained a solution to the inverse scattering problem it is now possible to extract the profiles $\rho(s)$ and $K(s)$ independently. This can be accomplished by using a data base of the form $r(k,\theta_n)$; n=1,2. From the structure of $\chi^2(s,\theta_n)$ we can then write

$$\rho(s) = \left[(\sin^2\theta_1 - \sin^2\theta_2)/(\chi_2^2 - \chi_1^2)\right]^{\tfrac{1}{2}} \qquad (79)$$

$$K(s) = \left\{(\chi_2^2\sin^2\theta_1 - \chi_1^2\sin^2\theta_2)/\left[(\chi_2^2 - \chi_1^2)(\sin^2\theta_1 - \sin^2\theta_2)\right]^{\tfrac{1}{2}}\right\}$$

$$(80)$$

where

$$\chi_n^2 \equiv \chi^2(s,\theta_n)$$

and from Eq.(49)

$$z = \int_0^s \frac{ds'}{\rho(s')} \qquad (81)$$

which gives the relationship between the 'apparent depth' s and the actual depth z. Note, that if measurements of the back-scattered field are made at normal incidence to the medium, the acoustic scatter generating function becomes the effective wave impedance. This technique may therefore be seen as an extention of impediography

(Leeman, 1979). Also, by repeating the process over a two dimensional surface, it will be possible to obtain three dimensional information about the scatter generating parameters.

CONCLUSION

We have explored the inverse scattering problem in acoustics for both three dimensional and one dimensional models. In both cases, we have devised inversion routines via the Jost and Kohn algorithm and outlined the requisite data bases necessary for quantitatively imaging the scatter generating parameters of the continuum model. The main weakness of this technique lies in its restriction to weak acoustic scatter generating functions and requires that the Born series converges. It has been shown by Prosser (1976) that the series generated by the Jost and Kohn algorithm actually converges in norm for all $\in$ for which

$$\in \|f_1\| \quad < \quad 3 - 2 \, 2^{\frac{1}{2}} = 0.172\ldots$$

and that the sum $f_\in$ is a scatter generating function which will yield the given data base $d_\in$. These constraints are ideally suited to medical imaging where the intrinsic scatter generating parameters of soft tissue are weak enough to allow convergence of the inverse scattering solutions.

The one dimensional analysis has been shown to yield a unique solution to the inverse problem based on the Liouville transform, used for generating the 1D Schrödinger equation from a given acoustic wave equation.

ACKNOWLEDGEMENTS

We would like to thank Prof. R. Burge for his help and encouragement. This work was done with the support of the science and engineering research council grant Ref. No. B/80303414.

REFERENCES

Ball, J., Johnson, S. A. and Stanger, F. 1979, Explicit inversion of the Helmholtz equasion for ultrasound insonification and spherical detection, Acoust. Im., 9:451, Plenum Press.

Blackledge, J. M., Fiddy, M. A., Leeman, S. and Zapalowski, L., 1982, Three dimensional imaging of soft tissue with dispersive attenuation, Acoust. Im., 12:423, Plenum Press.

Blackledge, J. M., 1983, Ph. D. Thesis, The theory of quantitative acoustic scatter imaging, London University.

Blackledge, J. M., Fiddy, M. A., Leeman, S. and Seggie, D., 1983, Reflectivity tomography in attenuating media, Ultra. Int., to be published.

Jones, J. P., Leeman, S. and Blackledge, J. M., 1982, Quantitative ultrasound scatter imeging, First Int. Symp. on Med. Im, and Im. Interp., 1:325.

Jost, R. and Kohn, W., 1952, Construction of a potential from the phase shift, Phys. Rev., 37:977.

Leeman, S., 1978, The impediography equations, Acoust. Im., 8, Plenum Press.

Mueller, R. K., Kaveh, M. and Wade, G., 1979, Reconstruction tomography and applications to ultrasonics, Proc. IEEE, 67(4):567.

Norton, S. and Linzer, M., 1981, Ultrasonic reflectivity imaging in three dimensions, Proc. IEEE, BME-28:202.

Pechenick, K. R. and Cohen, J. M. 1981, J. Math. Phys., 22:1513.

Prosser, R. T., 1979, Formal solutions of inverse scattering problems. II, J. Math. Phys., 17:1773.

Razavy, M., 1975, Determination of the wave velocity in an inhomogeneous medium from the reflection coefficient, J. Acoust. Soc. Am., 58(5):956.

A FILTERED BACKPROPOGATION ALGORITHM FOR FAN BEAM

DIFFRACTION TOMOGRAPHY

A. J. Devaney and G. Beylkin

Schlumberger-Doll Research
Ridgefield, CT 06877

ABSTRACT

The filtered backpropagation algorithm for parallel beam ultra-sound tomography (A. J. Devaney, Ultra. Imag. 4, 336-1982) is generalized to cases of fan beam insonification. For essentially two-dimensional objects (rod-like objects), the algorithm assumed cylindrical wave insonification and a receiver geometry consisting either of a fixed linear array of a semicircular array that partially surrounds the object and that rotates with the direction of insonification. For three-dimensional objects, point source insonification and either a fixed planar array or a semispherical array are employed. As is the case for the parallel beam algorithm, a partial reconstruction is obtained at each view angle via a process of convolutional filtering followed by backpropagation. The final reconstruction is then obtained by linearly superimposing the partial reconstructions over the available view angles. The under-lying theory for the algorithm will be presented for both the Born and Rytov approximations. Computer implementation of the algorithm will be discussed and the results of computer simulations will be presented.

FOURIER DOMAIN RECONSTRUCTION METHODS

WITH APPLICATION TO DIFFRACTION TOMOGRAPHY

M. Soumekh*, M. Kaveh+ and R.K. Mueller+

*Department of Electrical Engineering
Worcester Polytechnic Institute
Worcester, MA   01609

+Department of Electrical Engineering
University of Minnesota
Minneapolis, MN   55455

ABSTRACT

Presented are algorithms for reconstructing two-dimensional functions from discrete and finite information available about their Fourier transform. The procedure is based upon an interpolation technique in the Fourier domain. The concept is applied to straight-path and diffraction tomography where Fourier domain information is available on a finite number of rotated contours. Finally, the resultant Fourier domain based methods are compared with their corresponding spatial domain based methods, i.e., filtered backprojection and backpropagation in terms of accuracy of reconstruction and computational efficiency.

INTRODUCTION

An important issue in many imaging systems in general and tomography systems in particular is the reconstruction of a two-dimensional function (object function) on a uniform grid in the spatial domain from the knowledge of the discrete values of the spatial Fourier transform of the function on a finite number of known loci in the spatial frequency domain (observed function). We denote the object function in the spatial domain by $f(x,y)$. Let (μ,λ) represent the spatial Fourier domain and ~ indicate the spatial Fourier transform of a function, e.g., $\tilde{f}(\mu,\lambda) = F.T._{(x,y)}[f(x,y)]$. If $\tilde{O}(\alpha,\beta)$ is the observed function then:

$$\Upsilon(\mu',\lambda') = \eth(\alpha,\beta) \tag{1},$$

with

$$\begin{bmatrix} \mu' \\ \lambda' \end{bmatrix} = \underline{T}(\alpha,\beta) = \begin{bmatrix} T_1(\alpha,\beta) \\ T_2(\alpha,\beta) \end{bmatrix} \tag{2},$$

where T_1 and T_2 are known continuous functions. The loci defined by $\underline{T}$ i.e., the transform function are called object Reconstruction CONtours (RCONs). Hence, the task is to reconstruct $f(x_i, y_j)$ from the known values of $\eth(\alpha_n,\beta_m)$ (x_i, y_j, α_n and β_m belong to a subset of equally spaced discrete values in x, y, α and β domains respectively).

Since the imaging schemes result in the information about the Fourier transform of the object function, it is natural to search for a method that translates the available information into the distribution of the spatial Fourier transform of the object function on a uniform grid in the spatial frequency domain. In addition, the object function is always extended over a finite region in the spatial domain. We denote this finite region by D. Hence, for the partial derivatives of Υ, we have:

$$\left| \frac{\partial^{m+n}\Upsilon(\mu,\lambda)}{\partial\mu^m\partial\lambda^n} \right| = \left| \iint_D (-jx)^m(-jy)^n f(x,y)\exp[-j(\mu x+\lambda y)]dxdy \right| < \infty \tag{3}$$

for all finite values of m and n. In other words, all partial derivatives of Υ exist. Consequently, an algorithm which interpolates the components of $\Upsilon(\mu_i,\lambda_j)$ from the available data could yield accurate results. The two-dimensional discrete data in the spatial frequency domain, i.e., $\Upsilon(\mu_i,\lambda_j)$ is then translated to the object function on a spatial domain grid with the aid of discrete Fourier transform routines.

A well known approach of this class is the zero-order interpolation method in the spatial Fourier domain. The idea is to average the values of $\Upsilon(\mu',\lambda')$ that were found from (1) and (2) for which the corresponding (μ',λ') pairs fall into a grid pixel. The averaged value is then assigned to the center of the grid pixel.

Let (μ_i,λ_j) be the center of a pixel in the spatial frequency domain grid. We denote the domain of this pixel by S_{ij}. Thus, for the zero-order interpolation method we have:

$$\mathfrak{T}(\mu_i, \lambda_j) \simeq \frac{1}{N(\mu_i, \lambda_j)} \sum_{(\mu', \lambda') \varepsilon S_{ij}} \mathfrak{T}(\mu', \lambda')$$

$$\simeq \frac{1}{N(\mu_i, \lambda_j)} \sum_{(\alpha_n, \beta_m) \varepsilon T^{-1}(S_{ij})} \mathfrak{O}(\alpha_n, \beta_m) \tag{4}$$

where $N(\mu_i, \lambda_j)$ is the number of the available data points that fall in S_{ij}. However, one might be able to seek a more rigorous interpolation technique. In fact, a general model of the type:

$$\mathfrak{T}(\mu_i, \lambda_j) = \sum_n \sum_m \mathbb{W}(\mu_i, \lambda_j, \alpha_n, \beta_m) \cdot \mathfrak{O}(\alpha_n, \beta_m) \tag{5},$$

where $\mathbb{W}$ is some window function, might be more desirable. This is because the coefficients of the linear model of equation (5) could be chosen to be a functional of a measure of the distance between (μ_i, λ_j) and the location of the available data points in the spatial Fourier domain.

Our purpose is to search for a window function, i.e., $\mathbb{W}$ which would yield the "best" (as might be defined) approximation of $\mathfrak{T}(\mu_i, \lambda_j)$. The selection of the would be optimum window function depends upon the properties of the transform function, i.e., $\underline{T}$ and a priori knowledge available about the object function in the spatial and/or spatial Fourier domains. We formulate the problem for all classes of continuous transform functions. We then apply the subsequent results for specific forms of the transform functions.

FOURIER DOMAIN METHOD

As was mentioned earlier, the window function, i.e., $\mathbb{W}$ should ideally be based on information about the concentration of the available data in the spatial Fourier domain and physical properties of the object function. Our intention is to obtain an expression for the window function in the form of two multiplicative kernels that are related to the functional characteristics of the object function and the transform function. One of the kernels is the averaging function that accounts for the nonuniform distribution of the available data in the spatial Fourier domain. The other kernel carries a priori information about the object function.

It is obvious that the density of the available data on RCONs follows a pattern that depends upon the functional properties of the transform function. Hence, an appropriate window function should

incorporate the information available about the transform function.
To show this fact, consider a region in the spatial Fourier domain,
e.g., A_{ij} with $(\mu_i, \lambda_j) \varepsilon A_{ij}$ such that the following holds (by the mean
value theorem for the integrals):

$$\hat{f}(\mu_i \lambda_j) = \frac{1}{|A_{ij}|} \iint_{A_{ij}} \hat{f}(\mu', \lambda') d\mu' d\lambda' \qquad (6)$$

where $|A_{ij}|$ is the measure of the surface area of A_{ij}. With the help
of equations (1) and (2), we can make variable transformations in
equation (6) to obtain:

$$\hat{f}(\mu_i, \lambda_j) = \frac{1}{|A_{ij}|} \iint_{T^{-1}(A_{ij})} \eth(\alpha, \beta) \, \mathcal{J}(\alpha, \beta) d\alpha d\beta \qquad (7)$$

where $\mathcal{J}(\alpha, \beta)$ is the Jacobian of the transformation from (μ', λ') to
(α, β), i.e.,

$$\mathcal{J}(\alpha, \beta) = \left| \frac{\partial(\mu', \lambda')}{\partial(\alpha, \beta)} \right| \qquad (8),$$

For the available discrete and equally spaced data, equation (7) can
be approximated by the midpoint (or trapezoidal) rule of integration:

$$\hat{f}(\mu_i, \lambda_j) = C \sum_{(\alpha_n, \beta_m) \varepsilon T^{-1}(A_{ij})} \mathcal{J}(\alpha_n, \beta_m) \, \eth(\alpha_n, \beta_m) \qquad (9).$$

With $A_{ij} = S_{ij}$, the linear models of equations (4) and (9) are
identical except for the coefficients of the models. Note that the
zero-order interpolation method assigns the same weight, i.e.,
$1/N(\mu_i, \lambda_j)$ to all the available data that fall in the S_{ij} pixel. On
the other hand, the model of equation (9) exploits the information
available about the functional characteristics of the transform function to account for the density variations of the available data
within each segment of the S_{ij} pixel. Therefore, the latter model
should be more accurate than the zero-order interpolation method.

The above argument gives an intuitive reasoning for the use of the Jacobian function, i.e., $\mathcal{J}(\alpha,\beta)$ as the optimal averaging function. With the establishment of this concept, we can now search for the other multiplicative kernel in the window function which carry a priori information about the object function. A general form of that kernel was given in [1,2,3]. However, the following formulation utilizes a specific characteristic of the object function, namely, spatially limited to obtain an expression for the second kernel. This approach also automatically brings out the Jacobian function, the physical significance of which was shown earlier, as one of the multiplicative kernels.

Consider the inverse Fourier transform equation of:

$$f(x,y) = \int_{-\infty}^{\infty}\int_{-\infty}^{\infty} \tilde{f}(\mu',\lambda') \cdot exp\left[j(\mu'x+\lambda'y)\right]d\mu'd\lambda' \tag{10}$$

For simplicity, let us assume that the mapping of the transform function is one-to-one. Therefore, with the help of equations (1) and (8), we can make variable transformations in equation (10) to obtain:

$$f(x,y) = \int_{-\infty}^{\infty}\int_{-\infty}^{\infty} \tilde{O}(\alpha,\beta)\cdot exp\left[j\{T_1(\alpha,\beta)x+T_2(\alpha,\beta)y\}\right] \cdot \mathcal{J}(\alpha,\beta)d\alpha d\beta \tag{11}.$$

Hence, the Fourier transform of the left-hand side of (11) evaluated at (μ_i,λ_j) is:

$$\tilde{f}(\mu_i,\lambda_j) = \int_D\int \int_{-\infty}^{\infty}\int_{-\infty}^{\infty} \mathcal{J}(\alpha,\beta)\cdot\tilde{O}(\alpha,\beta)\cdot exp\left[j\{T_1(\alpha,\beta)x+T_2(\alpha,\beta)y\}\right]\cdot$$
$$d\alpha d\beta \cdot exp\left[-j(\mu_i x+\lambda_j y)\right]dxdy \tag{12},$$

where D is the finite support of the object function, i.e., $f(x,y)$ in the spatial domain. After some rearrangements, equation (12) can be written as:

$$\tilde{f}(\mu_i,\lambda_j) = \int_{-\infty}^{\infty}\int_{-\infty}^{\infty} \mathcal{J}(\alpha,\beta)\tilde{O}(\alpha,\beta) \cdot$$

$$\int_D\int exp\left[-j\{\mu_i-T_1(\alpha,\beta)\}x-j\{\lambda_j-T_2(\alpha,\beta)\}y\right]dxdy \;\; d\alpha d\beta \tag{13}$$

We define the indicator function for the support of the object function by the following:

$$
I(x,y) = \begin{cases} 1 & \text{if } (x,y) \varepsilon D \\ 0 & \text{otherwise} \end{cases} \tag{14}
$$

Hence, we have:

$$
\Upsilon(\mu,\lambda) = \int_{-\infty}^{\infty}\int_{-\infty}^{\infty} I(x,y) \cdot \exp\left[-j(\mu x + \lambda y)\right] dx dy
$$

$$
= \int_D\int \exp\left[-j(\mu x + \lambda y)\right] dx dy \tag{15}.
$$

With the aid of equation (15), we can rewrite equation (13) as follows:

$$
\Upsilon(\mu_i,\lambda_j) = \int_{-\infty}^{\infty}\int_{-\infty}^{\infty} \Im(\alpha,\beta) \cdot \eth(\alpha,\beta) \cdot \Upsilon[\mu_i - T_1(\alpha,\beta), \lambda_j - T_2(\alpha,\beta)] d\alpha d\beta \tag{16}.
$$

Finally, the double integral of equation (16) can be approximated with the help of the midpoint (or trapezoidal) rule of integration for equally spaced data:

$$
\Upsilon(\mu_i,\lambda_j) \simeq C \sum_n \sum_m \Im(\alpha_n,\beta_m) \cdot \eth(\alpha_n,\beta_m) \cdot
$$
$$
\Upsilon[\mu_i - T_1(\alpha_n,\beta_m), \lambda_j - T_2(\alpha_n,\beta_m)] \tag{17}
$$

We can further simplify the model of equation (17) by dropping the available data for which the functional value of their corresponding Υ is small in equation (17)[6]. This can be achieved through summing over only the (α_n,β_m) pairs for which the main-lobe of their Υ functions contain (μ_i,λ_j). Consequently, the model can be rewritten as follows:

$$
\Upsilon(\mu_i,\lambda_j) \simeq C \sum_{(\mu_i,\lambda_j) \varepsilon \text{main-lobe } \Upsilon} \Im(\alpha_n,\beta_m) \cdot \eth(\alpha_n,\beta_m) \cdot
$$
$$
\Upsilon[\mu_i - T_1(\alpha_n,\beta_m), \lambda_j - T_2(\alpha_n,\beta_m)] \tag{18}.
$$

In general, the exact extension of the object domain, i.e., D and/or the expression for the I function is not readily available. However, one can consider a more trivial and known domain that covers <u>at</u> <u>least</u> D as the object domain. A practical choice for such a domain is the smallest disk that covers the object domain. If the radius of such a disk is R, its corresponding indicator function will simply be:

$$I(x,y) = \begin{cases} 1 & \text{if} \quad x^2+y^2 \leq R \\ 0 & \text{otherwise} \end{cases} \tag{19}$$

It can be shown that the Fourier transform of the indicator function of equation (19) is [5]:

$$T(\mu,\lambda) = R^2 \frac{J_1(R\sqrt{\mu^2+\lambda^2})}{R\sqrt{\mu^2+\lambda^2}} \tag{20}$$

where J_1 is the Bessel function of the first kind, first order. The reconstruction results that are presented in the following section are based upon the application of this window function.

APPLICATION IN TOMOGRAPHY

The Fourier domain based reconstruction method that was outlined in the previous section can be applied in many imaging schemes such as multiple frequency, chirp frequency, diffraction tomography and straight-path tomography. In this paper, we present an empirical study of the Fourier domain reconstruction method in the latter two imaging schemes. The two classes of transform functions which define the loci of the available data, i.e., RCONs in the two tomography problems and their corresponding Jacobian of the transformation from (μ',λ') to (α,β) are given as follow [6]:

   i)   Straight-path Tomography

$$\underline{T}(\alpha,\beta) = \begin{bmatrix} \cos\alpha & \sin\alpha \\ -\sin\alpha & \cos\alpha \end{bmatrix} \begin{bmatrix} 0 \\ \beta \end{bmatrix} \quad \text{for} \quad \begin{cases} 0 \leq \alpha < 2\pi \\ 0 \leq \beta < \infty \end{cases} ,$$

$$J(\alpha,\beta) = |\beta| \tag{21}.$$

ii) Diffraction Tomography

$$\underline{T}(\alpha,\beta) = \begin{bmatrix} \cos\alpha & \sin\alpha \\ -\sin\alpha & \cos\alpha \end{bmatrix} \begin{bmatrix} -k_0 + \sqrt{k_0^2 - \beta^2} \\ \beta \end{bmatrix}$$

$$J(\alpha,\beta) = \frac{k_0|\beta|}{\sqrt{k_0^2-\beta^2}} \qquad \text{for} \begin{cases} 0 \leq \alpha < 2\pi \\ 0 \leq \beta \leq k_0 \end{cases}, \qquad (22)$$

where k_0 is a known constant (the wavenumber of the impinging field). In both cases, α is the object angular orientation and β is the spatial frequency domain of the observed field [2].   Note that in the diffraction case, the available data only produces information about a lowpass band of the object function in the Fourier domain.   The error that is introduced by neglecting the higher frequency components of $\Upsilon(\mu,\lambda)$ does not affect the basic formulation of the reconstruction method.

An alternative approach to the Fourier domain reconstruction method is the spatial domain reconstruction technique. The spatial domain methods in the straight-path tomography and diffraction tomography problems are called filtered backprojection and backpropagation (see Devaney [4]), respectively. These methods are practically based upon a rearranged form of equation (11) and an approximation to it [6].

Our task in this section is to study the relative merits of the Fourier and spatial domain reconstruction techniques in the two tomography problems. We introduce two classes of object functions for the purpose of the study of the reconstruction techniques:

1) Cylindrical object function of radius R_0

$$f(x,y) = \begin{cases} 1 & \text{if} \quad \sqrt{x^2+y^2} \leq R_0 \\ 0 & \text{otherwise} \end{cases},$$

$$\Upsilon(\mu,\lambda) = R_0^2 \, \frac{J_1(R_0\sqrt{\mu^2+\lambda^2})}{R_0\sqrt{\mu^2+\lambda^2}} \qquad (23)$$

2) Gaussian object function

$$f(x,y) = \exp\left[-\pi \frac{x^2+y^2}{\tau^2}\right],$$

where τ is a constant,

$$\Upsilon(\mu,\lambda) = A \exp\left[- \frac{\tau^2(\mu^2+\lambda^2)}{4\pi}\right] \tag{24}.$$

The examples that follow are for the object functions that are composed of a combination of cylindrical or Gaussian object functions. The object functions are reconstructed on a uniform 64x64 grid in the spatial domain. The number of the available discrete values in the β domain, i.e., the spatial frequency domain of the observed field is 32. The number of the available discrete values of α, i.e., the object angular orientation (call it K) is varied between 32 and 128 over the full 2π radian rotation. The sample spacing in the spatial domain is $\lambda_o/2$ (λ_o is the wavelength of the impinging field).

Figure (1) shows the sliced images, i.e., the distribution of the reconstruction of $f(x,0)$ for a Gaussian object function with $\tau = 8\lambda_o$. The available data lie on the RCONs of the diffraction case. The Fourier domain method was produced accurate reconstruction results for K=32 and K=64. At the same time, the spatial domain method reconstructions are greatly in error though the envelope of the images resemble the original Gaussian object function. The distortions that are seen in these reconstructions are mainly midband and high frequency errors.

Figure (2) shows the sliced images of an object function which is composed of two cylindrical functions. One of the cylinders is a positive disk of radius $4\lambda_o$. The other cylinder is a negative disk of radius $8\lambda_o$. The amplitude of the two cylinders are adjusted such that the DC value of the resultant object function is zero. The available data lie on the RCONs of the diffraction case. Again, the superior accuracy of the Fourier domain method to the spatial domain approach is evident. In the case of K=32, the midband and high frequency errors that are introduced by the backpropagation (spatial) method have badly distorted the $8\lambda_o$ cylinder. Note that lowpass-filtering the reconstructed object function for the purpose of the removal of some of these errors can result in smearing of the object function.

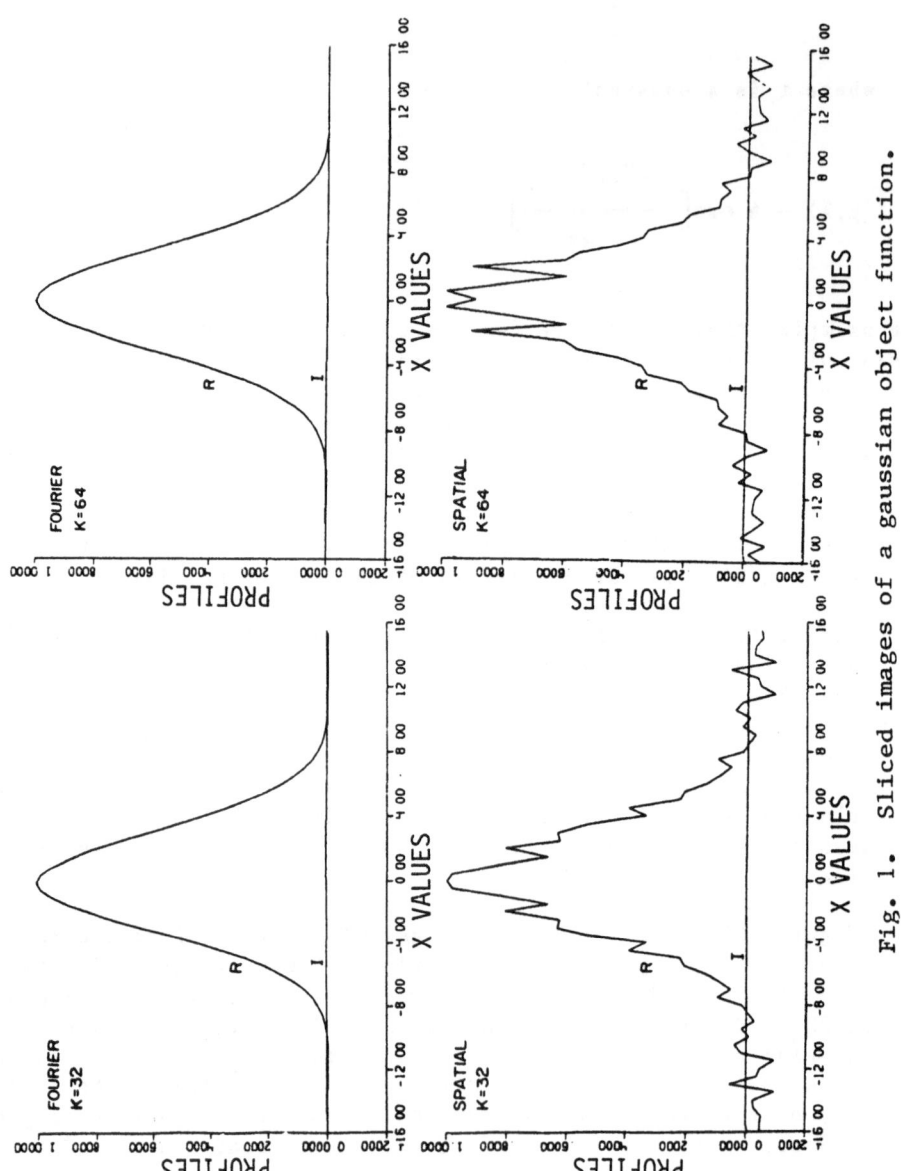

Fig. 1. Sliced images of a gaussian object function.

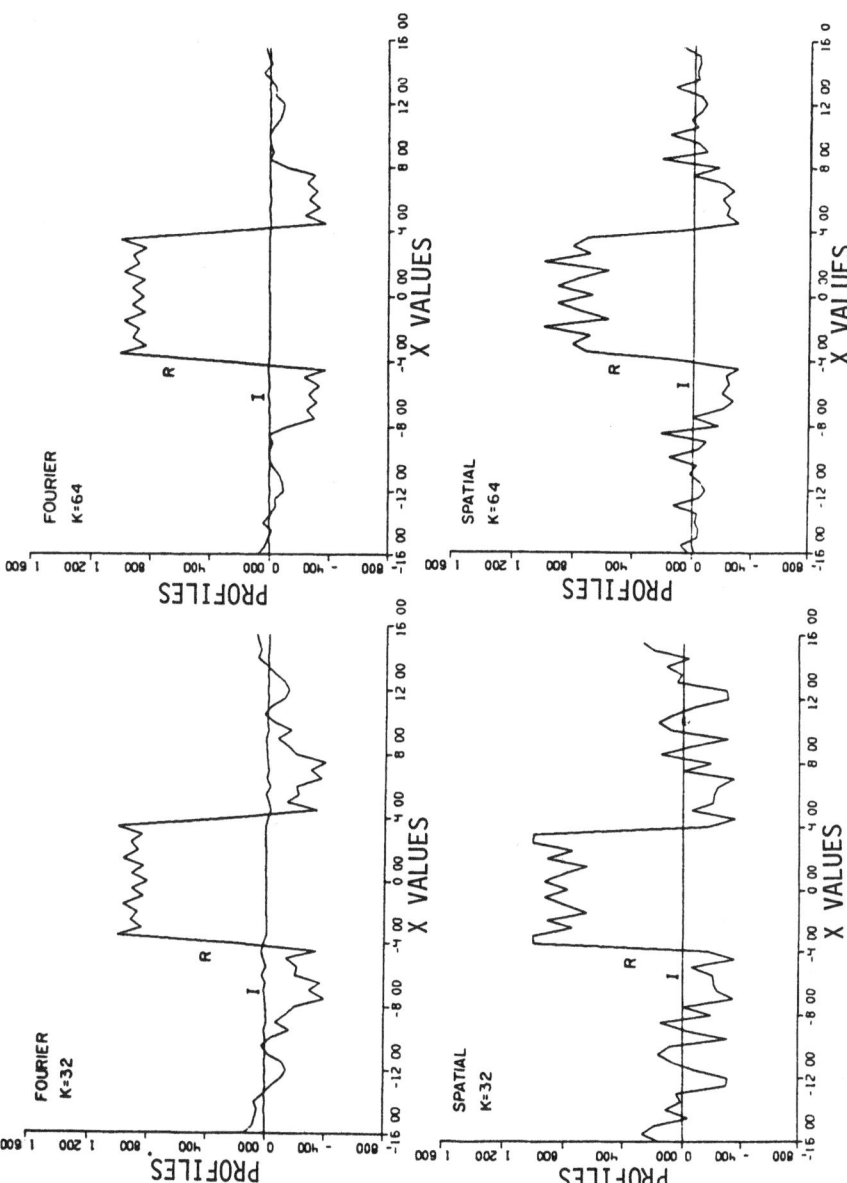

Fig. 2. Sliced images of the two cylinder object function.

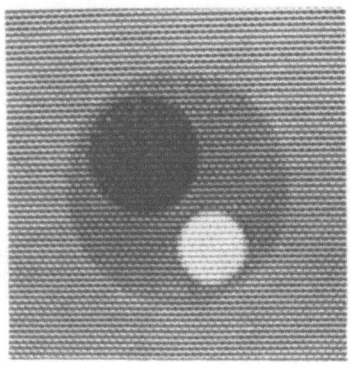

A. Fourier   B. Spatial

Diffraction Tomography Data

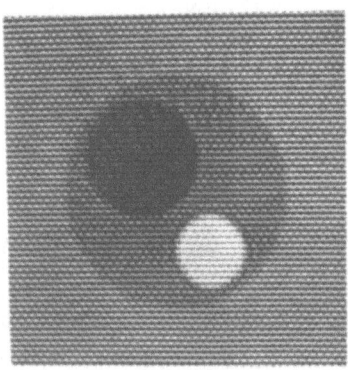 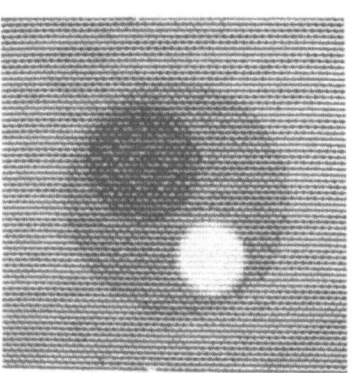

C. Fourier   D. Spatial

Straight-Path Tomography Data

Fig. 3. Reconstructions of the three cylinder
         object function (64 projections).

It has also been speculated that the spatial domain method produces better results as the number of object angular orientation, i.e., K decreases. Our study has proved otherwise. In fact, for $32 \leq K \leq 128$, the Fourier domain method has produced more accurate reconstruction results than the spatial domain method for the classes of Gaussian and cylindrical object functions.

Figure (3) shows the gray scale images of the reconstruction of an object function in diffraction tomography and straight-path tomography. The object function is a superposition of $3\lambda_o, 5\lambda_o$ and $10\lambda_o$ cylinders. The Fourier domain method reconstructed images have sharp contrast with relatively uniform distribution within the three cylinders. The pictures clearly show the outline of the boundary of the cylinders. However, the spatial domain method has produced blurred images. Overall, the spatial domain reconstruction results are inferior to the Fourier domain ones.

Finally, it should be recalled that the Fourier domain method usually requires substantially less computational time than the spatial domain method [1]. For example, the Fourier domain method is about an order of magnitude faster than the spatial domain method in the diffraction tomography problem. An approximation to the backpropagation method has been introduced to reduce the computational time [4]. However, this approximation results in further degradation of the midband and high frequency components of the object function. Obviously, this can not be tolerated when the spatial domain method itself performs poorly at those frequency bands.

CONCLUSION

We formulated a Fourier domain based reconstruction method when the available data lie on arbitrary contours. The knowledge of the transform function and physical properties of the object function were utilized in the derivation of the Fourier method. This concept was applied in diffraction tomography and straight-path tomography. In the tomography problems, our empirical study disclosed the superior accuracy of the Fourier domain approach to the spatial domain methods, also known as filtered backprojection and backpropagation, for two classes of object functions. In addition to superior accuracy, the computational efficiency and the ease of implementation of the Fourier method make this approach the more preferable reconstruction technique. Moreover, the Fourier domain method can exploit other forms of a priori information about the object function in design of the kernels of equation (5) for the purpose of image enhancement.

ACKNOWLEDGEMENT

This work was supported by The National Science Foundation under Grant No. ECS-7926008.

REFERENCES

1. M. Soumekh, M. Kaveh and R.K. Mueller, "Algorithms and Experimental Results in Acoustic Tomography Using Rytov's Approximation", ICASSP '83 Proceedings, Boston (1983).

2. M. Kaveh, M. Soumekh, Z.Q. Lu, R.K. Mueller and J.F. Greenleaf, "Further Results on Diffraction Tomography Using Rytov's Approximation", Acoustical Imaging, Vol 12 (1982).

3. M. Kaveh, M. Soumekh and R.K. Mueller, "Tomographic Imaging Via Wave Equation Inversion", ICASSP '82 Proceedings, Paris (1982).

4. A. Devaney, "A Filtered Backpropagation Algorithm for Diffraction Tomography", Ultrasonic Imaging, Vol. 4 (1982).

5. L. Rabiner and B. Gold, Theory and Application of Digital Signal Processing, Prentice-Hall (1975).

6. M. Soumekh, Ph.D. Dissertation, University of Minnesota (1983).

IMAGE RECONSTRUCTION FROM LIMITED PROJECTIONS
USING THE ANGULAR PERIODICITY OF THE FOURIER TRANSFORM
OF THE RADON TRANSFORMS OF AN OBJECT

M. Soumekh

Department of Electrical Engineering
Worcester Polytechnic Institute
Worcester, MA 01609

ABSTRACT

A noniterative method of image reconstruction from limited projections is presented. The approach is based on the periodicity
property of the Fourier transform of the Radon transforms of an
object (observed function) with respect to the projection angle.
This functional charactristic of the observed function is utilized
to interpolate the missing line integrals by a truncated Fourier
series expansion. Methods for finding the components of the truncated Fourier series from the available data are given. Use of a
priori information in formulating the solution of the truncated
Fourier series is also discussed.

INTRODUCTION

In many imaging systems, the number and range of the availability of the object angular projections is limited by the physical
constraints of the system. This results in regions of missing data
in the spatial Fourier domain which unable one to use conventional
algorithms to reconstruct the object function (a two-dimensional
function that carries certain information about the object under
study) from the observed function (the spatial Fourier transform of
the object angular projections).

The object function has a finite support in the spatial domain.
An immediate consequence of the space-limited property of the object
function is the existence and analytic continuity of the spatial
Fourier transform of the object function and all its partial derivatives [1]. As a result, an algorithm that extrapolates the missing
data in the spatial Fourier domain could yield accurate results.

There are two well known set of approaches to the limited projection problem. One is based on the application of the Whittaker-Shannon theorem in which the Fourier transform of the space-limited object function is interpolated with the aid of its analytic continuity property (e.g., [2]). These methods are susceptible to the system noise. The other set of approaches are practically variations of Gerchberg-Papouils iterative reconstruction algorithm. These methods impose certain constraints, e.g., space-limited and positivity (a priori information) on the object function to develop an iterative method of reconstruction from limited projections (e.g., [3]). In addition to their undesirable iterative nature, these methods are solely dependent on a priori information to enhance the quality of the reconstructed image.

This paper presents a reconstruction algorithm from limited projections which is noniterative and does not require any type of a priori information about the object function. However, the algorithm can incorporate certain a priori information about the object function for further enhancement of the reconstructed image.

Nevertheless, our main purpose is to bring out a special functional property of the object function, namely, its periodicity with respect to the angle of propagation which enables one to develop an efficient and accurate scheme for the recovery of the missing data in the spatial Fourier domain. Hence, the method that is developed here can serve as a first step or basis for any reconstruction algorithm from limited projections.

This concept is applicable in many imaging schemes such as straight-path tomography, diffraction tomography and fan beam tomography where the limited projection problem arises. We present the technique for the case of straight-path tomography. We start with the derivation of a scheme for interpolating the observed function with the aid of trigonometric polynomials (truncated Fourier series expansion). We then introduce two methods for finding the coefficients of the trigonometric interpolating polynomials from the available data. We also make use of a priori information about the object function to obtain a better solution for those coefficients. Finally, some examples are given.

RECONSTRUCTION FROM LIMITED PROJECTIONS

Randon transforms of a function are defined to be the integral over the straight lines that pass through that function. The objective in straight-path tomography is to reconstruct a two-dimensional function $f(x,y)$ (object function) on a uniform grid in the spatial domain with the aid of information available about the finite number of its Randon transforms. Let (μ,λ) represent the spatial Fourier domain and$\tilde{\ }$ indicate the spatial Fourier transform of a function, e.g., $\Upsilon(\mu,\lambda) = F.T._{(x,y)} [f(x,y)]$. If $\mho$ is the spatial Fourier transform of the Randon transforms then:

$$\Upsilon(\mu',\lambda') = \mho(\theta,\rho) \tag{1}$$

where θ is the projection angle, ρ is the spatial frequency domain of the Randon transforms and

$$\begin{bmatrix} \mu' \\ \lambda' \end{bmatrix} = \begin{bmatrix} \cos\theta & \sin\theta \\ -\sin\theta & \cos\theta \end{bmatrix} \cdot \begin{bmatrix} 0 \\ \rho \end{bmatrix} \quad \text{for} \quad \begin{cases} 0 \le \theta < 2\pi \\ 0 \le \rho < \infty \end{cases} \tag{2}.$$

In other words, (θ,ρ) is the polar representation of (μ',λ'). In practice, one can reconstruct the object function with the knowledge of the values of the observed function at discrete and equally spaced points in the (θ,ρ) domain with a suitable sampling rate [1].

It is obvious that the observed function can be viewed as a periodic function of θ with period $T=2\pi$. Therefore, the observed function can be represented by the following Fourier series expansion for each value of ρ:

$$\mho(\theta,\rho) = \sum_{n=-\infty}^{\infty} C_n(\rho) e^{j\frac{2\pi}{T}n\theta}$$

$$= \sum_{n=-\infty}^{\infty} C_n(\rho) e^{jn\theta} \tag{3},$$

In the limited projection problem, the observed function is not known for certain intervals in the θ domain. Let A_i's be the segments in each period of 2π in the θ domain where discrete values of the observed function are available at a suitable sampling rate and N be the total number of the available data in all the A_i intervals, e.g., $\theta_1, \theta_2, \ldots \theta_N$. Also, let B_i's be the segments in each period of 2π in the θ domain that correspond to the missing data (line integrals in straight-path tomography) (see Figure (1)). Let us assume that there exists a finite integer, e.g., M such that the following truncated Fourier series expansion is a good approximation of the observed function:

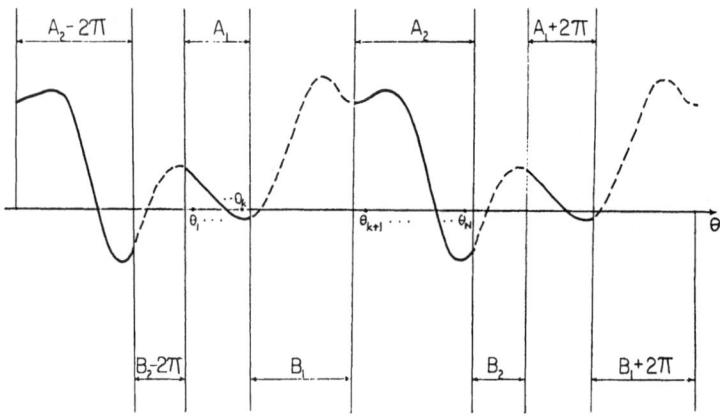

Fig. 1.   Depiction of a periodic function with its available data
          segments (A_i's) and missing data segments (B_i's).

$$\tilde{U}(\theta,\rho) \simeq \sum_{n=-M}^{M} C_n(\rho)e^{jn\theta} \tag{4}.$$

If the observed function is bandlimited in the Fourier domain of θ, an integer M could be found such that equation (4) becomes exact. However, in general the observed function is not bandlimited in that sense. Nevertheless, the spectrum of the observed function in the Fourier domain of θ could be assumed to drop sharply for $|n| > M$. This assumption is not equivalent to the space-limited constraint imposed in the Gerchberg-Papoulis iterative reconstruction methods.

Determination of the Truncated Fourier Series Coefficients

Since N values of the observed function are known at N distinct points, we can write the following set of N linear equations from (4):

$$\sum_{n=-M}^{M} \hat{C}_n(\rho)e^{jn\theta_m} = \tilde{U}(\theta_m,\rho) \quad \text{for} \quad m=1,2,\ldots,N \tag{5},$$

where $\hat{C}_n(\rho)$ is an estimate of $C_n(\rho)$. Equivalently, we can write:

$$\underline{E}(\theta) \cdot \hat{\underline{C}}(\rho) = \underline{\tilde{U}}(\theta,\rho) \tag{6},$$

where:

$\underline{E}(\theta)$ is a NxL (L=2M+1) matrix of the exponent coefficients $\{e^{jn\theta_m}\}$,

$\hat{\underline{C}}(\rho)$ is a column vector of size L of the estimates of the truncated Fourier series $\{C_n(\rho)\}$,

$\underline{\tilde{U}}(\theta,\rho)$ is a column vector of size N of the available data $\{\tilde{U}(\theta_m,\rho)\}$.

A well known approach for finding $\hat{\underline{C}}(\rho)$ from equation (6) is the method of least square [4]. Let $\underline{E}^+(\theta)$ be the generalized inverse of $\underline{E}(\theta)$. It can be shown [4] that $\hat{\hat{\underline{C}}}(\rho)$, where:

$$\hat{\hat{\underline{C}}}(\rho) = \underline{E}^+(\theta) \cdot \underline{\tilde{U}}(\theta,\rho) \tag{7},$$

is the least linear square estimate of $\widehat{C}(\rho)$. It should be mentioned that the $\underline{E}^+(\theta)$ matrix is invariant of ρ. Hence, the same generalized inverse matrix $\underline{E}^+(\theta)$ can be used for all values of ρ. This implies a significant saving in the computational time.

It can be shown that [4]:

i) $\underline{E}^+(\theta) = (\underline{E}^T(\theta) \cdot \underline{E}(\theta))^{-1} \underline{E}^T(\theta)$

for N>L provided that the rank of $\underline{E}(\theta)$ is L,

ii) $\underline{E}^+(\theta) = \underline{E}^{-1}(\theta)$

for N=L provided that the rank of $\underline{E}(\theta)$ is N,

iii) $\underline{E}^+(\theta) = \underline{E}^T(\theta)(\underline{E}(\theta) \cdot \underline{E}^T(\theta))^{-1}$

for N<L provided that the rank of $\underline{E}(\theta)$ is N.

A particular case of interest is when N>L (overdetermined system) which is known to reduce the noise effects. However, it should not be assumed that a better solution could be achieved by selecting a very large number of observations (N) within a limited range of projection. Such solutions could become susceptible to certain types of error caused by uncertainties in the imaging system (e.g., receiver jittering). One should also avoid using this method at high frequencies where signal to noise power ratio is small and the distance between adjacent points in the (μ, λ) domain is large.

Finally, the observed function can be approximated at any point in the θ domain, specifically, within the B_i segments with the knowledge of $\widehat{\widehat{C}}(\rho)$:

$$\eth(\theta, \rho) \simeq \sum_{n=-M}^{M} \widehat{\widehat{C}}_n(\rho) e^{jn\theta} \qquad (8).$$

Consequently, the observed function can be reconstructed over the full period of 2π in the θ domain. The object function can then be reconstructed from the observed function with the aid of conventional reconstruction tomography algorithms [1].

We can also reformulate the problem in terms of Lagrange interpolating polynomials by a transformation. Consider the case of N=L. We define:

$$\alpha = e^{j\Theta},$$

and $g(\alpha,\rho) = \widehat{U}(\Theta,\rho)$ (9).

With the help of (9), equation (4) can be rewritten in terms of α as follows:

$$g(\alpha,\rho) \simeq \sum_{n=-M}^{M} C_n(\rho)[e^{j\Theta}]^n = \sum_{n=-M}^{M} C_n(\rho)\alpha^n$$

$$= \alpha^{-M} \sum_{n=0}^{2M} C_{n-M}(\rho)\alpha^n$$ (10).

The values of α are known at N distinct points ($\alpha_m = e^{j\Theta_m}$ for $m=1,2,\ldots,N$). Therefore, we can interpolate the values of $g(\alpha,\rho)$ at all values of α by the following modified Lagrange interpolation scheme:

$$g(\alpha,\rho) \simeq \sum_{i=1}^{N} (\alpha/\alpha_i)^{-M} \prod_{\substack{j=1 \\ j\neq i}}^{N} g(\alpha_i,\rho) \cdot [(\alpha-\alpha_i)/(\alpha_j-\alpha_i)]$$ (11)

Note the presence of $(\alpha/\alpha_i)^{-M}$ multiplier in the coefficients of the Lagrange interpolating polynomails that accounts for the negative powers of α in equation (10).

USE OF A PRIORI INFORMATION

As was mentioned earlier, it is possible to exploit a priori information about the object function to achieve better estimates of the truncated Fourier series of equation (4). For example, in X-Ray tomography, the object function (the distribution of the absorption coefficient in the object under study) is purely real. An immediate consequence of this property can be shown to be:

$$\mathfrak{T}(\mu,\lambda) = \mathfrak{T}^*(-\mu,-\lambda) \tag{12}.$$

Equivalently, from (1) and (12), one can see that:

$$\eth(\theta,\rho) = \eth^*(\theta+\pi,\rho) \tag{13}.$$

From (3), one can also write:

$$\eth^*(\theta+\pi,\rho) = \sum_{n=-\infty}^{\infty} C_n^*(\rho) e^{-jn(\theta+\pi)}$$

$$= \sum_{n=-\infty}^{\infty} C_{-n}^*(\rho) e^{jn\pi} \cdot e^{jn\theta} \tag{14}$$

Substituting (3) and (14) in equation (13) gives:

$$\sum_{n=-\infty}^{\infty} C_n(\rho) e^{jn\theta} = \sum_{n=-\infty}^{\infty} C_{-n}^*(\rho) e^{jn\pi} \cdot e^{jn\theta} \tag{15}$$

Because of the uniqueness of Fourier series, one can conclude from equation (15) that:

$$C_n(\rho) = C_{-n}^*(\rho) e^{jn\pi} = (-1)^n C_{-n}^*(\rho) \tag{16}.$$

Let $C_n = a_n + jb_n$. After some mathematical manipulations, one can show from (3) and (16) that:

$$\text{Real } [\eth(\theta,\rho)] = \sum_{n=0}^{\infty} a_{2n} \cos(2n\theta) - b_{2n} \sin(2n\theta) \tag{17}$$

$$\text{Imaginary } [\eth(\theta,\rho] = \sum_{n=0}^{\infty} a_{2n+1} \sin[(2n+1)\theta] + b_{2n+1} \cos[(2n+1)\theta]$$

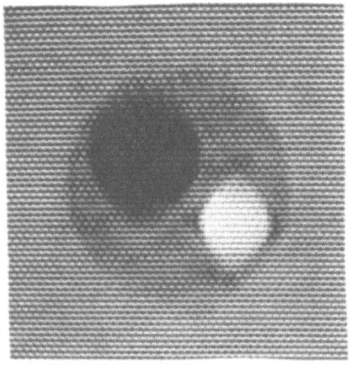 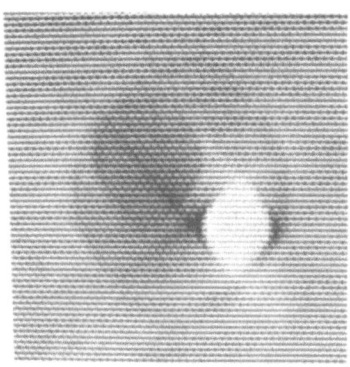

A. Interpolated   B. Limited Data

Fig. 2. Reconstructions from 16 projections in 90°
         angular interval (straight-path tomography
         data).

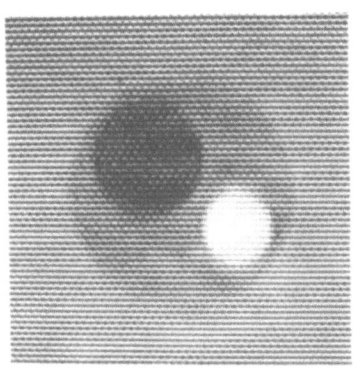 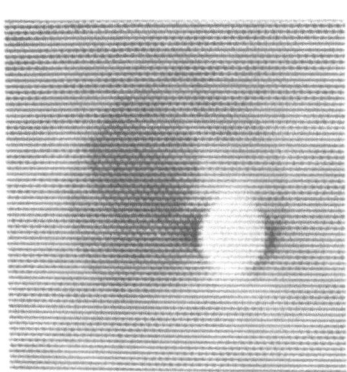

A. Interpolated   B. Limited Data

Fig. 3. Reconstructions from 16 projections in 90°
         angular interval (diffraction tomography
         data).

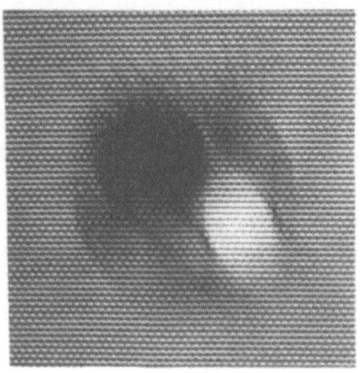 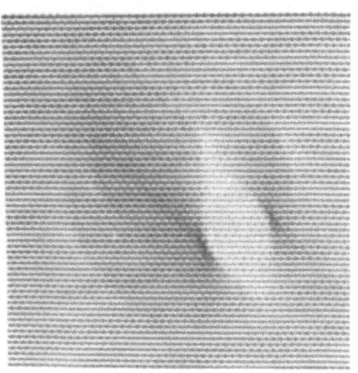

A. Interpolated   B. Limited Data

Fig. 4.   Reconstructions from 8 projections in
45° angular interval.

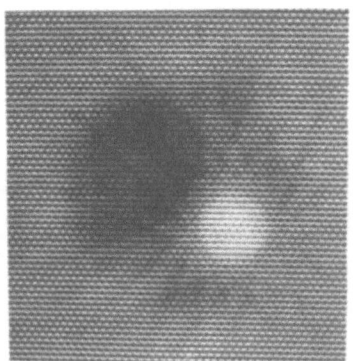 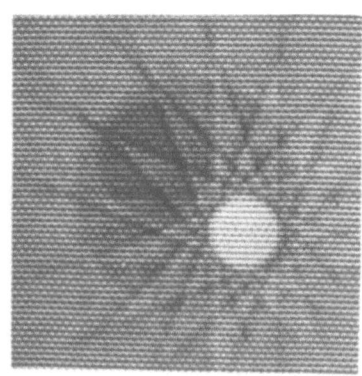

A. Interpolated   B. Limited Data

Fig. 5.   Reconstructions from 8 projections in
180° angular interval.

In other words, the real part of the observed function is a function of the even components of the C_n series and the imaginary part of the observed function is a function of the odd components of the C_n series. This implies that we are able to rearrange equation (5) to solve for twice as many Fourier series as indicated by equation (5). A more detailed version of this concept and the subsequent linear system of equations will be given in the future.

EXAMPLES

The examples that are presented are for the three cylinder structure that was also studied in [1]. The results of reconstructions by the truncated Fourier series method (interpolated) and filling zero in the spatial Fourier domain where projections are missing (limited data) are given. Figure (2) is for the case of 90° limited angle of projected (16 projections) which translates to 50% available projections for a real object function. The improvement by the truncated Fourier series is evident. Figure (3) is a similar situation in diffraction tomography (we mentioned earlier that the method is also applicable in diffraction tomography). Figure (4) shows the reconstructions for 45° limited angle of projection (8 projections). It can be seen that the limited data provides a totally washed out reconstruction. At the same time, the interpolated reconstruction clearly shows the object structure. The final example (figure (5)) shows the reconstruction results when 8 evenly spaced projections are available in 180° angle. This situation arises when it is desirable to minimize the object exposure to radiation. Again, the truncated Fourier method shows good fidelity of reconstruction.

CONCLUSION

We presented a method of reconstruction from limited projections which utilizes a special functional property of the Fourier transform of the Randon transforms of an object. The method was shown to be also able to evoke a priori information about the object under study to solve for the higher order components of the Fourier series and further enhancement of the resultant reconstructed image. An important feature of this method is its noniterative nature which is indicative of its computational efficiency. The method was shown to produce good results for limited angle of projections which are as small as 45°. A study of the method for the case of oversampled noisy observed functions is in order.

ACKNOWLEDGEMENT

Examples were processed in the Acoustical Imaging Center of the University of Minnesota with Professor Mostafa Kaveh's permission. Helpful discussions with Dr. Kaveh are also appreciated.

REFERENCES

1. M. Soumekh, M. Kaveh and R.K. Mueller, "Fourier Domain Recon-
    struction Methods with Application to Diffraction Tomography",
    Acoustical Imaging, Vol. 13 (1983).

2. J.W. Goodman, Introduction to Fourier Optics, McGraw-Hill, New
    York (1968).

3. T. Sato, et al., "Tomographic Image Reconstruction from Limited
    Projections Using Iterative Revisions in Image and Transform
    Spaces", Applied Optics, Vol. 20, No. 3 (1981).

4. B. Noble, Applied Linear Algebra, Prentice-Hall (1969).

MULTIFREQUENCY DIFFRACTION TOMOGRAPHY

James F. Greenleaf and Aloysius Chu

Biodynamics Research Unit, Department of Physiology and Biophysics, Mayo Clinic/Foundation, Rochester MN 55905

INTRODUCTION

In an attempt to obtain quantitative images of two-dimensional distributions of basic mechanical properties of tissue in a noninvasive manner, ultrasonic computer-assisted tomography has been developed[1,2,3]. Current techniques for applying ultrasonic computer-assisted tomography rely on mathematical methods of solution which require straight-line assumptions for the traversing trajectories of the ultrasonic rays[4]. Several authors have suggested more accurate methods of reconstructing distributions of parameters representing material properties such as attenuation and speed within tissue, using methods such as ray tracing[5], miniature profiles[6], phase insensitive apertures[7], and inverse scattering techniques.[8,9,10,11] The promise of such methods is to obtain images having higher fidelity than those obtained using simple straight-ray assumptions.

The purpose of this paper is to report experimental results in which multifrequency signals were transmitted, in a plane wave format, through scattering objects and the Rytov method of inverse scattering[9] was utilized to calculate the real and imaginary parts of the refraction index under the assumption that variations in density were absent. The use of multiple frequencies allows subsequent analysis for frequency dependent attenuation and absorption and ultimately may allow for the solution of not only compressibility but density variations[12]. Although in the experiment described by this paper, pseudo random noise was used as the multifrequency source, this technique represents only one

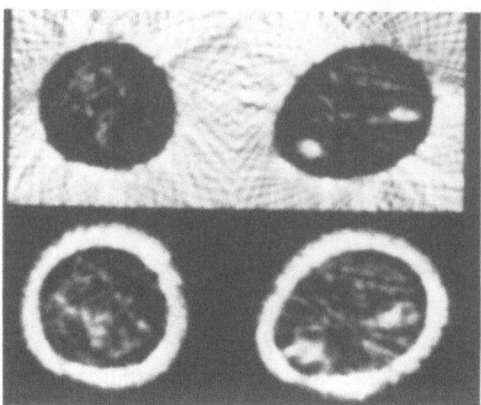

Fig. 1. Ultrasound reconstruction of speed (upper) and
         attenuation (lower) in breasts of a woman having grade IV
         adenocarcinoma, 1.6 cm in diameter, in the lower inner
         quadrant of the left breast. (Reproduced with permission
         from Greenleaf, et al.[14])

of several methods which could be used for generating multiple
and simultaneous frequencies.

     Fig. 1 illustrates a typical reconstruction of the distri-
bution of ultrasonic speed and attenuation within the breasts of a
patient in which straight-line assumptions for the reconstructions
were used.[13] A translate rotate scanner and associated reconstruction techniques described elsewhere, were utilized for obtaining
these results[14]. The system has been used in a preliminary clinical trial, in which the specificity, sensitivity and accuracy using
the straight ray technique in a series of some 80 individuals were
reported.[15] As can be seen from Fig. 1, various artifacts are present. The low resolution is a result of several effects, two of
which are 1) the fact that the rays do not travel in a straight
line and 2) the fact that first arrival time methods were used to
measure time-of-flight and thus the later arrival of energy was
ignored. The effects of diffraction and refraction on straight-
line tomography have been reported separately.[16] The artifact at
the edge of the breast and the incorrect size of these images are
the result of the diffraction of the ultrasonic energy and its
violation of the underlying assumptions of straight-ray trajectories used by the algorithms applied to the data obtained from

these patients. Although cancer and other diseases can be detected using this method, it seems clear that a higher fidelity technique for calculating the two-dimensional distributions of speed and attenuation is necessary to utilize textural, architectural and morphologic variations in the breast and thus obtain a greater level of accuracy.

The equation for wave propagation assumed to be applicable in this investigation is the Helmholz equation:

$$\nabla^2 \Psi + K_o^2 \eta^2 \Psi = 0, \tag{1}$$

where $\Psi = \Psi(\overline{x}, K)$ is the wave function, $\overline{x}$ is the position vector of an arbitrary point p in space, K_o is the wave number in the unperturbed medium and $\eta = \eta(\overline{x})$ is the refractive index.

Brekhovskikh[17] showed previously that this equation applies both to optics and electromagnetics as well as to acoustics under conditions of practical importance. Under the assumption that density variations are absent, we make the following substitutions: we take the field Ψ^I to be incident upon the scattering region and we define the total field $\Psi = \exp(\gamma)$ and $\Psi^I = \exp(\gamma^I)$, then $\gamma = \tilde{\gamma} + \gamma^I$. Substituting into Eq. 1 gives

$$\nabla^2 \tilde{\gamma} + 2 \nabla\gamma^I \cdot \nabla\tilde{\gamma} + \nabla\tilde{\gamma} \cdot \nabla\tilde{\gamma} + K_o^2(\eta^2 - 1) = 0. \tag{2}$$

In the Rytov procedure the term $\nabla\tilde{\gamma} \cdot \nabla\tilde{\gamma}$ is now neglected.

By making the definition

$$\phi = \Psi^I \gamma \tag{3}$$

and substituting into (2) we obtain

$$\nabla^2\phi + K_o^2 \phi = -K_o^2 (\eta^2 - 1)\Psi^I. \tag{4}$$

Following Morse and Ingard,[18] we then obtain the following equation:

$$\phi(\overline{x}) - \Psi^I(\overline{x}) = \iiint_V K_o^2(\eta^2(\overline{x}_o) - 1)\Psi^I(\overline{x}_o)G(\overline{x}|\overline{x}_o)d\overline{x}_o. \tag{5}$$

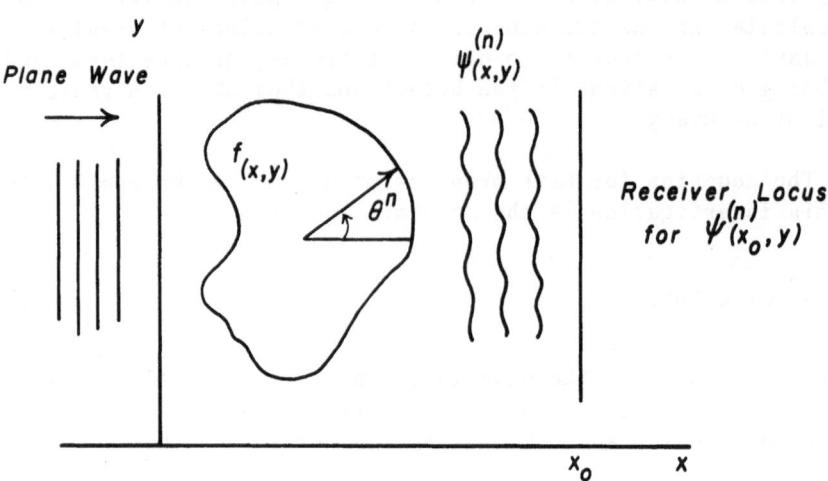

Fig. 2.   The object is insonified by a plane continuous wave. The
         scattered wave ψ is received along a line X_0 and its phase
         and amplitude are measured as a function of y. Many ψ^n
         (n=1,...M) are measured, each for a separate rotation of
         the object θ^n. (Reproduced with permission from J. F.
         Greenleaf, Methods of Experimental Physics-Ultrasound, P.
         Edmonds, ed., Academic Press, NY (1981))

This equation relates the source term on the right-hand side
of Eq. 4 through a convolution with the Green's function G to the
scattering from the object ϕ along with the incident distribution
of pressure ψ^I. Now following Iwata and Nagata[9] we solve this
equation for the geometry shown in Fig. 2 and obtained the
following equation:

$$\overline{\eta}(U,V) = \frac{1}{2\pi} \frac{u}{K_0} \exp[i(K_0 - u)x_0]\, \hat{\phi}(x_0, u, \theta_i)$$

where

$$\overline{\eta}(U,V) = \frac{1}{(2\pi)^2} \iint \eta(x,y)\exp[-i(Ux + Vy)]\, dx\, dy$$

$$U = (u - K_o) \cos \theta_i - v \sin \theta_i \qquad\qquad (6)$$

$$V = (u - K_o) \sin \theta_i + v \cos \theta_i$$

and

$$u^2 + v^2 = K_o^2,$$

and where $\overline{x} = x\overline{i} + y\overline{j}$.

This equation states that taking the Fourier transform of the complex phases $\phi(x_0, y, \theta_i)$ received along the receiver locus ($x = x_0$) in Fig. 2 for insonification angle θ_i gives $\hat{\phi}(x_0, u, \theta_i)$ which, except for a phase term, is equal to the values along a circular locus in the Fourier domain of the object as shown in Fig. 3. By rotating the object, one can obtain circular loci and at various angles through the Fourier space and fill in the Fourier plane as described by Kaveh, et al[19]. By changing the frequency the radius of the circular locus can be changed also helping to fill in the Fourier domain (assuming $\eta(\overline{x})$ is not a function of K).

It should be noted here that in the reflection mode the exterior half of the circle is measured (the crosses) and in the transmission mode the interior half of the circle shown in Fig. 3 can be measured (the closed circles). Therefore, in the transmission mode one can measure the low frequency ($\sqrt{U^2 + V^2} \leq \sqrt{2} K_o$) or central region of the Fourier Transform plane and the DC values. In the reflection mode the central region cannot be measured although an annular region ($\sqrt{2} K_o \leq \sqrt{U^2 + V^2} \leq 2 K_o$) of the Fourier Transform domain can be acquired. This explains the low frequency but quantitative nature of transmission tomography images and the differentiated but high frequency character of B-scans.

METHODS

An experiment was designed to test the Rytov method of inverse scattering. Fig. 4 is a schematic description of the experiment. Pseudo random noise signals were generated by loading a shift register with random numbers. The random signal, after being amplified, was transmitted by a cylindrical transducer made from PVDF. The cylindrical transducer produced a cylindrical wave focused at the center of rotation of the phantom in an attempt to confine as much energy as possible to the plane of interest. The scattered signal was received with a small hydrophone (1 mm diameter x 3 mm long cylinder) that

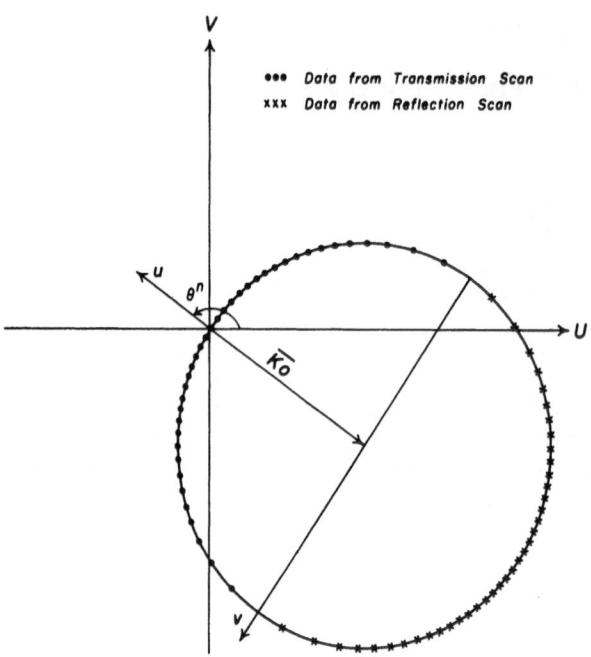

Fig. 3. The measured "profile" is Fourier transformed and
         after multiplying by a phase factor is placed along
         a circle in the two-dimensional Fourier plane. The
         circle is centered on the wave vector $\overline{K}_0$ and moves
         as the magnitude (wavelength) $\propto 1/|\overline{K}_0|$ as direction
         of K_0 varies. Thus, the Fourier plane can be filled
         in by rotating the object or by varying frequency.

detected the total scattered pressure field Ψ. The hydrophone
was scanned under computer control. After suitable amplification
the signal was A/D converted using a Biomation Transient Recorder
Model 8100 digitizing at 10 MHz. The data received from each of
50 angles of view were spectral analyzed by Fast Fourier Transform
and specific frequencies having high energy were selected for reconstruction using the Fourier Transform inversion method described
by Iwata and Nagata[9] and later by Kenue[12] and Kaveh[19] and
Greenleaf[20].

    The modulus of the Fourier transform of the signal received
through water is shown in Fig. 5 and indicates that, although randomly distributed, there were some amplitude peaks which could be
used for reconstruction. By selecting four separate frequencies,
we applied the Rytov reconstruction technique[9] to the resulting
data and obtain reconstructions shown in Fig. 6.

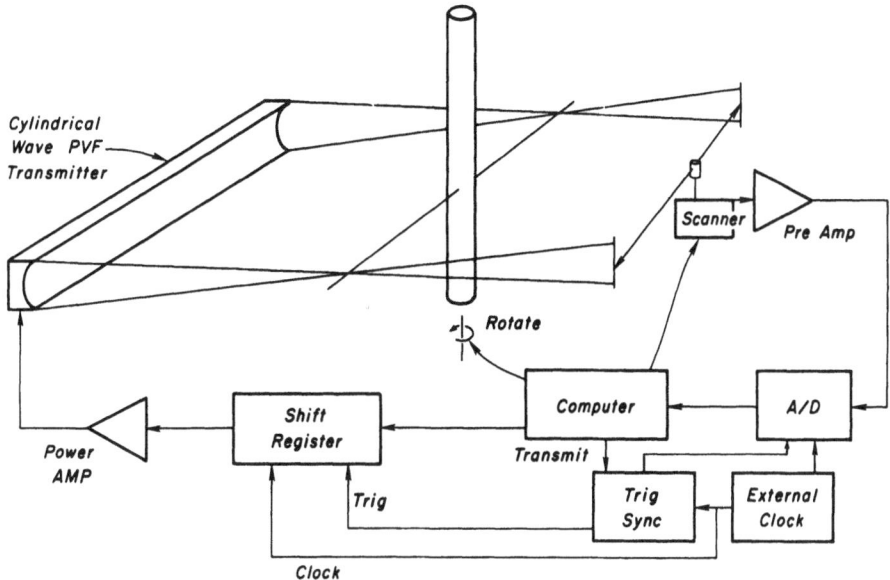

Fig. 4.   Schematic of experimental arrangement.   Random 8 bit
          numbers are shifted from register into amplifier.
          Cylindrical wave insonifies object and scattered wave
          is measured by a hydrophone scanned by computer in
          synchrony with transmit triggers.   A/D converted
          signals are stored in computer for analyses and re-
          construction.

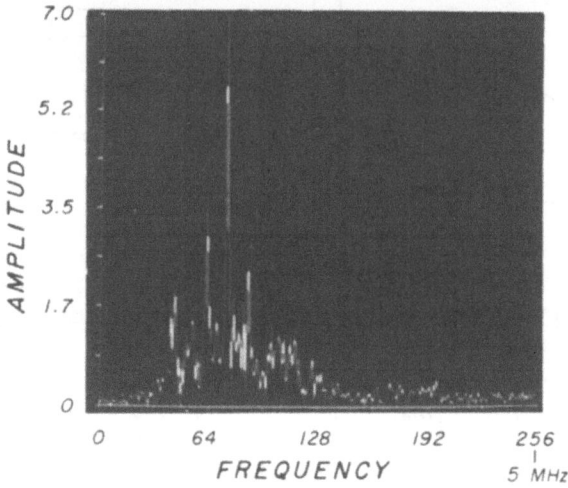

Fig. 5. Spectrum of signal received through water. Peaks in
energy were used for reconstruction.

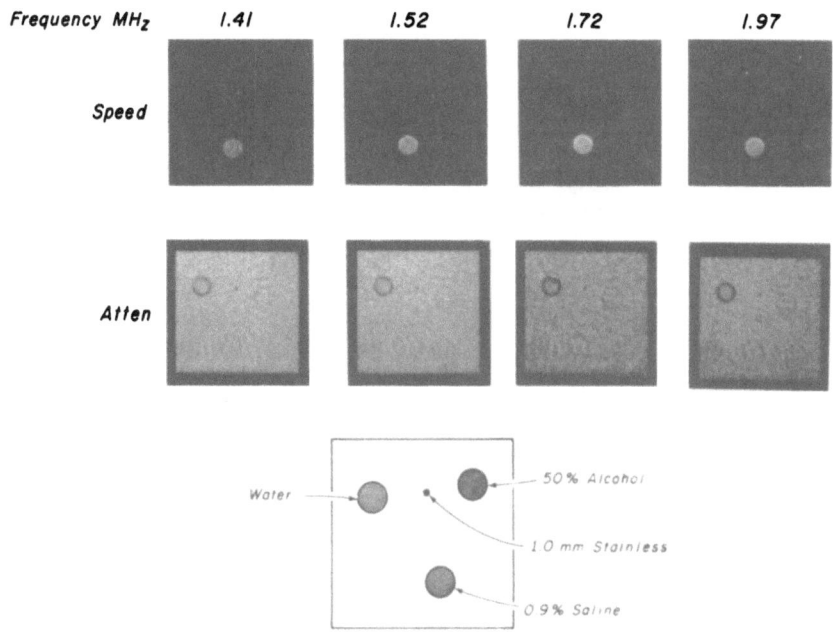

Fig. 6. Reconstructions of speed and attenuation at different
         frequencies using Rytov approximation.

         The attenuation of the water filled finger cot was
         apparently due to microbubbles on the surface of rubber.

         Frequencies were selected from peaks of energy shown in
         Fig. 5.

         Fig. 7 illustrates images of slope and intercept of speed
and attenuation vs. frequency obtained from values in the
reconstructions shown in Fig. 6. From these images one can see
the frequency dependent and frequency independent components of
speed and attenuation associated with the phantom.

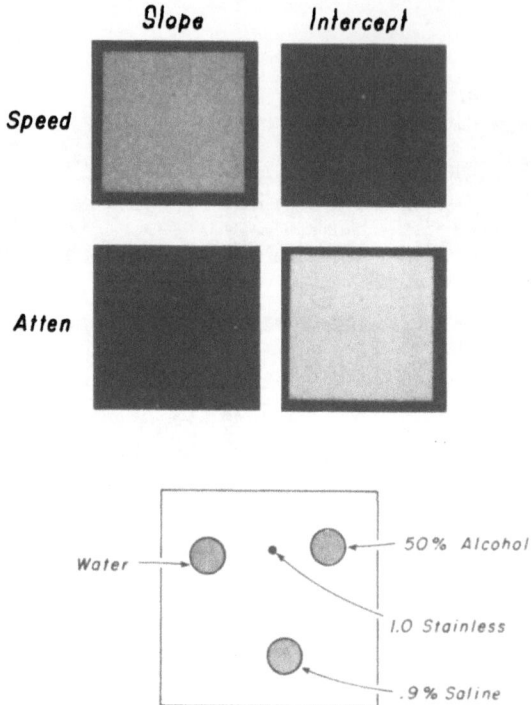

Fig. 7.   Images of slope and intercept from attenuation and speed
          reconstructions shown in Fig. 6. Wire can be seen in
          slope images indicating scattering phase and amplitude
          are both function of frequency. Flat attenuation slope
          images indicate very little frequency dependent
          attenuation (e.g., absorption) was present.

          Fig. 8 illustrates averages of speed and attenuation,
calculated from values of the four separate figures of Fig. 6.

DISCUSSION

          The separate reconstructions of the real and imaginary, that
is, speed and attenuation components of refraction index, respectively, are remarkably consistent over a range of virtually 2 to
1 in frequency. Notice that the artifacts associated with reconstructions shown in Fig. 1 are absent and that Fig. 6 illustrates
apparently high resolution and accurate reconstructions.

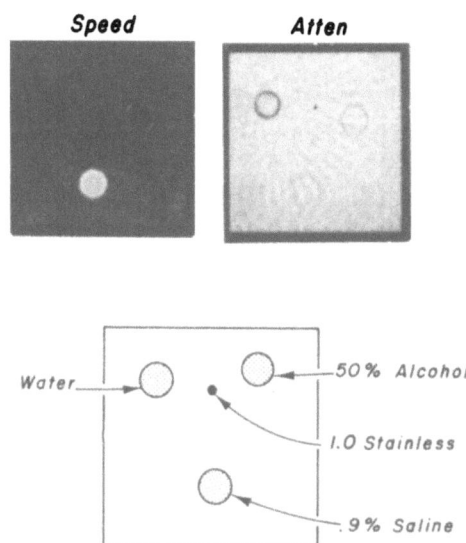

(1.41, 1.52, 1.72, & 1.97 MH₃)

Fig. 8. Average of speed and attenuation values calculated from
image shown in Fig. 6. Some pattern remains in
background, apparently due to interpolation in Fourier
space.

The images of slope and intercept (Fig. 7) illustrate that
the energy scattered into the receiver must be a function of
frequency for the wire but not for the finger cots. As can be
seen from the lower left panel, this apparently has to do with
the complex scattering associated with objects having a product
of radius times wave number (K) of 2.

The image of the wire in the slope of the speed images is
curious. Apparently, the relative phase shift due to the wire
for the the various frequencies was not commensurate with the
speed of sound in the wire. This is not surprising since the
impulse response of the imaging method probably varies with

frequency and as a result, the image of wire is different at different frequencies (Fig. 6).

The frequency independent images (intercept) show the speed but not the attenuation terms indicating most of the attenuation was reflective and not absorptive.

CONCLUSION

Whether multifrequency reconstructions will allow us to solve for the frequency dependence of scattering independent of absorption is yet to be determined in tissues.

However, it appears that multifrequency reconstructions can be obtained using pulses or other broadband signals and that frequency independent and frequency dependent speed and attenuation images can be obtained in simple models using the Rytov method of reconstruction.

ACKNOWLEDGMENTS

The authors thank E. C. Quarve for secretarial assistance and S. J. Richardson and J. L. Hanson for graphics and P. Thomas for technical assistance.

This work was supported in part under Grants CA 24085 from the National Cancer Institute, GM 24994 from the National Institutes of Health, and ECS 7926008 from the National Science Foundation.

REFERENCES

1. J. F. Greenleaf, S. A. Johnson, S. L. Lee, G. T. Herman, and E. H. Wood, Algebraic reconstruction of spatial distributions of acoustic absorption within tissue from their two-dimensional acoustical projections, in: "Acoustical Holography," P. S. Green, ed., Plenum Press, New York (1974).
2. G. H. Glover and J. C. Sharp, Reconstruction of ultrasound propagation speed distributions in soft tissue: Time of flight tomography, IEEE Trans on Sonics and Ultrasonics 24:229 (1977).
3. P. L. Carson, T. V. Oughton, and W. R. Hendee, Ultrasound transaxial tomography by reconstruction, in: "Ultrasound

in Medicine," D. N. White and R. W. Barnes, eds., Plenum Press, New York (1976).

4. J. F. Greenleaf and R. C. Bahn, Clinical imaging with transmissive ultrasonic computerized tomography, IEEE Trans Biomed Eng 28:177 (1981).

5. S. A. Johnson, J. F. Greenleaf, B. Rajagopalan, and R. C. Bahn, Ultrasound images corrected for refraction and attenuation: A comparison of new high resolution methods, in: "Computer-Aided Tomography and Ultrasonics in Medicine," J. Raviv, J. F. Greenleaf, and G. T. Herman, eds., North-Holland, Amsterdam (1979).

6. E. J. Farrell, Tomographic imaging of attenuation with simulation correction for refraction, Ultrasonic Imaging 3:144 (1981).

7. J. G. Miller, J. R. Klepper, G. H. Brandenburger, L. J. Busse, M. O'Donnell, and J. W. Mimbs, Reconstructive tomography based on ultrasonic attenuation, in: "Computer Aided Tomography and Ultrasonics in Medicine," J. Raviv, J. F. Greenleaf, and G. T. Herman, eds., North-Holland, Amsterdam (1979).

8. R. K. Mueller, M. Kaveh, and G. Wade, Reconstructive tomography and applications to ultrasonics, Proc IEEE 67:567 (1979).

9. K. Iwata and R. Nagata, Calculation of refractive index distribution from interferograms using the Born and Rytov's approximation, Jpn J Appl Phys 14:379 (1975).

10. D. Nahamoo and A. F. Kok, Ultrasonic diffraction imaging. Technical Report TREE-82-20, School of Electrical Engineering, Purdue University (1982).

11. A. J. Devaney, A filtered backpropagation algorithm for diffraction images, Ultrasonic Imaging 4:336 (1982).

12. S. K. Kenue and J. F. Greenleaf, Limited angle multifrequency diffraction tomography, IEEE Trans on Sonics and Ultrasonics 29:213 (1982).

13. J. F. Greenleaf and R. C. Bahn, Clinical imaging with transmissive ultrasonic computerized tomography, IEEE Trans Biomed Eng 28:177 (1981).

14. J. F. Greenleaf, J. J. Gisvold, and R. C. Bahn, A clinical prototype ultrasonic transmission tomographic scanner, in: "Acoustical Imaging," E. A. Ash and C. R. Hill, eds., Plenum Press, New York (1982).

15. J. F. Greenleaf, R. C. Bahn, J. J. Gisvold, and J. S. Schrieman, Evaluation of interpretation methods for transmission ultrasonic tomography of the breast, Proceedings of the International Congress on the Ultrasonic Examination of the Breast, Tokyo, Japan (1983) - In Press.

16. J. F. Greenleaf, P. J. Thomas, and B. Rajagopalan, Effects of diffraction on ultrasonic computer-assisted tomography,

      in: "Acoustical Imaging," J. P. Powers, ed., Raven Press,
      NY (1982).
17.   Brekhovskikh, Waves in layered media, Academic Press (1960).
18.   P. M. Morse and K. V. Ingard, Theoretical acoustics,
      McGraw-Hill (1968).
19.   M. Kaveh, M. Soumekn, Z. Q. Lu, R. K. Mueller, and J. F.
      Greenleaf, Further results on diffracton tomography using
      Rytov's approximation, in: "Acoustical Imaging," E. A.
      Ash and C. R. Hill, eds., Plenum Press, New York (1982).
20.   J. F. Greenleaf, Computerized tomography with ultrasound,
      Proc IEEE 71:330 (1983).

THE INVERSE ACOUSTICAL SCATTERING PROBLEM FOR LAYERED MEDIA IN THE PRESENCE OF BROAD-BAND ACOUSTIC NOISE AND WITH LIMITED TRANSDUCER BANDWIDTH

P.C. Pedersen[+], O.J. Tretiak[+], and J. Bai[*]

[+]Dept. of Electrical and Computer Engineering
[*]Dept. of Physics and Atmospheric Science,
Drexel University, Philadelphia, PA 19104

INTRODUCTION

The inverse problem in acoustics (or acoustical imaging) is based on the theory of acoustic wave propagation, absorption, and scattering. The techniques for processing and display of the medium parameters, based on scattered acoustic energy may be divided into 2 general categories: (1) techniques based upon transmitted, or forward scattered ultrasound energy; (2) techniques based upon reflected, or backscattered, ultrasound energy. The reconstruction techniques, discussed in this paper, are based on backscattered acoustic energy.

There are several categories of inverse methods. One category reconstructs the acoustic impedance profile directly by solving an integral equation using approximation techniques, such as the optimization technique [1] or the forward scattering approximation (FSA) method [2]. The optimization technique makes use of the technique of minimizing a cost function which is related to the difference between the measured field and the field scattered by a given medium, to solve the integral equation of the system. The FSA method makes use of Fourier transformation to convert the integral equation into a differential equation before solving it. Other inverse methods utilize the reflection impulse response to derive the properties of the unknown medium [3-5]. Here the acoustic impedance profile of the medium may be reconstructed in two steps: first a deconvolution is performed to remove the effect of the transmitter and receiver characteristics so that the reflection impulse response can be obtained from the incident and reflected waves, and then the acoustic impedance profile is reconstructed from the impulse response by certain approximation techniques. In this paper we will examine reconstruction techniques which utilizes the reflection impulse response.

In our recent work [6-8], we have evaluated numerically two techniques for reconstructing the acoustic impedance profile of an unknown medium, based on the reflection impulse response of the medium. One reconstruction technique, the so-called impediography method [3], is based on the assumption that the higher order reflections are negligible, and is therefore only valid when the total impedance change is small. The impediography method is a non-recursive method which therefore does not have any stability problems. The other technique, which is based on a recursive algorithm by Goupillaud [4,5], (and is thus called Goupillaud's method), utilizes first order as well as higher order reflections, and is therefore valid for acoustic impedance changes of arbitrary magnitude. Goupillaud's method may become unstable in the presence of significant noise due to its recursive nature.

Our previous investigation was based upon a number of simplifying assumptions: (1) The interrogating wave is a plane wave; (2) The interrogating wave is a one-dimensional longitudinal wave; (3) The interrogating wave has an infinite bandwidth; (4) All signals are noise free; (5) The medium has no attenuation and is isotropic; (6) The acoustic impedance of the unknown medium varies in only one direction.,

The purpose of this paper is to investigate, both theoretically and by computer simulation, the reconstruction under more practical conditions: in the presence of noise, especially the acoustic noise; and assuming limited bandwidth of the transmitting and receiving transducers. We will show simulation results and give a theoretical estimation of the effect of noise alone, and then present simulation results for the effect of limited bandwidth alone, and the effect of both noise and limited bandwidth together with compensation for the effect of the limited bandwidth by means of deconvolution.

LIST OF SYMBOLS AND GIVEN ACOUSTIC IMPEDANCE PROFILES
  $h(t)$: impulse response of the transducer
  $H(f)$: frequency response of the transducer
  $g(t)$: combined impulse response of transmitting and receiving transducer $=h(t)*h(t)$
  $G(f)$: combined frequency response of transmitting and receiving transducer $=H^2(f)$
  $r_o(t)$: reflection impulse response of the medium
  $r(t)$: $r(t)=r_o(t)*g(t)+n(t)*h(t)$
  $\hat{r}(t)$: reflection impulse response of the medium after deconvolution
  $R_o(f)$: reflection frequency response of the medium
  $R(f)$: $R(f)=R_o(f)H^2(f)+N(f)H(f)$, or $R(f)=R_o(f)G(f)+N(f)H(f)$

$\hat{R}(f)$:   the Fourier transform of $\hat{r}(t)$

$z_g(x)$:   given acoustic impedance profile of the medium

$z_o^g(x)$:   reconstructed acoustic impedance profile without effects of noise and transducers

$z(x)$:   reconstructed acoustic impedance profile under the effect of noise or transducer and without deconvolution

$\hat{z}(x)$:   reconstructed acoustic impedance profile with the aid of deconvolution

$\sigma(n)$:   standard deviation of acoustic noise signal, $n(t)$

σ_z :   standard deviation of reconstructed acoustic impedance profile, due to noise

Given Acoustic Impedance Profiles

$z_1(x) = \exp(0.0953x)$ $0 \leqslant x \leqslant 1 \text{cm}$

$z_2(x) = 1.05 + 0.05\sin(4\pi x - 0.5\pi)$ $0 \leqslant x \leqslant 1 \text{cm}$

$z_3(x) = \exp(2.3x)$ $0 \leqslant x \leqslant 1 \text{cm}$

$z_4(x) = 5.5 + 4.5\sin(4\pi x - 0.5\pi)$ $0 \leqslant x \leqslant 1 \text{cm}$

THEORY

a. The Direct Problem

In order to perform computer simulation of the acoustic impedance profile reconstruction, we need to determine the reflection impulse response, $r_o(t)$ for the given acoustic impedance profile. (When simulating realistic measurement conditions, only $r(t)$ or $\hat{r}(t)$ is available). To obtain $r_o(t)$, we start with a given acoustic impedance profile for the inhomogeneous layer, $z_g(x)$, $0 \leqslant x \leqslant d$. This lossless layer is bounded by two semi-infinite homogeneous regions whose acoustic parameters are known. Since there does not exist a closed-form solution for the reflection and transmission properties of $z_g(x)$ in general, we first divide $z_g(x)$ into a large number of thin layers. The acoustic impedance profile within each layer is then approximated to an exponential impedance profile as a function of travel time. (For constant sound velocity in the inhomogeneous layer, this can be written as a function of x). Based on the wave equation, we have derived a transmission matrix for an exponential impedance profile by means of which the transmission properties of each of the thin layers can be described. By multiplying together the transmission matrices for the thin layers, the transmission properties of the total inhomogeneous layer is obtained from which the reflection frequency response, $R_o(f)$, can be calculated. Finally, $r_o(t)$ is obtained as the inverse Fourier transform of $R_o(f)$ [6,8].

b. Reconstruction Algorithms for the Inverse Problem

The two reconstruction algorithms which will be utilized in this paper have both been described in detail in the literature [4,6,7]. Hence, only a brief outline of these algorithms is given below.

For the Impediography method, the relation between the reflection coefficients r(t) and the acoustic impedance profile is:

$$z(t)=z(0)\exp(2\int_0^t r(t')dt') \qquad (1)$$

where z(t) and r(t) represent the acoustic impedance and the reflection coefficient, respectively, at the point, x=ct/2 reached by the wave at time t/2. z(0) is the acoustic impedance of reference medium at incident side of the inhomogeneous medium.

The Goupillaud's method is obtained by means of the stratified medium model which is divided in such a way that the travel time through each layer is the same. This leads to the following relation between reflection coefficents r and acoustic impedance profile z :

$$z(k+1)= \prod_{i=0}^{k}(1+f_i(r))/(1-f_i(r)) \qquad (2)$$

where

$$f_0(r)=r_0$$
$$f_1(r)=r_1(1-f_0^2(r))^{-1}$$
$$f_2(r)=(r_2+f_0 f_1 r_1)[(1-f_0^2)(1-f_1^2)]^{-1}$$
e.t.c.

Here, z(k) represents the impedance of the kth layer. r_i represents the reflection coefficient of the ith layer, where i is an integer from 0 to N, and N is the total number of layers.

Whenever the total relative impedance change is small (less than 50%) both the Impediography method and Goupillaud's method will give nearly identical and correct results. However, with large relative impedance changes only Goupillaud's method can reconstruct the acoustic impedance profile with small error.

c. Simulation Model of Physical System

The physical model of the problem we will consider is: The transmitting transducer is excited by an electrical impulse, $\delta(t)$, and produces an ultrasonic pulse, h(t) which will propagate into the medium under investigation (h(t) is the impulse response of the transducer). During propagation through the inhomogeneous medium, part of the transmitted acoustic energy is reflected back to the transducer which now functions as the receiving transducer. If the reflection impulse response of the medium is $r_0(t)$, the backscattered energy is $h(t)*r_0(t)$. In addition, acoustic noise, n(t), may be added, so that the total acoustic energy, returned to the receiving transducer is $[h(t)*r_0(t)+n(t)]$. The receiving transducer produces an electrical signal, y(t), which is equal to the total received acoustic energy, convolved with the impulse response, h(t), of the transducer in receiving mode. The output signal, y(t), is the input signal to the reconstruction algorithms, either the Impediography method, or Goupillaud's method. The signal degradation due to the transmitting and receiving transducers may be partially compensated for by inverse filtering (deconvolution) of y(t), before feeding the signal to the reconstruction algorithms.

While the physical model of the acoustic measurement system is most realistically described in the time domain, the simulation model is carried out in the frequency domain, except for the actual impedance profile reconstruction. In the simulation model, shown in Figure 1, the reflection frequency response, $R_o(f)$, of the inhomogeneous medium, is calculated by means of a transmission matrix method, as described earlier. The frequency responses of the transmitting and receiving transducers are lumped together as $G(f)=H^2(f)$, and since the added noise is acoustic noise, the noise spectrum, $N(f)$, is multiplied with the frequency response of the receiving transducer, $H(f)$, before being added to $R(f)=R_o(f)G(f)$. This combined signal, $R_o(f)G(f)+N(f)H(f)$, can be seen to correspond to the Fourier transform of the output signal, y(t), in the physical model.

In Figure 1, signal restoration in terms of deconvolution (inverse filtering) is carried out before taking the inverse Fourier transform to give $\hat{r}(t)$, and applying $\hat{r}(t)$ to the impedance reconstruction algorithms. However, in this paper, we will first evaluate the acoustic impedance profile reconstruction due to noise alone, i.e. $G(f)=1$, and of no deconvolution. Then we will examine the impedance profile reconstruction when no acoustic noise is present, i.e. $N(f)=0$, and no deconvolution. Finally, we will carry out the impedance profile reconstruction, as shown in Figure 1.

d. Transducer Model

In order to model the effect of limited transducer bandwidth, we approximate the transducer function by $H(f)$.

$$H(f) = \frac{jff_o/Q}{(f_o^2-f^2+jff_o/Q)} \tag{3}$$

where f_o represents the resonance frequency of the transducer. The shape of this transducer response curve is similar to what has been obtained experimentally with free-field radiation from ultrasound transducers [9-10].

Recalling that $G(f)=H^2(f)$ and that g(t) and h(t) are the corresponding impulse responses, we have, in the absence of acoustic noise,

$$R(f)=R_o(f)G(f) \text{ and } r(t)=r_o(t)*g(t)=r_o(t)*h(t)*h(t)$$

The duration of r(t) is at most the sum of the durations of the impulse responses in the convolution. With the help of Laplace transformation tables one finds that

$$h(t)=-2b[\cos(at)+(b/a)\sin(at)]\exp(bt) \tag{4}$$

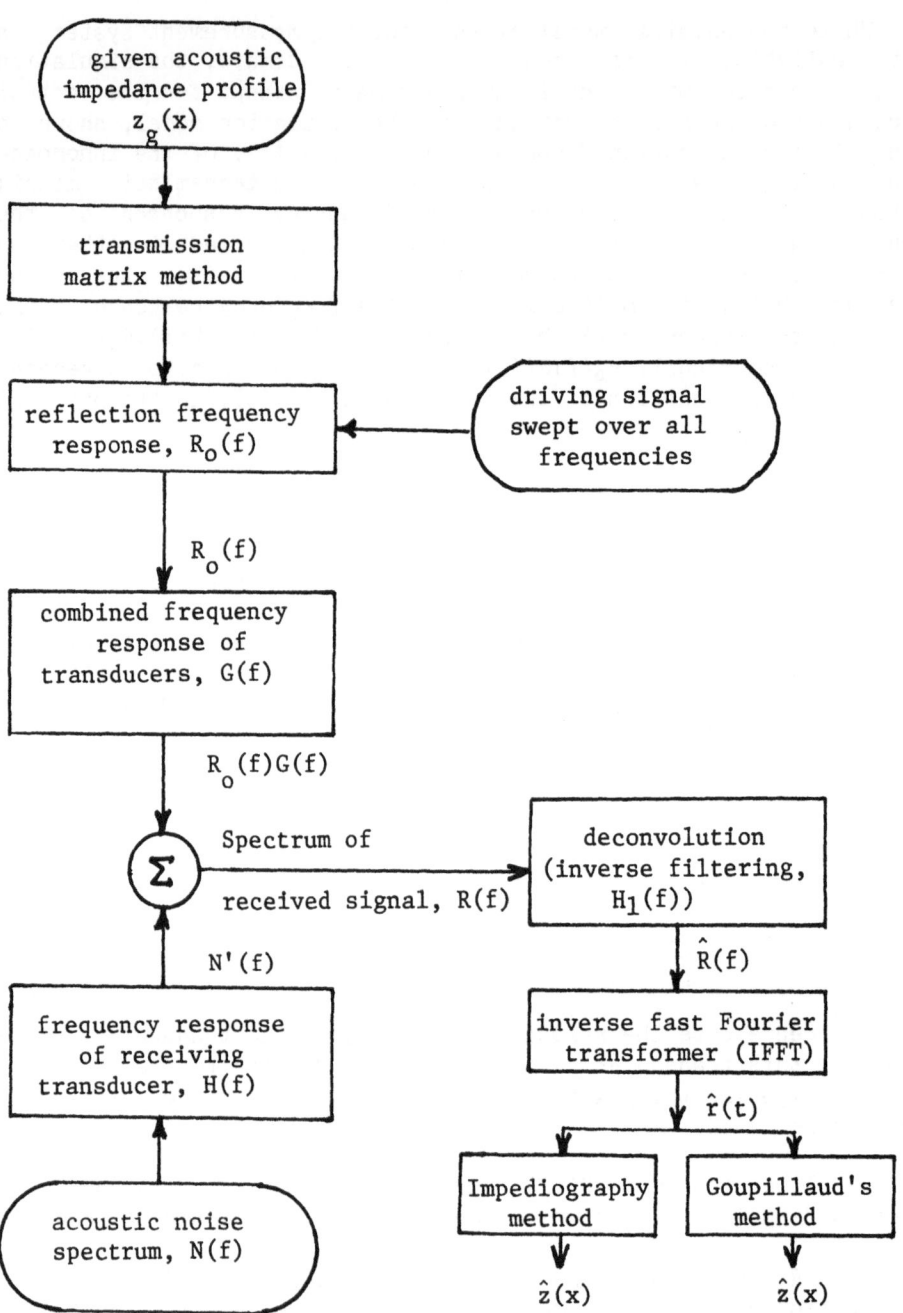

Figure 1.   Simulation Model of Acoustic Measurement System

where $b=-\pi f_o/Q$, $a=2\pi f_o\sqrt{1-1/4Q^2}$, Hence, h(t) and therefore r(t) increases in duration when Q is increased and f_o is decreased. This is of importance in the simulation since the maximum frequency interval between consecutive values of R(f) must be less than the reciprocal of the duration of r(t), in order to prevent undersampling. As is seen in Figure 1, r(t) (or $\hat{r}(t)$ with deconvolution) is found as IFFT{R(f)}.

e. Noise Model

While we will assume that the physical noise, n(t) is white, we will use bandlimited white noise in the simulation model. The characteristics of bandlimited white noise can be summarized briefly as follows [11]: (1) the mean value of the noise is zero; (2) it approaches white noise when the bandwidth goes to infinity.

Based on the assumption of white noise for the physical model, we have

$$R_n(t)=N_o\delta(t), \quad S_n(f)=N_o$$

where $R_n(t)$ is the noise autocorrelation function, $S_n(f)$ is the noise power density spectrum, and N_o is the noise power per unit frequency. Let n'(t) be the noise at the receiving transducer terminals. Then

$$n'(t)= \int_0^\infty h(t')n(t-t')dt' \tag{5}$$

$$S_n'(f)=N_o|H(f)|^2 \tag{6}$$

$$\sigma^2(n')=N_o \int_{-\infty}^\infty |H(f)|^2 df \tag{7}$$

where $\sigma(n')$ is the standard deviation of n'.

We will define the signal-to-noise ratio, (S/N), of the input to the reconstruction algorithms as

$$(S/N) \text{ [dB]}=20 \log[\widetilde{peak}(y)/\sigma(n')]=20 \log[m] \tag{8}$$

where $\widetilde{peak}(y)$ is peak(y) for n(t)=0, and where peak(y) is the peak value of the output signal, as described earlier. In other words, the S/N ratio, m, is defined as the ratio of the peak value of the noise free received signal to standard deviation of the received noise. The received signal is the reflection impulse response of the unknown medium, after convolution with the impulse response of the transmitting and receiving transducers, and the received noise is the acoustic noise signal, convolved with the impulse response of the receiving transducer.

From (7) and (8) we have

$$N_o=\sigma^2(n')/ \int_{-\infty}^\infty |H(f)|^2 df=[\widetilde{peak}(y)]^2/m^2 \int_{-\infty}^\infty |H(f)|^2 df \tag{9}$$

In formulating the noise model to be used for computer simulation, n(t) is generated by taking a sequence of independent pseudorandom numbers, uniformly distributed in [-0.5, 0.5], and multiplying them by an appropriate constant k which is related to the maximum value of the acoustic noise. n'(t) is then computed by convolving n(t) with h(t). In the formulation, given below, $S_n(f)$ is redefined to represent the noise power density spectrum for the simulation model. T is the time interval between sampling points in r(t) (or $\hat{r}(t)$). As seen from Figure 1, $\hat{r}(t)$ (r(t) without deconvolution) is obtained as the inverse Fourier transform of $\hat{R}(f)$ (or R(f)). Therefore, $T=1/2\Delta f$ where $-\Delta f \leqslant f \leqslant \Delta f$ is the frequency range over which $\hat{R}(f)$ (or R(f)) is evaluated.

The pseudorandom sequence has the power spectrum as follows

$$S_n(f) = \begin{cases} N_i, & \text{for } -1/2T \leqslant f \leqslant 1/2T \\ 0 & \text{otherwise} \end{cases} \tag{10}$$

where N_i is the noise power per unit frequency for the simulation model. To find N_i, we note that [11]

thus
$$\int_{-\infty}^{\infty} S_n(f)\,df = k^2/12 = N_i/T \tag{11}$$

$$N_i = k^2 T/12 \tag{12}$$

For the physical model, an expression for the noise power per unit frequency, N_o, is given in (9). (12) gives the corresponding expression for the simulation model. In order to simulate the physical model, we must equate (9) and (12), giving

$$k^2 = 12 \ \widetilde{peak}^2(y)/[Tm^2 \int_{-\infty}^{\infty} |H(f)|^2 df] \tag{13}$$

In this paper, we will deal with the problem under a S/N ratio varying between 10dB to 80dB.

f. Deconvolution Model and Theory

To experimentally obtain the reflection impulse response of the unknown medium one would need transmitting and receiving transducers of infinite bandwidth which is of course physically unrealizable, and a noise-free environment. The actual received reflection response is modified due to the bandlimited transducers and due to noise. Based on the method of least squares [12], we can improve the degraded signal by restoration processing. This method is based on the concept of making the square of difference between the restored signal and the signal without distortion be a minimum. The procedure of the deconvolution is indicated in Figure 1 where $H_i(f)$ is the frequency response of the restoration filter. From Figure 1, it is seen

$$\hat{R}(f) = R(f)H_i(f) = [R_o(f)H(f) + N(f)]H(f)H_i(f) \tag{14}$$

We will define ε as the expected value of the difference squared between $\hat{r}(t)$ and $r_o(t)$

$$\varepsilon = E[\int (\hat{r}(t) - r_o(t))^2 dt]$$

where $E[x]$ represents the expectation value of x.

Since the noise is random and, on the other hand, $R_o(f)$, $H(f)$, $H_I(f)$ are all deterministic functions, we have,

$$\varepsilon = 2\pi \int_{-\infty}^{\infty} [|R_o(f)|^2 [|H^2(f)H_I(f) - 1|^2] + N_2 |H(f)H_I(f)|^2] df \qquad (15)$$

where we define $N_2 = Tt\sigma^2 = t_o N_o$. $t_o = NT$ is the time duration of the bandlimited white noise, $n(t)$. N is the number of the sampling points, and $\sigma(n)$ is the standard deviation of the bandlimited white noise.

To follow the standard steps for minimization [12], we let $H_I(f) = A + jB$. Then by calculating $\partial\varepsilon/\partial A = 0$ and $\partial\varepsilon/\partial B = 0$, we find that ε will assume a minimum when

$$H_I(f) = |R_o|^2 H^{2*}/(|R_o H|^2 + N_2)|H|^2 \qquad (16)$$

If $N(f) = 0$, then $H_I H^2 = 1$, and $\hat{R}(f) = R_o(f)$. Furthermore, when $N_2 = Tt\sigma^2(n) = t_o N_o$ is a real function, $H_I H^2$ is also real. Therefore, the term $H_I H^2$ has no effect on the phase of $R_o(f)$ but can only cause magnitude distortion.

EFFECT OF NOISE ALONE

a. Theory of Reconstruction Error due to Noise

In this part, we will evaluate the effect of noise alone on the reconstruction algorithms. With reference to Figure 1, this means that $H(f) = 1$, and $H_1(f) = 1$, and thus in the time domain we have as input function to the reconstruction algorithms

$$r(t) = r_o(t) + n(t) \qquad (17)$$

where $r_o(t)$ is the reflection impulse response of the medium.

Considering the Impediography method, we have from (1) and (17),

$$z(t) = \exp(2\int_o^t r(t')dt') = \exp(2\int_o^t r_o(t')dt' + 2\int_o^t n(t')dt')$$
$$= \exp(2\int_o^t r_o(t')dt') \exp(2\int_o^t n(t')dt')$$
$$= z_o(t)\exp(2\int_o^t n(t')dt') \qquad (18)$$

where we use $z_o(t)$ to represent the reconstructed acoustic impedance by the Impediography method without noise. Thus from (18) we see that the effect of noise upon the Impediography method functions as a multiplying factor. Introducing a new variable, $n_I(t) = \int_0^t n(t')dt'$, we have

$$z(t) = z_0(t)\exp(2n_I(t)) \tag{19}$$

In the case of $n_I(t)$ being very small, we can approximate the exponential function by series expansion and neglect higher order terms. Then, to a good approximation, we have
$$\exp(2n_I(t)) \approx 1 + 2n_I(t)$$
and therefore (19) becomes,
$$z(t) = z_0(t) + 2n_I(t)z_0(t) \tag{20}$$

We can consider $z(t)$ as the sum of the actual value, $z_0(t)$, and an additive term.
Since $n(t)$ has zero mean, $E[n_I(t)] = \int_0^t E[n(t')]dt' = 0$

We assume that $n(t)$ is bandlimited white noise, as discussed previously. In other words,
$$S_n(f) = \begin{cases} T\sigma^2(n) & \text{for } |f| \leqslant 1/2T \\ 0 & \text{otherwise} \end{cases}$$
Then, $E[n^2(t)] = \sigma^2(n)$, and whenever $t \gg T$,

$$\begin{aligned}
\text{Var}[n_I(t)] = E[n_I^2(t)] &= E\left[\int_0^t\int_0^t n(t')n(t'')dt'dt''\right] \\
&= \int_0^t\int_0^t E[n(t')n(t'')]dt'dt'' \\
&= tT\sigma^2(n)
\end{aligned} \tag{21}$$

Let us define σ_z as the standard deviation of reconstructed acoustic impedance profile. From (20), we have $\Delta z = z(t) - z_0(t) = 2n_I(t)z_0(t)$,

and $\sigma_z^2(t) = \text{Var}[z] = E[\Delta z^2] = 4z_0^2(t)\ \text{Var}[n_I(t)]$

Then the standard deviation of the reconstructed impedance is

$$\sigma_z = 2z_0(t)\sqrt{\text{Var}[n_I(t)]} = 2\sigma(n)\sqrt{T}(z_0(t)\sqrt{t}) \tag{22}$$

This shows that the standard deviation of reconstructed impedance is proportional to the standard deviation of the noise, and its time dependence is proportional to the product of the acoustic impedance function and the square root of the travel time.

The meaning of the standard deviation of the acoustic impedance profiles, $\sigma_z(t)$, may be clarified as follows: When an impedance profile is reconstructed many times from an impulse response to which noise with constant standard deviation $\sigma(n)$ has been added, the reconstructed acoustic impedance profile differs from reconstruction to reconstruction. The mean value is $z_0(x)$ which is the reconstructed acoustic impedance profile without the effects of noise and transducers while the spread of the reconstructed impedance profiles is given in terms of the standard deviation, $\sigma_z(t)$.

In (8), we have an expression for the S/N ratio of input data to the reconstruction algorithm, with the input data being $r_0(t)+n(t)$. As we in this part are evaluating the effect of noise alone upon the reconstruction, and thus are ignoring the effect of the transducers, $\widetilde{peak}(y)=peak(r_0(t))$, and $\sigma(n')=\sigma(n)$. (8) is thus changed to:

$$S/N_r \text{ [dB]}=20 \log[peak(r_0(t))/\sigma(n)] \qquad (23)$$

The S/N ratio for the reconstructed acoustic impedance profiles is defined as:

$$S/N_z \text{ [dB]}=20 \log[z_0(t)/\sigma_z(t)] \qquad (24)$$

Using (22), this reduces to:

$$S/N_z \text{ [dB]}=-20 \log[2\sigma(n)\sqrt{Tt}] \qquad (25)$$

The S/N_z ratio is evaluated at the far side of the inhomogeneous layer, i.e., t corresponds to the double travel time through the layer. Using c=1500m/s, and a 1cm thick layer, t=13.3 μsec.

Figure 2 shows the relationship between S/N_r from (23) and S/N_z from (25), calculated for the four impedance profiles, z_1 to z_4, listed under LIST OF SYMBOLS AND GIVEN ACOUSTIC IMPEDANCE PROFILES. From (22), we have shown that except for z_4, $\sigma_z(t)$ is maximum and S/N_z is a minimum when t corresponds to the double travel time, as is used in Figure 2.

We have tested the validity of our theoretical analysis of noise effect by computer simulation. It was done for 8 noise levels, which means that the computer simulation was performed for S/N ratios from 10dB to 80dB with intervals of 10dB. Figure 3 shows the standard deviation, $\sigma_z(x)$ in reconstruction of the impedance profile, z_2(small sinusoidal impedance variation) as determined from computer reconstructions. $\sigma_z(x)$ is calculated from ten actual impedance profile reconstructions, using ten sets of random noise samples, as follows: first we use the pseudorandom number generator function to generate ten sets of noise sequences, $n_i(t)$ for i from 1 to 10; second, by adding these sets of noise sequences to the noise free impulse response, $r_0(t)$, we get ten sets of noisy impulse responses, $r_i(t)=r_0(t)+n_i(t)$ for i=1 to 10; then, with these $r_i(t)$, we reconstruct ten sets of acoustic impedance profiles, $z_i(x)$; and finally, we calculate the standard deviation, σ_z, with following formula

$$\sigma_z =\sqrt{\sum_{i=1}^{10}(z_i(x)-z_0(x))^2/10} \qquad (26)$$

where $z_0(x)$ is the acoustic impedance reconstructed without noise.

As is seen in Figure 3, the error due to noise in the reconstruction of $z_2(x)$ generally increases with the depth of the

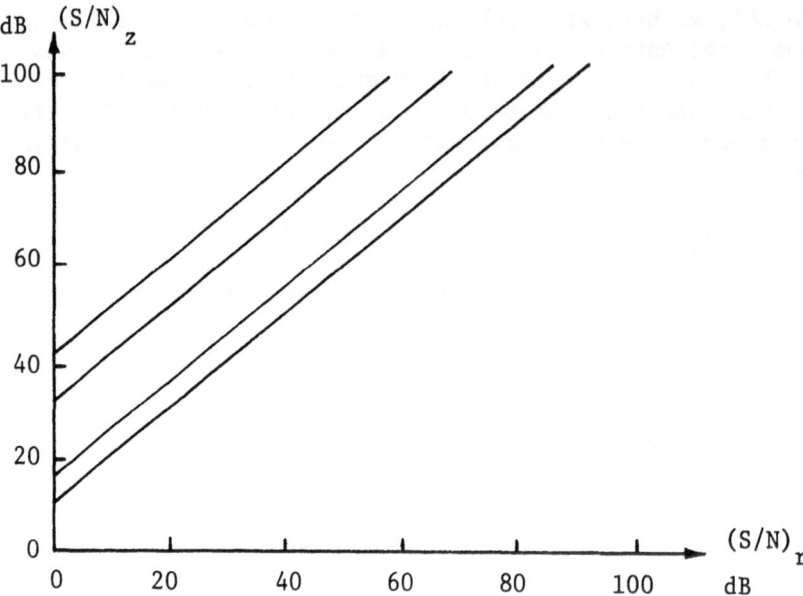

Figure 2.   Calculated S/N of Reconstructed Impedance Profile, Using the Impediography Method, Versus S/N of Reflection Impulse Response.   From Top to Bottom the Lines are for $z_1(x)$, $z_2(x)$, $z_3(x)$,
and $z_4(x)$, respectively. All Curves are Calculated at x=1cm.

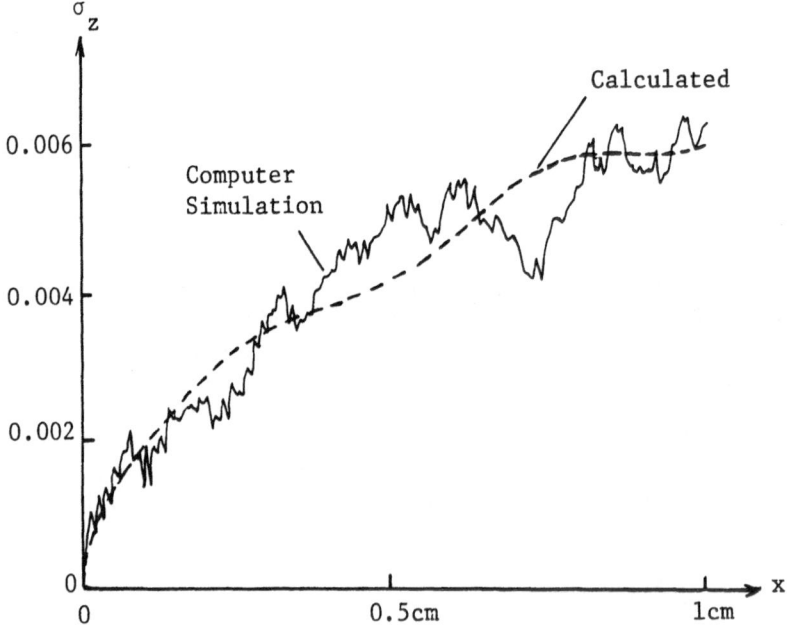

Figure 3.   Standard Deviation from Computer Simulation Data of
Reconstructed Acoustic Impedance Profile, $z_2(x) = 1.05 +
0.05\sin((4x-0.5)\pi)$, S/N = 10dB.   The Calculated Standard Deviation
is included for Comparison.

medium. A theoretical curve, obtained from (22), is plotted in Figure 3 for comparison. Even though the experimental curve is based on only 10 runs, there is a good agreement between the theoretically predicted standard deviation and the actually obtained standard deviation. Performing the same experiment as shown in Figure 3 for the impedance profiles z_1, z_3, and z_4 gave similarly good agreements except for some deviation near x=1cm for profiles z_3 and z_4.

EFFECT OF LIMITED BANDWIDTH ALONE

Here we will evaluate the effect of limited bandwidth alone on the reconstruction algorithms. Referring to Figure 1, N(f)=0, $H_1(f)$=1, H(f) is given in (3), and the input to the reconstruction algorithms is IFT{R_o(f)G(f)} where G(f)=H^2(f). If G(f)=|G(f)|$\underline{/\text{arg}G(f)}$, we have found that if either IFT{R_o(f)|G(f)|} or IFT{R_o(f) $\underline{/\text{arg}G(f)}$} is used as input to the reconstruction algorithms, there will in both cases be serious reconstruction errors. The reconstruction errors will increase with increased Q of the transducer and will decrease with increased overlap between the curve for the power spectrum of the reflected acoustic energy (proportional to $|R_o(f)|^2$) and the curve for the amplitude of the combined transducer function, |G(f)|.

Figure 4 and 5 show the reconstruction of the acoustic impedance profile of $z_2(x)$ and $z_4(x)$, respectively. In both figures, the given acoustic impedance profile is included for comparison. Since $z_2(x)$ exhibits only a small relative impedance variation, both the Impediography method and Goupillaud's method give the same result. In contrast, $z_4(x)$ has a large relative impedance variation, and hence only Goupillaud's method has been used in Figure 5. A low transducer Q value, Q=1/4 has been used, but even in this case the distortion is very serious. The effect of transducer resonance frequency is evaluated by carrying out the reconstruction at three different resonance frequencies, 100, 200, and 300khz. A resonance frequency of 100khz seems to provide a sligtly better result.

The curves in Figure 5 clearly show that without an adequate deconvolution technique, the distortion in the reflected signal, caused by even low Q transmitting and receiving transducers, can make impedance profile reconstruction technique useless.

DECONVOLUTION

As the examples of impedance profile reconstruction, shown in Figure 4 and 5, have demonstrated, the distortion in the reconstructed profiles due to the limited transducer bandwidth is unacceptable. We will now investigate the complete computer simulation system, including deconvolution, as depicted in Figure 1.

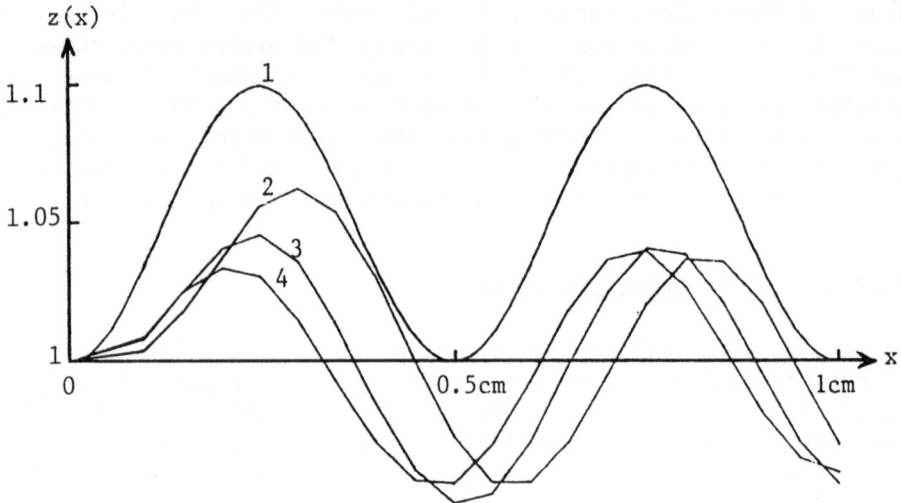

Figure 4. The Reconstructed Impedance Profile, $z_2(x)$, Distorted by the Total Transfer Function of the Combined Transducer. Transducer Q = 1/4. 1: given impedance profile. 2,3 and 4: f_o = 100khz, 200khz, and 300khz.

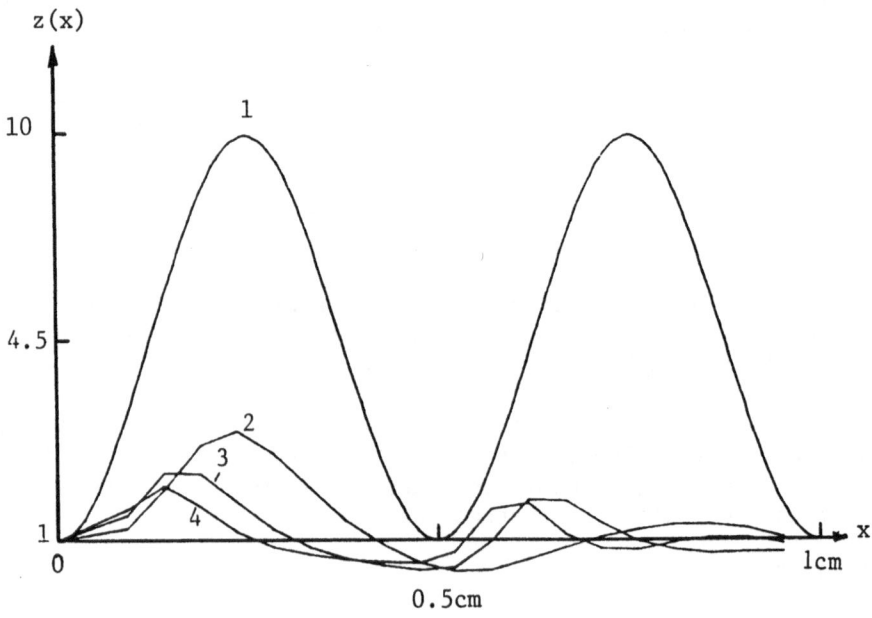

Figure 5. The Reconstructed Impedance Profile, $z_4(x)$, Based on Goupillaud's Method, Distorted by the Total Transfer Function of the Combined Transducer. Transducer Q = 1/4. 1: given impedance profile. 2,3, and 4: f_o = 100khz, 200khz, and 300khz.

The effectiveness of the deconvolution is dependent upon how closely $H_I H^2$ approaches unity over the frequency range where $R_o(f)$ contains significant spectral components, and where $R_o(f)$ is above the noise level. $H_I(f)$ is the frequency response of the inverse filter. As shown in (16), the function $H_I H^2$ is real. For $N(f)=0$, $H_I H^2 = 0$ whenever $R_o(f)=0$. For a low noise level, $H_I H^2$ approaches unity except for frequencies where $R_o(f)=0$ or $H(f)=0$. For $N(f)=0$, we have $H_I H^2 = 1$ except when $H(f)=0$. Therefore, if we ignore the isolated frequencies for which $R_o(f)=0$, the function $G(f)H_I(f)$ acts as an ideal low-pass filter.

We have generated a number of curves of $H_I(f)H^2(f)$ for different impedance profiles, S/N levels, transducer Q's, and transducer resonant frequencies, and we have found that even for poor S/N ratios (1 to 5dB) and moderately high transducer Q's (5 or higher), the function $H_I H^2$ is very close to one, except in the immediate vicinity of those frequencies where $R_o(f)=0$.

Figure 6 and 7 show 2 examples of acoustic impedance profile reconstruction using $H_I(f)$ as defined in (16). Figure 6 shows the reconstruction of $z_2(x)$, for a transducer Q of 5, S/N ratio=40dB, and 3 different transducer resonance frequencies. The reconstructed profiles, $\hat{z}_2(x)$, are in very good agreement with the given profile. This result should be contrasted to the result in Figure 4 where $z_2(x)$ is reconstructed without the aid of deconvolution, but with a much lower Q of 0.25. Since $z_2(x)$ has only a small relative impedance variation, both the Impediography method and Goupillaud's method provide the same result. Figure 7 displays the result of the reconstruction of $z_4(x)$ along with the given profile, for a transducer with a Q of 1 and a S/N=40dB. The good reconstruction result, obtained with Goupillaud's method, should be contrasted with the result, shown in Figure 5 where $z_4(x)$ is reconstructed without employing deconvolution. Since the relative impedance variation is large, the Impediography method does not provide an acceptable reconstruction.

DISCUSSION

In this paper, we have shown the effects of more realistic measurement conditions on the reconstruction of acoustic impedance profiles by the Impediography and Goupillaud's methods. The effect of noise alone is negligible, especially for the Impediography method. The effect of limited bandwidth of the transducers alone is serious for both methods. If noise is negligible, the effect of transducer distortion can be eliminated perfectly with the help of deconvolution. When acoustic noise and limited bandwidth transducers exist at the same time the reconstruction of acoustic impedance profile will be good when the S/N ratio is greater than 30dB. Although the Impediography method is less sensitive to noise than Goupillaud's method, Goupillaud's method is still far better

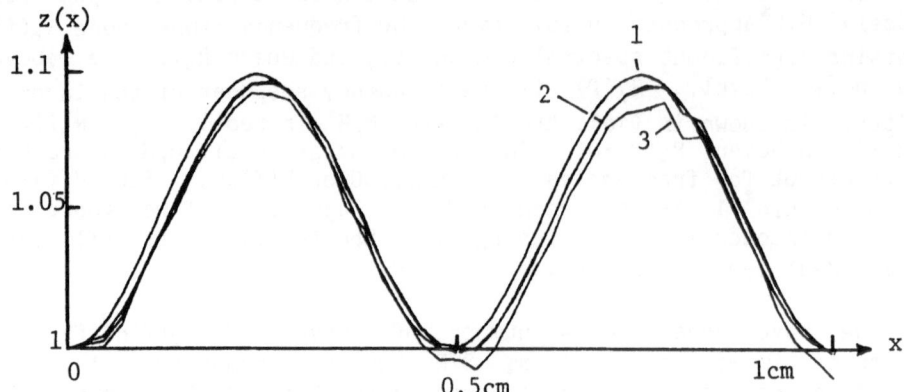

Figure 6. The Reconstructed Acoustic Impedance Profile After
Inverse Filtering for $z_2(x)$.
$Q = 5$, $S/N = 40dB$, and f_o = 50khz, 150khz, and 250khz
1: $z_g(x)$, 2: f_o = 50khz and 250khz, 3: f_o = 150khz.

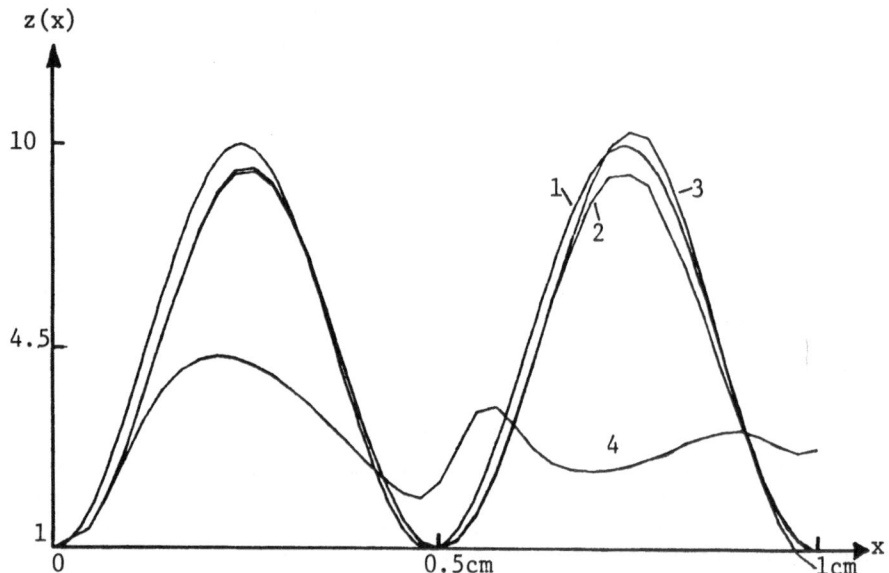

Figure 7. The Reconstructed Acoustic Impedance Profile After
Inverse Filtering for $z_4(x)$. $Q=1$, $S/N=40dB$, and f_o= 150 khz and
300khz.
1: $z_g(x)$; 2: f_o=150khz, Goupillaud; 3: f_o=300khz, Goupillaud;
4: by Impediography Method

than the Impediography method for reconstruction of the acoustic impedance profile with large and rapid impedance change. With deconvolution, either a priori knowledge of signal and noise spectrum is required or iteration method should be used to determine the reflection frequency response of signal and the character of noise.

ACKNOWLEDGEMENT

The work presented in this paper was supported by National Science Foundation Grant ECS-802531, and by Drexel University.

REFERENCES

[1]D. Lesselier, Optimization techniques and inverse problems: Probing of acoustic impedance profile in time domain. J. Acou. Soc. Am. 1276-1284 (1982).
[2]S. M. Candel, F. Defillipi and A. Launary, Determination of the inhomogeneous structure of a medium from its plan wave reflection response, Part II: A numerical approximation. J. Sound and Vibration 583-595 (1980).
[3]J. P. Jones, Impediography: a new ultrasonic technique for diagnostic medicine. "Ultrasound in Medicine", Vol. 1. Proceedings of the 1974 Meeting of the American Institute of Ultrasound in Medicine, Plenum Press, 499-508 (1975).
[4]J. A. Ware and K. Aki, Continuous and discrete inverse-scattering problems in a stratified elastic medium. I. Plane waves at normal incidence, J. Acou. Soc. Am. 911-921 (1969).
[5]P. L. Goupillaud, An approach to inverse filtering of near-surface layer effects from seismic records, Geophysics 754-760 (1961).
[6]P. C. Pedersen, O. J. Tretiak, and P. He, Numerical techniques for the inverse acoustical scattering problem in layered media. Acoustical Imaging, 443-457 (1982).
[7]P. He, Direct and inverse acoustic scattering problem for a medium with continuous impedance profile. M. S. Thesis, Drexel University, (1981).
[8]P. C. Pedersen, O. J. Tretiak, and P. He, Impedance-matching properties of an inhomogeneous matching layer with continuously changing acoustic impedance, J. Acou. Soc. Am., 327-336 (1982).
[9]M. G. Silk, Predictions of the effect of some constructional variables on the performace of ultrasonic transducers, Ultrasonics 27-33 (1983).
[10]M. O'Donnell, L. J. Busse, and J. G. Miller, Chapter 1. Piezoelectric transducers, "Ultrasonics," Academic Press, New York, 29-64 (1981).
[11]R. E. Ziemer and W. H. Tranter, "Principle of Communications," Houghton Mifflin Company, Boston, (1976).
[12]D. Slepian, Linear least-squares filtering of distorted imaging, J. Optical Soc. Am. 918-922 (1967).

FAST ITERATIVE ALGORITHMS FOR INVERSE SCATTERING SOLUTIONS

OF THE HELMHOLTZ AND RICCATI WAVE EQUATIONS

S. A. Johnson[1], Y. Zhou[2], M. K. Tracy[2],
M. J. Berggren[1], and F. Stenger[3]

Departments of Bioengineering[1], Electrical Engineering[2]
and Mathematics[3]
Salt Lake City, Utah 84112

I. INTRODUCTION

Solving the inverse scattering problem for the Helmholtz wave equation without employing the Born or Rytov approximations is a challenging problem, but some slow iterative methods have been proposed [1, 2]. One such method suggested and demonstrated by us is based on solving systems of nonlinear algebraic equations that are derived by applying the method of moments to a sinc basis function expansion of the fields and scattering potential [2, 3]. In the past, we have solved these equations for a 2-D object of n by n pixels in a time proportional to n^5 [1-3]. We now describe further progress in the development of new methods based on FFT convolution and the concept of backprojection [4, 5], which solves these equations in time proportional to $n^3 \cdot \log(n)$.

Two important equations for modeling the propagation of sound through an inhomogeneous fluid-like body immersed in a homogeneous fluid bath are the Helmholtz wave equation (the inhomogeneous form) and the Riccati wave equation. Their inverse scattering solutions may provide a high-spatial resolution, quantitative image of the body. They are related by the transform $p(\underline{x}) = p_o(\underline{x})e^{w(\underline{x})}$ and are given, respectively, by

$$\nabla^2 p(\underline{x}) + k_o^2 p(\underline{x}) = k_o^2 \gamma(\underline{x})p(\underline{x}), \tag{1}$$

$$\nabla^2 w(\underline{x}) + \nabla w(\underline{x}) \cdot \nabla w(\underline{x}) + 2\nabla \ln p_o(\underline{x}) \cdot \nabla w(\underline{x}) - k_o^2 \gamma(\underline{x}) = 0, \tag{2}$$

where $k_o^2 = \omega^2/c_o^2$, c_o^2 is a constant speed of sound (usually taken in the fluid bath), ω is constant angular frequency, $p(\underline{x})$ is acoustic pressure, and $p_o(\underline{x})$ acoustic pressure in the incident field. Here, γ is the scattering potential given by $\gamma = 1 - c_o^2/c^2(\underline{x}) - i2c_o^2\alpha(\underline{x})/[\omega\ c(\underline{x})]$, where $c(\underline{x})$ is the speed of sound in the body. Here α contains absorption information and is essentially equal to the linear absorption coefficient for weak scattering, e.g. in soft tissues.

The equations above may be modified to include density variations within the body by a transformation to a form equivalent to Eq. (1), except the field p in Eq. (1) is now replaced by a new field $p(\underline{x})\left(\rho(\underline{x})\right)^{1/2}$ where $\rho(\underline{x})$ is the density [2]. Using that new field, Eq. (1) may be used to find a new $\gamma(c(\underline{x}), \rho(\underline{x}), \alpha(\underline{x}),\omega)$. Parameters $\rho(\underline{x})$ and $c(\underline{x})$ may be found from $\gamma(c, \rho, \alpha, \omega)$ by the use of multiple frequencies by the method given in [2].

Eqs. (1) and (2) have different mathematical properties and physical interpretations. If p is the total pressure field and if p_o, the incident pressure field, is defined to be the field in the absence of the body, then w/i is the complex phase difference between the total field and the incident field. The quantity w/ik_oc_o = w/iω has units of time and may be considered to be an average time of arrival of the scattered field. The imaginary part of w is the ratio of the moduli of incident and the total fields. For ultrasound computed transmission tomography (UCTT) in soft tissues, γ is usually small, e.g., $|\gamma| \leqslant 0.1$ and, therefore, w is a more slowly changing function than p. Thus, it is not suprising that for UCTT, linearizing Eq. (2), i.e., the Rytov approximation, produces superior images than obtained from linearizing Eq. (1), i.e., the Born approximation [6]. We next describe methods for solving Eqs. (1, 2) without the Born or Rytov approximations.

Let f represent some field such as p or $p\rho^{1/2}$. Then the Helmholtz wave equation (1) in f may be transformed to the Lippmann-Schwinger equation [2] that in turn may be transformed to a set of detector or measurement equations given by [1-3]

$$f_{\phi m}^{(sc)} = \sum_{j=1}^{N} \gamma_j f_{\phi j} D_{mj}, \qquad \begin{array}{l} m = 1, \ldots, M \\[4pt] \phi = 1, \ldots, \Phi \end{array}, \qquad (3)$$

and a set of field constraint equations given by [1-3]

$$f_{\phi\ell} - f_{\phi\ell}^{(in)} = \sum_{j=1}^{N} \gamma_j f_{\phi j} C_{\ell j}, \qquad \begin{array}{l} \ell = 1, \ldots, N \\[4pt] \phi = 1, \ldots, \Phi \end{array}, \qquad (4)$$

where $f_{\phi\ell} \triangleq f_\phi(\underline{x}_\ell)$ is the actual field, $f_{\phi\ell}^{(in)} \triangleq f_\phi^{(in)}(\underline{x}_\ell)$ is the incident field, $f_{\phi m}^{(sc)} \triangleq f_\phi^{(sc)}(\underline{x}_m)$ is the measured values of the scattered field on detector point $\underline{x}_m$, and ϕ is the source location. Although $D_{mj} = C_{mj}$, we use a separate symbol D_{mj} to correspond to measurement point $\underline{x}_m$ on a "detector." C_{mj} and $C_{\ell j}$ are constants given by

$$C_{\mu j} \triangleq \int \psi(\underline{x}' - \underline{x}_j')g(|\underline{x}' - \underline{x}_\mu|)d^Q\underline{x}', \quad \begin{array}{l} \mu = m \text{ or } \ell \\ m = 1, \ldots, M. \\ \ell = 1, \ldots, N \end{array} \quad (5)$$

Here the functions $\psi_j(\underline{x}') \triangleq \psi(\underline{x}' - \underline{x}'_j)$ belong to a basis set $\{\psi_j(\underline{x}')\}$ and $g|\cdot|$ is the outward going Green's function. In [3], we describe the evaluation of constants $C_{\mu j}$ via Eq. (5) for a sinc function basis set.

Four different iterative algorithms are given in [1-3] for the solution of Eqs. (3) and (4). Although these algorithms are quite robust, they are slow. If a square 2-dimensional image is sought, then $N = n^2$ where n is the number of pixels per side of the image. If $M\phi > N$, then over n^4 and n^5 computations must be performed to evaluate Eqs. (3) and (4), respectively. Since all algorithms for solving Eqs. (3) and (4) which we described [1-3] must compute the sums in these equations, a faster means of performing these sums would be helpful for increasing the speed of solution.

II. FAST INVERSE SCATTERING SOLUTIONS FOR THE HELMHOLTZ EQUATION

A fast iterative method which computes a trial solution in order $n^3 \cdot \log(n)$ time has been found [4, 5]. This speedup is made possible by application of two principles: (1) Fast Fourier transform (FFT) convolution and (2) backprojection. The derivation of these fast methods follow.

If γ is small enough, a fixed point method may be used to solve Eq. (4) by the following iterative scheme [2-5, 7-9]:

$$f_{\phi\ell}^{(k+1)} = \beta\left[f_{\phi\ell}^{(in)} + \sum_{j=1}^{N} C_{\ell j}f_{\phi j}^{(k)}\gamma_j^{(n)}\right] + (1-\beta)\,f_{\phi\ell}^{(k)}, \quad 0 < \beta < 1. \quad (6)$$

Here $\gamma^{(n)}$ is a trial guess for γ, and $f_{\phi j}^{(k)}$ is the k-th guess for γ. If the spectral radius of the matrix $\{\beta C_{\ell j}\gamma_j + (1-\beta)\delta_{\ell j}\}$ is less than unity, Eq. (6) will converge [4, 5, 7, 8]. A nearly optimum β is found by a search method. Convergence may be verified by examining the residuals of Eq. (6) as per Eq. (18) of [2] as $k \to \infty$.

The sum in Eq. (6) is a convolution of $C_{\mu j}$ with $\left(\gamma_j f_{\phi j}\right)$, since $C_{\mu j}$ is a function of $\left(\left|\underline{x}_\mu - \underline{x}_j\right|\right)$ by inspection of Eq. (5). The convolution in Eq. (6) may be performed in order $n^2 \cdot \log(n)$ operation by use of the convolution theorem and using the FFT [9, 10]. Several iterations of Eq. (6) with fixed $\gamma^{(n)}$ will lead to a best fit of f_ϕ to a particular $\gamma^{(n)}$. But $\gamma^{(n)}$ is not necessarily correct and must be updated. This suggests an algorithm where f_ϕ is updated for a given fixed $\gamma^{(n)}$ and then $\gamma^{(n)}$ is updated while holding fixed the new value of f_ϕ; the process is repeated until convergence is acceptable. In short, the process is a sequence of solving the detector and field equations iteratively.

Since we have a fast method for solving Eq. (4) for f_ϕ, we desire a fast way to solve Eq. (3) for γ. In x-ray computed tomograph (x-ray CT), backprojecting x-ray CT data directly produces an image which is a blurred version of the original object [11]. The original object may be recovered by convolving with a deblurring function. The back projection operation spreads the detector density values along the ray paths back to the detector. The sum of this operation for all source positions is the backprojected image. Equation (3) is a sum over all scattering points in the body each of which radiate an outward going spherical wave whose initial modulus and phase at point j are given by the product $\gamma_j f_{\phi j}$, and whose complex phase at detector m is given by multiplying this product by the Green's-function-like propagation coefficient D_{mj}.

Therefore, the step that is analogous to backprojection into pixel ℓ would be to reradiate the detected signal of each detector m with a phase corresponding to the negative of the shift in scattering from pixel ℓ to detector m. Thus, the round trip phase shift, from pixel ℓ to detector m to pixel ℓ, would be zero and the sum over all m into ℓ would add constructively. Two possible formulas come to mind; we might multiply $f_{\phi m}^{(s)}$ by $D_{m\ell}^*$ (here, * means complex conjugate) and sum over m and ϕ, or we might divide $f_{\phi m}^{(s)}$ by $D_{m\ell}$ and sum over m and ϕ. Thus for each ϕ, backprojection, as defined above, produces a blurred image of $\left(f_\phi \gamma\right)$.

Backprojection of γ can be achieved by dividing the backprojected image of $\left(f_\phi \gamma\right)$ by f_ϕ for each ϕ, and then sum all such modified images over ϕ as in [4]. Using the $D_{m\ell}^{*}$ form, we may summarize the process by a formula for $\tilde{\gamma}$, the backprojected approximation to γ :

$$\tilde{\gamma}_{\ell}^{(k)} = K_{\ell} \sum_{\phi=1}^{\Phi} (1/f_{\phi\ell}^{(k)}) \sum_{m=1}^{M} D_{m\ell}^{*} f_{\phi m}^{(sc)} \stackrel{\Delta}{=} B[f_{\phi m}^{(sc)}]. \qquad (7)$$

Here K_{ℓ} is a normalizing factor which we take to be $2\pi/\Phi$ and B is the backprojection operator. The sum over M already has a renormalization factor built into the D^{*}, so K does not depend on M.

   The closeness of approximation to γ by $\tilde{\gamma}$ is a question we have investigated by considering the much simplified problem of a single weak scattering point at x_{-s} (so that the Born approximation is valid), plane wave sources incident sequentially from all angles and a circular ring detector at a very large radius (where the asymptotic form of Green's function is valid). In this case, $\tilde{\gamma}$ obtained from the integral form of Eq. (7) may be evaluated analytically and is found to be proportional to $J_{o}^{2}(k_{o}|x - x_{-s}|)$. For the three-dimensional case of a spherical detector of a very large radius, $\tilde{\gamma}$ is proportional to $\text{sinc}^{2}(k_{o}^{2}|x - x_{-s}|)$. For larger objects, stronger scattering, or other detector configurations, the blurring in $\tilde{\gamma}$ will be different from the above examples and not necessarily convolutional.

   The above image of $\tilde{\gamma}^{(k)}$ must be deblurred to find $\gamma^{(k)}$. Consider the intuitive notion of adding to $\tilde{\gamma}^{(k)}$ the correction obtained by operating with $B[\cdot]$ on the difference of the true scattered field $f_{\phi}^{(sc)}$ and the predicted scattered field $P[\tilde{\gamma}^{(k)}]$. Here $P[\gamma_{j}] = \sum_{j} D_{mj}f_{\phi j}\gamma_{j}$ is a "projection" (actually scattering) operator. This may be written

$$\tilde{\gamma}^{(k)(1)} = B[f_{\phi}^{(sc)} - P\tilde{\gamma}^{(k)(o)}] + \tilde{\gamma}^{(k)(o)}, \qquad (8)$$

where $\tilde{\gamma}^{(k)(o)}$ is defined to be $\tilde{\gamma}^{(k)}$ from Eq. (7). Of course, the operation of deblurring may be repeated and the general iteration formula, upon reordering and introducing an underrelaxation factor α is

$$\tilde{\gamma}^{(k)(n+1)} = \alpha B f_{\phi}^{(sc)} + [I - \alpha BP]\tilde{\gamma}^{(k)(n)}. \qquad (9)$$

This process converges only if $(I - \alpha BP)$ has positive eigenvalues which lie within the unit circle [4, 5, 8]. In this case, optimum convergence of Eq. (9) occurs when $\alpha = 2/(\lambda_{max} + \lambda_{min})$, where λ_{max} and λ_{min} are the maximum and minimum eigenvalues, respectively, of BP. Further discussion of deblurring formulas is given in [4, 5].

To complete our derivation of a fast method for solving Eq. (3), we note that both B and P are convolution operations. The convolution $Bf_\phi^{(sc)}$ can be done by FFT only if $f_\phi^{(sc)}$ is defined on a rectangular grid containing field points or is interpolated onto such a rectangular grid. One such scheme for interpolation is proposed in [4].

In closing this section, we make a remark concerning the possibility of $f_{\phi\ell}^{(k)}$ vanishing at some pixel ℓ. In such cases, a modification of Eq. (7) is suggested which still may be valid:

$$\tilde{\gamma}_\ell^{(k)} = K_\ell \sum_{\phi=1}^{\Phi} (f_{\phi\ell}^{*(k)} / \sum_{\theta=1}^{\Phi} f_{\theta\ell}^{*(k)} f_{\theta\ell}^{(k)}) \sum_{m=1}^{M} D_{m\ell}^* f_{\phi m}^{(sc)} = B[f_{\phi m}^{(sc)}]. \quad (10)$$

It is instructive to insert the scattered field from a single scattering point, $f_{\phi m}^{(sc)} = \sum_j \delta_{sj} f_{\phi j} D_{mj}$, into either Eq. (7) or Eq. (10) and verify that a sharply peaked distribution of $\tilde{\gamma}$ values occur around pixel s (here, $\delta_{sj} = 1$ for s = j and 0 for s $\neq$ j). In practice, any K_ℓ is made workable by the correct choice of α in Eq. (9). The sums over $f_{\phi\ell}^* f_{\phi\ell}$ and $D_{m\ell}^* D_{m\ell}$ can be completed in n^3 operations. Thus, Eq. (10) preserves the $n^3 \cdot \log(n)$ time dependence.

In summary, the fast method for solving Eq. (1) for γ consists of the steps:

1. Pick initial γ and f_ϕ, for example, 0 and $f_\phi^{(in)}$, respectively.
2. Solve for γ via Eq. (7) or Eq. (10).
3. Deblur by applying Eq. (9) iteratively until change is small.
4. Stop if convergence is satisfactory.
5. Solve for f_ϕ by applying Eq. (6) iteratively until change is small.
6. Go to step 2.

III. SUGGESTED FAST INVERSE SCATTERING SOLUTIONS TO THE RICCATI EQUATION

The Helmholtz partial differential equation (PDE), Eq. (1), may be transformed to the Riccati PDE, Eq. (2) as described in the introduction. By use of Green's theorem, the PDE (2) may be transformed [12] to the following integral equation, which we shall call the Riccati integral equation:

$$w_\phi(\underline{x}) = -k_o^2 \int \gamma(\underline{x}')G_\phi(\underline{x}, \underline{x}')d^Q\underline{x}' - \int (\nabla w_\phi \cdot \nabla w_\phi)(\underline{x}')G_\phi(\underline{x}, \underline{x}')d^Q\underline{x}'.$$

(11)

For incident plane waves $e^{ik_o(\phi)}$, $G_\phi = e^{ik_o(\phi)(x-x')} \cdot g(|\underline{x} - \underline{x}'|)$, where g is the same as in Eq. (5). For 3-dimensions, $Q = 3$ and $g = e^{ik_o|\underline{x}-\underline{x}'|}/4\pi(|\underline{x} - \underline{x}'|)$. For 2-dimensions, $Q = 2$ and $g = H_o^{(2)}(k_o|\underline{x} - \underline{x}'|)$ where $H_o^{(1)}(\cdot) = J_o(\cdot) - iY_o(\cdot)$ and where J_o and Y_o are the well known Bessel functions [13]. See [12] for the form of G_ϕ when the source at ϕ produces spherical waves.

By the method of moments, w_ϕ, γ, and $(\nabla w_\phi \cdot \nabla w_\phi)$ may be expanded in special sinc basis functions $\{\psi_j\}$ defined in [2, 3]. For example, $w(\underline{x}) = \sum_j w(\underline{x}_j)\psi_j(\underline{x})$. $\{\psi_j\}$ has the special feature that $\psi_j(\underline{x}_\ell) = \delta_{j\ell}$ for $\underline{x}_j$ and $\underline{x}_\ell$ located at the nodes of a rectangular coordinate system. Following the method described in [2], upon substituting the basis expansions of w_ϕ, γ, and $(\nabla w_\phi \cdot \nabla w_\phi)$ into Eq. (11), interchanging integration and summation, and evaluating $\underline{x}$ at rectangular node points, the following algebraic equations are obtained, respectively, for points $\underline{x}_m$ on the detector and for points $\underline{x}_\ell$ in the Q-dimensional support window containing nonzero values of γ.

$$W_{\phi m} + \sum_{j=1}^{N} D'_{\phi m j}(\nabla w_\phi \cdot \nabla w_\phi)_j \triangleq W'_{\phi m} = -\sum_{j=1}^{N} k_o^2 D'_{\phi m j}\gamma_j \qquad \begin{matrix} m = 1, \ldots, M \\ \phi = 1, \ldots, \Phi \end{matrix},$$

(12)

$$w_{\phi\ell} = -k_o^2 \sum_{j=1}^{N} D'_{\phi\ell j}\gamma_j - \sum_{j=1}^{N} D'_{\phi\ell j}(\nabla w_\phi \cdot \nabla w_\phi)_j \qquad \begin{matrix} \ell = 1, \ldots, N \\ \phi = 1, \ldots, \Phi \end{matrix}. \qquad (13)$$

The constants $D'_{\phi m j}$ and $D'_{\phi\ell j}$ in Eqs. (12) and (13) are obtained from

$$D'_{\phi\mu j} = \int \psi_j(\underline{x}')G_\phi(\underline{x}' - \underline{x}_j)d^Q\underline{x}' \quad , \quad \mu = m, \ell. \tag{14}$$

Here $W_{\phi m}$ is defined to be the measured value of w_ϕ at detector point $\underline{x}_m$. Also, here γ_j, $w_{\phi j}$, and $\left(\nabla w_\phi \cdot \nabla w_\phi\right)_j$ are the values of γ_j, $w_{\phi j}$, and $\left(\nabla w_\phi \cdot \nabla w_\phi\right)_j$ at point x_j. Equations (12) and (13) are the detector and field equations analogous, respectively, to Eqs. (3) and (4). Inspection of Eqs. (12-14) and the definition of G_ϕ shows that the sums in Eqs. (15) and (13) are convolutions and can thus be done in $n^3 \cdot \log(n)$ time by the FFT.

In comparing the nonlinear properties of the Helmholtz and Riccati models, we see that in Eqs. (3) and (4), the field f_ϕ and potential γ are coupled as a product $f_\phi\gamma$; in Eqs. (12) and (13), the nonlinearity is restricted only to the field like variable $\left(\nabla w_\phi \cdot \nabla w_\phi\right)$. This new field-like quantity, when divided by $w_\phi \cdot w_\phi$, is large in regions producing diffraction or refraction effects. These observations suggest several iterative schemes for determining γ that are analogous to those employed in the previous section. Consider the following algorithm:

1. Set iteration index n = 0.

2. Define $\left(\nabla w_\phi \cdot \nabla w_\phi\right)_j^{(n)} = 0$ for n = 0,

3. Solve for $\gamma^{(n+1)}$ (as, for example, per [6, 14]) from

$$\ell_{hs}(\phi, m) \overset{\Delta}{=} W_{\phi m} + \sum_{i=1}^N D'_{\phi m j}\left(\nabla w_\phi \cdot \nabla w_\phi\right)_j^{(n)}$$

$$= -k_o^2 \sum_{j=1}^N D'_{\phi m j}\gamma_j^{(n+1)} \overset{\Delta}{=} r_{hs}(\phi, m), \tag{15}$$

4. If $\sum_\phi \sum_m \left|\ell_{hs}(\phi, m) - r_{hs}(\phi, m)\right| \leqslant \varepsilon_1$ for some ε_1, then go to step 8, or else,

5. Set $\gamma^{(n+1)} = \gamma'$, evaluate $w_{\phi\ell}^{(n+1)}$ from

$$L_{hs}(\phi, \ell) \overset{\Delta}{=} w_{\phi\ell}^{(n+1)} = -\sum_{j=1}^N D'_{\phi m j}\left(\nabla w_\phi \cdot \nabla w_\phi\right)_j^{(n)} - k_o^2 \sum_{j=1}^N D'_{\phi m j}\gamma_j'$$

$$\overset{\Delta}{=} R_{hs}(\phi, \ell). \tag{16}$$

6. Compute $\left(\nabla w_\phi \cdot \nabla w_\phi\right)^{(n+1)}$ from $w_\phi^{(n+1)}$,

7. If $\sum_{\phi\ell} \left| R_{hs}(\phi, \ell) - L_{hs}(\phi, \ell) \right| > \varepsilon_2$ for some ε_2, go to step 5, else go to step 3,

8. Stop, store results.

A second suggested algorithm would omit step 7 and would replace step 5 with the forward propagation of a plane wave through $\gamma^{(n+1)}$ by a finite element method such as given in [15]. This constitutes the second algorithm of this section.

A third variation would subtract a trial computed $W_{\phi m}$, i.e. $W_{\phi m}^{(n)}$, from the measured $W_{\phi m}$ and compute a correction $\Delta\gamma$ from

$$W_{\phi m} - W_{\phi m}^{(n)} = -\sum_j D_{\phi m j}\left[\left(\nabla w_\phi \cdot \nabla w_\phi\right)_j^{(\infty)} - \left(\nabla w_\phi \cdot \nabla w_\phi\right)_j^{(n)}\right] - \sum_j k_o^2 D_{\phi m j}\Delta\gamma_j^{(n)}$$

(17)

Assuming the differences in $\nabla w_\phi \cdot \nabla w_\phi$ in the first term are smaller than the second term in $\Delta\gamma_j^{(n)}$, Eq. (17) is solved for $\Delta\gamma_j^{(n)}$ via [14]. Here, $(\cdot)^{(\infty)}$ means the true value of $(\cdot)$. Next, $\gamma^{(k)}$ is updated by

$$\gamma^{(n+1)} = \gamma^{(n)} + \Delta\gamma^{(n)}.$$

(18)

Next, using $\gamma^{(n+1)}$, a new $w_{\phi\ell}^{(n+1)}$ is computed via Eq. (16). Then, using $w_{\phi\ell}^{(n+1)}$, a new $W_{\phi m}^{(n+1)}$ is computed via

$$W_{\phi m}^{(n+1)} = -\sum D_{\phi m j}\left[\left(\nabla w_\phi \cdot \nabla w_\phi\right)^{(n+1)} + k_o^2 \gamma^{(n+1)}\right].$$

(19)

Next, a new $\Delta\gamma_j$ is computed via Eq. (17) and the process is repeated until $\sum_j \left| \Delta\gamma_j^{(n)} \right|$ becomes sufficiently small.

A fourth variant would consist of replacing Eq. (16) for computing $w_{\phi j}$ and $\left(\nabla w_\phi \cdot \nabla w_\phi\right)_j$ by a finite element method such as in [15]. The first and third algorithms use Eq. (16) and are as accurate as the band limited image of γ_j. The second and fourth algorithms use finite element approximations which model the forward scattered component only, and thus may be less accurate but faster.

The predicted advantages of the Riccati equation model (for plane incident waves) are:

1. An accurate, fast, exact method for solving for γ, given $W'_{\phi m}$, is known [14].

2. Iterative methods for solving for γ and $w_{\phi j}$ from $W_{\phi m}$ probably converge faster than the iterations involving γ and $f_{\phi j}$ because $w_{\phi j}$ contains the additional information of absolute phase, while $f_{\phi j}$ contains only relative phase, i.e., adding 2π radians of phase to $f_{\phi j}$ does not change $f_{\phi j}$, but such an addition of phase will change w.

The possible disadvantages of the Riccati equation model are:

1. Determination of the absolute phase of $W_{\phi m}$ is difficult (has not been achieved at this writing to our knowledge).

2. The coefficients $D'_{\phi m j}$ and $D'_{\phi \ell j}$ require more storage, because of source dependence ϕ, than do the $D_{m j}$ and $C_{\ell j}$ coefficients from the Helmholtz model.

The first objection does not seem fundamental and may yield to interpolation schemes in the $(\omega, \underline{x}_{m}, \phi)$ domain. The second objection does not seem to be serous, since a single set of coefficients $D'_{m j}$, $D'_{\ell j}$ for the source at a standard location $\phi = \phi_{o}$ could be stored. Thus, computation of w_{ϕ} at some other source position $\phi \neq \phi_{o}$ could be performed in a properly rotated coordinate system using $D'_{m j}$ and $D'_{\ell j}$. Passing $\gamma^{(n)}$, $w_{\phi}^{(n)}$, and $(\nabla w_{\phi} \cdot \nabla w_{\phi})^{(n)}$ images between Eqs. (15) and (16) would then be done by interpolating between the fixed and rotated coordinate systems.

The above four algorithms based on the Riccati model have not been tested. However, in the next section, we give computer experimental results for the fast methods based on the Helmholtz equation.

IV. EXPERIMENTAL RESULTS

Investigation was undertaken of an iterative algorithm that solves Eq. (6) alternately with Eqs. (7) and (9). This process converged to the correct value of γ, when $f_{\phi}^{(in)}$ and 0 were taken as the initial values of f_{ϕ} and γ, respectively. The process some-

Table I

Percent error versus number of iterations to solve complete problem
for simulated data generated from Eqs. (3, 4) with γ chosen as
$(.1 + i\ .01)\exp\left(-r^2/3.91\right)$, r is distance from center in pixels.

Deblurring Matrices $\left[M_1\ ;\ M_2\right]$	Number of Iterations for 7×7 Pixel Image					
	1	2	4	6	8	10
$[B\ ;\ BP]$	10.0	1.4	0.06	≈ 0	≈ 0	≈ 0
$\left[P^\dagger\ ;\ P^\dagger P\right]$	10.0	1.4	0.4	0.4	0.4	0.4
$\left[(BP)^\dagger B\ ;\ (BP)^\dagger BP\right]$	11.0	1.3	0.8	0.7	0.7	0.7
$\left[(SBP)^\dagger SB\ ;\ (SBP)^\dagger SBP\right]$	23.0	21.0	21.0	21.0	21.0	21.0

Table I shows the results of solving Eqs. (3) and (4) by
finding f_ϕ from Eq. (6), and γ from applying Eq. (7) followed by a
deblurring operation $\tilde{\gamma}^{(k)(n+1)} = \alpha M_1 f^{(sc)} + \left[I - \alpha M_2\right]\tilde{\gamma}^{(k)(n+1)}$.
The particular deblurring matrices $\left[M_1\ ;\ M_2\right]$ used are shown in the
left column. For example, $[B\ ;\ BP]$ gives Eq. (9).

times diverged with P replacing B and $P^\dagger P$ replacing BP. The process of choosing α by searching different α values for optimum
convergence is effective, but should be made more rapid if possible. In all of these experiments reported in this paper, the
convolutions were done directly and not by use of the FFT.

The four deblurring methods are tested for error in the computed image of γ at the end of 1, 2, 4, 6, 8, 10 iterations.
Matrices $[B\ ;\ BP]$, i.e., Eq. (9), gave the best results. The fixed
point step Eq. (6) was applied eight times and the deblurring step
was applied 50 times in each iteration through Eqs. (3) and (4) to
obtain a measure of the global iteration effectiveness. The optimal balance of deblurring steps, Born approximation iterative solution steps, and global iteration steps was not studied. Neither
do we know if the number of global iterations to produce an acceptable image remains constant as the size of the image is increased.

The error listed in Table I is computed by the expression
$\|\tilde{\gamma}^{(k)} - \gamma_T\|/\|\gamma_T\|$ using the one norm, where $\tilde{\gamma}^{(k)}$ is γ after k iterations and γ_T is the true value of γ.

The accuracy and the speed of the fast algorithms described here when implemented with FFT convolution and appropriate hardware represent a new class of methods which look very promising for producing high resolution (0.5 to 1.0 mm), quantitative, medical ultrasound images.

ACKNOWLEDGMENTS

Discussions with David T. Borup on the use of reported [8] and other FET methods for solving Eq. (6) are appreciated. Discussions with J. F. Greenleaf concerning finite element wave propagation methods [15] and with A. J. Devaney concerning backpropagation algorithms for UCTT [16, 17] are also appreciated. We appreciate discussions with the editor of *Ultrasonic Imaging* to digest [4] and to duplicate Table I from [4]. In the present paper, Sections I, II, and IV are digested from [4], while Section III contains new explicit formulas which have not been submitted previously for publication. The editing and typing help of the Department of Electrical Engineering is appreciated. Support in part from grants PDT-110B from the American Cancer Society and R01-CA1-29728 from the National Cancer Institute, Public Health Service, are appreciated.

REFERENCES

[1]   Johnson, S. A., Yoon, T. H., and Ra, J. W., "Inverse scattering solutions of the scalar Helmholtz save equation by a multiple source, moment method," *Electronics Lett.* *19*, 130-132 (1983).

[2]   Johnson, S. A., and Tracy, M. L., "Inverse scattering solutions by a sinc basis moment method -- Part I: theory," *Ultrasonic Imaging 5*, 361-375 (1983).

[3]   Johnson, S. A. and Tracy, M. L., "Inverse scattering solutions by a sinc basis, moment method -- part II: numerical evaluation," *Ultrasonic Imaging 5*, 376-392 (1983).

[4]   Johnson, S. A., Zhou, Y., Tracy, M. K., Berggren, M. J., and Stenger, F., "Inverse scattering solutions by a sinc basis, multiple source, moment method -- Part III: fast algorithms," submitted

[5]   Johnson, S. A., Zhou, Y., Berggren, M. J., and Tracy, M. L., "Acoustic inverse scattering solutions by moment methods and back propagation," submitted to *Proceedings of the SIAM Special Issue on Inverse Scattering*, Tulsa, Oklahoma, May 1983.

[6] Kaveh, M., Soumekh, M., Lu, Z. Q., and Mueller, R. K., "Further results on diffraction tomography," in *Acoustical Imaging 12*, E. A. Ash and C. R. Hill, eds., 599-608 (Plenum Press, New York, 1982).

[7] Dahlquist, D., *Numerical Methods* (Prentice-Hall, Englewood Cliffs, New Jersey, 1974).

[8] Borup, D. T. and Gandhi, O. P., "Fast-Fourier-transform method for calculation of SAR distributions in finely-discretized inhomogeneous models of biological bodies," *IEEE Trans. Microwave Theory Techniques* (in press).

[9] Bojarski, N. N., "K-Space Formulation of the electromagnetic scattering problem," Technical Report AFAL-TR-71-5, March 1971, (U.S. Air Force, 1971).

[10] Brigham, E. O., *The Fast Fourier Transform* (Prentice-Hall, Englewood Cliffs, New Jersey, 1974).

[11] Herman, G. T., *Image Reconstruction from Projections, The Fundamentals of Computerized Tomography* (Academic Press, New York, 1980).

[12] Ball, J. S., Johnson, S. A., and Stenger, F., "Perturbation methods for reconstructing 3-D complex acoustic velocity and impedance in circular geometries with spherical waves," in *Acoustical Imaging 9*, K. Y. Wang, ed., 451-461 (Plenum Press, New York, 1980).

[13] Abromowitz, M. and Stegun, I. A., *Handbook of Mathematical Functions* (Dover Publications, New York, 1972).

[14] Soumekh, M., Kaveh, M., and Mueller, R. K., "Fourier domain reconstruction methods with application to diffraction tomography," *Acoustical Imaging 13*, submitted to Plenum Press, New York.

[15] Greenleaf, J. F., "Computed tomography from ultrasound scattered by biological tissues," submitted to *American Math. Society Symposium of Inverse Problems*, April 12-13, 1983, New York, N. Y.

[16] Devaney, A. J., "A filtered back propagation algorithm for diffraction tomography," *Ultrasonic Imaging 4*, No. 4, 336-350 (1982).

[17] Devaney, A. J. and Oristaglio, M. L., "Inversion procedure for inverse scattering within the distorted-wave Born approximation," *Physical Review Letters 51*, 237-240 (1983).

ACOUSTICAL IMAGING/NDE OF COMPLEX GEOMETRIES USING REFRACTION AND

REFLECTION TRANSMISSION TECHNIQUES IN ACOUSTIC MICROSCOPY

Michael Oravecz, Carol Vorres and Lawrence W. Kessler

Sonoscan, Inc.
530 East Green Street
Bensenville, IL 60106

ABSTRACT

The application of scanning laser acoustic microscopy (SLAM) for flaw detection, metallurgical characterization and evaluation of processing variable influence has been limited to components (ceramic turbine blades, plastic IC packages, metal cylinders, etc.) which allow access to two surfaces. Such specimens were tested based only on amplitude variations of <u>transmitted</u> 10 to 500 MHz sonic energy. This paper presents practical SLAM applications where parts are tested based on amplitude variations of <u>refracted</u> and/or <u>reflected</u> 10 to 100 MHz sonic energy. This distinction allows testing of components that restrict access to only one surface and of components where the area of interest is masked by intervening material. For example, miniature welded or brazed corner and T joints, thin wall tubing and the bonding of small electronic parts on thick heat sinks can be nondestructively tested using refraction and reflection SLAM imaging.

COMPUTER-ASSISTED TOMOGRAPHIC ACOUSTIC

MICROSCOPY FOR SUBSURFACE IMAGING

Zse-Cherng Lin, Hua Lee,[*]
Glen Wade and Carl F. Schueler[**]

Department of Electrical & Computer Engineering
University of California
Santa Barbara, CA 93106

ABSTRACT

Computer-assisted ultrasonic tomography has received much attention in recent years. Acoustic microscopy is an important branch of non-destructive evaluation which provides high-resolution imaging of the detailed structure of an object. STAM (Scanning Tomographic Acoustic Microscope) is a system capable of producing tomographic images by scanning the source or rotating the specimen to generate a sequence of tomographic projections. This system has advantages over conventional approaches, especially for complex objects with planar structure such as integrated-circuit chips. An earlier paper provided the analysis and the reconstruction algorithms for the signal processing of planar tomographic systems including an algorithm for "back-and-forth" propagation. This paper summarizes the modifications and revisions of this latter algorithm as applied to microscopic digital imaging. Two different schemes of rotation to acquire STAM data and the corresponding reconstruction procedures are described. Simulations are presented which demonstrate the resolving capability of the STAM system.

---

[*] Hua Lee is now with the Department of Electrical Engineering, University of Illinois at Urbana-Champaign, IL 61801.

[**] Carl F. Schueler is with the Santa Barbara Research Center, Goleta, CA 93117.

INTRODUCTION

There has been a surge of interest in the use of ultrasound for nondestructive evaluation and computer-assisted tomography in recent years [1-3]. Acoustic energy can often give a view of a cross section not available with X-rays or other types of radiation. Acoutic microscopy is an important branch of non-destructive evaluation which provides high-resolution micrographs of specimens to be evaluated. When an acoustic microscope operates in the transmission mode, the micrograph is simply a shadowgraph of all the structure encountered by the paths of the acoustic rays passing through the object. The resultant image shows a two-dimensional mapping of three-dimensional internal structure and is particularly difficult to comprehend in the case of specimens of substantial thickness and structural complexity.

STAM (Scanning Tomographic Acoustic Microscope) is a system capable of producing microscopic tomograms (cross-sectional images) to overcome this difficulty. This system has advantages over conventional approaches, especially for complex objects with planar structure. It makes use of elements from an existing ultrasonic microscope known as SLAM (Scanning Laser Acoustic Microscope) in which the acoustic wavefield is detected by a scanning laser beam. STAM incorporates digital signal processing to obtain tomograms of the internal structure of microscopic specimens to be evaluated.

The analysis and reconstruction algorithms for the signal processing of planar tomographic systems were presented in an earlier paper [4]. In STAM, a suitably modified "back-and-forth propagation" method [4] is employed for use with a plane-wave source. Elements of SLAM are used in such a way as to acquire the data for the system. Both amplitude and phase information are needed and problems associated with obtaining this information are discussed in another paper [5]. With this instrument, projections are generated, one after another, for the tomographic image reconstruction. Two different schemes of rotation for data acquisition are discussed. Simulations show that these techniques can significantly eliminate the blurring and overlapping of various layers in an object and give well-differentiated tomograms. In this paper, we summarize the modifications and revisions of this algorithm for digital reconstruction of the image. We also present results from computer simulations.

BACK-AND-FORTH PROPAGATION

The three-dimensional structure of an object can be determined from holographic data [6]. In general, many holograms, corresponding to different directions of illumination of the object, are needed to reconstruct the three-dimensional structure. Back-and-forth propagation is a holographic approach for tomographic image reconstruction [4]. Assume we have an object of planar structure as shown in Fig. 1 and that it is illuminated by an ultrasonic wave from below. The object is homogeneous to the ultrasound except at plane 2 which contains object elements of different elastic composition. The acoustic transmission through plane 2 is denoted by $t(x,y)$.

The wavefield at plane 1 is $u_1(x,y)$. The acoustic waves travel through the object and the wavefield just below plane 2 is $u_2(x,y)$. If we neglect the evanescent waves, the relation between $u_1(x,y)$ and $u_2(x,y)$ can be expressed by

$$U_2(f_x,f_y) = U_1(f_x,f_y) \exp\{K(z_2-z_1)(1-f_x^2\lambda^2-f_y^2\lambda^2)^{\frac{1}{2}}\}, \qquad (1)$$

with

$$U_1(f_x,f_y) = F\{u_1(x,y)\},$$

$$U_2(f_x,f_y) = F\{u_2(x,y)\},$$

and

$$K = -\alpha + jk,$$

where z_1 and z_2 denote the positions in the Z-direction for plane 1 and plane 2 respectively, F stands for the two-dimensional Fourier Transform, K is the acoustic propagation constant, α is the attenuation coefficient and λ is the wavelength in the object region [7].

At plane 2, $u_2(x,y)$ is modified by the distribution $t(x,y)$ and gives

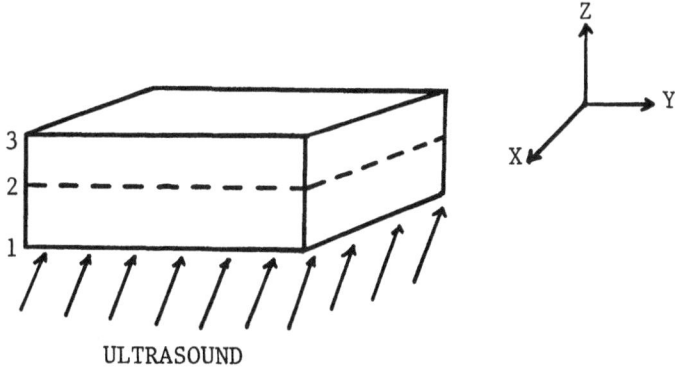

ULTRASOUND

Fig. 1   An object of planar structure irradiated by ultrasound coming from below.

$$\hat{u}_2(x,y) = u_2(x,y)t(x,y). \tag{2}$$

The modified wave propagates through the object to plane 3 (the receiving plane) and becomes $u_3(x,y)$ with

$$U_3(f_x,f_y) = \hat{U}_2(f_x,f_y) \exp\{K(z_3-z_2)(1-f_x^2\lambda^2-f_y^2\lambda^2)^{\frac{1}{2}}\}, \tag{3}$$

where

$$U_3(f_x,f_y) = F\{u_3(x,y)\}$$

and

$$\hat{U}_2(f_x,f_y) = F\{\hat{u}_2(x,y)\}.$$

The distribution $t(x,y)$ will have its effect on the wavefield $u_3(x,y)$. If there are additional distributions in other planes, the actual wavefield $v_3(x,y)$ detected at plane 3 will be different from the above calculated $u_3(x,y)$ so as to include contributions from all planes. If we back-propagate $v_3(x,y)$ to plane 2 by using

$$V_2(f_x,f_y) = V_3(f_x,f_y) \exp\{-K(z_3-z_2)(1-f_x^2\lambda^2-f_y^2\lambda^2)^{\frac{1}{2}}\} \tag{4}$$

with

$$V_2(f_x,f_y) = F\{v_2(x,y)\}$$

and

$$V_3(f_x,f_y) = F\{v_3(x,y)\},$$

the wavefield $v_2(x,y)$ will not be the same as $\hat{u}_2(x,y)$. We can express the relation between $v_2(x,y)$ and $\hat{u}_2(x,y)$ by

$$v_2(x,y) = \hat{u}_2(x,y) + n(x,y) = u_2(x,y)t(x,y) + n(x,y), \tag{5}$$

where $n(x,y)$ represents an undesired signal caused by higher-order scattering due to the wave interaction within the object. When the object is illuminated by a plane wave all terms in equation (5) are functions of the incident angle of the acoustic source except for $t(x,y)$. Equation (5) can be rewritten to include the incident angle ϑ

$$v_2(x,y;\vartheta) = \hat{u}_2(x,y;\vartheta) + n(x,y;\vartheta) = u_2(x,y;\vartheta)t(x,y) + n(x,y;\vartheta). \tag{6}$$

As the angle ϑ varies, $n(x,y;\vartheta)$ changes both in magnitude and phase. We assume that $n(x,y;\vartheta)$ has little correlation with respect to ϑ. Therefore, it can be regarded as a zero-mean noise term because of the random phase. Then a least-square estimate [8] of the distribution $t(x,y)$ can be found

$$\hat{t}(x,y) = \frac{\sum_i v_2(x,y;\vartheta_i) \; u_2^*(x,y;\vartheta_i)}{\sum_i u_2(x,y;\vartheta_i) \; u_2^*(x,y;\vartheta_i)}, \tag{7}$$

where the summation is carried out over all different angles of view. Since plane waves are used, $u_2(x,y;\vartheta_i)u_2^*(x,y;\vartheta_i)$ remains constant for different incident angles ϑ_i. Equation (7) can be further simplified, i.e.,

$$\hat{t}(x,y) = \kappa \sum_i v_2(x,y;\vartheta_i) \; u_2^*(x,y;\vartheta_i), \tag{8}$$

where κ is a scaling factor which depends on how many projections are used for image reconstruction.

For each angle ϑ_i, the received wavefield $v_3(x,y;\vartheta_i)$ is back-propagated to obtain $v_2(x,y,;\vartheta_i)$ and the source field $u_1(x,y;\vartheta_i)$ is forward-propagated to find $u_2^*(x,y;\vartheta_i)$. Finally, $\hat{t}(x,y)$ is computed as the least-square estimate of $t(x,y)$. In this way, a tomogram at plane 2 can be obtained.

SYSTEM DESIGN AND DIGITAL IMAGE RECONSTRUCTION

The SLAM system is capable of operating both in the holographic and direct-imaging mode [9]. An acoustic transducer generates plane waves to illuminate the specimen under test. As shown in Fig. 2, a scanning laser beam detects the acoutic wavefield transmitted through the object to constitute the data acquisition and provide the required phase and amplitude information. In the holographic mode, the micrographs are in the form of interferograms which retain phase information concerning the transmitted wavefield. In the direct-imaging mode, the micrographs are images which give only the intensity distribution of the wavefield. The data set acquired when the acoustic plane wave propagates at a specific incident angle is termed a projection. A number of projections in sequence can be generated by rotating the acoustic transducer or the specimen through an angular increment at each step. The projections are recorded and stored on video tape for digital image reconstruction.

Two different schemes of rotation are currently considered.

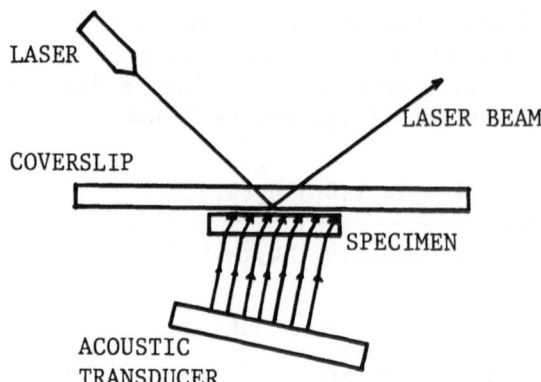

Fig. 2.   Simplified   schematic   diagram   of   SLAM   (Scanning   Laser
           Acoustic Microscope).

One of them is shown in Fig. 3(a). The transducer is rotated
around an axis which, as shown, is the center line of the top surface of the transducer and points perpendicularly out of the figure. The distance between the object and the rotation axis must
be kept constant to avoid phase errors caused by the rotation.
With this rotation scheme, the algorithm described in the previous
section can readily be applied. Fig. 3(b) shows the other method
of rotation. The object is circularly rotated around a vertical
axis passing through its center. The transducer remains stationary. This is equivalent to rotating the transducer so that the
wave vectors of the generated plane waves are rotating circularly.
Because of the rotation of the object, we have to perform coordinate transformation on the received wavefield and incident wavefield in order to obtain the same data as acquired by rotating the
transducer. The back-and-forth propagation algorithm can then be
applied to reconstruct the images. One important point is that
the coordinate transformation can be performed after the wave propagation. This saves one coordinate transformation for each angle
of view and also allows the forward-propagated incident wave to be
used for all different angles of view. Simplifications in the
algorithm greatly reduce the computation-time required for image
reconstruction. Flow charts of algorithms for these two different
schemes of rotation are shown in Fig. 4.

     The first scheme requires rotating the transducer. Detri-
mental phase errors will be introduced if the rotation axis is not
kept at the specified position to within a fraction of a wavelength. It is extremely difficult to achieve this stability.
Another difficulty is that the detection sensitivity of SLAM is

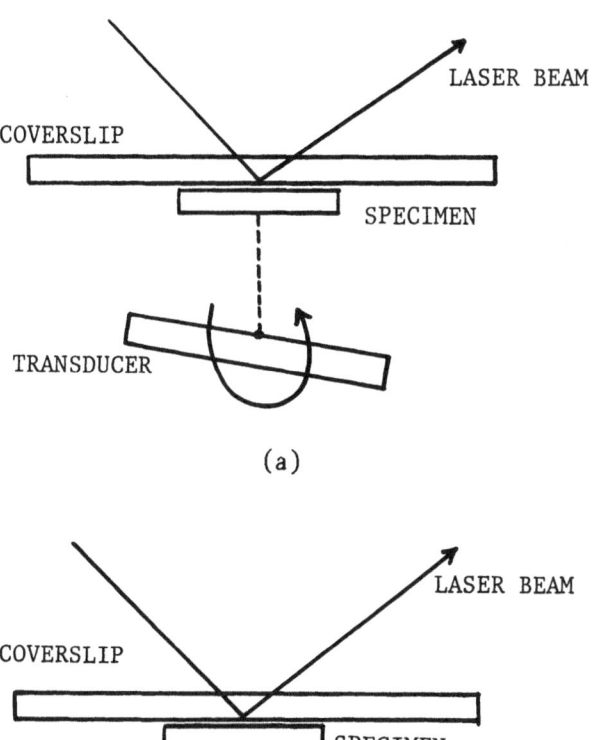

(a)

(b)

Figure 3. Two different schemes of rotation to generate projections for STAM (Scanning Tomographic Acoustic Microscope). (a) Rotation of the transducer. (b) Rotation of
the specimen.

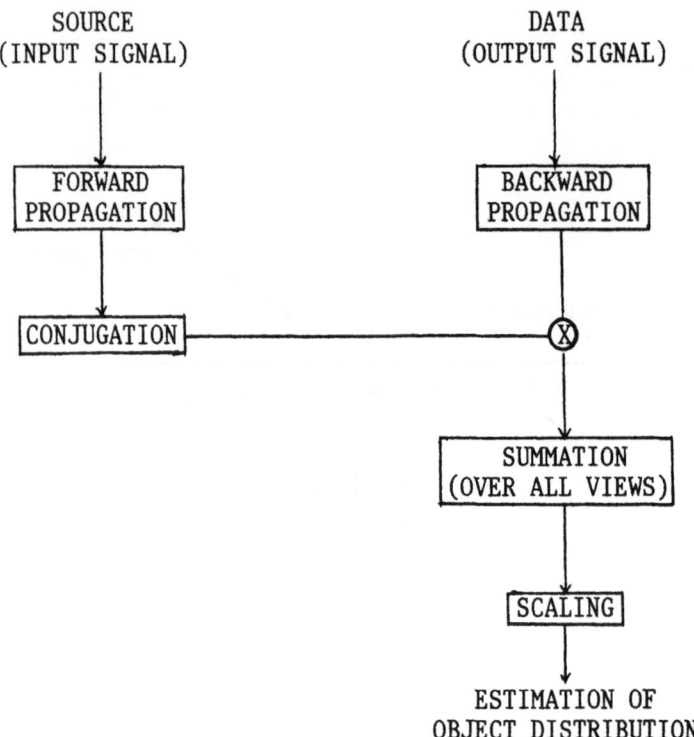

Fig. 4(a) Flow chart of the reconstruction algorithm for the
scheme of rotating the transducer.

not uniform with rotation because of the knife-edge technique [9].
Compensation of this nonuniformity is therefore necessary to
obtain compatible projections. The normalization factor for each
projection is hard to determine.

The second scheme does not have the problem of nonuniform
sensitivity because the transducer is stationary during the rotation process. Nevertheless, the specimen still has to be kept
flat and at the same horizontal level as it is rotated to avoid
phase errors. From practical experience, we have noted that it is
easier to stabilize the object in this sense than the transducer.

Because of the availability of high-speed computers and the
techniques of digital signal processing, it is appropriate to
modify this back-and-forth propagation method for digital treatment. In order to calculate wave propagation, the measured wavefield is decomposed into spatial frequency components by the Fast
Fourier Transform. Each frequency component is back- or forward-
propagated by the transfer function

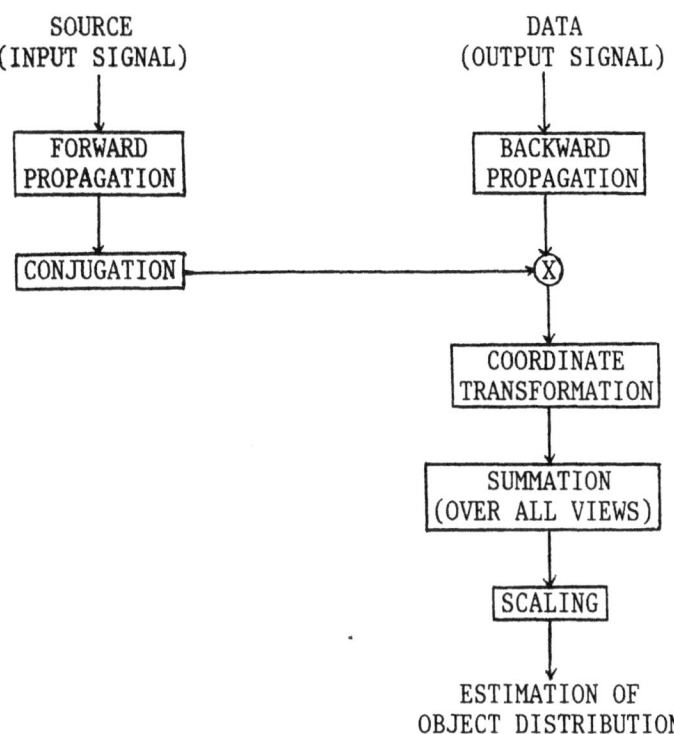

Fig. 4(b) Flow chart of the reconstruction algorithm for the scheme of rotating the object.

$$H(f_x, f_y) = \begin{cases} \exp\{Kz(1-f_x^2\lambda^2-f_y^2\lambda^2)^{\frac{1}{2}}\}, & f_x^2+f_y^2 < \lambda^{-2} \\ \\ 0 & \text{otherwise.} \end{cases} \qquad (9)$$

where z is the propagation distance. The transfer function is low-pass since the evanescent waves have been neglected. Because the cutoff frequency of the transfer function is $1/\lambda$, the sampling spacing must be smaller than $\lambda/2$ so that the Nyquist criterion is satisfied [10].

The object distribution can be computed, coded into gray level values and displayed on the CRT terminal. The resulting tomograms can be regarded as a three-dimensional image of the object structure.

SIMULATION RESULTS

       To illustrate the resolving capability of the STAM system, we
have performed computer simulations. A three-dimensional object
with planar structure was used as the simulated specimen and irradiated by acoustic plane waves. The object was assumed to be
attenuation-free with respect to the acoustic waves except for two
horizontal layers ten wavelengths apart. The top layer was eight
wavelengths away from the receiving plane. Different patterns
including structure 50 percent transparent to the acoustic wave
were contained in those two layers.

       For the first rotation scheme, eighteen simulated projections
of the acoustic wave transmitted through the object were generated
and used for signal processing. The projections were generated by
assuming that the incident angle of the acoustic wave inside the
specimen was linearly increased from negative 45 degrees to positive 45 degrees. The amplitude images of these projections are
equivalent to SLAM images. One of them is shown in Fig. 5.
Obviously the image is ambiguous due to the overlapping and diffraction. The shaded region near the left edge is caused by the
assumption of the limited input aperture.

Fig. 5.   One of the simulated amplitude images from a projection
           (i.e., a SLAM image), with illuminating waves of inci-
           dent angle $\vartheta = 45°$.

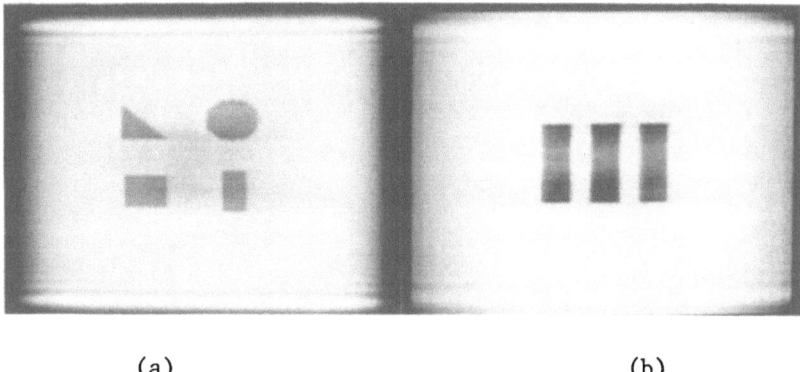

(a)   (b)

Fig. 6.   Tomograms reconstructed from eighteen projections gener-
          ated by rotating the transducer. (a) Top layer. (b)
          Bottom layer.

The projections were then processed by the method described
in utilizing various numbers of projections. These tomograms show
the gradual differentiation and deblurring of these two layers as
the number of projections increases. Fig. 6 shows the tomograms
of these two layers reconstructed from eighteen projections. Note
that the blurring and overlapping of the two layers are significantly removed.

For the second rotation scheme, the transducer was assumed to
be stationary and tilted so that the incident angle of the acoustic wave inside the specimen was 45 degrees. The object was
rotated circularly with a constant angular increment and going
through a full 360 degrees. When the coordinate transformation
for the rotation was performed, zero order interpolation was used
to fill out all points on the new coordinate system. Figures
7(a-c) show the results of images reconstructed from 18, 36 and 72
projections respectively. The results demonstrate the differentiation and deblurring capability of this algorithm. As the number
of projections increases, the reconstructed image becomes
smoother.

A simulation was also performed with an object of three
layers. The layers were still ten wavelengths apart and the top
layer was eight wavelengths away from the receiving plane. The
incident angle of the acoustic wave was 30 degrees. Only nine
projections were generated and used for image reconstruction. The
results are shown in Fig. 8.

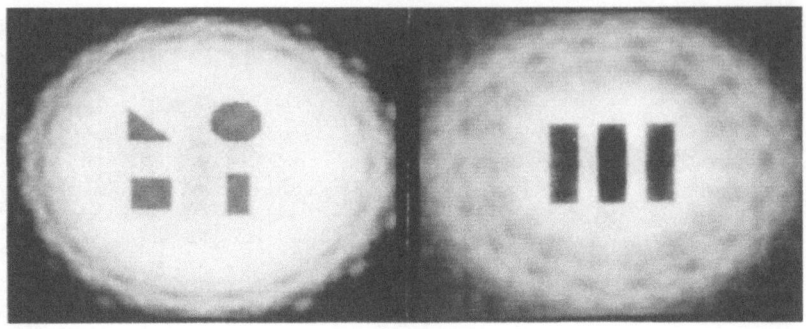

(a)

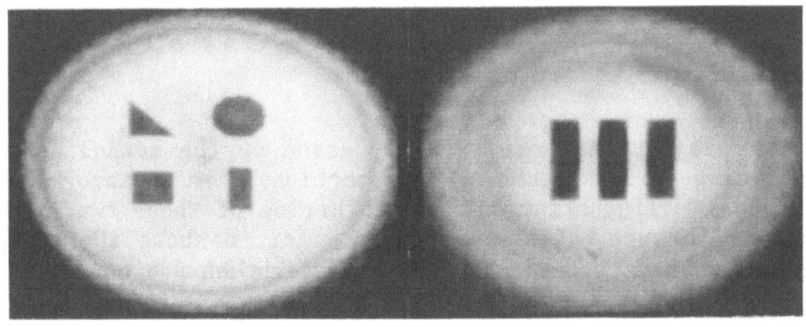

(b)

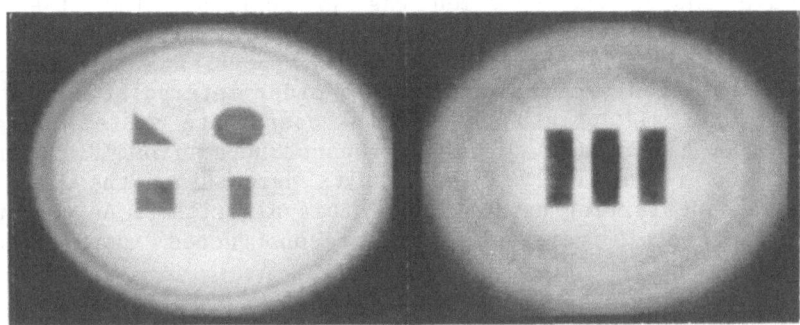

(c)

Fig. 7. Tomograms reconstructed from various numbers of projections generated by circularly rotating the specimen. (a) 18 projections. (b) 36 projections. (c) 72 projections.

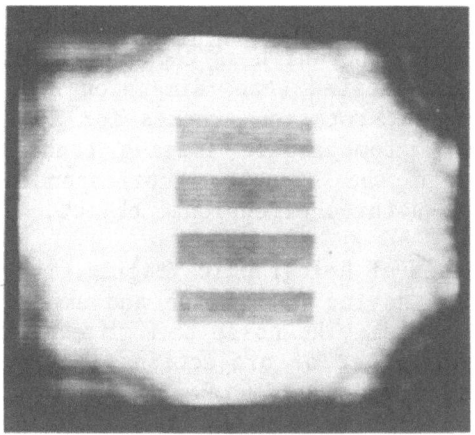

(a)

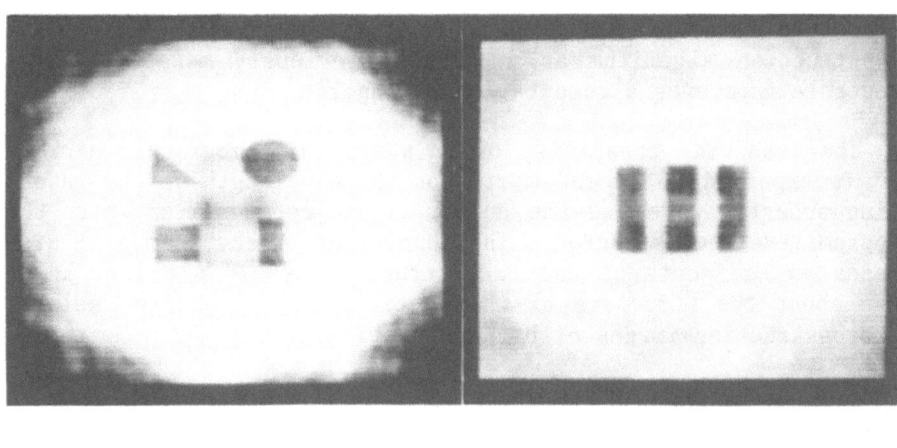

(b)   (c)

Fig. 8.   Tomograms reconstructed from 9 projections generated by
           circularly rotating the specimen of three layers.   (a)
           Top layer.   (b) Middle layer.   (c) Bottom layer.

CONCLUSION

The long-range purpose of this project is to produce a tomographic acoustic microscope which is capable of imaging the internal structure of complicated three-dimensional objects. We have
simulated two different rotation schemes for acquiring the data.
Both show that this tomographic imaging technique is able to
cleanly differentiate the structure corresponding to different
horizontal layers in a three-dimensional object.

As in other types of holographic imaging, the success of this
technique depends on having both phase and amplitude information
concerning the wavefield. The noise term in equation (6) tends to
cancel out when the number of projections increases. Therefore,
equation (8) gives a good estimate of the object distribution.
According to past experience with data acquisition, the first
scheme of rotation raises two serious practical problems. One is
that it is hard to keep the axis of rotation stationary during the
rotating process. The other is that a SLAM system has a different
system response for different angles of incidence [9]. The second
scheme of rotation avoids these problems. Although it requires a
coordinate transformation on the data, the simplification in the
reconstruction algorithm as described previously makes it quite
acceptable from even a computation standpoint.

The resolving capability of this type of tomographic microscopy is expected to depend mostly on the wavelength of the insonifying acoustic wave and the number of projections from which the
tomograms are reconstructed. The resolution of each tomogram also
depends on the depth of the reconstructed layer. A priori knowledge about the structure of the specimen is, of course, helpful
in reconstructing images of high resolution.

ACKNOWLEDGEMENT

The authors are greatly indebted to the Santa Barbara
Research Center for its encouragement and financial support of
this project, to Lawrence W. Kessler and Michael G. Oravecz of
Sonoscan for their valuable discussions of the practical problems
associated with acoustic microscopes and to Robert Adler of Zenith
for suggesting object rotation instead of transducer rotation.

The authors would like to acknowledge Steve Silverman who
helped prepare pictures and Chin-Hsing Chen who proofread and participated valuably in the manuscript's preparation. Special
thanks are due to Michael Galindo for typing this manuscript for
publication.

REFERENCES

[1] G. Wade, R.K. Mueller and M. Kaveh, "A Survey of Techniques for Ultrasonic Tomography," Computer-Aided Tomography and Ultrasonics in Medicine, Ed. J. Raviv et al., North Holland Publishing Co., Amsterdam, 1979, pp. 165-215.

[2] J.F. Greenleaf, S.A. Johnson, S.L. Lee, G.T. Herman and E. H. Wood, "Algebraic Reconstruction of Spatial Distribution of Acoustic Absorption within Tissue from their Two-Dimensional Acoustic Projections," Acoustical Holography, Vol. 5, Ed. P.S. Green, Plenum Press, New York, 1974, pp. 591-603.

[3] Takuso Sato, Osamu Ikeda and Kazuyuki Endo, "Combined Spectral and Aperture Synthetic Ultrasonic Imaging System," IEEE Transactions on Sonics and Ultrasonics, Vol. SU-28, No. 2, March 1981, pp. 64-69.

[4] H. Lee, C. Schueler, G. Flesher and G. Wade, "Ultrasonic Planar Scanned Tomography," Acoustical Imaging, Vol. 11, Ed. John Powers, Plenum Press, New York, 1982, pp. 309-323.

[5] Z. Lin, H. Lee, G. Wade, M.G. Oravecz and L.W. Kessler, "Data Acquisition in Tomographic Acoustic Microscopy," to be published in Proceedings of the 1983 Ultrasonics Symposium.

[6] E. Wolf, "Three-Dimensional Structure Determination of Semi-Transparent Objects from Holographic Data," Optics Communications, Vol. 1, No. 4, Sept./Oct. 1969, pp. 153-156.

[7] J.W. Goodman, Introduction to Fourier Optics, McGraw-Hill, 1968, pp. 54.

[8] A.D. Whalen, Detection of Signal in Noise, Academic Press, 1971, pp. 335-336.

[9] L.W. Kessler, "Imaging with Dynamic-Ripple Diffraction," Acoustic Imaging, Ed. G. Wade, Plenum Press, New York, 1976, Chap. 10.

[10] Alan V. Oppenheim and Ronald W. Schafer, Digital Signal Processing, Englewood Cliffs, NJ:Prentice-Hall, 1975.

THE ORIGIN OF GRAIN CONTRAST IN THE SCANNING ACOUSTIC MICROSCOPE

M.G. Soumekh, G.A.D. Briggs and C. Ilett

Department of Metallurgy and Science of Materials
University of Oxford, Parks Road, Oxford OX1 3PH

INTRODUCTION

The contrast in the scanning acoustic microscope may be described by the V(z) curve (1,2). The potential of the SAM lies in the fact that the V(z) curve is a strong function of the elastic properties of the material. It is thus possible to image elastic properties with spatial resolution of the order of a micron or better provided the incident radiation has a sufficiently small wavelength.

V(z) curves obtained both experimentally and theoretically from isotropic surfaces show a periodic sequence of peaks and troughs whose separation is dependent upon the velocity of "leaky" Rayleigh waves. The theory describing the V(z) response of the microscope on an isotropic surface has been extensively studied using either an approximate ray model (3,4) or more rigorously by Fourier optics (5,6). The ray approach requires that the existence of a surface wave is assumed. On the other hand the Fourier optical formulation merely requires that a reflectance function be generated that satisfies the wave equations in the specimen and coupling fluid and also the boundary conditions between them. The existence of surface wave activity arises naturally from the solution of these equations. A further advantage of the Fourier optics formulation is that the complete v(z) curve including the amplitude and phase may be predicted rather than just the periodicity of the peaks and troughs.

Work carried out upon anisotropic specimens by Kushibiki et al. (7,8) has demonstrated that the response of the acoustic microscope upon anisotropic specimens is qualitatively different from the response on isotropic surfaces because two surface waves are excited on certain planes along certain directions.

In this paper the formulation of Fourier Optics is applied to anisotropic specimens and the full V(z) curve is predicted from the independent elastic constants. The presence of two propagating surface waves is inferred from the calculated reflectance functions. Finally the Fourier Optics approach is employed to explain the contrast from grains of different orientations.

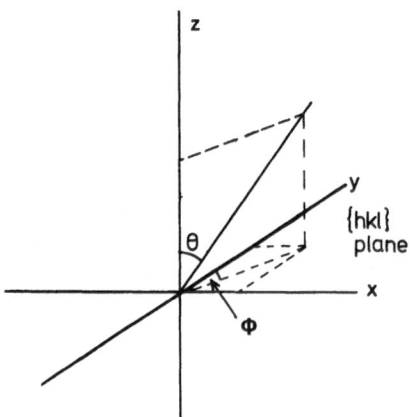

Figure 1   Coordinate system defining the azimuthal angle ϕ

CYLINDRICAL LENS SYSTEMS

The work of Chubachi et al. has been carried out with a cylindrical lens. The advantage of such a system for specimen identification is that the radiation from the lens is incident at only one azimuthal angle, ϕ, as shown in figure 1 below.

The $V(z)$ response of a cylindrical acoustic lens on an anisotropic specimen is given by Sheppard and Wilson (6) as

$$V(z) = \int_0^\alpha P(\theta) \ R(\theta) \ \exp(2jkz\cos\theta) \ \cos\theta \ d\theta \qquad (1)$$

where $P(\theta)$ is the pupil function of the lens so defined account for the two-way passage through the lens.
α is the semi-angle of the lens.
k is the wave number of the incident sound in the coupling medium.
$R(\theta)$ is the reflection coefficient as a function of the angle θ.

For an isotropic substrate $R(\theta)$ is independent of ϕ so that rotation of the cylindrical lens relative to the specimen will not effect the $V(z)$ response of the microscope.

For an anisotropic specimen the formulation of Sheppard and Wilson for a cylindrical lens can be modified thus

$$V(\phi,z) = \int_0^\alpha P(\theta) \ R_{hkl}(\phi,\theta) \ \exp(2jk\cos\theta) \ \cos\theta \ d\theta \qquad (2)$$

where the subscript hkl indicated the Miller indices of the specimen plane. When the specimen is anisotropic the $V(z)$ response is a function of both the plane of incidence and the incident azimuthal angle, ϕ. The method of calculation of $R_{hkl}(\theta,\phi)$ is described in detail in another work by the authors (9).

The form of the Anisotropic Reflectance function

Figure 2 shows four reflectance functions from a 100 nickel/water interface. The elastic properties used were those employed by Farnell (10) that is : c_{11}=2.61 X 10^{-11} Nm^{-2}

$\quad c_{12}$=1.51 X 10^{-11} Nm^{-2}

$\quad c_{44}$=1.309 X 10^{-11} Nm^{-2}

$\quad$ density= 8900 kg m^{-3}

$\quad$ velocity of sound in water = 1551 m s^{-1}

The parameter ϕ represents the deviation of the incident azimuthal angle from the (100) direction.

Figure 2a gives the reflection function when ϕ = 0 degrees. This reflectance function looks similar in form to reflectance functions obtained from isotropic specimens, with features that have immediate counterparts. Point A represents the quasi-longitudinal wave criticality. The point B does not, however, correspond to the quasi-shear wave velocity along (100) in nickel but represents the incident angle above which no propagating quasi-shear wave may be transmitted into the solid. The phase velocity for shear waves in

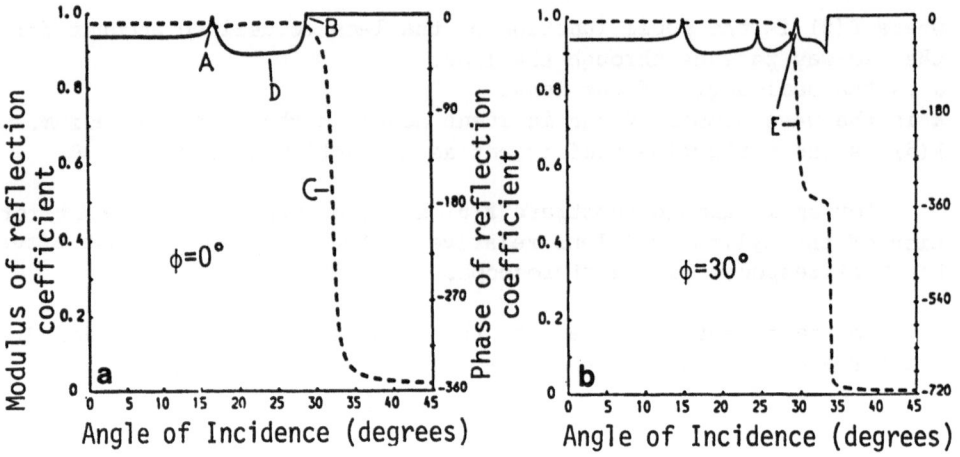

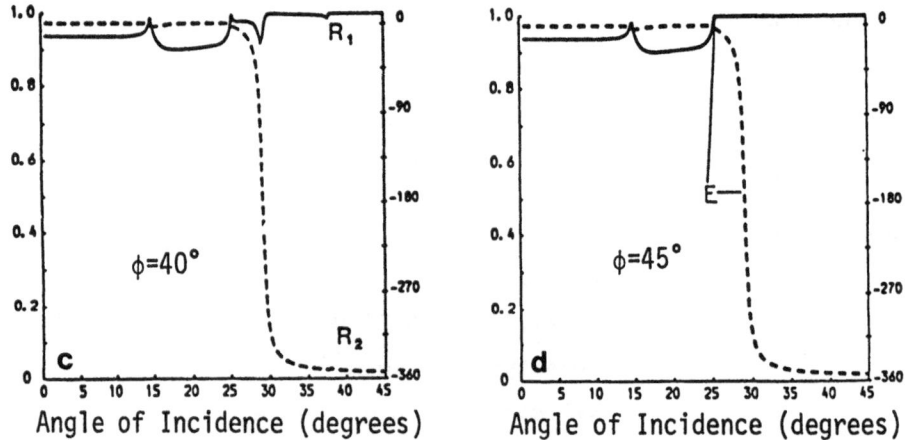

Figure 2   Reflectance functions between {100} nickel and water
           a) Reflectance function when ϕ = 0 degrees.
           b) Reflectance function when ϕ = 30 degrees, note the
              change in scale of the phase ordinate.
           c) Reflectance function when ϕ = 40 degrees.
           d) Reflectance function when ϕ = 45 degrees.

nickel along this direction is 3834 ms^{-1} corresponding to a criticality at point D, a wave incident at θ_s (= arcsin 1551/3834) will
generate a shear wave whose wave vector is parallel to the surface.
For an isotropic surface this would represent the incident angle
above which the reflection coefficient was unity, the fact that the
slowness surface is not spherical for an anisotropic surface means
that this is not the situation obtaining here. In the case we are
considering here the slowness surface bulges out for a wave whose
wave vector is inclined below the surface, the maximum projection of
the k vector upon the surface is thus obtained when this vector is
directed into the bulk of the specimen. Point B therefore does not
correspond to the velocity of a shear wave along the 100 direction
in the 100 plane. We see then that the phase velocity corresponding
to point B of 3199 ms^{-1} is a consequence of the slowness surface
deviating from a spherical form. The point C corresponds to a phase
velocity of 2909 ms^{-1}, a value very close to the Rayleigh velocity
upon a stress-free surface (10). Point C represents the Rayleigh
angle criticality, with the familiar rapid phase change through $\simeq 2\pi$
around this angle.

Figure 2b shows a qualitative change in appearance from any
reflectance function obtained with any isotropic substrate. There
are now two phase changes through the order of 2π (note the change
in scale of the ordinate of figure 2b). The phase change occurring
around θ_R = 34 degrees corresponds to the true surface wave, the Rayleigh wave. At this incident angle the modulus of the reflection
coefficient is unity implying that no energy is leaked or absorbed
into the solid. The phase change occurring at θ_{ps} has no counterpart with isotropic materials both because the phase change corresponds to a velocity greater than the slowest quasi-shear wave, and
because the modulus of the reflection coefficient is less than unity.
At angles close to θ_{ps} a wave is excited which appears to behave
rather like a surface wave but since the modulus of the reflection
coefficient is somewhat less than one a proportion of the incident
radiation must be converted to a bulk wave propagating into the body
of the solid. Note that the possibility of energy absorption is
excluded because the viscosity tensor is zero. A wave of this type
is called a pseudo-surface wave.

Figure 2c shows the reflectance function when ϕ = 40 degrees.
The points R_1 and R_2 show that the Rayleigh wave has a very small
effect upon the reflectance function whereas the predominant surface
wave activity is due to the pseudo-surface wave.

When ϕ =45 degrees the trend seen in figure 2c is completed.
The wave associated with the pseudo-surface wave has acquired the
characteristics of a true surface wave by which we mean that the
modulus is equal to one close to its critical angle. For propagation
along a (110) direction upon the {100} plane the wave associated with
the slowest bulk wave and the Rayleigh wave are degenerate, this wave

is not coupled into the solid from the liquid and so does not in any
way perturb the reflectance function (i.e. the bumps R_1 and R_2 in
figure 2c have dissappeared). We have a situation then when ϕ = 45
degrees which has no counterpart when the specimen is isotropic, that
is the Rayleigh wave has degenerated into a bulk wave which does not
effect the reflectance function. This bulk wave is,however, <u>slower</u>
than the surface wave which gives the 2¶ phase change of figure 2d.

Predicted V(z) curves from an anisotropic surface

   In order to predict the shape of the V(z) curves obtained with
a cylindrical lens the reflectance functions of figure 2 were substituted into equation 2. The pupil function was defined such that
$P(\theta)$ = 1 for $|\theta|$ < 45 degrees and $P(\theta)$ = 0 when $|\theta|$ > 45 degrees.
Figure 3 a-d shows the predicted V(z) curves. The curves for ϕ = 0,
40 and 45 degrees show a V(z) periodicity which agrees well with the
formula given by Atalar (11) viz

$$\Delta z = \frac{v_{water}}{2 f (1 - \cos\theta_s)} \tag{3}$$

where Δz is the extra defocus distance between successive nulls
f is the frequency of the acoustic radiation
v_{water} is the velocity of sound in the coupling medium (water) and
θ_s is the optimum angle for the excitation of surface waves.

   When ϕ = 40 degrees the value of θ_s that predicts the periodicity Δz is that corresponding to the pseudo-surface wave, this is to
be expected from inspection of figure 2c. The V(z) curves for values
of ϕ close to 30 degrees show rather more complicated behaviour.
This is because at and around these ϕ values both Rayleigh and
pseudo-surface waves are strongly excited. The spacing between the
nulls in figure 3b is not regular although Δz is approximately equal
to the value one would obtain if the critical angle for the excitation of pseudo-surface waves were substituted into equation 3. This
indicates that when ϕ = 30 degrees the pseudo-surface wave is more
important to SAM response than the Rayleigh wave.

   The Fourier optical approach we have described has the advantage
over the ray approach that the shape of the V(z) curve and the relative contribution of each type of surface wave can be obtained without recourse to arbitary coupling constants between the surface waves
and the incident wave from the liquid.

SPHERICAL LENS SYSTEMS

   Cylindrical lenses have proved very useful for quantitative
materials measurements, but they have not been useful for true acoustic imaging because they do not have spatial resolution in the

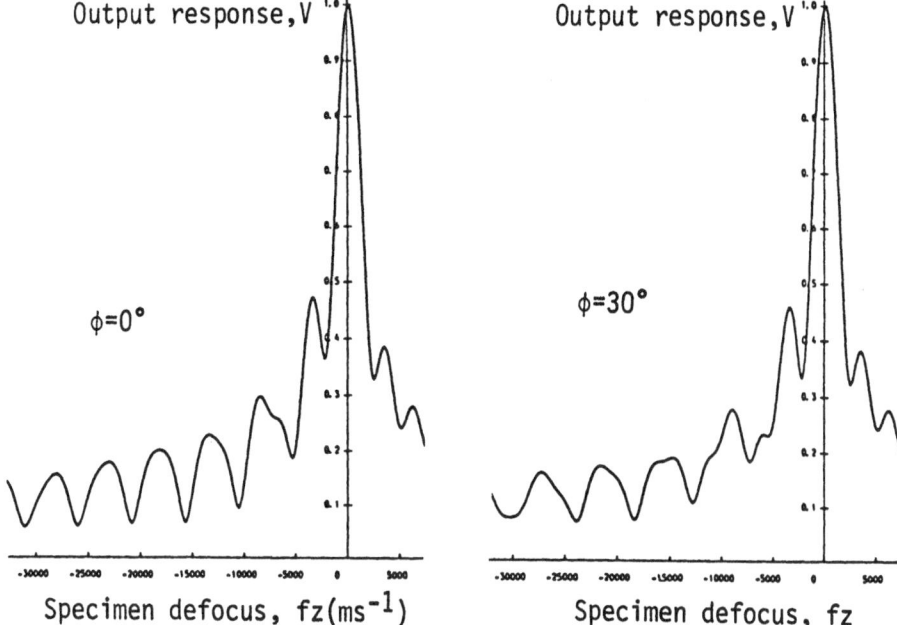

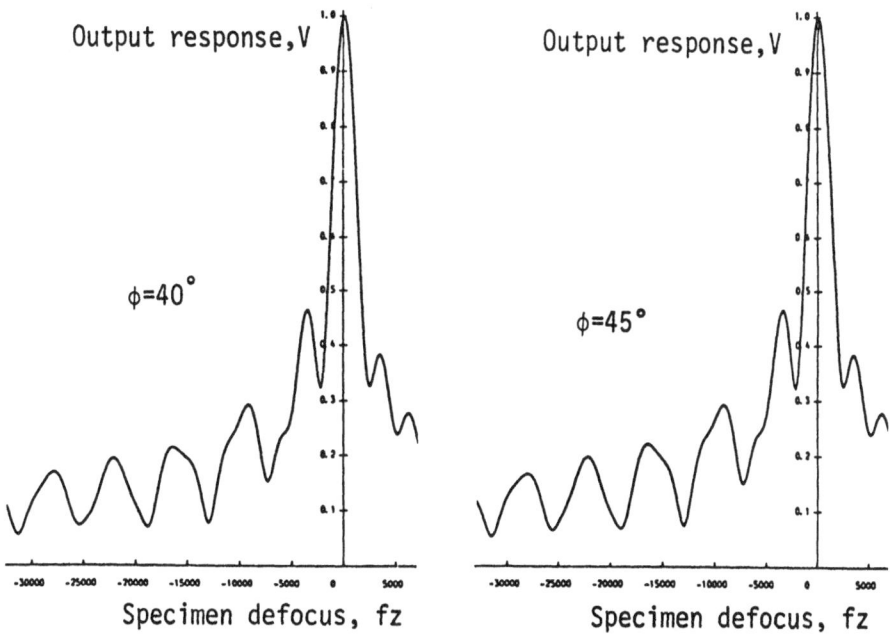

Figure 3. Theoretical V(z) curves for {100} plane of nickel using a "top hat" pupil function. The azimuthal angles are shown on each curve.

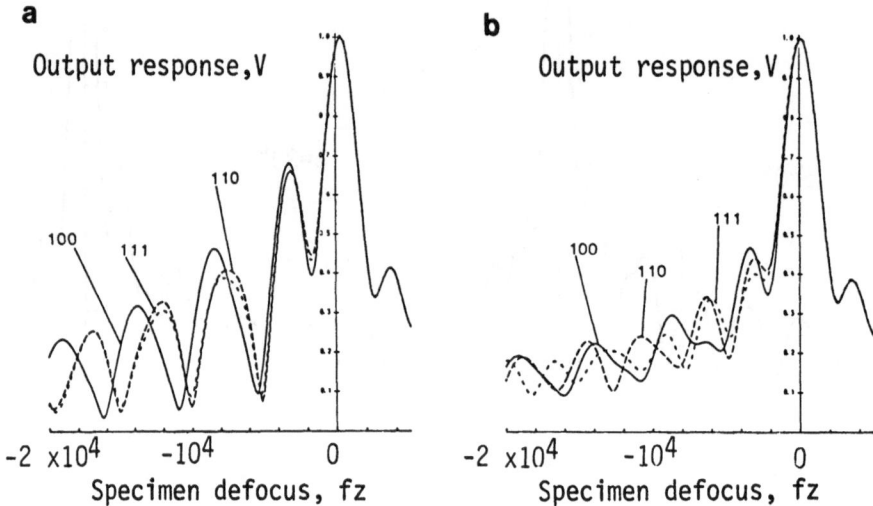

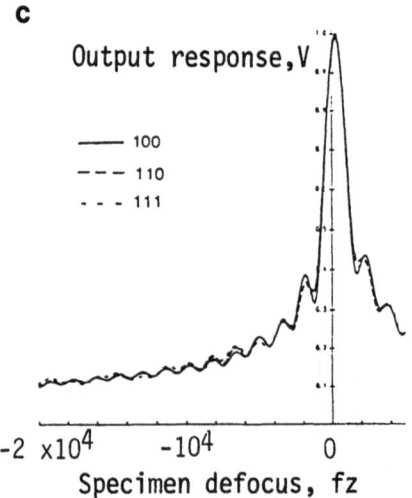

Figure 4. Theoretical V(z) curves from a spherical lens, for the
principal orientation of various materials.
a) aluminum
b) nickel
c) copper

direction parallel to the axis of the cylinder. In order to explain
the images obtained in the SAM for polycrystalline materials we must
take into account the fact that the specimen is excited with sound
from every azimuthal angle. The V(z) formulation of Sheppard and
Wilson for a spherical lens and an isotropic specimen is

$$V(z) = \int_0^\alpha R(\theta) \; P(\theta) \; \exp(2jkz\cos\theta) \; \cos\theta \; \sin\theta \; d\theta \qquad (4)$$

For an isotropic substrate the reflectance function is independent
of ϕ. The expression for V(z) in the presence of an anisotropic
substrate is modified to

$$V_{hkl}(z) = \int_0^{2\pi}\int_0^\alpha R_{hkl}(\theta, \; \phi) \; P(\theta) \; \exp(2jkz\cos\theta) \; \cos\theta \; \sin\theta \; d\theta d\phi \qquad (5)$$

Changing the order of integration we can write

$$V_{hkl}(z) = \int_0^\alpha R'_{hkl}(\theta) \; P(\theta) \; \exp(2jkz\cos\theta) \; \cos\theta \; \sin\theta \; d\theta \qquad (6)$$

where

$$R'_{hkl}(\theta) = 1/2\pi \int_0^{2\pi} R_{hkl}(\theta,\phi) \; d\phi$$

$R'_{hkl}(\theta)$ is the complex mean reflectance function presented by a
circularly symmetrical (i.e. a spherical) lens. The complex mean
reflectance function serves the same role in the generation of the
V(z) response an anisotropic specimen as does the reflectance function in isotropic materials. This function will henceforth be
abbreviated to CMR. The CMRs of three common metals, aluminium,
nickel and copper have been computed for the three principal crystallographic planes, these CMRs were then used to calculate the V(z)
response from these planes. The results of these computations are
shown in figure 4a-c.

These theoretical findings confirm experimental results, that is
that there is very little contrast between grains of aluminium and
copper but that there is considerable grain contrast in nickel (both
experimentally and theoretically iron gives very similar results to
nickel). One would expect there to be very little grain contrast
from polycrystalline aluminium because this metal is only slightly
anisotropic, with an anisotropy factor $,\eta(= 2c_{44}/c_{11} - c_{12})$ of 1.22,
compared to $\eta = 1$ for an isotropic material. Nickel has an anisotropy
factor of 2.38 so it is not suprising that the acoustic images show
good contrast between grains in flat specimens. Copper on the other
hand has the greatest anisotropy factor of all the three metals
studied, that is 3.2, and yet there is very little contrast between
grains of different orientations. This is partly due to the fact that
surface waves propagate rather slowly on copper, so that the critical
angles are rather large. An extremely wide angle lens is thus needed
to detect the re-radiated surface waves. In our computations this
effect was to some extent allowed for by using a "top-hat" pupil function allowing 100 per cent transmission for waves of all angles up to

60 degrees but nevertheless our calculations confirm the experimental difficulty in observing contrast between grains of copper. Further measurements and calculations on copper specimens using different coupling fluids such as liquid metals and methanol are needed in
order to see whether changing the acoustic coupling and wave velocity
of the sound from the coupling medium will enhance grain contrast.

CONCLUSION

   In this section we will show how the theoretical calculations
of V(z) may be used to aid characterisation of anisotropic surfaces.
Figure 5a shows theoretical V(z) curves calculated for the principal
planes of nickel. This curve differs from figure 4b because the
pupil function employed was calculated to correspond to the geometry
of our 730 MHz lens,the pupil function is not known exactly as the
size and position of the transducer are not completely characterised.
This curve predicts the grain contrast as the defocus is varied.

   Figure 5b-5d shows acoustic micrographs of unetched nickel at
0,-4 and -7 μm defocus. Looking at the theoretical V(z) curve we
note that there should be no contrast between grains at focus; at
defocuses greater than about a couple of microns considerable
is expected between grains. As the focus is increased grains of
{100} orientation will first appear dark then light relative to
grains of {110} and {111} orientations. It is not possible with
the information presented to assign crystallographic orientations to
each grain, especially as the grains are unlikely to correspond
exactly to principal orientations and the pupil function has not
been perfectly characterised. The images shown are,however, consistent with the following orientations : α {110},β {100} and γ {111}.

   The theoretical V(z) curves for anisotropic materials for both
cylindrical and spherical lenses show that there are features
present that are not observed in isotropic specimens due largely
to the possibility of excitation of pseudo-surface waves. Finally
we gave a simple illustration of how the theory could be used to
indicate the orientation of the grains. More sophisticated
techniques probably using both amplitude and phase of V(z) in conjunction with the theory developed here should eventually allow
determination of orientation with good spatial resolution for
materials such as iron and nickel which give good grain contrast.

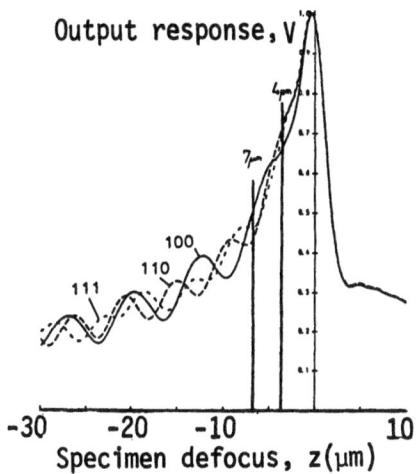

Figure 5a.

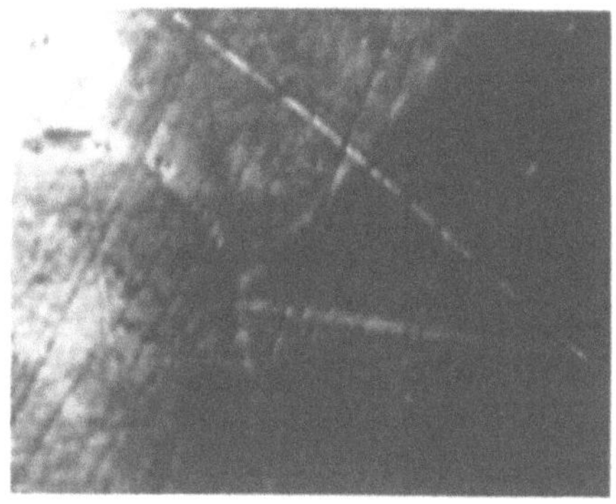

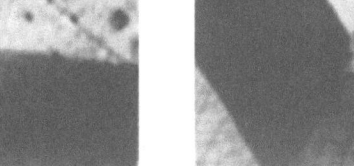

Figure 5b.

Figure 5c.    Figure 5d.

Figure 5

a) Theoretical V(z) curve on the principal planes of nickel using
    a pupil function corresponding to our 730 MHz lens.
b) Acoustic micrograph of nickel taken at focus
c) Acoustic micrograph of same specimen, z=-4μm.
d) Acoustic micrograph of the same area, z=-7μm.

REFERENCES

1) R D Weglein and R G Wilson 1978, Electronics Letters, 14, 352.

2) A Atalar 1978, J. Appl. Phys., 49, 5130.

3) R D Weglein and R G Wilson 1979, Appl. Phys. Lett., 34, 179.

4) W Parmon and H L Bertoni 1979, Electronics Letters, 15 684.

5) H K Wickramasinghe 1978, Electronics Letters, 14, 305.

6) C J R Sheppard and T Wilson, 1981, Appl. Phys. Lett 38(11) 858.

7) J Kushibiki, K Horii and N Chubachi, 1983 Electronics Letters, 19 359.

8) J Kushibiki, K Horii and N Chubachi, 1983 Electronics Letters, 19 404.

9) M G Somekh, G A D Briggs and C Ilett to be published in Phil. Mag.

10) G W Farnell 1970 "Physical Acoustics Principles and Methods", Vol Vol.VI edited by W D Mason and R N Thurston (London, New York: Academic Press) Chapter III.

11) A Atalar 1979, J. Appl. Phys., 50, 8237.

ACKNOWLEDGEMENTS

    We wish to thank AERE Harwell for providing support in both resources and manpower, in particular we wish to thank Mr Roger Martin for his work on the electronics of the microscope.
    It is a pleasure also to thank the following bodies who provided financial support for the authors the SERC (MGS), the Royal Society(GADB) and SERC and NPL for CI. MGS would particularly like to acknowledge the SERC who have provided funds for him to attend this conference.
    We wish to thank Dr Paul Bloch and Mr Nick Doe for helpful discussions over some aspects of the computations. Finally we wish to thank Prof Sir Peter Hirsch for support and guidance throughout this project.

DETECTION OF SURFACE-BREAKING CRACKS IN THE ACOUSTIC MICROSCOPE

M.G. Soumekh, G.A.D. Briggs and C. Ilett

Department of Metallurgy & Science of Materials
University of Oxford, Parks Rd, Oxford OX1 3PH

INTRODUCTION

The prime advantage of the scanning acoustic microscope lies
in its ability to image the interaction of elastic waves with the
specimen. Over the past two years a body of information has been
accumulated which shows that the scanning acoustic microscope (SAM)
is a powerful instrument for detection of cracks and elastic discontinuities. This ability is due to the dominant role played by leaky
surface waves in the contrast of the SAM. When a Rayleigh wave propagates along the surface of the specimen it may be strongly scattered
by defects much less than a wavelength thick. Fine cracks and grain
boundaries which would not be resolvable by conventional criteria can
thus be seen in acoustic micrographs.

The subject of our other paper in these proceedings (1) is an
example of how the response of the SAM is determined by the elastic
properties of the material. In that paper, however, we assumed that
the elastic properties were invariant in a direction parallel to the
surface. We were thus able to assign a reflectance function for each
plane wave incident upon the specimen, multiplication of the incident
plane wave spectrum by the reflectance function gives the returned out
output angular spectrum. A specimen whose properties vary across the
surface is more difficult to treat, because an incident plane wave
will no longer be reflected at the same angle of incidence with modified amplitude and phase but will scatter into a spectrum of plane
waves propagating in a range of directions. A tightly closed crack
would not be expected to significantly scatter incident plane waves
unless these waves couple into surface waves propagating along the
specimen which then hit the crack broadside on. This is the situation that obtains in the SAM.

The aim of this paper is to discuss the contrast in the acoustic microscope from surface-breaking elastic discontinuities.

ACOUSTIC IMAGES OF CRACKS

In this section we will discuss some of the acoustic micrographs which illustrate the ability of the SAM to detect cracks.

Figure 1a shows an acoustic micrograph of a hard metal cutting tool which has been coated with TiN a few microns thick to improve wear resistance. The film was sputter coated at high temperatures; subsequent differential contraction has caused compressive thermal stress relief cracks to form in the coating. The cracks appear in the acoustic micrograph as bright lines, the width of which appear to be approximately 8μm. The optical image (b) was taken using a Reichart Univar microscope at its highest magnification, with Nomarski interference, the cracks are barely visible and cannot easily be distinguished from surface scratches. Examination of the SEM micrograph (c) shows that the cracks are approximately 200nm wide at the surface.

Several important points arise from comparison of the three micrographs.
1) The acoustic microscope distinguishes between shallow scratches and surface breaking cracks presumably because surface waves propogate around scratches with little perturbation.
2) The width of the cracks in the acoustic images appear to be about 8μm which corresponds to approximately twice the wavelength of surface waves. It appears then that the SAM is detecting cracks much smaller than the wavelength of the interrogating radiation, because of the distinctive nature of the contras the SAM "displays" the cracks with a thickness determined by the acoustic wavelength.
3) Despite the fact that the cracks appear broad compared to their true extent further examination in the SEM confirmed that the network of cracks had been reproduced faithfully. This is a great strength of the SAM. One is able to detect cracks orders of magnitude smaller than the incident radiation but yet can scan much larger areas than is possible with other techniques (cf figures 1a and 1c).

Fatigue cracks are also revealed by this mechanism. Figue 2a-c shows a series of acoustic images of an Al20%Si plane bearing alloy which has failed in fatigue. The alloy is in the centre of the pictures and the steel substrate is on the left. We see that changing the the defocus determines whether the crack appears bright or dark relative to the background. Throughout this specimen there are regions where the fatigue cracks are visible both optically and acoustically and others where they are only detected with the SAM. We note again that the thickness of the cracks is determined by the Rayleigh wavelength

Figure 3 shows two experimental V(z) curves taken directly over

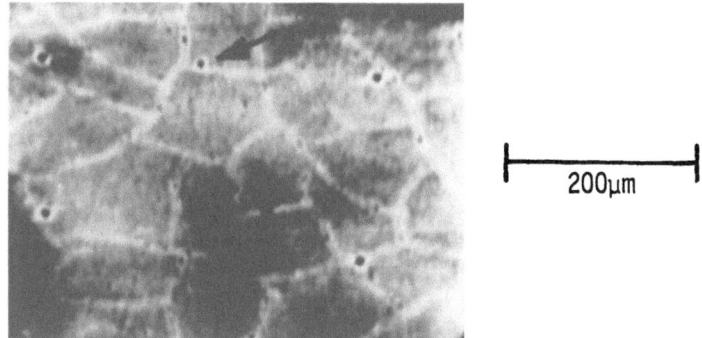

Figure 1a Acoustic image of cutting tool, the arrow marks the microhardness indent around which the optical and SEM micrographs were taken.

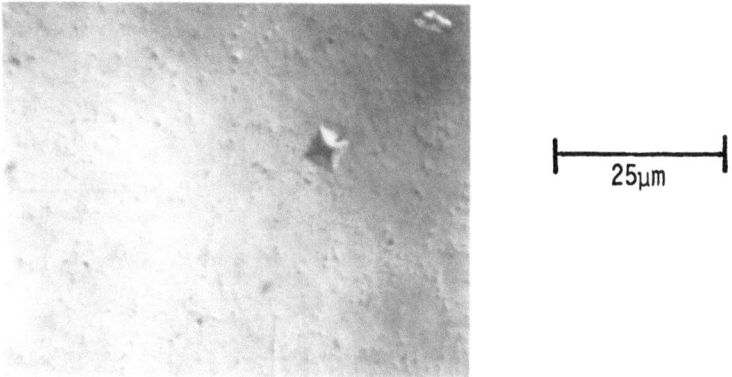

Figure 1b Optical micrograph of a portion of the cutting tool taken in Nomarski

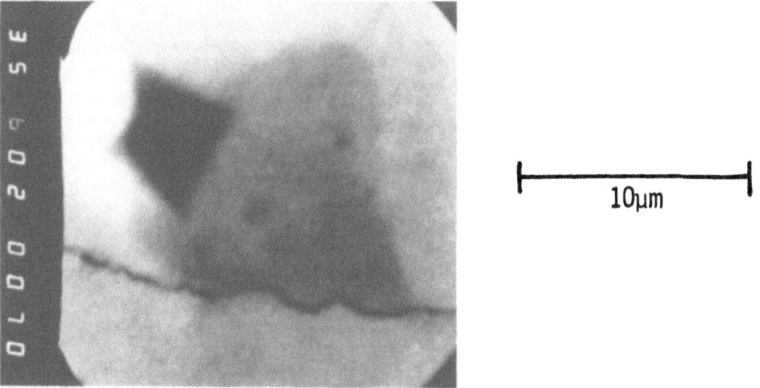

Figure 1c SEM micrograph of the cutting tool indicating the true width of the crack.

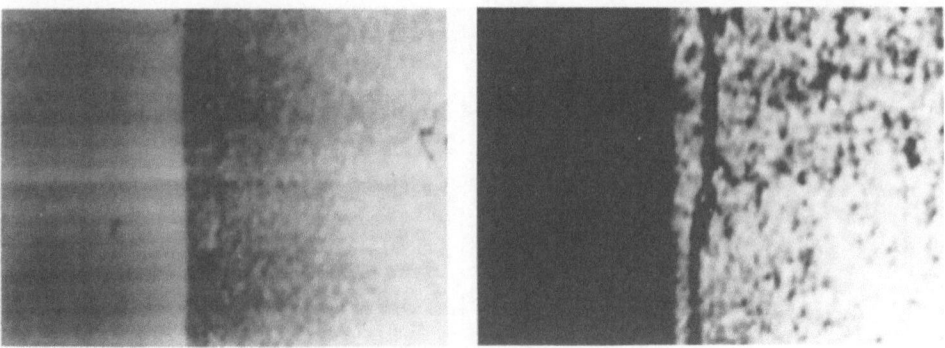

Figure 2a Acoustic micrograph of bearing alloy showing a fatigue crack, taken at focus

Figure 2b Acoustic micrograph of the same area with defocus of 5μm.

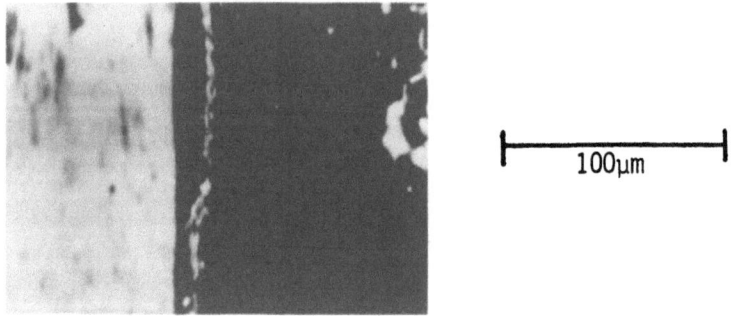

Figure 2c Acoustic micrograph of the same area with defocus of 11μm.

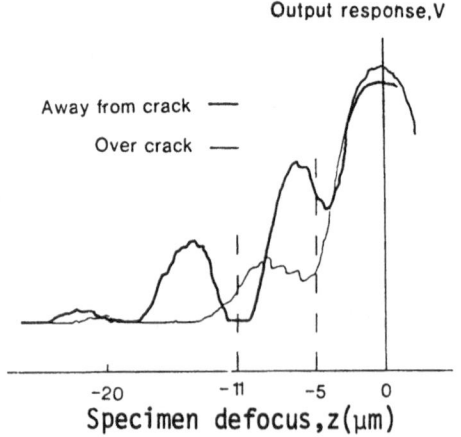

Figure 3 Experimental V(z) curve taken on the bearing alloy, the thick line was taken over sound material and the fine line over the crack.

the crack shown in figure 2 and over in a region where the material
is sound. These V(z) cuves explain the contrast available in the
pictures. At focus the crack can be seen but the contrast is patchy
with the crack sometimes appearing light and sometimes dark, when
specimen is defocused by 5µm the crack appears dark because the
output voltage from the crack is less than that from the adjoining
material. When the specimen is moved still further towards the lens
(11µm above the focus) the crack appears to be light because the output voltage over the crack is greater than that from the good material
ial.

The effect of the crack on the V(z) curve is to reduce the ripple due to interference between axially reflected sound and re-radiated surface waves. The crack scatters the Rayleigh wave propagating
along the specimen which prevents its re-radiation into the liquid.
The V(z) curve over the crack seems to assume a mean level relative
to the strong V(z) oscillations on the sound material.

Acoustic images of polycrystalline materials often reveal strong
contrast at the grain boundaries, since a grain boundary is only a
few atomic spacings thick the contrast which again appears as a line
a few microns wide is due to a surface wave hitting the boundary
broadside on.

A SIMPLIFIED MODEL FOR CONTRAST AT A CRACK

In this section we develop a simplified model which takes into
consideration the factors discussed in the previous section. To
simplify the problem and to bring out the salient features of SAM
contrast over a crack we make the following simplifying approximations.
i) We assume that the lens is cylindrical with its axis running
parallel to the crack, so that any surface waves incident on the
crack are reflected normally.
ii) In order to introduce symmetry into the case we assume that
the crack is placed centrally under the acoustic beam.
We consider the Rayliegh waves to be excited from the points a and a'
and to propagate towards the crack. They are reflected normally from
the crack with a reflection coefficient R_c. The symmetrical placing
of the crack has the effect that the reflected Rayliegh wave produces
the same effect upon the V(z) response of the lens as if it had
propagated undisturbed from a to a' but attenuated by a value equal
to the reflection coefficient R_c. The effect of a central crack can
thus be regarded as being similar to that produced by a lossy medium.
The difference between a crack and an absorptive medium is that the
amplitude of a surface wave propagating along such a lossy medium
decreases with the distance of propagation whereas the effect of the
crack is to provide a fixed loss independent of the path length aa'.
(see figure 4)

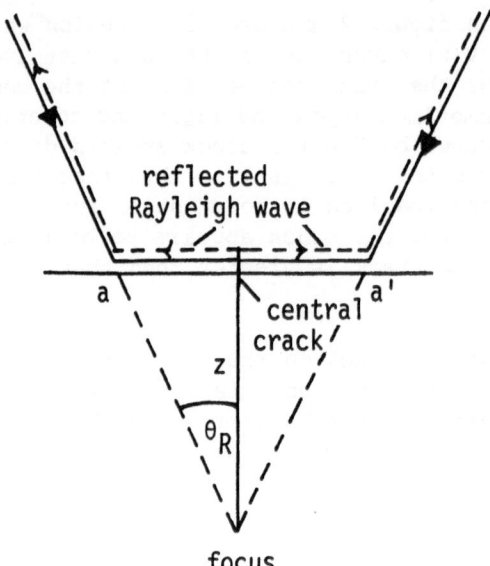

Figure 4 Diagram indicating the theoretical model for the V(z) over a crack.

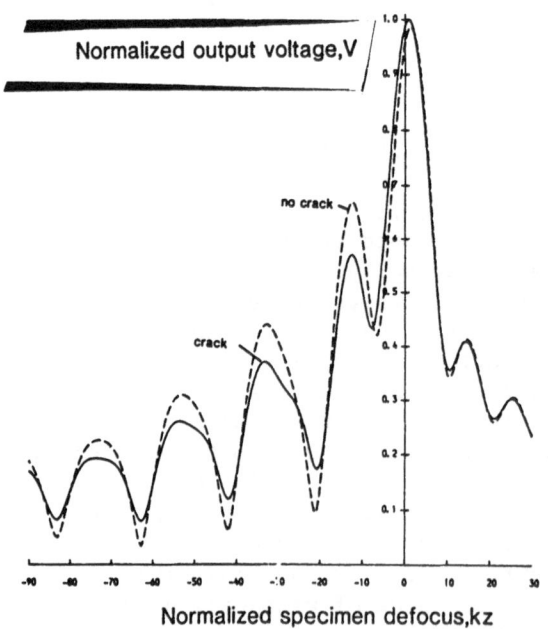

Figure 5   Theoretical V(z) curves calculated for a line focus lens showing the relative contrast for cracked and good material.

We will express this situation more formally

Path length aa' = $2zt\tan\theta_R$

Let the dissipative loss in the solid be α_d, so that

$$\exp(-\alpha_d 2zt\tan\theta_R) = R_c \qquad (1)$$

which gives $\alpha_d = \log_e R_c / 2|z| \tan\theta_R$
(note that we are interested in V(z) for z<0.)
We thus wish to insert a reflectance function corresponding to an absorptive loss which is inversely proportional to the defocus. The V(z) formulation of Sheppard and Wilson (2) is modified in the presence of a crack thus

$$V(z) = \int_0^\alpha P(\theta) \ R(\theta,z) \ \exp(2jkc\cos\theta) \ \cos\theta \ d\theta \qquad (2)$$

where k is the wave vector of the incident sound in the water
P(θ) is the two-way pupil function
α is the semi-angle of the lens
R(θ,z) is the reflectance function which is now being considered to be a function of the defocus z.

The reflection coefficient is calculated for a given defocus by substituting in the appropriate Rayleigh wave loss factor α_d. In order to calculate the reflectance function given the Rayleigh wave loss factor the corresponding complex velocities of the shear and longitudinal waves were computed using the formulation given by Viktorov (3).

Assuming the losses are not too great

$$\alpha_d = A\alpha_{long} + (1-A)\alpha_{shear}$$

that is to say the loss factor for Rayleigh waves is a linear sum of the loss factors for the shear and longitudinal waves. The factor α_d can be calculated from the elastic constants of the material but it is sufficient for our purposes to say that for most common materials with Poisson's ratio between .3 and .4 that A is less than .1. To within the accuracy of this approximate model we can say that the loss factor for the shear wave is equal to that of the Rayleigh wave. This model was employed to calculate V(z) over a crack in aluminium using a value for R_c = .25 (4), the results of this calculation are shown along with the theoretical V(z) for sound material in figure 5. We note that the crack has the effect of reducing the ripple associated with Rayleigh wave interference. Precise comparison with the experimental curves of figure 3 should be avoided because of the assumptions made. Firstly we have assumed a cylindrical lens in the theory compared to the spherical lens used for imaging and secondly when |z| is small the point of excitation of Rayleigh waves is unlikely to correspond exactly to the points a and a'.

Similar calculations were carried out using the parameters appropriate to nickel and the contrast predicted from a crack was not as great as that we would expect in a cracked specimen of aluminium, the reasons for this are discussed in the next section.

DISCUSSION

In this paper we have shown that the scanning acoustic microscope is a powerful instrument for the examination of surface-breaking elastic discontinuities. The distinctive contrast mechanism involving excitation and re-radiation of leaky Rayleigh waves is responsible for this ability. One would therefore expect that detection of cracks was more sensitive in materials in which leaky Rayleigh waves were strongly excited from the coupling fluid. Bertoni and Tamir (5) have shown that the reflectance function for an isotropic material may be separated into two components, the first part representing the so-called geometric reflection which depends upon all the acoustic wave activity except for the contribution of the leaky Rayleigh wave pole, the second term represents the effect of that pole.
Bertoni and Tamir's separation of the reflectance function is shown below.

$$R(\theta) = R_o(\theta) \left\{ 1 + \frac{k_p^2 - k_o^2}{k_x^2 - k_p^2} \right\}$$

where $R_o(\theta)$ is the geometrical reflection coefficient, k_p is the pole of the reflection coefficient, k_o is the zero and k_x is the wave vector of the incident wave. The first term in the brackets represents the geometrical reflection and the second term gives the effect of the Rayleigh wave. Each term is substituted into the formulation for $V(z)$ (2), the partial reflection coefficient will thus yield a partial $V(z)$ in which the contribution from the leaky wave is separated from the $V(z)$ due to geometrical reflection.
That is

$$V(z) = V_G(z) + V_R(z) \tag{3}$$

Curves showing these three components are shown in figure 6a-c, it should be noted that the values of $V_G(z)$ and $V_R(z)$ as shown on the graph do not add to $V(z)$ because we have plotted the modulus of the transducer voltage and the equality of equation 3 only holds when the $V(z)$ contributions are added vectorially. Figure 6a shows why aluminium may be expected to give good contrast over cracks, the $V(z)$ contribution is very large compared to the Rayleigh contribution to the $V(z)$ in nickel, shown in figure 6b. The difference can be attributed to the imaginary part of k_p which is determined primarily by the relative density of the substrate and coupling fluid. Figure 6c shows three $V(z)$ curves for nickel and an acoustic coupling fluid whose acoustic velocity is identical with that of water but whose density is three times as great, we can see that the $V(z)$ curve due to Rayleigh wave excitation is much enhanced. We would thus expect more sensitive crack detection if a denser coupling fluid were used.

In conclusion we have shown acoustic micrographs which demonst-

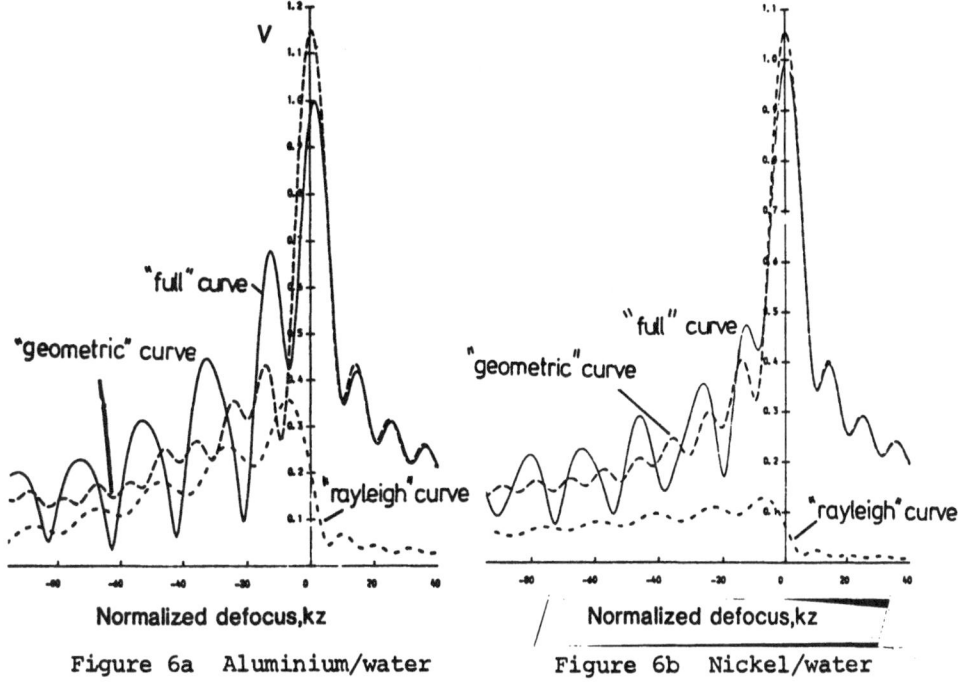

Figure 6a   Aluminium/water

Figure 6b   Nickel/water

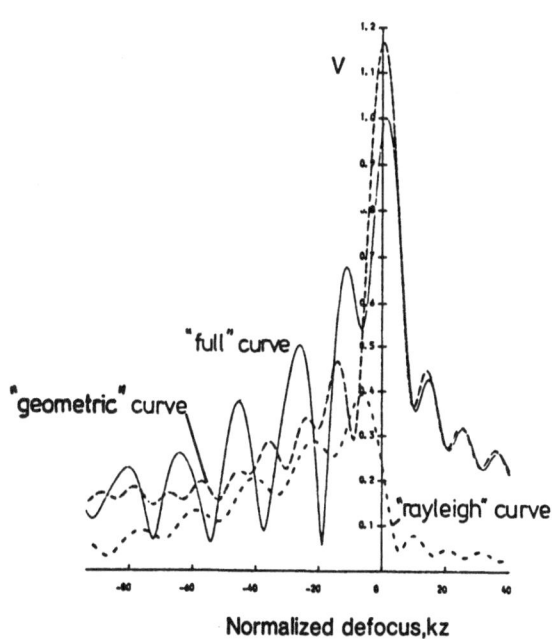

Figure 6c   Nickel/ "dense" liquid interface.

Figure 6 Calculated curve showing two "partial" V(z) curves and the "full" V(z) curve.

rate that the SAM can detect cracks orders of magnitude smaller than the wavelength of the incident sound. These micrographs show the network of cracks very clearly and yet large areas of the specimen may be imaged. The sensitive detection of cracks is due to excitation of Rayleigh waves along the surface of the specimen. A theory which although approximate suggests the important points of the crack detection mechanism. We have made some calculations to assess the magnitude of the Rayleigh wave contribution to the V(z) curve, the results of which suggest that cracks will be detected with greater sensitivity when the ratio of the density of the specimen to the density of the coupling fluid is comparitively low. It seems possible then that the sensitivity of the SAM for detecting cracks may be still further improved by a suitable choice of coupling fluid.

ACKNOWLEDGEMENTS

We wish to express our gratitude to AERE Harwell for financial support. It is also a pleasure to thank the SERC for provision of a post-doctoral research fellowship for MGS, the Royal Society for a research fellowship in the physical sciences for GADB, and the SERC and NPL for a CASE studentship for CI.
Gratitude is due Prof H **L Bertoni** for fascinating discussions, and to Dr E D Boyes for the SEM photograph. Finally we wish to thank Dr J A Champion and Prof Sir Peter Hirsch for support and guidance throughout this project.

REFERENCES

1) M G Somekh, G A D Briggs and C Ilett in these proceedings.
2) C J R Sheppard and T Wilson, 1981, Appl. Phys. Lett. 38(11) 858.
3) I A Viktorov "Rayleigh and Lamb waves" (Plenum, New York,1967).
4) Y C Angel and J D Achenbach, we acknowkedge receipt of their paper prior to publication.
5) H L Bertoni and T Tamir, 1973, Appl. Phys. 2 157.

DESCRIPTION OF THE PULSED ACOUSTIC FIELD OF A LINEAR DIAGNOSTIC ULTRASOUND ARRAY USING A SIMPLE APPROXIMATE EXPRESSION

M.Pappalardo, N.Denisenko, G.Scarano and M.Matteucci

Istituto di Acustica "O.M.Corbino" CNR
Via Cassia, 1216 Roma Italy

INTRODUCTION

The most successful technique used to investigate the properties of the matter by means of elastic wave interaction is the well known pulsed echo system. The fundamental reason is that this method has the best intrinsic axial resolution as compared to the quasi-optical, holographic and tomographic techniques. As far as ultrasound medical diagnostic is concerned, in recent years the transducer design and technology have made considerable progress and presently single or array transducers with very large bandwidths are available and employed in sophisticated imaging systems.

In order to predict the transient acoustic field generated by transducers able to emit very short pulses, the harmonic solution is totally inadequate. For the case of a baffled planar piston the transient acoustic field theory has recently been reviewed by Harris[1]. The most useful solution, which has a simple physical significance and can be applied to any geometrical vibrating surface, is the convolution/impulse response approach originally proposed by Stephanishen[2]. This approach relates the acoustic transient field to the derivative of the radiating source velocity waveform v(t) by means of convolution with a function h(x,t). This function takes into account the geometrical shape of the vibrating surface, the observation point space position and the propagation medium characteristics. The so called spatial impulse response h(x,t) was determined in closed form by Stephanishen[2] for a circular piston, by Lockwood and Willette[3] for a rectangular surface, and by Lasota and Salamon[4] for a linear source.

129

The convolutional approach gives an exact solution and can be applied to any transducer of practical interest in the field of medical diagnostics, including ultrasonic arrays, if the superposition principle is applied. Nevertheless, an easy application of this convolutional algorithm for practical transducer design is prevented by the time and storage requirements needed for its implementation on a computer.

For practical purposes a sufficiently accurate prediction of the transient field would be adequate in order to design the transducers geometry. Some characteristics of the acoustic field , such as beam-width, depth of focus and diffraction lobes amplitude are of major interest in order to optimize the performances of the whole imaging system. In a previous work[5], following a completely different approach in the time domain, we presented a simple approximate closed-form expression able to describe the space-time acoustic pressure distribution generated by a linear source excited by a short sinusoidal burst.

In the present work we extend this approach to a more realistic excitation, e.g. amplitude modulated sinusoidal burst, and, by applying the superposition principle, we give some examples for the case of arrays of linear sources. The proposed expression is only valid within the limits of paraxial approximation but the computational complexity and the time and storage requirements are considerably reduced compared to the exact convolutional solution. Therefore, this method can be advantageously used for array design and optimization.

APPROXIMATE THEORY

With reference to Fig.1 consider a linear transducer of aperture 2a which is driven with a short pulse of duration τ . We suppose that the transducer is imbedded in a half-space with acoustically soft boundary. These are the conditions which exist in medical diagnostic ultrasonics as shown by Archer-Hall and Gee[6], and Delannoy et al.[7].The transient acoustic field $p(x_o,z_o,t)$ can be described by the following integral obtained by extending the classical scalar diffraction theory to the case of wide band excitation:[4,8]

$$p(x_o,z_o,t) = \frac{1}{2\pi c} \int_{-\infty}^{\infty} \frac{\partial}{\partial t} p(x,0,t-\tfrac{r}{c}) \frac{\cos\varphi}{r} dx \qquad (1)$$

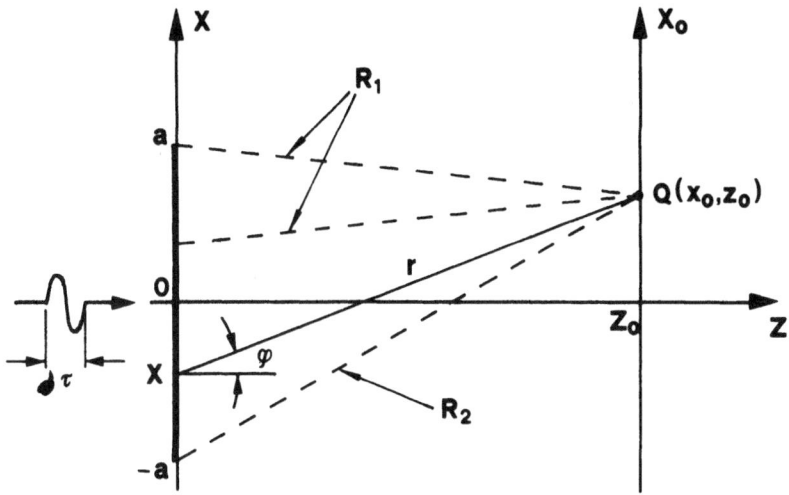

Fig.1. Linear source geometry.

where c is the velocity of the sound in the medium,and:

$$\cos\varphi = z_o/r \quad ; \quad r = \sqrt{z_o^2 + (x-x_o)^2}$$

Relation (1) states that the acoustic pressure at an arbitrary point $Q(x_o, z_o)$ at time t can be found by summing all the elementary contributions emitted by the transducer which arrive at the point Q at the same time t.

Our approach for calculating this expression is based on limiting the integration only to that part of the transducer surface which, at the time instant t, actually contributes to the pressure at the point Q, i.e. to vary the integration limits as a function of time t and the coordinates of point Q.

We assume the transducer is uniformly excited by a sinusoidal pulse of frequency $f = \omega/2\pi$ and duration τ, and the pressure is zero outside the transducer, i.e.:

$$p(x,0,t) = A(x)\,\Pi(t)\,\sin\omega t \tag{2}$$

where $\Pi(t)$ is the unitary square pulse of duration τ and $A(x)$ is the aperture function of the source with:

$$\Pi(t) = \begin{cases} 1, & 0 \leqslant t \leqslant \tau \\ 0, & \text{elsewhere} \end{cases} \quad \text{and} \quad A(x) = \begin{cases} 1, & |x| \leqslant a \\ 0, & \text{elsewhere} \end{cases}$$

Suppose that the duration τ of the acoustic pulse on the transducer face is an integer multiple of the half-period of the excitation burst $\tau = nT/2$. In this case substituting equation (2) in (1) we obtain the following expression:

$$p(x_0, z_0, t) = \frac{\omega}{2\pi c} \int_{-\infty}^{\infty} A(x) \, \Pi(t - \frac{r}{c}) \cos\omega(t - \frac{r}{c}) \frac{\cos\varphi}{r} \, dx \qquad (3)$$

or equivalently where, for convenience, the exponential form is employed:

$$\overline{p}(x_0, z_0, t) = \frac{\omega}{2\pi c} \int_{-\infty}^{\infty} A(x) \, \Pi(t - \frac{r}{c}) e^{-j\omega(t - \frac{r}{c})} \frac{\cos\varphi}{r} \, dx \qquad (4)$$

For a broad band transducer, which is able to generate one or two cycles of a sinusoidal wave, n will be 2 or 4, while if we want to approach the quasi-harmonic case n will take a large value.

It is easy to see that, at the generic point Q (x_0, z_0) and at the time instant t, arrive only the elementary contributions which fulfill the following conditions:

$$c(t - \tau) \leqslant r \leqslant ct \qquad (5)$$

i.e. only those which arrive from points of the transducer ranging within the limits:

$$(x'_- ; x''_-) \quad \text{and} \quad (x'_+ ; x''_+)$$

where (see Fig.2):

$$x'_\pm = x_0 \pm \sqrt{c^2 t^2 - z_0^2}$$
$$x''_\pm = x_0 \pm \sqrt{c^2(t-\tau)^2 - z_0^2} \qquad (6)$$

For the time t given by relation (5) we have $\Pi(t-r/c)=1$ and relation (4) can be divided in the following two integrals:

$$\overline{p}(x_0, z_0, t) = \frac{\omega e^{-j\omega t}}{2\pi c} \left[\int_{x'_-}^{x'_+} A(x) e^{j\omega \frac{r}{c}} \frac{\cos\varphi}{r} \, dx - \int_{x''_-}^{x''_+} A(x) e^{j\omega \frac{r}{c}} \frac{\cos\varphi}{r} \, dx \right] \qquad (7)$$

where the time dependence is now only in the integration limits.

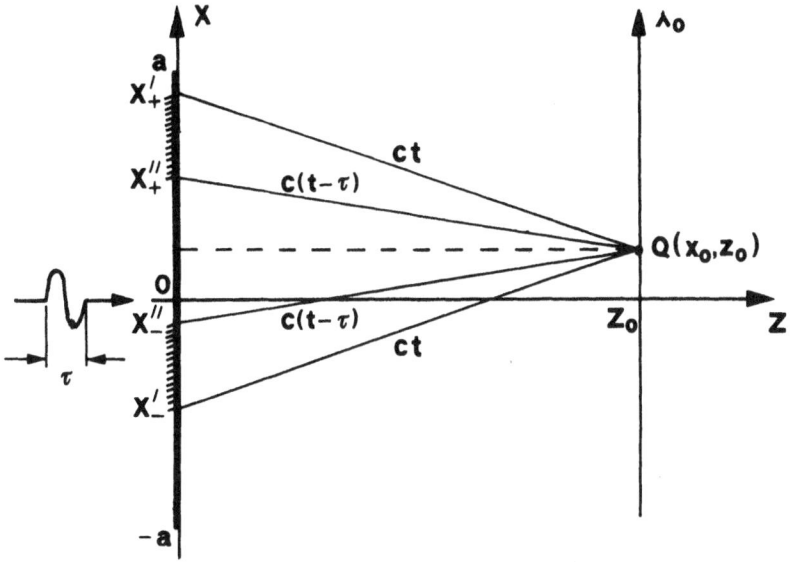

Fig.2. Graphical explanation of the time dependent integration limits.

Applying the classical paraxial approximation:

$$\frac{\cos\varphi}{r} \approx \frac{1}{z_o} \quad ; \quad r \approx z_o + \frac{(x-x_o)^2}{2z_o}$$

and making the following substitutions:

$$\xi = M(x-x_o) \quad ; \quad M = \sqrt{\omega/\pi z_o c}$$

$$\xi'_\pm = \pm M\sqrt{c^2 t^2 - z_o^2} \quad ; \quad \xi''_\pm = \pm M\sqrt{c^2(t-\tau)^2 - z_o^2}$$

we have:

$$\overline{p}(x_o, z_o, t) = \frac{M}{2} e^{-j\omega(t-\frac{z_o}{c})} \left[\int_{\xi'_-}^{\xi'_+} A(\xi) e^{j\frac{\pi}{2}\xi^2} d\xi - \int_{\xi''_-}^{\xi''_+} A(\xi) e^{j\frac{\pi}{2}\xi^2} d\xi \right] \quad (8)$$

where:

$$A(\xi) = \begin{cases} 1, & -M(a+x_o) \leq \xi \leq M(a-x_o) \\ 0, & \text{elsewhere} \end{cases}$$

Because $A(\xi)$ is a 0-1 step function it is easy to verify that relation (8) can be simplified as follows:

$$\overline{P}(x_0, z_0, t) = \frac{M}{2} e^{-j\omega(t-\frac{z_0}{c})} \left[F(\eta_1) - F(\eta_2) - F(\eta_1') + F(\eta_2') \right] \quad (9)$$

where:

$$F(\eta) = C(\eta) + j S(\eta)$$

and

$$C(\eta) = \int_0^\eta \cos\frac{\pi}{2}\xi^2 d\xi \quad ; \quad S(\eta) = \int_0^\eta \sin\frac{\pi}{2}\xi^2 d\xi$$

are the classical Fresnel integrals[9].

The values of η can be obtained comparing the integration limits $\xi_{\pm}'$ and $\xi_{\pm}''$ with the limits of the function $A(\xi)$. Table I shows the values of η_1 and η_2 as functions of time t and the coordinates of the point Q. In order to find the values of η_1' and η_2' it is sufficient to substitute t-τ for t. Because the spatial pressure distribution is symmetrical, only values of η relative to points with coordinates $x_0 \geqslant 0$ are considered in Table I. The time t is considered to range over four different intervals, while the space is divided in two regions ($x_0 < a$; $x_0 \geqslant a$). A physical interpretation of this classification is given by Fig.1.

Finally, the expression of the transient field takes the following form:

$$p(x_0, z_0, t) = \frac{M}{2} \left[C(\eta)\cos\omega(t-\frac{z_0}{c}) + S(\eta)\sin\omega(t-\frac{z_0}{c}) \right] \quad (10)$$

where:

$$C(\eta) = C(\eta_1) - C(\eta_2) - C(\eta_1') + C(\eta_2')$$
$$S(\eta) = S(\eta_1) - S(\eta_2) - S(\eta_1') + S(\eta_2')$$

and the pressure field envelope is:

$$\left| p(x_0, z_0, t) \right| = \frac{M}{2} \sqrt{C(\eta)^2 + S(\eta)^2} \quad (11)$$

We have, up to now, assumed a sinusoidal burst excitation[10,11] in order to be able to integrate relation (1). A more realistic excitation can

Table 1. Values of η_1 and η_2 as functions of time t and of the coordinates of the point Q.

	$x_0 < a$		$x_0 \geq a$	
	η_1	η_2	η_1	η_2
$t < z_0/c$	0	0	0	0
$z_0/c \leq t < R_1/c$	$M\sqrt{c^2t^2 - z_0^2}$	$-M\sqrt{c^2t^2 - z_0^2}$	0	0
$R_1/c \leq t \leq R_2/c$	$M(a - x_0)$	$-M\sqrt{c^2t^2 - z_0^2}$	$M(a - x_0)$	$-M\sqrt{c^2t^2 - z_0^2}$
$t > R_2/c$	$M(a - x_0)$	$-M(a + x_0)$	$M(a - x_0)$	$-M(a + x_0)$
$R_1 = \sqrt{z_0^2 + (a - x_0)^2}$			$R_2 = \sqrt{z_0^2 + (a + x_0)^2}$	

be considered if we approximate any modulated pulse with a temporal sequence of properly weighted half cycles of time duration τ as shown in Fig.3. In this case the acoustic pressure on the transducer face can be described by:

$$p(x,0,t) = A(x) \sum_{i=1}^{N} \beta_i \, \Pi_i(t) \sin \omega t \qquad (12)$$

where β_i is a weighting factor and $\Pi_i(t)$ is defined as follows:

$$\Pi_i(t) = \begin{cases} 1, & (i-1)\tau \leq t \leq i\tau \\ 0, & \text{elsewhere} \end{cases}$$

Substituting relation (12) in (1) we have:

$$p(x_0,z_0,t) = \frac{\omega}{2\pi c} \sum_{i=1}^{N} \int_{-\infty}^{\infty} A(x) \, \Pi_i(t - \tfrac{r}{c}) \cos \omega(t - \tfrac{r}{c}) \frac{\cos \varphi}{r} \, dx \quad (13)$$

and the pressure can be easily computed as previously described.

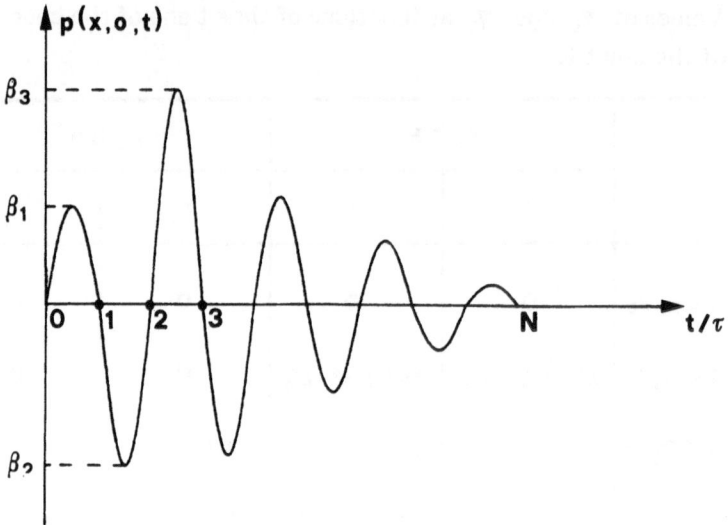

Fig.3 Approximate excitation pulse.

Finally the transient field of the array of linear sources shown in Fig.4 can be computed using the time-space invariance of the system:

$$P_{ar}(x_0, z_0, t) = \sum_{i=1}^{L} P_i(x_0, z_0, t) = \sum_{i=1}^{L} p(|x_0 - x_i|, t - \Delta t) \qquad (14)$$

where $p_{ar}(\,.\,)$ is the pressure generated by the array in the point (x_0, z_0) at the time t, L is the number of the elements and $p_i(\,.\,)$ is the contribute of the generic element i.

This calculation can be easily performed storing only the non-zero part of the field $p(\,.\,)$ of a single element centered on the array axis and looking-up the values, properly shifted in position and delayed in time, corresponding to each element of the array. Appropriate time delays can easily simulate electronic focalization.

COMPUTATIONAL RESULTS

All graphical results shown were obtained with a small minicomputer of only 32 K memory (Digital PDP11/04).

Fig.5 shows the evolution in the time of the envelope of the

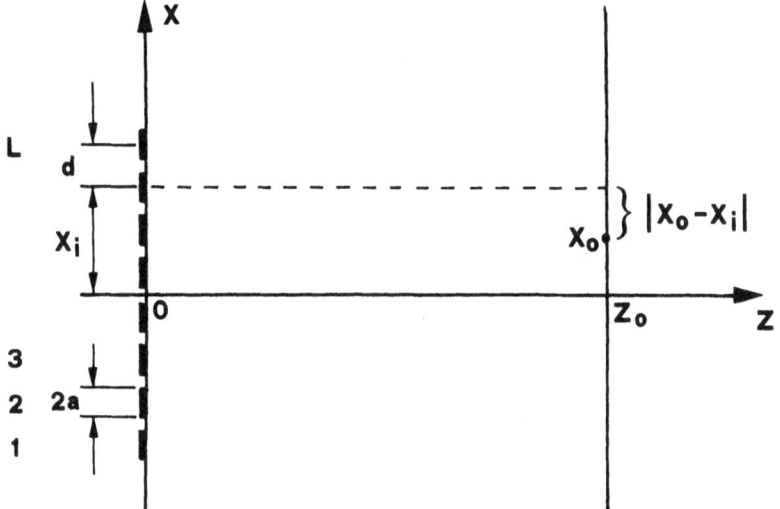

Fig.4. Geometry of a linear array of L elements.

acoustic pulse generated by a linear source at 2mm at different depths; the excitation is a sinusoidal burst of only one cycle.

In order to check our approximate expression we simulated a quasi-harmonic excitation applied to the same transducer of Fig. 5 (about 100 cycles). Fig.6 shows the expected results e.g. the classical "sinc" function shape. It must be noted that in the far field region the behaviour of the field is not exactly equal to that predicted by the harmonic theory. In fact the minimum of the pressure between two successive lobes is not exactly zero, nevertheless the agreement is quite good.

Fig.7 shows the field at the same depths of the previous cases for an array of 32 linear sources whose geometry is specified in the figure for a burst excitation of two cycles.

Fig.8 shows the field of the same array of Fig.7 focused at a distance of 20 mm. Depth of focus, beam width, depth resolution can be easily evaluated and used for transducer design. In order to show the fine structure and the details of the transient field we reported in Fig. 9 the same results as in Fig. 8 but from an opposite point of observation.

Finally we compare the transient field in the focal region for the

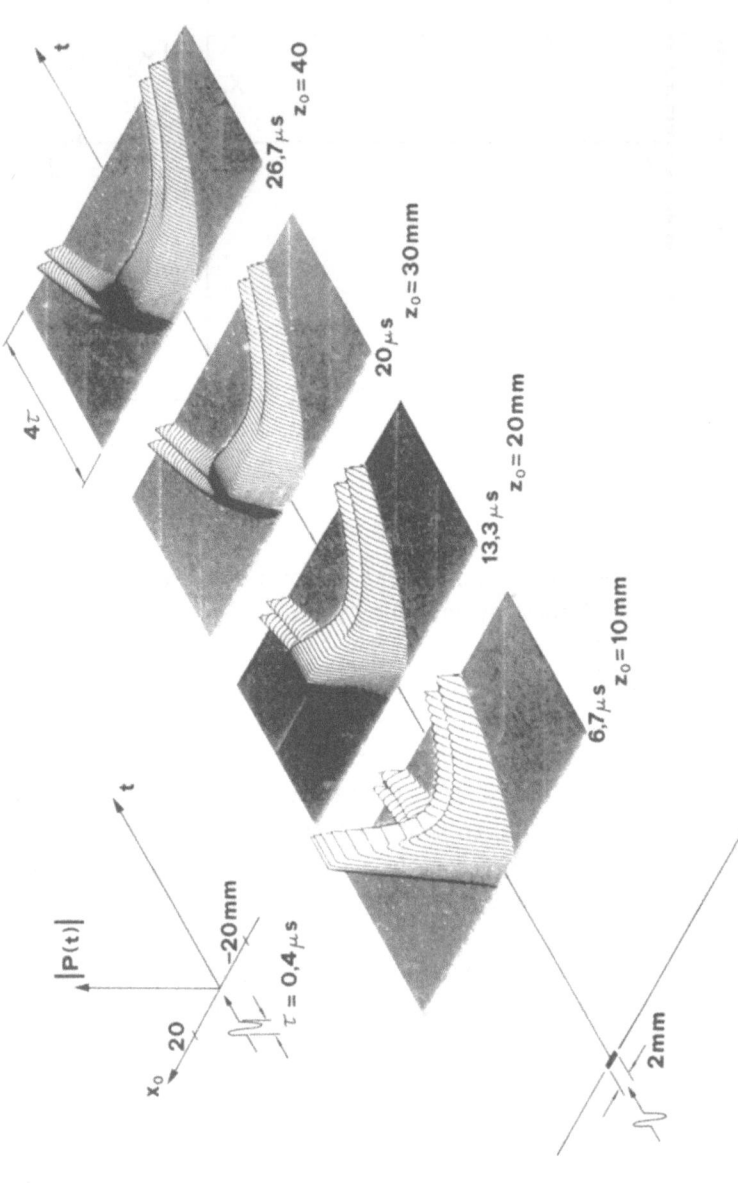

Fig.5.   Normalized time-space distributions of the envelope of the acosutic pulse generated by a linear source of aperture 2mm at four different distance z_o. The transducer is excited with a one-cycle 2.5 MHz sinusoidal pulse. The displayed time duration of every plot is 1.6 μs.

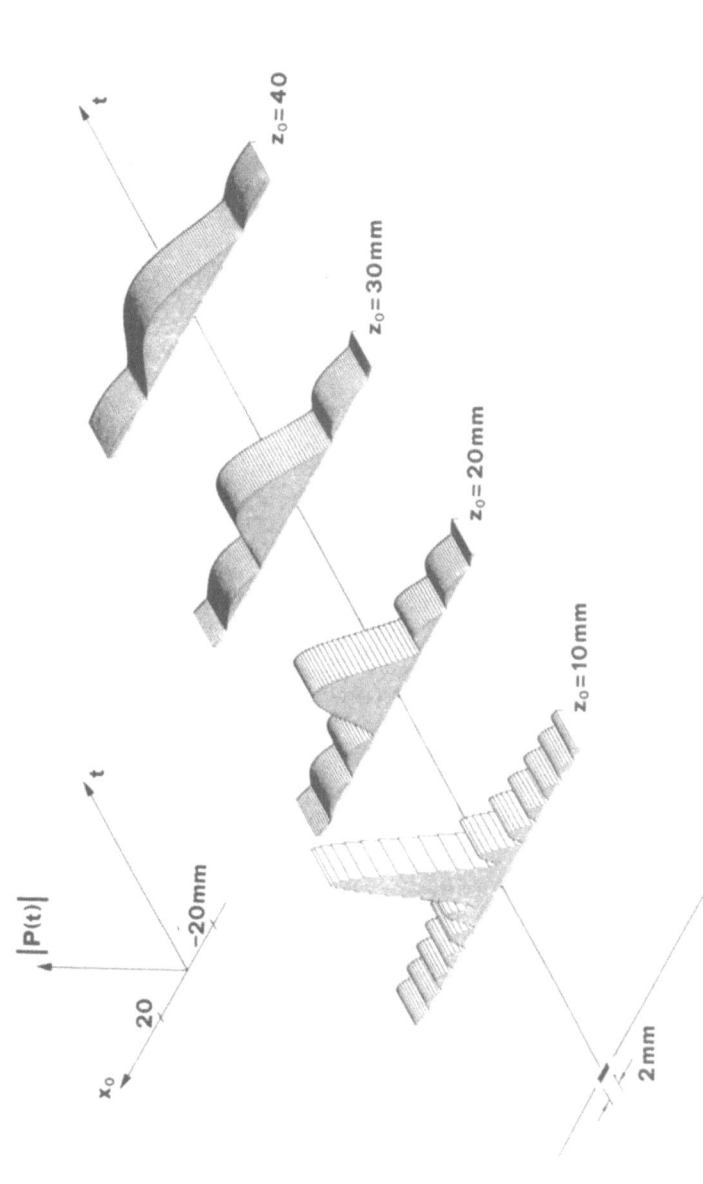

Fig.6. Normalized time-space distributions of the envelope of the acoustic field generated by a linear source of aperture 2mm at four different distances. Quasi-harmonic excitation (100 cycles; frequency 2.5 MHz). Timing: after 26.7 μs, 33.3 μs, 40 μs,46.7 μs.

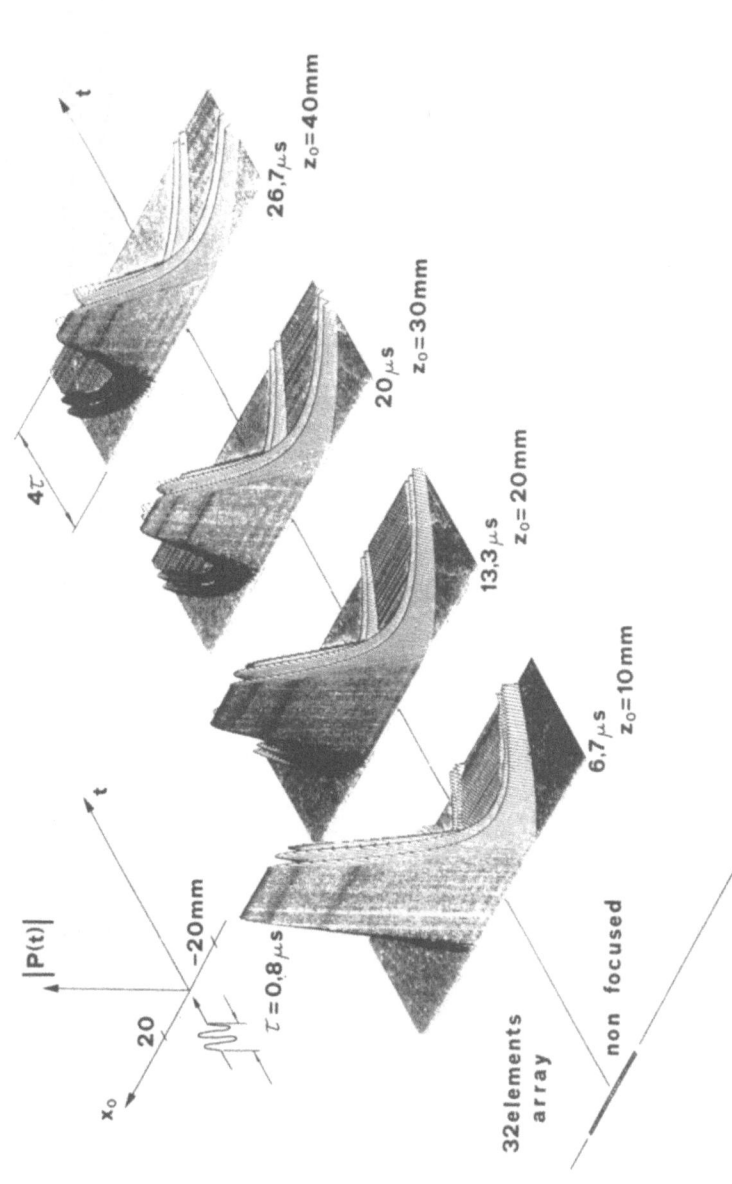

Fig.7. Normalized time-space distributions of the envelope of the acoustic pulse, generated by an unfocused array of 32 elements, with 2a=0.3mm, d=0.35mm (see Fig.4). The array is excited with a two-cycle 2.5 MHz sinusoidal pulse.The displayed time duration of every plot is 3.2 μs.

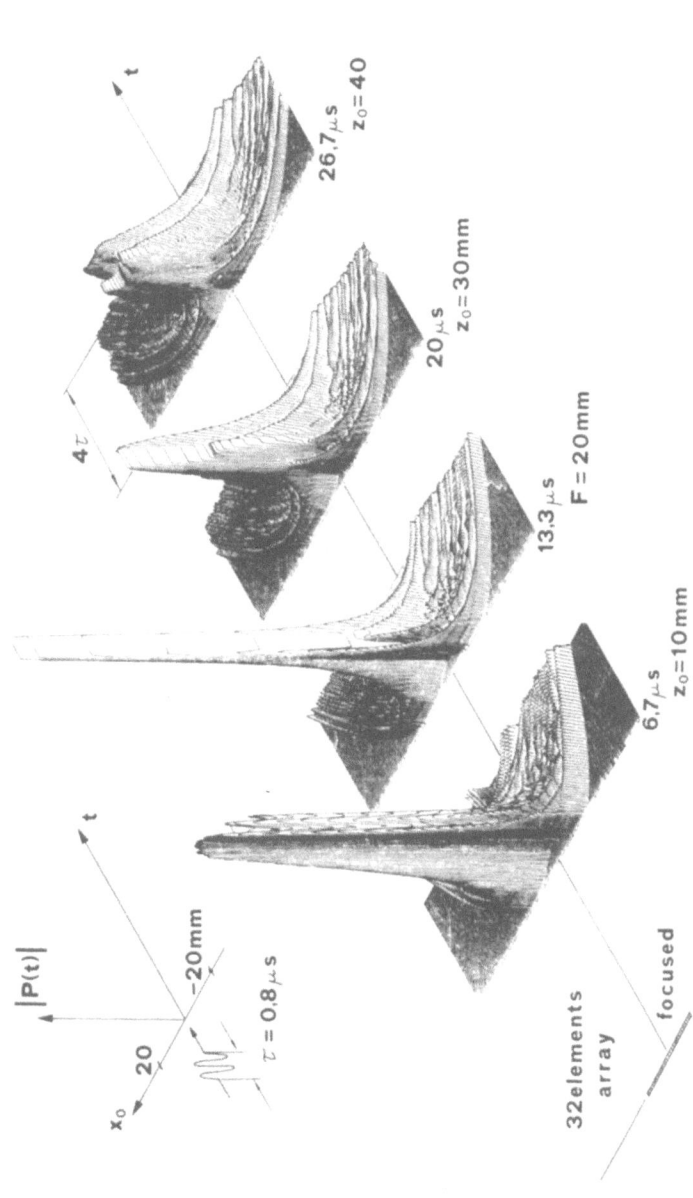

Fig.8. Normalized time-space distribution of the envelope of the acoustic pulse generated by a same array as in Fig.7, focused at 20mm and excited with a two-cycle 2.5 MHz sinusoidal pulse. The displayed time duration of every plot is 3.2 μs.

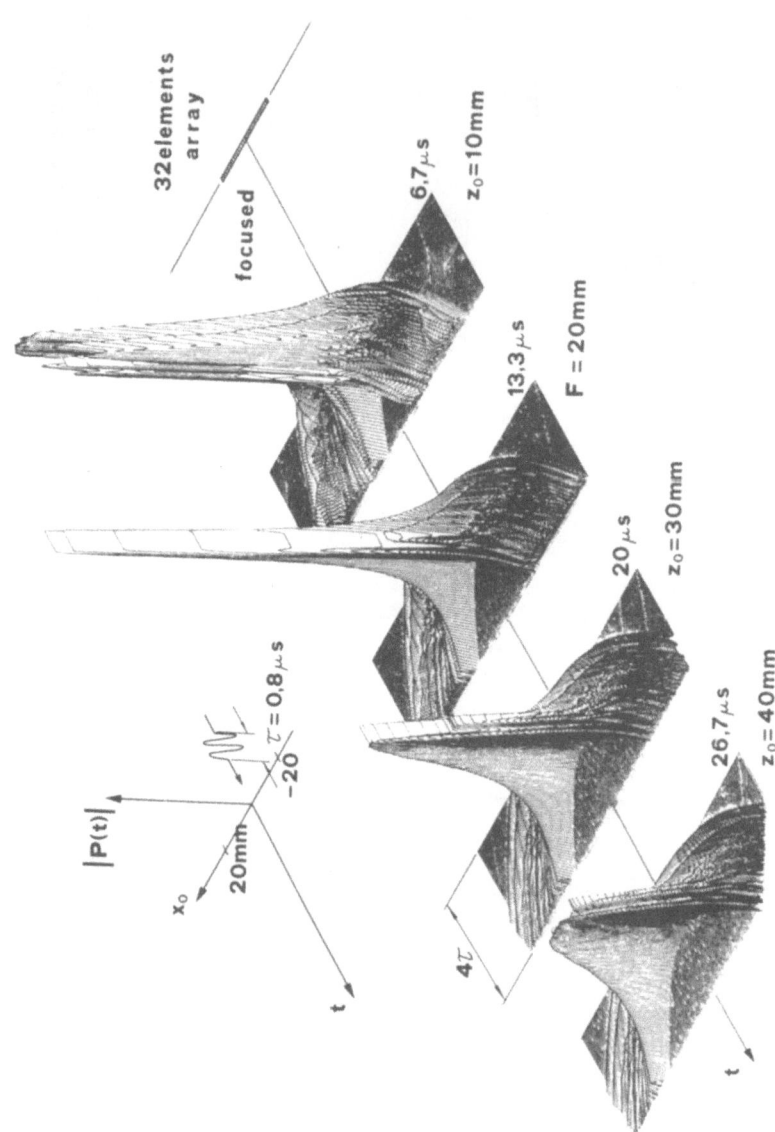

Fig.9.   Same transient field as in Fig.8 but from the oppposite point of observation.

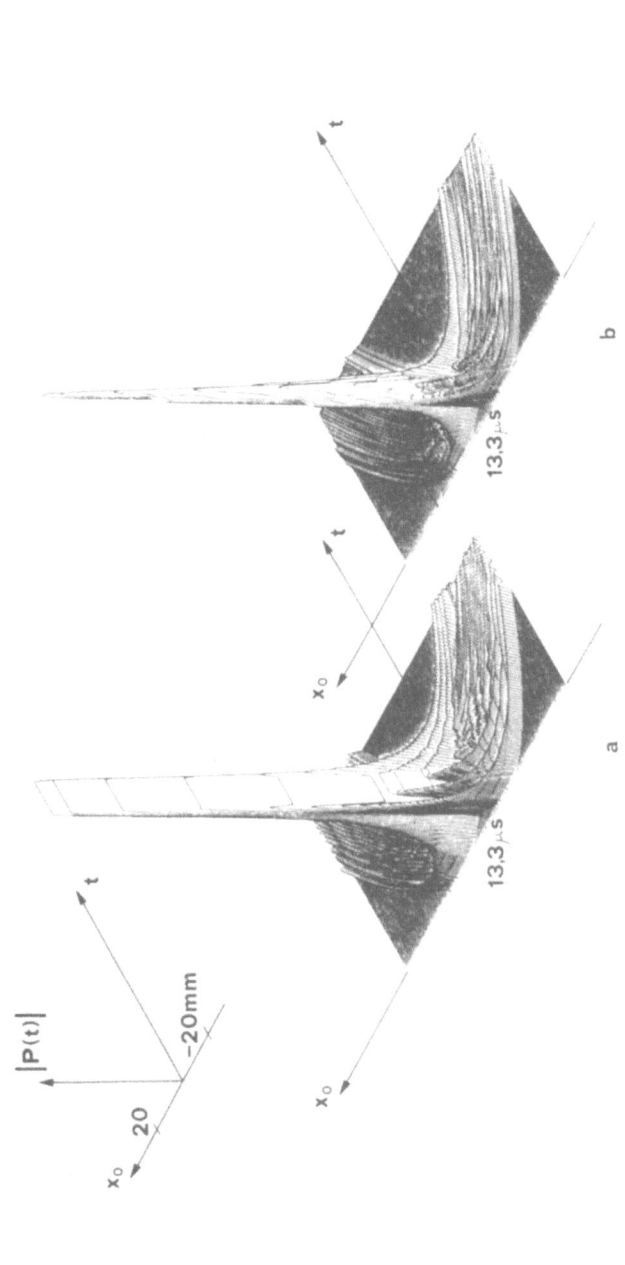

Fig.10. Normalized time-space distributions of the envelope of the acoustic pulse at distance 20mm, generated by a same array as in Fig.7, focused at 20mm and excited: a) with a three-cycle 2.5 MHz sinusoidal pulse; b) with a approximate modulated pulse ($\beta_1 = 0.4$, $\beta_2 = 0.75$, $\beta_3 = 1.$, $\beta_4 = 0.5$, $\beta_5 = 0.3$, $\beta_6 = 0.1$). The displayed time duration of every plot is 4.8 μs.

same array as in Fig.8 with a burst excitation of 3 cycles (Fig.10a) and for an approximate modulated pulse (Fig.10b).

CONCLUSIONS

The results obtained with our approximate expression are similar to those presented by other authors[11,12] obtained with a pure computer simulation.

Important parameters of the transient field like beam width, axial resolution and depth of focus can be easily extracted and used in transducer design.

The present approach does not need filtering, FFT processing or numerical computation of the convolution integral, drastically reducing computer storage and time requirements. These considerations make the proposed method attractive for practical optimization problems,as compared to the exact convolutional solution.

Work sponsored by S. P. on Biomedical and Clinical Engineering of CNR.

REFERENCES

1.   G. R. Harris, Review of transient field theory for a baffled planar piston, J.Acoust.Soc.Am. 70:10 (1981).

2.   P. R. Stephanishen, Transient radiation from pistons in an infinite planar baffle, J.Acoust.Soc.Am. 49:1629 (1971).

3.   J. C. Lockwood and J.G. Willette, High-speed method for computing the exact solution for the pressure variations in the nearfield of a baffled piston, J.Acoust.Soc.Am. 53:735 (1973).

4.   H. Lasota and R. Salamon, Application of time-space impulse responses to calculations of acoustic fields in imaging systems, Acoustical Imaging 10:493 (1980).

5.   N. Denisenko, M. Pappalardo, E. D'Ottavi and M. Matteucci, An approximated closed form of the transient acoustic pressure distribution generated by a linear source,(submitted for publication).

6.   J.A. Archer-Hall and D. Gee, A single integral computer method for axisymmetric transducers with various boundary conditions, NDT Int. 13:95 (1980).

7. B. Delannoy, H. Lasota, C. Bruneel, R.Torguet and E. Bridoux, The infinite planar baffles problem in acoustic radiation and its experimental verification, J.Appl.Phys. 50:5189 (1979).

8. J. W. Goodman, "Introduction to Fourier Optics", McGraw-Hill, New York (1968).

9. C. Toraldo di Francia, "La diffrazione della luce", Edizioni Scientifiche Einaudi, Torino (1958).

10. A. Weyns, Radiation field calculations of pulsed ultrasonic transducers, Ultrasonics 18:183 (1980).

11. A.J. Duerinckx, Matched Gaussian apodization of pulsed acoustic phased arrays, Acoustical Imaging 10:455 (1980).

12. J. Souquet, P.Stonestrom, M.Nassi, Ultrasound simulator, Acoustical Imaging 10:23 (1980).

CROSS-SECTIONAL MEASUREMENTS AND EXTRAPOLATIONS OF ULTRASONIC

FIELDS

R. C. Waag, J. A. Campbell, J. Ridder* and P. R. Mesdag*

Department of Electrical Engineering
University of Rochester, Rochester, NY 14627

*Department of Applied Physics
Delft University of Technology, Delft, The Netherlands

ABSTRACT

Ultrasonic fields produced by a disk transducer have been measured in two planes normal to the transducer axis and then extrapolated from one measurement plane to the other and also back to the face of the transducer. The measurements were performed with a small PVF hydrophone and a quadrature detector which yielded kthe complex values of the pressure at regularly spaced points in the scanned planes. Wave field extrapolation was carried out digitally using Fourier transformation methods which characterize the propagation of a field from one plane to another as a linear filtering operation. The filtering process was implemented as a product of the angular spectrum of the measured pressure field and a propagation matrix comprised of phase factors computer from parameters which included the wavelength of the measured field, the distance over which the field was to be extrapolated, and a maximum spatial frequency. The extrapolated field was then obtained by inverse Fourier transformation of the angular spectrum-phase matrix product. Comparisons of extrapolation results with measured data and with theoretical predictions show reasonable agreement and also demonstrate the sensitivity of the extrapolation process to the extent and shape of the spatial frequency window employed to determine the non-zero elements in the phase matrix. The results indicate fields can be extrapolated more accurately by sampling higher than the Nyquist rate and extending the size of the plane over which the measurements are made.

Supported in part by NSF Grant ECS8017683 and NATO Grant 09381.

SPECTRUM OF ECHO SCATTERED BY SIMPLE OBJECT PLACED

AT FOCAL PLANE OF ULTRASONIC TRANSDUCER

Mitsuhiro Ueda

Tokyo Institute of Technology
Research Laboratory of Precision Machinery and
Electronics
Nagatsuta, Midori-ku, Yokohama 227, Japan

A better understanding of scattering process is essential to interpret images obtained by ultrasonic pulse echo system. Scattering process in plane or spherical waves sound field has been analyzed so far, but it has hardly known about the scattering in focused beam sound field. In this paper spectrum of echo scattered by a simple object in focused sound field is analyzed by using the mathematical expressions of echo presented at the previous symposium. It becomes clear that frequency dependence of spectrum changes due to the geometrical figures of scatterers as well as the degree of focusing of the sound beam and some of the respects are verified experimentally.

INTRODUCTION

Recently an ultrasonic pulse echo system has been widely used in medical diagnosis and nondestructive testing. Roughly speaking images of scattering strength are observed in the system. The scattering of waves in plane or spherical sound field has been analyzed so far[1~3] but it has hardly known about the scattering in focused sound field. Since almost all pulse echo systems use focused sound beam, the analysis of echo signal scattered by an object in focused sound field is essential if we want to get more information about acoustic characteristics of the scatterers other than the locations of their discontinuities from the echo signal.

New mathematical expressions of the echo signal have been reported at the previous symposium which work for arbitrary objects in arbitrary sound field.[4] By using these expressions the spectrum of echo scattered by a simple object, that is, a point, line, film, and plane object, is analyzed and the ratios of spectrums of the

point and line objects with that of the plane object is calculated
and compared with experimentally obtained spectrums. It becomes
clear that the frequency dependence of spectrum changes by the
geometrical figure of scatterers as well as the degree of focusing
of the sound beam and some of these respects are verified experimentally.

PRINCIPLES

Expressions of echo signal[4]

    According to the previous report, the frequency component $E(\omega)$
of echo signal $e(t)$ can be expressed as follows

$$E(\omega) = j2\omega\rho_0 G(\omega)\int_{V'} R(\mathbf{r})\ (jk\Phi(\omega,\mathbf{r}))^2 dv \tag{1}$$

where ρ_0 shows the mean density of the medium, k shows the wave
number of ultrasonic waves ($k = \omega/c_0$, c_0 shows the sound velocity
of the medium.), $G(\omega)$ shows electrical characteristics of an ultrasonic transducer and it is assumed that the same transducer is used
as a transmitter and a receiver. $R(\mathbf{r})$ shows the distribution of
amplitude reflection coefficient of the scatterer and is given by

$$R(\mathbf{r}) = (\ z(\mathbf{r}) - z_0\ )/(\ z(\mathbf{r}) + z_0\ ) \tag{2}$$

where $z(\mathbf{r})$ shows the distribution of specific acoustic impedance of
the scatterer, z_0 shows the mean specific acoustic impedance of the
medium, and $\mathbf{r}$ shows the position vector. In Eq.(1) V' shows volume
of the scatterer after the effect of sound velocity difference is
compensated. If the inside of the scatter is uniform, Eq.(1) can be
expressed by a surface integral on the scattering volume as follows

$$E(\omega) = j\omega\rho_0 G(\omega)R_0\int_{S'}\ \Phi(\omega,\mathbf{r})\ \mathbb{N}\cdot\nabla\Phi(\omega,\mathbf{r})ds \tag{3}$$

where R_0 shows the amplitude reflection coefficient of the uniform
scatterer, $\mathbb{N}$ shows outer normal to S', and S' shows the surface of
the scatterer which is modified in order to compensate the sound
velocity difference. In Eqs.(1) and (3) $\Phi(\omega,\mathbf{r})$ shows a velocity
potential and is given as follows at the focal plane of a concave
circular transducer

$$\Phi(\omega,\mathbf{r}) = (a^2/2r_0)f(ka\sin\theta)e^{-jkr} \tag{4}$$

$$f(x) = 2J_1(x)/x \tag{5}$$

where (r,θ) shows a spherical coordinates system whose origin is
taken at the center of the transducer (Fig. 1), r_0 shows a focal
length of the tranducer, and J_1 shows a Bessel function of 1st order.

In this analysis it is assumed that the scatterers placed at the
focal plane is so thin compared with the wavelength of ultrasonic
waves that the velocity potential can be assumed as Eq.(4) all over
the scatterer.

Spectrum of echo scattered by a point object

As seen in Fig.1 a point object whose volume is V is placed on
the axis of the transducer at its focal plane. If the dimension of
the point object is much smaller than the wavelength of ultrasonic
waves, Eq.(1) becomes

$$E_s(\omega) = j2\omega\rho_0 G(\omega)R_s V'(jk)^2(a^2/2r_0)^2 e^{-j2kr_0}, \tag{6}$$

where $E_s(\omega)$ shows the frequency component of echo scattered by the
point object and V' shows the modified volume of the scatterer and
is given by

$$V' = Vc_0/c_1 , \tag{7}$$

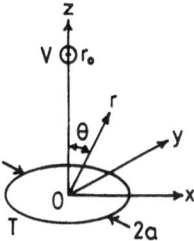

Fig.1   Arrangement of a point object and
an ultrasonic transducer (T).

where c_1 shows the sound velocity of the scatterer.   Then the spectrum is given by

$$S_s(\omega) = k^6(a^8/r_0^4)|z_0 G(\omega)R_s V'/2|^2 . \tag{8}$$

As seen in Eq.(8), the spectrum is proportional to k^6 .

Spectrum of echo scattered by a line object

As seen in Fig.2, a line object whose cross sectional area is
S is placed at the focal plane parallel to x-axis on the axis of
transducer. If the dimension of the wire is much smaller than the
wavelength of ultrasonic waves, Eq.(1) can be written as follows

$$E_\ell(\omega) = j2\omega\rho_0 G(\omega)R_\ell s'(jk)^2(a^2/2r_0)^2 e^{-j2kr_0}\int_{-\infty}^{\infty} e^{-jkx^2/r_0}f^2(kax/r_0)dx$$

$$(9)$$

where $E_\ell(\omega)$ shows the frequency component of echo scattered by the
wire, R_ℓ shows the amplitude reflection coefficient of the wire, S'
shows the modified cross sectional area and is given by

$$S' = S c_0/c_1 \qquad\qquad (10)$$

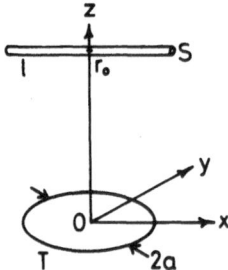

Fig.2   Arrangement of a line object and an ultrasonic transducer.

The following approximations are used in Eq.(9)

$$r = r_0 + x^2/2r_0 \qquad\qquad (11)$$

$$\sin\theta = x/r_0 \qquad\qquad (12)$$

Since it is rather difficult to perform the integral in Eq.(9)

analytically, let us approximate the radiation pattern $f^2(x)$ by a gaussian pattern $\exp(-x^2/\alpha_\ell)$ under the condition that the line integrals of the radiation patterns are equal, that is,

$$\int_{-\infty}^{\infty} f^2(x)dx = \int_{-\infty}^{\infty} e^{-x^2/\alpha_\ell}dx \tag{13}$$

then α_ℓ is given by (see appendix A)

$$\alpha_\ell = 3.67 \quad . \tag{14}$$

Then the integral part of Eq.(9) can be approximated as

$$\int_{-\infty}^{\infty} e^{-jkx^2/r_0} \ f^2(kax/r_0)dx$$

$$= (r_0/ka)\int_{-\infty}^{\infty} e^{-jx^2/F} \ f^2(x)dx$$

$$\doteqdot (r_0/ka)\int_{-\infty}^{\infty} e^{-(j/F + 1/\alpha_\ell)x^2} \ dx \quad , \tag{15}$$

where F shows a Fresnel number and is given by

$$F = ka^2/r_0 \quad . \tag{16}$$

By using the following relation

$$\int_{-\infty}^{\infty} e^{-jnx^2}dx = (\pi/n)^{\frac{1}{2}} \tag{17}$$

where n shows a complex number and it is assumed that $\mathrm{Re}(n) > 0$ and x shows a real number. Then the spectrum of echo scattered by the line object is given by

$$S_\ell(\omega) = k^4(a^6/r_0^2)|z_0G(\omega)|^2 \ (S'R_\ell)^2 \ (\pi\alpha_\ell/4)/(1 + (\alpha_\ell/F)^2)^{\frac{1}{2}}. \tag{18}$$

As seen from Eq.(18), if the focusing of beam is strong, that is,

$$F \gg \alpha_\ell \tag{19}$$

then the spectrum is given by

$$S_\ell(\omega) = k^4(a^6/r_0^2)|z_0G(\omega)|^2 \ (S'R_\ell)^2 \ (\pi\alpha_\ell/4) \quad . \tag{20}$$

Thus the spectrum is proportional to k^4. If the focussing of beam is weak, that is,

$$F \ll \alpha_\ell \tag{21}$$

then the spectrum is given by

$$S_\ell(\omega) = k^5(a^8/r_0^3)|z_0G(\omega)|^2 \ (S'R_\ell)^2 \ (\pi/4) \quad . \tag{22}$$

In this case the spectrum is proportional to k^5. Thus the spectrum of echo is influenced by the focusing of beam. As to the deviation due to the gaussian approximation of the beam, see appendix B.

Spectrum of echo scattered by a thin layer

As seen in Fig.3, a thin layer whose amplitude reflection coefficient is R_t and a thickness d is placed at the focal plane perpendicular to the axis of transducer. If the thickness of the layer is much smaller than the wavelength of ultrasonic waves, Eq.(1) can be written as follows

$$E_t(\omega) = j2\omega\rho_0 G(\omega)R_t d'(jk)^2(a^2/2r_0)^2 e^{-j2kr_0}\int_0^\infty e^{-jkx^2/r_0}f^2(kax/r_0)2\pi x dx$$

$$(23)$$

where $E_t(\omega)$ shows the frequency component of echo scattered by the layer, d' shows the modified thickness and is given by

$$d' = d\, c_0/c_1 . \tag{24}$$

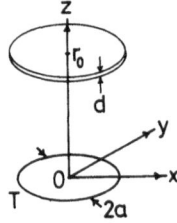

Fig.3 Arrangement of a film and an ultrasonic transducer.

It is also difficult to perform the integration in Eq.(23) analytically, comsequently let us approximate the radiation pattern by a gaussian pattern under the condition that the surface integrals of the radiation patterns are equal, that is,

$$\int_0^\infty f^2(x)2\pi x dx = \int_0^\infty e^{-x^2/\alpha_p}\, 2\pi x dx . \tag{25}$$

Then α_p is given by (see appendix C)

$$\alpha_p = 4 . \tag{26}$$

The integral part of Eq.(23) becomes

$$\int_0^\infty e^{-jkx^2/r_0} \, f^2(kax/r_0) 2\pi x dx$$

$$\doteq 2\pi(r_0/ka)^2 \int_0^\infty e^{-(j/F + 1/\alpha_p)x^2} x dx$$

$$= \pi(r_0/ka)^2/(j/F + 1/\alpha_p) \quad . \tag{27}$$

Then the spectrum of echo reflected by the thin layer is given by

$$S_t(\omega) = k^2 a^4 |z_0 G(\omega)|^2 \, (R_t d')^2 \, 4\pi^2/(1 + (\alpha_p/F)^2) \tag{28}$$

As seen in Eq.(28), if the focusing of beam is strong, the spectrum becomes

$$S_t(\omega) = k^2 a^4 \, |z_0 G(\omega)|^2 \, (R_t d')^2 \, 4\pi^2 \quad , \tag{29}$$

and if the focusing is weak, the spectrum becomes

$$S_t(\omega) = k^4 (a^8/r_0^2) |z_0 G(\omega)|^2 \, (R_t d')^2 (\pi^2/4) \tag{30}$$

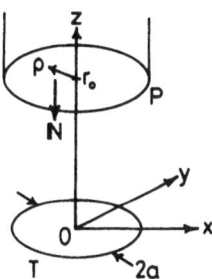

Fig.4   Arrangement of a plane reflector and
         an ultrasonic transducer.

Thus the spectrum of echo is also influenced by the focusing of
beam. As to the deviation due to the gaussian approximation of
beam, see appendix D.

Spectrum of echo scattered by a plane reflector

As seen in Fig.4, a plane reflector whose amplitude reflection coefficient is R_p is placed at the focal plane perpendicular to the axis of transducer. In this case the frequency component is calculated by using Eq.(3) and the following relations hold

$$N \cdot \nabla \Phi = \partial \Phi / \partial r \ N \cdot e_r + (1/r) \partial \Phi / \partial \theta \ N \cdot e_\theta$$

$$= (jk + 1/r) \Phi \cos \theta + (a^2/2r^2) e^{-jkr} \partial f / \partial \theta \sin \theta \quad (31)$$

If the following conditions hold

$$kr_0 \gg 1 \quad \text{and} \quad ka \gg 1 \quad (32)$$

Eq.(31) can be approximated as follows

$$N \cdot \nabla \Phi \simeq jk\Phi \quad (33)$$

Then Eq.(3) becomes

$$E_p(\omega) = j\omega\rho_0 G(\omega) R_0 (jk) (a^2/2r_0)^2 \int_0^\infty e^{-jkx^2/r_0} f^2 (kax/r_0) 2\pi x dx \quad (34)$$

and the spectrum is given by (Eq.(27))

$$S_p(\omega) = a^4 |z_0 G(\omega)|^2 \ R_0^2 \ \pi^2 / (1 + (\alpha_p/F)^2) \quad (35)$$

As seen in Eq.(35), if the focusing is strong, Eq.(35) becomes

$$S_p(\omega) = a^4 |z_0 G(\omega)|^2 \ R_0^2 \ \pi^2 \quad (36)$$

and if the focusing is weak, it becomes

$$S_p(\omega) = k^2 (a^8/r_0^2) |z_0 G(\omega)|^2 \ R_0^2 \ (\pi^2/16) \ . \quad (37)$$

Dependence of spectrum

Let us denote the spectrum as follows

$$S(\omega) = \{Ⓐ\} \cdot \{Ⓑ\} \cdot \{Ⓒ\} \cdot |z_0 G(\omega)|^2 \quad (38)$$

where the first braces show the dependence of spectrum on the wave number and it is shown in Table 1. As seen in Table 1 the dependence changes due to the geometry of scatterers as well as the focusing of beam. This shows the difficulty of deconvolution processing which is applied to the echo signal scattered by different kinds of objects. If the frequency band of transducer is sufficiently narrow, then the spectrum is mainly determined by the fourth braces in Eq.(38) and in this case the deconvolution will

work well. The broad band transducer, however, is used, then the
spectrum is mainly determined by the type of scatters and the
degree of focusing. In this case the deconvolution will not work
well.

The second braces in Eq.(38) show the dependence on the transducer geometry and is shown in Table 1. The third braces in Eq.(38)
show the dependence on the sizes and reflection coefficient of the
scatterers.

EXPERIMENTAL RESULTS

In the experiments a circular concave transducer of a = 8.3 mm
and r_0 = 100 mm is used and the ultrasonic frequency is 2.25 MHz.
In Fig.5(a) a waveform of echo reflected from a plane reflector
made of acrylic acid resin (R_p = 0.36) and its spectrum are shown.
The reflector is placed at the focal plane of transducer. In
Fig.6(a) a waveform of echo reflected from a wire made of nylon
(ρ_1 = 1.1, c_1 = 2680 m/s, and R_ℓ = 0.325) whose diameter is 94 μm
is shown. Compared with Fig.5(a) the enhancement of higher frequency components is clearly seen. The ratio of spectrum is shown
in Fig.6(c) and in this figure the theoretically estimated ratio
for strongly focused beam is shown by a dotted line. Two lines
show good agreement over the frequency interval 0.8~3.0 MHz which
corresponds to frequency band of transducer as seen from Fig.5(b).

Table 1 Dependence of Spectrum on Frequency and Transducer Geometry.

	(A) Frequency		(B) Transducer geometry	
focusing	strong	weak	strong	weak
point	k^6		a^8 / r_0^2	
line	k^4	k^5	a^6 / r_0^2	a^8 / r_0^3
film	k^2	k^4	a^4	a^8 / r_0^2
plane	k^0	k^2	a^4	a^8 / r_0^2

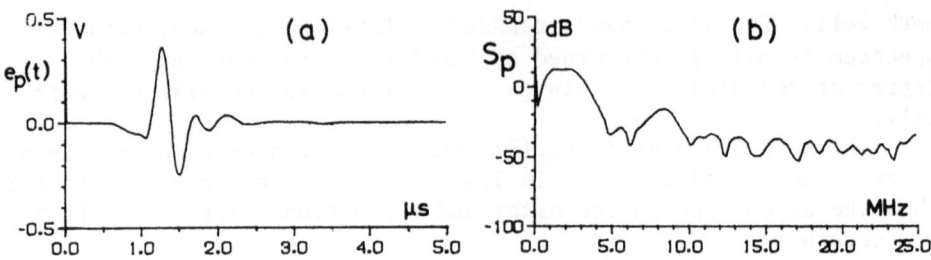

Fig.5   Waveform of echo scattered by a plane reflector (a)
        and its spectrum (b).

There appears two peaks at about 5 MHz and at this frequency the
assumption that the diameter of wire is much smaller than the wavelength of ultrasonic waves does not hold. Consequently the origin
of these peaks cannot be explained by the theory given in this
paper, but it may be considered to show some possibility to extract
acoustic characteristics (possibily sound velocity) of the scatterer by analyzing the echo signal.

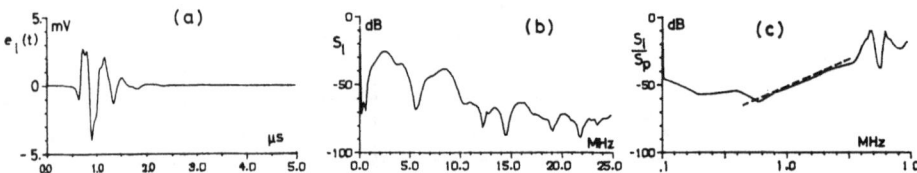

Fig.6   Waveform of echo scattered by a wire (a), its
        spectrum (b) , and ratio of spectrum (c). The
        dotted line shows theoretical estimate of ratio.

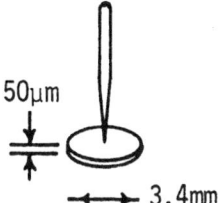

50μm

3.4mm

Fig.7 A disk and a needle used as a point object.

As to the echo reflected from a point object, it is rather
difficult to get the echo experimentally since its amplitude becomes
very small. Consequently a disk made of polyethylene (c_1 = 1950
m/s, R_s = 0.077) attached to a needle as shown in Fig.7 is used as
a point object. If the thickness of disk is much smaller than the
wavelength of ultrasonic waves and the diameter of disk is much
smaller than the beam width, it can be regarded as a point object.
The waveform of the echo reflected from the disk is shown in Fig.
8(a) and the ratio of spectrums is shown in Fig.8(c). In this

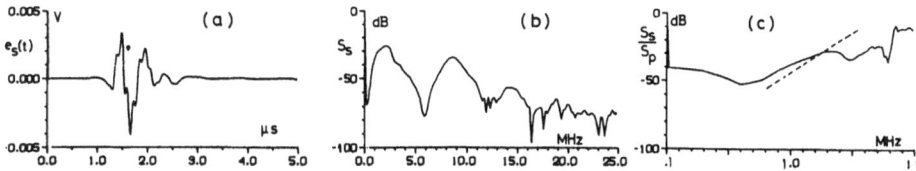

Fig.8 Waveform of echo scattered by a point object (a),
       its spectrum (b), and ratio of spectrum (c). The
       dotted line shows theoretical estimate of ratio.

figure the theoretically estimated ratio for strongly focused beam
is shown by a dotted line. As seen from Fig.8(c), the frequency
dependence of k^6 is observed at the rising of the experimental
curve but it disappeared as the frequency becomes higher. It may
be due to the deformation of disk surface and/or the interference
with the echo reflected from the needle.

CONCLUSIONS

The spectrum of echo reflected by a simple object in focused
sound field is analyzed in this paper. As results of the analysis
it makes clear that the frequency dependence of spectrum changes
by the geometrical figure of scatterer as well as the degree of
focusing of ultrasonic beam and some respects of the results are
verified experimentally. By extending the analysis to more complex
objects, it may become possible to estimate some quantitative
information about acoustic characteristics of scatterer from the
waveform of echo signal.

APPENDIX A : Derivation of Eq.(14)

According to ref.5, the following relation holds

$$\int_{-\infty}^{\infty} (2J_1(x)/x)^2 dx = 32/3\pi \tag{A1}$$

and the following relation is well-known

$$\int_{-\infty}^{\infty} e^{-x^2/\alpha_\ell} dx = \sqrt{\pi\alpha_\ell} \tag{A2}$$

From Eqs.(A1) and (A2), α_ℓ is given by

$$\alpha_\ell = 1024/9\pi^3 \doteqdot 3.67 . \tag{A3}$$

APPENDIX B : Evaluation of deviation due to gaussian approximation
            (line object)

As seen from Eq.(15) and (17), the integral I

$$I = \int_{-\infty}^{\infty} e^{-jx^2/F} f^2(x) dx \tag{B1}$$

is approximated as follows

$$I_a = \int_{-\infty}^{\infty} e^{-(j/F + 1/\alpha_\ell)x^2} dx$$

$$= (\pi/(j/F + 1/\alpha_\ell))^{0.5} . \tag{B2}$$

Consequently let us evaluate the deviation d_ℓ as follows

$$d_\ell = |I/I_a|^2 \tag{B3}$$

where I is computed numerically and the result is shown in Fig. B1 as a function of the Fresnel number F. As seen in this figure the approximation is very accurate.

APPENDIX C : Derivation of Eq.(26).

According to ref.7, the following relation holds

$$\int_{-\infty}^{\infty} (2J_1(x)/x)^2 x\,dx = 4 \tag{C1}$$

and the following relation is obtained easily

$$\int_{-\infty}^{\infty} e^{-x^2/\alpha_p}\, x\,dx = \alpha_p \tag{C2}$$

then α_p is given by

$$\alpha_p = 4 .$$

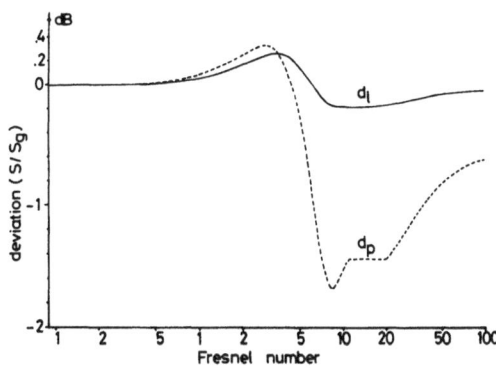

Fig. B1 Deviation due to gaussian approximation of beam.

APPENDIX D : Evaluation of deviation due to gaussian approximation
(plane objects)

As seen in Eq.(27), the integral I

$$I = \int_0^\infty e^{-jx^2/F} f^2(x) \, 2\pi x dx \qquad (D1)$$

is approximated as follows

$$I_a = \int_0^\infty e^{-(j/F + 1/\alpha_p)x^2} 2\pi x dx$$

$$= \pi/(j/F + 1/\alpha_p) . \qquad (D2)$$

Then the deviation is defined as

$$d_p = |I/I_a|^2$$

and d_p is shown in Fig.B1. In this case the maximum deviation of
-1.7 dB occurs at $F \doteqdot 10$ and it may be necessary to correct the
data by using the curve in Fig.B1 in the case of precision measurement.

REFERENCES

1. A. Freedman: Acoustica 12(1963) 61.
2. D.M. Johnson: J. Acoust. Soc. Am. 59(1976) 1319.
3. J. Saneyoshi: Handbook of ultrasonic technique (Nikankogyosha,
   Tokyo, 1978) p.53 (in Japanese).
4. M. Ueda: Acoustical Imaging, Plenum Press, New York 1982, vol.12,
   p.391.
5. S. Moriguti, K. Uda, and S. Hitotumatu: Mathematical formulas
   (Iwanamishoten, Tokyo, 1973), vol.3, p.196 (in Japanese).
6. ibid., vol.1, p.233.
7. ibid., vol.3, p.189.

MEASUREMENT AND SIMULATION OF THE SCATTERING OF ULTRASOUND BY PENETRABLE CYLINDERS

Brent S. Robinson and James F. Greenleaf

Biodynamics Research Unit, Department of Physiology
and Biophysics, Mayo Clinic/Foundation, Rochester
MN 55905

INTRODUCTION

Scattering of ultrasound occurs when a propagating wave encounters variations in some intrinsic property, such as density compressibility or absorption, of the propagation medium. Scattering is manifested as a change in the spatial distribution (and, in the case of absorption, the total amount) of the ultrasonic energy. There are four distinct components to be considered in any scattering problem. These are:
(i) The propagation process. A description of the physics of the propagation process is embodied in the wave equation which can be regarded as the propagation model.
(ii) The spatial distribution of acoustic properties of the medium, i.e., the 'object', which in this treatment we call F.
(iii) The insonifying wave, Ψ_0. In the absence of the object Ψ_0 is is the only field existing. However, in the presence of the object scattering causes an additional field called
(iv) the scattered field, Ψ_s. The actual field present at any point is termed Ψ and is the sum

$$\Psi = \Psi_0 + \Psi_s \quad . \tag{1}$$

In the so-called direct scattering problem, the scattered field is related to the other components by:

$$\Psi_s = G\{\Psi_o, F\} \quad , \tag{2}$$

where G is an operator derived from the propagation model. In comparison, ultrasound diffraction tomography is an application of inverse scattering in which information about the object is gained from knowledge of the incident and scattered fields using an inverse operator, G^{-1}, also derived from the propagation model. So for inverse scattering we have

$$F = G^{-1}\{\Psi_o, \Psi_s\} \quad . \tag{3}$$

This paper presents some results of a study on the direct scattering problem (Eq. 2). However, due to the intimate relationship between direct and inverse scattering as apparent from Eqs. (2) and (3), the results also have relevance for inverse scattering. In fact, the motivation for the study came from a need to test some of the assumptions we make in diffraction tomography experiments, i.e., most importantly, to verify that we can indeed faithfully measure in the laboratory the data predicted by our theoretical model. To achieve this aim, the scattering of plane ultrasound waves by penetrable circular cylinders was investigated both theoretically and experimentally. This particular geometry was chosen since it is possible to compute an exact solution for the direct scattering problem.[1,2] Similar comparisons already appear in the literature.[3,4] However, the agreement between theory[4] and experiment[3] was poor and furthermore, the early work investigated the magnitude of the scattering in the far-field only.

For our results to have relevance for diffraction tomography, we investigated the complex amplitude (i.e., both magnitude and phase) of forward scattering in the near field. Although scattering from a single cylindrical scatterer immersed in an otherwise homogeneous medium is a somewhat specialized case, it is perhaps not too far removed from, say, a tumor embedded in soft tissues for the results to have practical significance. Accordingly, we studied cylinders whose properties were close to what might be encountered in clinical imaging. The acoustic properties did not differ greatly from water (i.e., the scatterers were tenuous) and the diameters were a few dozen wavelengths across. To further extend previous work[3,4] we studied the two-dimensional distribution of scattering in a plane normal to the cylinder axis, the field within the cylinder, and the effects of absorption. Note that we assume the sign convention $\exp(-j\omega t)$ for temporal waveforms meaning that the phase shifts we report are inverted with respect to 'reality' in all results.

METHOD OF SIMULATION

The symmetry inherent in the scattering of a plane incident wave by a cylinder allows an exact direct solution to be computed comparatively simply. Details of the derivation are available in the literature[1-4] and are not repeated here. However, some comments on practical aspects of the simulation method are made. We begin with the time-independent wave equation[5]

$$\nabla^2 \Psi - \nabla\Psi \cdot \nabla\rho/\rho + k^2 \Psi = 0 \quad , \tag{4}$$

as our propagation model. Ψ, the excess pressure, ρ the density, and k the wave number are all functions of space. The second term on the left hand side involving the gradient of the density is zero in homogeneous regions (i.e., everywhere except at the cylinder boundary). The derivation[5,6] of Eq. (4) assumes linear wave motion and isotropy, neglects shear waves, and makes no attempt to model absorption mechanisms. However, provided that the absorption per wavelength is small, the propagation speed and absorption of the wave can be separately specified by the real and imaginary parts, respectively, of k, i.e.:

$$k = \omega\sqrt{\rho\kappa} + j\alpha \quad , \tag{5}$$

where ω is the angular temporal frequency, κ is the compressibility, α is the absorption coefficient and $j = \sqrt{-1}$.

The derivation of the simulation method involves separating the wave equation expressed in cylindrical coordinates (r, θ) for regions within and without the scattering cylinder.[1] The solution of the resulting simultaneous differential equations has the form of a product of weighted cylindrical harmonics and Bessel functions, i.e.,

$$\Psi_s(r,\theta) = \sum_{m=0}^{\infty} A_m H_m(kr) \cos(m\theta) \quad ; \quad r \geq a \tag{6}$$

and

$$\Psi(r,\theta) = B_0 J_0(k_c r) + \sum_{n=1}^{\infty} \varepsilon_n B_n J_n(k_c r) \cos(n\theta) \quad ; \quad r \leq a \tag{7}$$

where r is the distance from the cylinder center, θ is the scattering angle, a is the radius of the cylinder, the subscript c denotes a parameter inside the cylinder, and the Hankel function of the first kind (i.e., $H_m = J_m + jY_m$) is used. The complex coefficients, A_m and B_m, are characteristic of the particular cylinder at the frequency being considered and are obtained by equating separately the values of pressure and radial particle velocity across the cylinder boundary. This results in the expressions:

$$A_m = \frac{\varepsilon_m k J_m(k_c a) \dot{J}_m(ka)/\rho - \varepsilon_m k_c J_m(ka) J'_m(k_c a)/\rho_c}{k_c H_m(ka) J'_m(k_c a)/\rho_c - k J_m(k_c a) H'_m(ka)/\rho} \quad (8)$$

where

$$\varepsilon_m = 1 \text{ for } m = 0; \ \varepsilon_m = 2j^m \text{ for } m > 0 \quad (9)$$

and

$$B_n = \frac{J_n(ka) H'_n(ka) - H_n(ka) J'_n(ka)}{J_n(k_c a) H'_n(ka) - k_c \rho H_n(ka) J'_n(k_c a)/k \rho_c} \quad . \quad (10)$$

Computational Considerations

The derivatives Q' (where Q represents J or H) of the Bessel functions are evaluated by the relationship[7]

$$Q'_m(ka) = m Q_m(ka)/ka - Q_{m+1}(ka) \quad . \quad (11)$$

Our experience has shown that the series of Eqs. (6) and (7) are well behaved and both converge rapidly after a certain number of terms, N say. Thus, the series computations are terminated after the first N terms have been added rather than by using error stopping criteria. N can be determined by inspection of the 'A' coefficients, the magnitudes of which range between 0 and 2 decreasing rapidly to zero when $m > N$ (in contrast, the 'B' coefficients are sometimes observed to diverge for $n > N$). Alternatively, N can be determined by the conservative emperical formula

$$N = \text{Nearest Integer } \{3 + 1.2 \text{ ka }\} \quad . \quad (12)$$

Eq. (12) indicates that for cylinders with sub-wavelength diameters, the first three terms are sufficient. For large cylinders N is proportional to the ka value which is effectively the number of 'free-space' wavelengths around the circumference

of the cylinder. It is interesting to note that N is substantially independent of the acoustic parameters within the cylinder. However, as the cylinder becomes larger, the scattering
becomes increasingly detailed and higher order cylindrical harmonics are consequently required to characterize the scattering.

Bessel functions of complex arguments are required in
Eqs. (7), (8) and (10) to model absorption within the cylinder.
These are computed using standard library routines.[8] When the
observation point is far from the cylinder the argument of the
Bessel function in Eq. (6) is large enough for the asymptotic
expression[7] to be used accurately. Since this is most accurate
for low orders, recurrence relations[7] are used to compute the
higher order Bessel functions. In fact, since the form of the
simulation computation is a series evaluation, use may be made
of Bessel function recurrence relations to improve computational
efficiency.

Fig. 1 shows an example of the first 80 complex coefficients
and corresponding convergence of the scattering computation at a
point on the cylinder boundary. Each series of coefficients took
about 12 seconds to compute on a 32-bit (Perkin-Elmer 7/32) mini-
computer. When the library routines were used to evaluate all
Bessel functions, the field values computed by Eqs. (6) and (7)
agreed to 5 significant figures (normalized magnitude = 0.16297,
phase = 2.5981 radians). Computation time was 6 seconds for each
complex value. When the asymptotic expression was used in Eq.
(6) as outlined in the preceeding paragraph, the computed field
values differed by less than 0.5% and computation time was
reduced to 0.5 seconds. The agreement between Eqs. (6) and (7)
at the boundary gives us confidence in the field values computed
within the cylinder. These would otherwise be difficult to
verify expermentally.

MEASUREMENT TECHNIQUE

The advent of PVF piezoelectric films has allowed
comparatively easy fabrication of large transducers. To match
the simulation exactly, we should have insonified with a perfect
plane wave. However, such insonification is impossible to
achieve. Instead we used a large, cylindrically shaped, concave
PVF transducer with the scattering cylinder located in the narrow
(about $\frac{1}{2}$ cm vertical extent, refer to Fig. 2) focal region. In
clinical imaging, cylindrical wave insonification may prove preferable to plane wave insonification since it localizes the
energy. Scattering was sensed by a small, effectively omnidirectional, hydrophone (a "Mediscan microprobe") which was
scanned in two orthogonal directions as shown in Fig. 2.

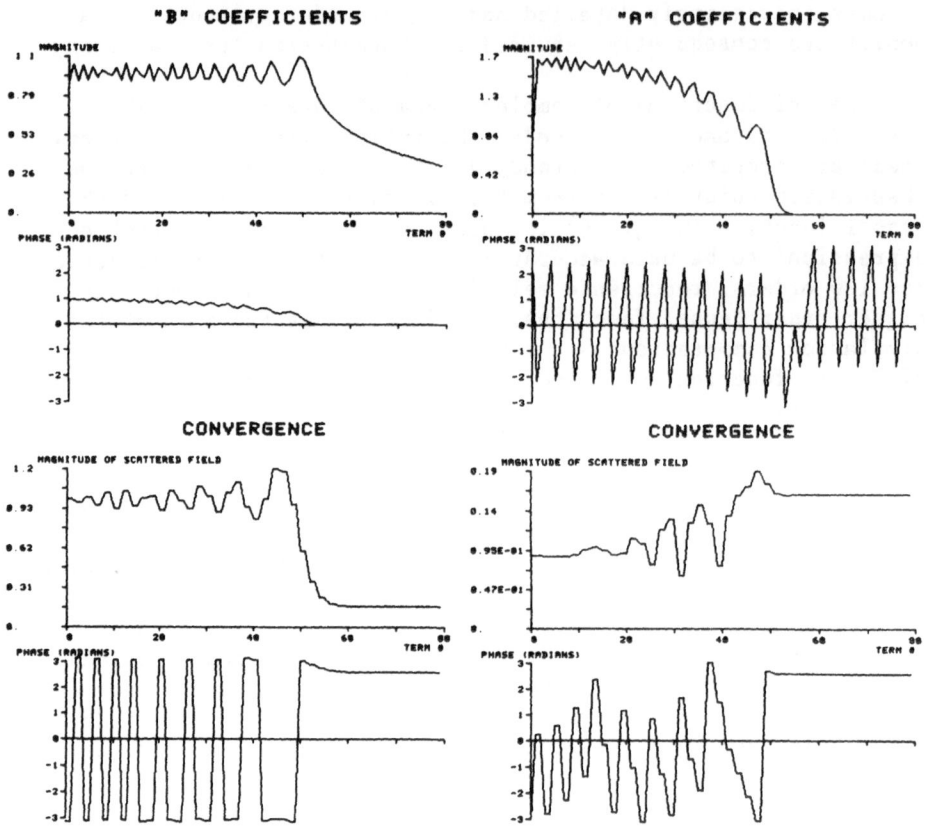

Fig. 1.   Simulation example.   Above - complex coefficients;
          below - convergence of series expression for scattered
          fields.   On left is result for internal field expression
          (Eqs. (7) and (10)); on right is result for external
          field expression (Eqs. (6) and (8)).   Scattering at 90°
          angle to a point on the cylinder surface is computed.
          Parameters of simulation are specified in Table 1.

        To ease mechanical and electrical precision requirements,
the experiments were performed at relatively low ultrasonic frequencies (0.7 and 1.5 MHz).   Latex rubber condoms were filled
with fluids of known acoustic speed, density and absorption and
immersed in water to act as cylindrical scattering objects.   We
found condoms to be ideal for these experiments since they are
virtually transparent acoustically and, when suspended under

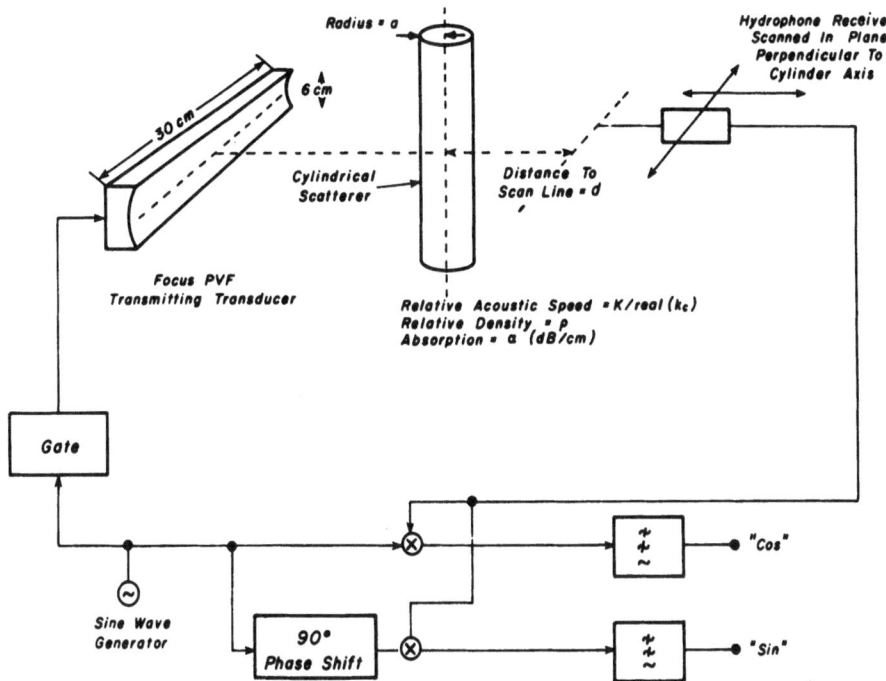

Fig. 2. Configuration of measurement system and signal
processing.

Fig. 3. Water tank containing PVF transmitting transducer on
left, cylindrical scattering object at center and
receiver probe to right. One of the stepper motor scanning drives is in place above the receiving transducer.

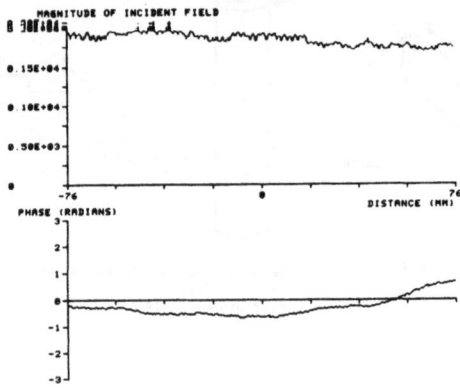

Fig. 4.   Measured magnitude and phase profiles of the incident
          field. Measurement conditions are specified in Table 1.
          This particular profile was used to normalize the field
          shown in Fig. 11. Horizontal scale is +/-76 mm.

tension, they provide large circular cross sections. A photograph of the experimental apparatus is shown in Fig. 3.

The transmitter was excited by narrow-band (75 μsec
duration) tone bursts to ensure effectively steady-state conditions and yet eliminate undesirable tank reverberations.
Standard phase coherent quadrature detection was used to measure
the complex amplitude of the pressure field. Measured values
were stored in a computer for subsequent display. When scanning
parallel to the direction of propagation of the incident wave a
second receiving transducer (not depicted in Figs. 2 or 3) was
employed to monitor, and compensate for the varying propagation
delay of the tone burst. The incident fields were also measured
in each experiment by scanning with the scatterer removed. A
typical incident field measured at 1.5 MHz tone burst center frequency is plotted in Fig. 4. The scan was normal to the direction of propagation of the incident wave at a distance d = 48 mm
from the nominal cylinder center. Measurements of the incident
fields were used to normalize the total fields measured with the
cylinder in place in all following results.

RESULTS

A comparison between simulated and measured scattering from
approximately 46% isopropyl alcohol solution is shown in Fig. 5.

Table 1. Acoustic parameters used in simulations of various
          scattering experiments. The radius (a) of the cylinder
          was 18 mm in all cases. Concentrations of isopropyl
          alcohol solutions (Figs. 5-10) are by weight. Figs. 4
          and 11 obtained at 1.5 MHz. All other results were at
          0.6796 MHz.

Fig. #	d (mm) Refer to Fig. 2	Relative Velocity $\frac{k}{\text{real } \{k_c\}}$	Relative Density $\rho c/\rho$	Absorption (dB/cm)	Medium Within Cylinder
1	0	.982	.9	0	---
4	48	1	1	0	Water
5	-30 to 213	.982	.9	0	46% Alchl.
6	-32 to 32	.982	.9	0	46% Alchl.
7	31	.974	.9	0	46% Alchl.
8	-30 to 213	1.032	.92	0	43% Alchl.
9	20	1.014	.9	0.1	44% Alchl.
10	24	1	.91	0	45% Alchl.
11	48	1.021	.95	1.64	Castor Oil

The horizontal grain in the experimental result is thought to
arise from imperfections in the normalization process described
in the preceeding section. There is also one missing scan at
about 190 mm displayed as a grey horizontal line. Nevertheless,
the qualitative agreement between simulation and experiment is
very good. The acoustic speed for this cylinder was 1.8% lower
than for water. As would be expected from acoustical ray theory,
the cylinder acts as a converging lens, focusing energy. However, also evident in Fig. 5 is considerable high spatial frequency detail close to the cylinder as well as streaks appearing
to emanate from the cylinder edges. These features, which are
perhaps better illustrated in the simulation 'close-up' (Fig. 6)
are thought to be due to interference of diffracted waves. The
simulations show that most of the scattering is in the forward
direction and that there is a gradual progressive phase shift
through the cylinder with relatively little spatial detail in the
phase distribution. Good quantitative agreement between simulation and experiment is evident from Fig. 7 which is for a profile
at about the uppermost level of Fig. 6 but for a slightly lower
speed cylinder (see Table 1).

An interesting point arises here. If one was doing
computed tomography based on the straight ray assumption,[9] then
the magnitude of Fig. 7 would be interpreted as an attenuation

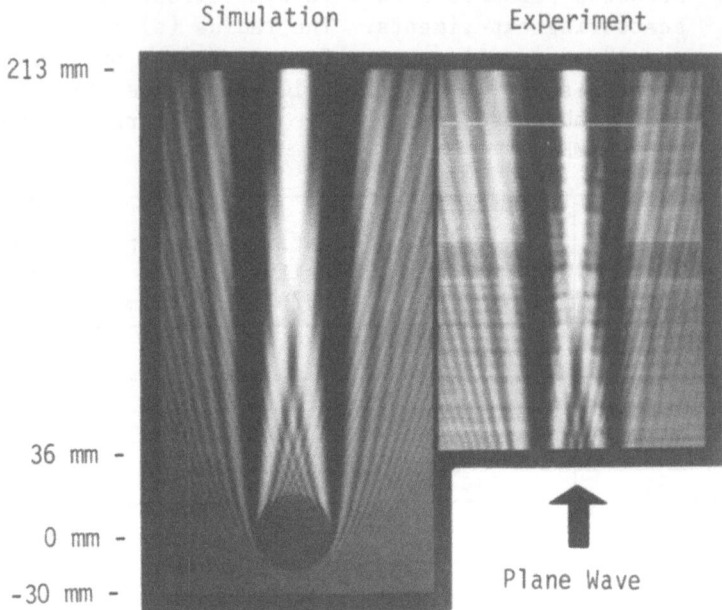

Fig. 5.   Two-dimensional distribution of magnitude of
          normalized total field for a low acoustic speed
          cylinder. Parameters are listed in Table 1. The
          field was measured in .508 mm horizontal by 1.016
          mm vertical increments; simulation was on a .508
          by .508 mm grid. However, displays are on iden-
          tical scales. As in the following figures, the
          grey scale is adjusted so that the minimum value
          of normalized field (0.55 in this simulation) is
          black, maximum (1.72) is white.

projection. Obviously this would be incorrect for a cylinder.
On the other hand, the phase shift corresponds to a time-of-
flight projection and would be a reasonably faithful projection.
This observation further explains why speed reconstructions tend
to be quantitatively more accurate than attenuation reconstructions in ultrasound computed tomography,[10] although in practice
directional transducers with localized beams are used.[9]

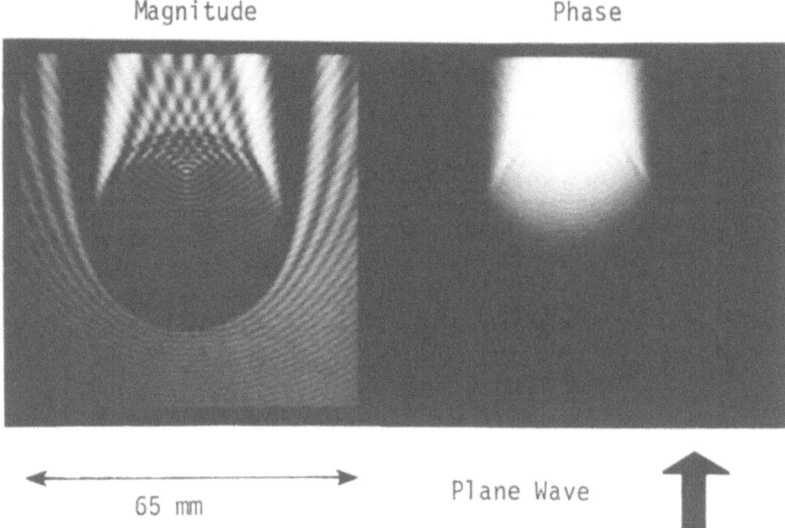

Fig. 6. Simulated normalized magnitude and phase of total
field due to scattering by cylinder of Fig. 5.
Magnitude ranges from 0.69 to 1.27, phase shift
ranges from -0.16 to 2.01 radians. The normalization procedure removes the progressive phase
shift due to propagation of the incident wave from
this picture.

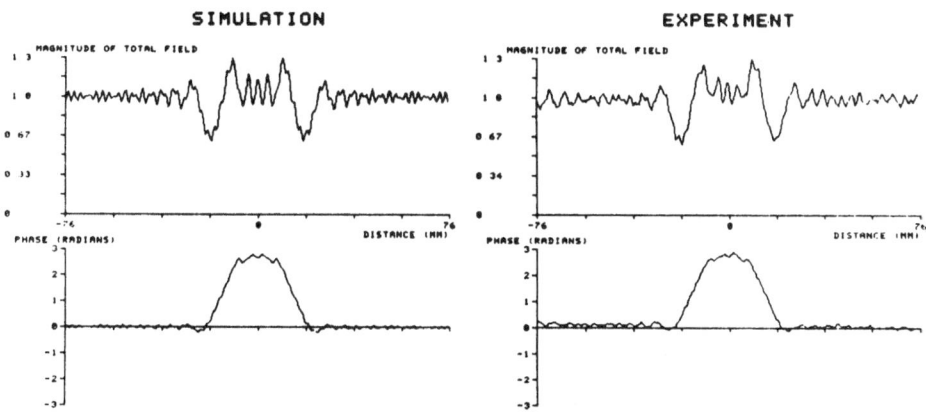

Fig. 7. Profile of normalized total field for low acoustic speed
cylinder. Experimental result shows a high degree of
symmetry as expected from geometry.

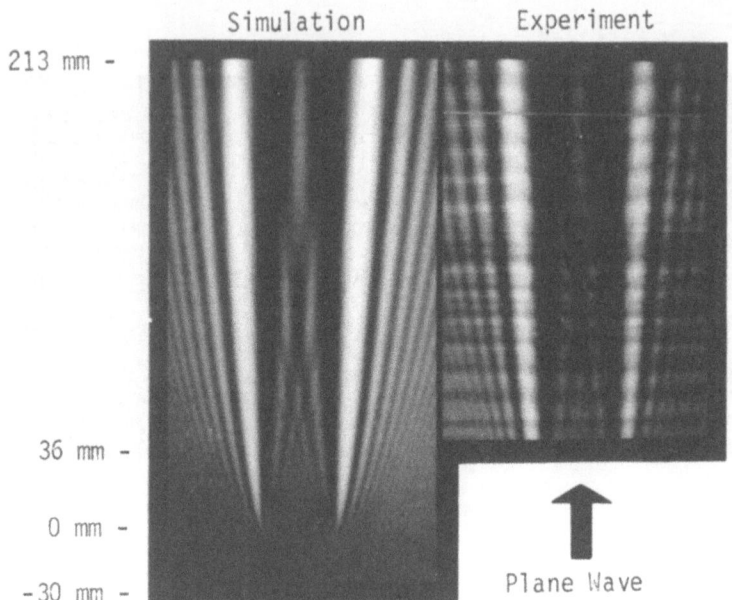

Fig. 8.   Two-dimensional distribution of magnitude of
          normalized field for high acoustic speed cylinder.
          Magnitude of simulation ranges from .32 to 1.54.

In Figure 8 the two-dimensional field distribution arising
from a higher speed cylinder is presented. The characteristics
of the scattering are similar to those of Figs. 5 and 6, although
there is no focal region evident in Fig. 8. Fig. 9 is also for a
cylinder of greater acoustic speed but the profile is measured
only 2 mm away from the cylinder surface. The phase shift is
inverted with respect to Fig. 7. Once again, fine details in
the magnitude seem to arise from the cylinder edges.

The final two figures show the results of experiments and
simulations designed to examine the effects of the various
acoustic parameters on scattering. This is, of course, convenient to do in simulation. It is possible to adjust the concentration of isopropyl alcohol solutions so that the propagation
speed matches that of water. Since absorption of isopropyl alcohol is negligible, the cylinder essentially becomes a density-
only scatterer. As seen in Fig. 10, there is very little forward
scatter even for a density difference of 9%. The slight residual
phase shift in the experimental result arises from an imperfect
speed match.

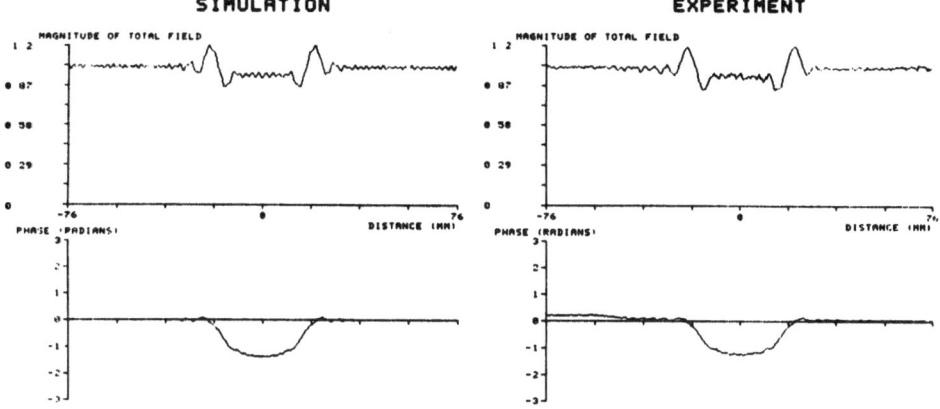

Fig. 9.   Profile of normalized total field for high acoustic
          speed cylinder.

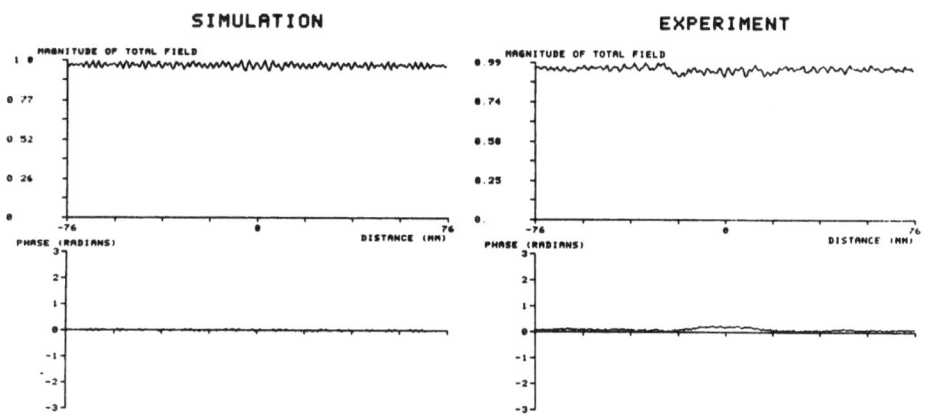

Fig. 10. Profile of normalized total field for cylindrical
          scatterer having only variations in density.

    The effect of absorption was tested by filling the
cylinder with castor oil with the result shown in Fig. 11.
Absorption is handled phenomenologically in the simulation by
making the wave number inside the cylinder complex. Strictly
speaking, absorption is not modelled correctly by Eqs. (8) and
(10). This is because the expression for the radial particle
velocity used to equate the boundary values and hence obtain
the A and B coefficients does not include the phase shift between pressure and velocity resulting from absorption. However, in view of the very close agreement between the results
of experiment and simulation, it appears that this aspect of
the simulation is not critical.

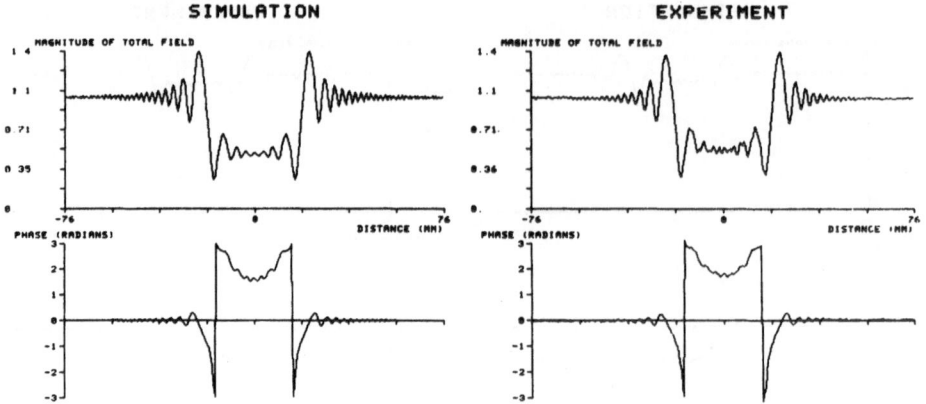

Fig. 11.   Profile of normalized total field for cylinder with
            absorption. The phase in the central portion is
            actually less than $-\pi$ radians but has been displayed
            in "wrapped" form to emphasize the modulo 2π nature
            of both measurement and simulation.

DISCUSSION

In all cases, the agreements between the experimental
measurements and simulations were very good. As is common,[6] a
simplified wave equation was used as the starting point of the
simulation for reasons of tractability. However, these results
suggest that Eq. (4) is a sufficiently accurate model of reality
in this application and presumably, therefore, a valid basis for
diffraction tomography. Just as important is the conclusion that
our measurement techniques allow us to measure what we presume we
measure. This is perhaps somewhat surprising when it is remem-
bered that we transmitted an imperfect cylindrical wave rather
than an ideal plane wave. Evidently the normalization procedure
described in the section on measurement techniques largely
corrects for the imperfections (indicated in Fig. 4) in flatness
of the incident wave. Also, insonification with cylindrical
waves apparently results in essentially the same scattering in
the central plane as for plane waves.

The results indicate that quite small changes in acoustic
speed cause significant scattering, particularly in the forward
direction. Both the magnitude and the phase are affected. This
is consistent with the concept that the wave progressively accu-
mulates phase shift as it propagates through regions of different
acoustic speed. In comparison, equivalent variations in density
contribute little to scatter. Although it can be shown[11] that

the density gradient term in Eq. (4) acts as a delta function-
like source at the cylinder boundary, its integrated contribution to scattering is evidently quite small. The low
contrast of the forward scattered field arising from density
variations may make attempts to separately image density[12]
difficult, at least for large scatterers. Absorption reduced
the gross level of the field in the 'shadow' of the cylinder
but did not significantly alter the detailed shape of the field
distribution or the phase. The scattering reported here cannot
be explained in terms of refraction effects alone. It appears
that diffraction from the edges of the cylinders is also significant. Of course, this study is restricted to fluid media
with abrupt, regular interfaces and it is not certain how well
these observations apply to scattering by soft tissues.

Finally, we remark that Eqs. (2) and (3) are rather
idealized. It is often necessary to invoke linearizing approximations, such as those due to Born or Rytov, in order to
obtain operators G and G^{-1} in the general case.[13] The simulation method described here, which is an exact solution of Eq.
(4), is apparently consistent with reality. Thus, it provides
a useful means for testing other direct and inverse scattering
methods against a 'gold standard'. The simulation can be
easily extended to handle wide temporal bandwidth signals by
Fourier synthesis. Similarly, non-plane wave insonification
could be synthesized from a sum of plane waves.

ACKNOWLEDGMENTS

We wish to thank Dr. Paul Thomas for assistance in making
the measurements and Drs. Kaveh and Soumekh of the University
of Minnesota for suggestions regarding the simulation. The
manuscript was prepared by Elaine Quarve and graphics by James
Hanson. These studies were supported in part by Grants CA
24085 (NCI) and ECS 7926008 (NSF).

REFERENCES

1. P. M. Morse and K. U. Ingard, "Theoretical Acoustics,"
    McGraw-Hill, New York, p. 464 (1968).
2. J. J. Bowman, T. B. A. Senior, and P. L. E. Uslenghi,
    "Electromagnetic and Acoustic Scattering by Simple
    Shapes," North-Holland, Amsterdam, p. 125 (1969).
3. P. Tamarkin, Scattering of an underwater ultrasonic
    beam from liquid cylindrical obstacles, J. Acoust. Soc.
    Am. 21:612 (1949).

4. S. J. Bezuszka, Scattering of underwater plane ultrasonic
    waves by liquid cylindrical obstacles, J. Acoust. Soc.
    Am. 25:1090 (1953).
5. E. J. Skudrzyk, "The Foundation of Acoustics,"
    Springer-Verlag, New York (1971).
6. S. A. Johnson, F. Stenger, C. Wilcox, J. Ball, and M. J.
    Berggren, Wave equation and inverse solutions for soft
    tissue, in: "Acoustical Imaging," J. P. Powers, ed.,
    Plenum, New York, 11:409 (1982).
7. M. Abramowitz and I. A. Stegun, "Handbook of Mathematical
    Functions," Dover, New York, p. 355 (1970).
8. A. H. Morris, Jr., NSWC/DL Library of Mathematics
    Subroutines, Naval Surface Weapons Center, Virginia,
    Report #NSWC TR 81-410 (1981).
9. J. F. Greenleaf and R. C. Bahn, Clinical imaging with
    transmissive ultrasonic computerized tomography, IEEE
    Trans. BME-28:177 (1981).
10. J. F. Greenleaf, P. J. Thomas, and B. Rajagopalan, Effects
    of diffraction on ultrasonic computer-assisted
    tomography, in: "Acoustical Imaging," J. P. Powers, ed.,
    Plenum, New York, 11:351 (1982).
11. R. H. T. Bates and F. L. Ng, Polarization-source for-
    mulation of electromagnetism and dielectric-loaded
    waveguides, Proc. IEEE 11:1568 (1972).
12. S. J. Norton, Generation of separate density and
    compressibility images in tissue, Ultrasonic Imaging
    5:240 (1983).
13. M. Kaveh, M. Soumekh, Z. Q. Lu, R. K. Mueller, and J. F.
    Greenleaf, Further results on diffraction tomography
    using Rytov's approximation, in: "Acoustical Imaging,"
    E. A. Ash and C. R. Hill, eds., Plenum, New York, 12:599
    (1982).

FRESNEL APPROXIMATION IN THE NON PARAXIAL CASE

AND IN ABSORBING MEDIA

P. Alias and P. Cervenka

Institut de Mécanique Théorique et Appliquée
Université Pierre et Marie Curie, Paris, France

INTRODUCTION

The Fourier theory of harmonic radiation is now a classical tool in Optics or in Acoustics {1}. It is in general limited to an isotropic non absorbing medium and most of its applications are obtained in the frame of the Fresnel (or Fraunhofer) paraxial approximation which remains valid for the great majority of optical or acoustical imaging or detection devices. It implies essentially a 2-D Fourier analysis in planes Π_z normal to a mean propagation direction Oz. The purpose of this article is to recall {2,3,4,5} that such an analysis may be extended straight forward to absorbing media just by using adequate plane modes expressing a phase variation dictated by a classical real wave vector and a an attenuation dictated by an imaginary wave vector parallel to Oz in such a way that the equiamplitude planes remain parallel to the reference plane z = O ; the signature of such a "generalized plane wave" remains adequate for a Fourier analysis. It will be shown that the Fresnel approximation has also a straight forward extension in this case, so that all the classical results of the Fourier theory in the Fresnel frame may easily be corrected. Furthermore, the Fresnel approximation consists in using a limited development of the z component of the wave vector in terms of the spatial frequency $\vec{f}$ valid for small values of $\vec{f}$. It may be extended to non paraxial situations important in NDE or in submarine detection of objects buried in sediments occurring when a beam is focused through a plane interface at large incidence. The mean direction of the transmitted beam may encounter a large deflexion, but corresponds in fact to the same spatial frequency $\vec{f}_0$ in planes parallel to the interface than the mean incident direction in the first medium. It is then possible to extend the Fresnel approximation to limited developments in terms of $\vec{f}' = \vec{f} - \vec{f}_0$, an approximation which remains valid if the original beam has a reasonable angular aperture. The main aberrations encountered by the transmitted beam may be shown easily when assuming that the incident focus beam has a gaussian structure.

PLANE FOURIER ANALYSIS OF A LINEAR HARMONIC RADIATION IN AN HOMOGENEOUS MEDIUM

If the radiation field may be described by means of a scalar function $\phi(M,t) = U(M)e^{-j\omega t}$, a Fourier analysis of the complex amplitude U may be developed inside planes Π_z normal to a chosen direction Oz, with the family of functions U_z :

$$U_z(\vec{m}) \;=\; U(\vec{M}) \quad , \quad \vec{M} = \vec{m} + z\vec{z} \in \Pi_z$$

For the Fourier expression of $U_0(\vec{m})$ in Π_0 :

$$U_0(\vec{m}) \;=\; \int A_0(\vec{f}) \; e^{2j\pi \vec{f} \cdot \vec{m}} \; d\vec{f} \tag{1}$$

the solution for the half space $z > 0$ in the absence of sources may be written :

$$U(\vec{M}) \;=\; \int A_0(\vec{f}) \, u \,(\vec{M}, \vec{f}) \; d\vec{f} \tag{2}$$

where $u(\vec{M},\vec{f})$ is the solution of the propagation equation which has the signature $e^{2j\pi\vec{f}\cdot\vec{m}}$ in Π_0 .

For an homogeneous medium, the solution u is a "plane" wave $e^{j\vec{k}\cdot\vec{M}}$ where $\vec{k} = 2\pi\vec{f} + k_z \, \vec{z}$ must satisfy the dispersion equation associated to the propagation equation and the relation (2) gives :

$$U_z(\vec{m}) \;=\; \int A_0(\vec{f}) \; e^{j k_z z} \; e^{2j\pi \vec{f} \cdot \vec{m}} \; d\vec{f} , \tag{3}$$

which means that the radiation in Π_z may be expressed from its expression in Π_0 through the operations in the real space and in the Fourier space :

$$A_z(\vec{f}) = A_0(\vec{f}) \, H_{0z}(\vec{f}) \;, \qquad H_{0z}(\vec{f}) = e^{j k_z z}$$

$$U_z(\vec{m}) = U_0(\vec{m}) * h_{0z}(\vec{m}) \qquad h_{0z}(\vec{m}) = TF^{-1}(H_{0z}) \tag{4}$$

In the classical Fourier theory of optics or acoustics, the propagating medium is isotropic and non absorbing and the propagation equation is the ordinary wave equation :

$$\square \, \varphi = 0 \quad , \quad \square \equiv \frac{1}{c^2} \frac{\partial_t}{\partial_t^2} - \nabla^2 \qquad (5)$$

The plane waves $e^{j\vec{K}\cdot\vec{M}}$ are ordinary plane waves :

$$K = \frac{\omega}{c} = \frac{2\pi}{\lambda} \quad , \quad K_z = \frac{2\pi}{\lambda} \left(1 - \lambda^2 f^2 \right)^{1/2} . \qquad (6)$$

The Fresnel paraxial approximation $(\lambda f < 1)$ uses the classical expressions for the operators H_{oz} and h_{oz} :

$$H_{oz} \approx e^{2j\pi \frac{z}{\lambda}} \, e^{-j\pi \lambda z \, f^2}$$

$$h_{oz} \approx \frac{1}{j\lambda z} e^{2j\pi \frac{z}{\lambda}} . \, e^{j\pi \frac{m^2}{\lambda z}} \qquad (7)$$

THE CASE OF AN ATTENUATING MEDIUM

The preceding formalism is extended straightforward to the case of an absorbing medium, for example the acoustical radiation in a viscous fluid : the propagation equation may be written for the velocity potential ϕ :

$$\square' \varphi = 0 \quad , \quad \square' \equiv \frac{1}{c^2} \frac{\partial_t}{\partial_t^2} - \left(1 + \frac{\eta}{\rho_o c^2} \frac{\partial}{\partial t} \right) \nabla^2 \qquad (8)$$

The dispersion equation :

$$K'^2 \left(c^2 - j\omega \eta / \rho_o \right) = \omega^2 \qquad (9)$$

leads simply to attenuated plane wave modes $e^{j\vec{K}''\cdot\vec{r}}$, where $\vec{K}'' = k''\vec{n}$ and $k'' \cong k + j\alpha$:

$$K = \frac{\omega}{c} = \frac{2\pi}{\lambda} \quad , \quad \alpha = \frac{\eta \omega^2}{2\rho_o c^3} ,$$

α being the classical attenuation coefficient : an approximate solution valid for $\alpha\lambda \ll 1$.

These modes are not appropriate to a Fourier analysis and must be replaced by modes $u(\vec{f},\vec{M}) = e^{j\vec{K}'\cdot\vec{M}}$ of signature $e^{2j\pi\vec{f}\cdot\vec{m}}$ in Π_0 with a complex component k'_z of the wave vector $\vec{k}'$. These modes (Fig. 1) have equiphase planes normal to the real part of k' and equiamplitude planes normal to $0z$, and are determined from the dispersion equation according to :

$$K'^2_z = \frac{\omega^2}{c^2 - j\omega \eta / \rho_o} - 4\pi^2 f^2 \qquad (9')$$

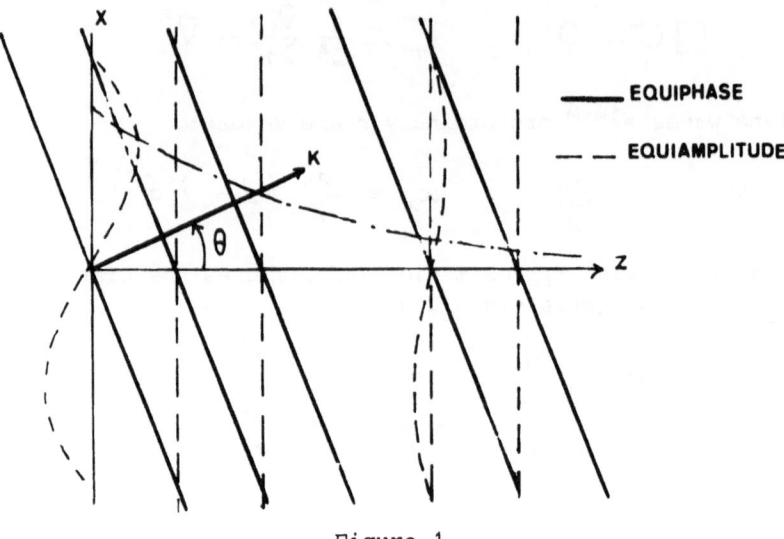

Figure 1

If $\alpha\lambda \ll 1$, the adequate solution may be written :

$$\vec{k}' = \frac{2\pi}{\lambda}\,\vec{n} + j\,\frac{\alpha}{\cos\theta}\,\vec{z} \quad , \qquad \theta = \left(\vec{z},\vec{n}\right) \tag{10}$$

and the corresponding Fresnel paraxial approximation ($\lambda f \ll 1$) to :

$$k'_z \simeq \frac{2\pi}{\lambda} - \pi\lambda f^2 + j\alpha\left(1 + \frac{\lambda^2 f^2}{2}\right) \tag{11}$$

The propagation operators from Π_0 to Π_z are then modified from (7) according to :

$$H'_{0z}(\vec{f}) = H_{0z}A_{0z}\ , \quad A_{0z} = e^{-\alpha z}\,e^{-\pi\left(\frac{\alpha z \lambda^2}{2\pi}\right)f^2}\ ,$$

$$h'_{0z}(\vec{m}) = h_{0z} \ast a_{0z}\ , \quad a_{0z} = \frac{2\pi}{\alpha z \lambda^2}\,e^{-\alpha z}\,e^{-\pi\left(\frac{2\pi}{\alpha z \lambda^2}\right)m^2}\ , \tag{12}$$

which means that the effect of the attenuation is in addition of a global attenuation $e^{-\alpha z}$ a smoothing by a gaussian function $e^{-\pi(m/\delta)^2}$. This smoothing effect is in general of negligible importance in practical imaging devices because δ remains very small in front of the focused spot size $\varepsilon = \lambda/\theta_0$, where θ_0 is the angular aperture of the focusing system. The equality $\varepsilon = \delta$ requires $\alpha z = 2\pi/\theta_0^2$ and a global attenuation of 55 dB for an aperture of $\theta_0 = 1$, which seems to be of consideration only for acoustical microscopy where both high angular apertures and high attenuations may be encountered {5}.

EXTENSION OF THE FRESNEL APPROXIMATION TO NON PARAXIAL BEAMS

It is well known that a focused beam oriented obliquely towards a plane interface separating two media of different velocities encounters strong aberrations. This situation interests the submarine detection of objects buried in sediments or, in N D E, the detection of defects not parallel to the surface. In this case, the Fourier analysis in planes Π_z parallel to the interface Π_0 remains very useful but the paraxial approximation is no longer valid. However, an equivalent approximation may be developed considering that the mean direction of the beam corresponds to plane wave modes associated to the same spatial frequency $\vec{f}_0$ in both media so that the spectrum $A_z(\vec{f})$ of a beam of reasonable aperture may be developed around $\vec{f}_0$, i.e. in terms of the spatial frequency $\vec{f}' = \vec{f} - \vec{f}_0$. The equivalent Fresnel approximation consists then in developing k_z in terms of $\vec{f}'$ and using the associated expressions for the propagation operators H_{oz} and h_{oz}, the dispersion equation associated to an isotropic medium gives through an elementary calculation, choosing $\vec{f}_0 = f_0 \vec{x}$:

$$K_z'^2 + 4\pi^2 f^2 = F(\omega) = K_{z_0}'^2 + 4\pi^2 f_0^2 \,,$$

$$K_z' \approx K_{z_0}' - 2\pi \left(\frac{2\pi f_0}{K_{z_0}}\right) f_x' - \pi \left(\frac{2\pi}{K_{z_0}'}\right)\left[\left(1+\frac{4\pi^2 f_0^2}{K_{z_0}'^2}\right) f_x'^2 + f_y'^2\right] \,, \quad (13)$$

or, for the classical isotropic non absorbing medium :

$$K_z \approx K_{z_0} - 2\pi \, tg\,\theta_0 \, f_x' - \pi \frac{\lambda}{\cos\theta_0} \left[\frac{f_x'^2}{\cos^2\theta_0} + f_y'^2 \right] \,, \quad (13')$$

where θ_0 is the incidence associated to the spatial frequency $\vec{f}_0$.

Obviously, the propagation operator $H_{oz} = e^{jk_z z}$, associated to the "Fresnel" expression of k_z, will deliver for a gaussian beam focused in Π_{z_0} ($z = z_0$) :

$$A_{z_0}\left(f_x',f_y'\right) = C_0 \; e^{-\pi\left(b_{ox}'' f_x'^2 + b_{oy}'' f_y'^2\right)}$$

$$U_{z_0}(x,y) = \frac{C_0}{\sqrt{b_{ox}'' b_{oy}''}} \; e^{-\pi\left(\frac{x^2}{b_{ox}''} + \frac{y^2}{b_{oy}''}\right)} \; e^{2j\pi f_0 x} \quad (14)$$

a solution in Π_z of the form :

$$A_z\left(f_x',f_y'\right) = C \, e^{-2j\pi a f_x'} \, e^{-j\pi\left(b_x f_x'^2 + b_y f_y'^2\right)} \,, \quad (15\text{-}a)$$

$$U_z(x,y) = \frac{C}{\sqrt{b_x b_y}} \; e^{-j\pi\left[\frac{(x-a)^2}{b_x} + \frac{y^2}{b_y}\right]} \cdot e^{2j\pi f_0 x}, \qquad (15\text{-}b)$$

where a, b_x, b_y are complex quantities, and $\sqrt{b_x b_y}$ the square root with a positive real part.

Noting $a, b_{x,y} = a', b'_{x,y} - ja'', b''_{x,y}$, one may check that the $b''_{x,y}$ must remain > 0, and that the focus F_0 corresponds to the nulling of the $b'_{x,y}$. If now this beam is focused through an interface located in Π_0 (z = 0) so that F_0 is a virtual focus, this formalism permits to attain easily the nature of the beam transmitted through the different material disposed in z > 0 and noted with the subscript 1 ;

$$A'_z(\vec{f}') = A_{z_0} \cdot H_{z_0 0} \cdot t \cdot H^1_{o z} , \qquad (16)$$

where $t(\vec{f}')$ is the coefficient of transmission for generalized plane modes associated to the signature $\vec{f}'$ in both materials.

A problem arises from the transmission term $t(f')$ which depends on the physical nature of the problem. For beams of low angular aperture, a reasonable approximation is to assume $t(\vec{f}) \cong t(\vec{f}_0)$ because $t(\vec{f})$ has slow variation in phase in front of $A_0(\vec{f}')$. We have also used matched developments of $t(\vec{f}')$ which have the mathematical form given in (15) for the A_z so that the gaussian structure of the beam remains preserved. This last operation affords negligible corrections on the numerical results obtained in the great majority of the problems that we have studied. In both cases, the solution $U^1_z(x,y)$ remains identical to the solution (15). An elementary calculation shows that the maximum radiation is located in Π_z at :

$$x = a' - a''(b'_x / b''_x) , \quad y = 0, \qquad (17)$$

and the width of the beam at 6 dB in the directions x or y is :

$$\ell_x = .47 \sqrt{b''_x + b'^2_x / b''_x} , \quad \ell_y = .47 \sqrt{b''_y + b'^2_y / b''_y} \quad (18)$$

THE CASE OF NON ABSORBING MEDIA

This case interests many optical or acoustical situations. In acoustics, an interesting example is in N D E the detection of defects oriented obliquely or even normally to the surface of a solid. Such a detection may be done using an immersion technique and a focusing transducer oriented obliquely towards the surface with an adequate angle of incidence, so that the transmitted beam of transverse (or longitudinal) waves has sufficient incidence to give enough reflection from the target, specularly or by edge diffraction. In the solid medium, the shear wave (of the SV kind) may be described through a vector potential $\vec{\psi}^1 = \psi^1\,\vec{y}$ which gives the velocity through the relation $V = \overline{\mathrm{Rot}\ \vec{\psi}^1}$ so that the velocity potential in water (first medium, $c = 1500$ m/s) is associated to the scalar ψ^1 in the second medium (1) (steel for example : $C_T^1 = 3200$ m/s) and the transfer coefficient $t(\vec{f})$ must be expressed in term of ψ^1/ϕ for a plane wave of signature $\vec{f}$. For longitudinal waves, the retained scalar is obviously the velocity potential ϕ^1 (in steel , $C_L^1 = 5900$ m/s).

In a first approach, it is reasonable to neglect the attenuation for both media, and referring to the expression (13') of k_z for non absorbing media, it may be checked that the solution (15) associated to a gaussian beam (14) is modified by the propagation operators H_{oz} and H_{oz}^1 according to (16) in keeping constant values for $b_{x,y}'' = b_{ox,y}''$. Then, from (18), it appears that the minimum width of the beam, i.e. focusing, is obtained in the x (or y) direction for a value of z such that b_x' (or b_y') = 0, and from (17), the focus point is located at the corresponding value of $x = a'(z)$. If we assume $t(\vec{f}) = t(\vec{f}_0) = t_0$, the solution A_z^1 given by (16) corresponds to the values of $a, b_{x,y}$ taking in account the expression (13') of k_z :

$$
\begin{cases}
a(z) = -z_o\,tg\,\theta_o + z\,tg\,\theta_o^1 \\[2mm]
b_x(z) = -\dfrac{\lambda z_o}{\cos^3\theta_o} + \dfrac{\lambda^1 z}{\cos^3\theta_o^1} - j\,b_{ox}'' \\[3mm]
b_y(z) = -\dfrac{\lambda z_o}{\cos\theta_o} + \dfrac{\lambda^1 z}{\cos\theta_o^1} - j\,b_{oy}''
\end{cases}
\qquad (19)
$$

There are two focal points F_x and F_y where the beam width is minimized in the x and y directions respectively. They are located on the mean refracted ray $x = a(z)$ at respective depths :

$$
\begin{cases}
F_x : \quad z_x = z_o\,\dfrac{\lambda}{\lambda^1}\,\dfrac{\cos^3\theta_o^1}{\cos^3\theta_o} , \\[3mm]
F_y : \quad z_y = z_o\,\dfrac{\lambda}{\lambda^1}\,\dfrac{\cos\theta_o^1}{\cos\theta_o} .
\end{cases}
\qquad (20)
$$

Taking in account that $f_0 = \sin\theta_0/\lambda = \sin\theta_0^1/\lambda^1$, it is easy to check that F_y is on Oz like F_0 and that the focal distances IF_x and IF_y are in the ratio :

$$
IF_x / IF_y = \left(\cos\theta_o^1 / \cos\theta_o\right)^2 ,
\qquad (21)
$$

ratio which is < 1 if C^1 > C. This aberration is well known and may be explained using a ray theory {6}.

Another interesting remark is that at the focuses F_x and F_y, the beam exhibits in the x and y directions respectively exactly the same gaussian variation as the virtual beam at F_0.

NUMERICAL EXPERIMENTS

In the proposed numerical applications, we point out the characteristics of the refracted acoustic field for two kinds of interfaces with different patterns of the incident beam.

Results are exposed through equiamplitude lines in the incident plane (X,Z) and in the perpendicular cylindrical (quasi planar) surface which contains the line of maxima levels. This second network is projected upon a vertical plane (Y,Z), and also upon the interface (X,Y) if the refracted mean angle is large. The corresponding 6 dB X and Y resolutions and the maximal pressure level are drawn as functions of depth. All distances are given in wave lengths λ in the first medium.

The first example concerns the interface water/steel with such an angular incidence that only shear waves are generated in metal. We assume a constant transmitting coefficient in those computations and attenuation is neglected, except when displaying the X Y resolutions and maximal level in Figs. 2-B and 3-B. In this case, $C/C^1 = 0.459$, $\rho/\rho^1 = 0.127$. We retain for the incident angle the value $\theta_0 = 18.94°$ (the critical angle $\theta_c = 27.33"$). 6 dB-half width at virtual focused points are $\ell_x = \ell_y = 2\lambda$.

The figures 2-A and 2-B illustrate the case of an incident gaussian beam which is focused at the depth $Z_x = Z_y = 200 \lambda$. The figures 3-A and 3-B show the effects fo the aberration correction obtained using a bifocused beam with $Z_x = 358 \lambda$ and $Z_y = 200 \lambda$ ($Z_x/Z_y = \cos^2 \theta_0/\cos^2 \theta_0^1$). In the displays 2-B and 3-B of levels and resolutions, dashed lines (b) and (c) show the influence of the absorption in the second medium for respective attenuations of 0.5 and 1 dB/λ^1. These results show that, as mentioned earlier for a normal incidence, the resolution is very weakly affected by the attenuation.

The second example recovers characteristics of a model water/ marine sediments (modelled as an absorbing liquid):

$C/C^1 = 0.754$; $\rho/\rho^1 = 0.667$; $\alpha^1 = 0.5$ dB/λ^1

We simulate a square projector with sides of 11.8 λ located 20 λ above the interface. Computations concern two different incident angles : 48° and 40° (the value of the critical angle is $\theta_c = 49.84°$). In each case, the depth of the focus F_y is settled so that lateral resolution keeps a maximal value at a depth of 20 λ. We obtain :

$Z_y = 36.4 \lambda$ with $\theta_0 = 48°$ ($\ell_y = 3.2 \lambda$)

and $Z_y = 18.3 \lambda$ with $\theta_0 = 40°$ ($\ell_y = 1.9 \lambda$)

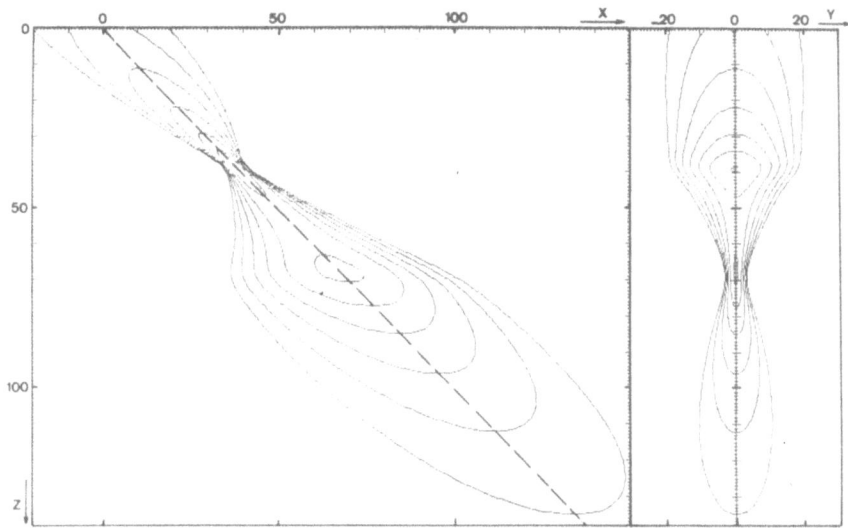

Figure 2-A - Equiamplitude lines : - 0.1, - 3, - 6, - 9, - 12 and
          - 15 dB ref. maxi pressure in the refracted field.

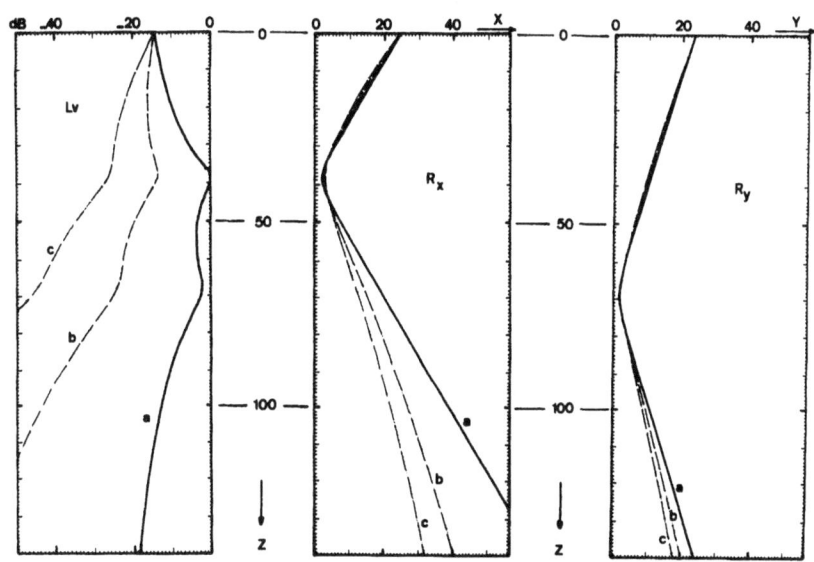

Figure 2-B - Maximal level (Lv) and X-Y resolutions (R_x and R_y)
          versus depth :
          (a) $\alpha^1 = 0$ dB/λ^1 ; (b) $\alpha^1 = 0,5$ dB/λ^1 ; (c) $\alpha^1 = 1$ dB/λ^1

                    INTERFACE WATER/STEEL
      $c/c^1 = 0.459$; $\rho/\rho^1 = 0.127$; $\theta_0 = 18.94°$
      Unique virtual focus : $Z_x = Z_y = 200\ \lambda$; $\ell_x = \ell_y = 2\ \lambda$

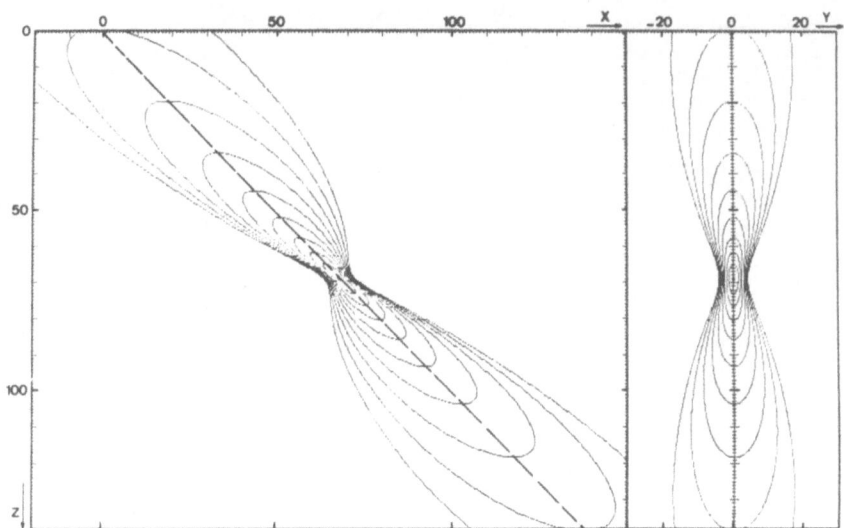

Figure 3-A - Equiamplitude lines : - 0.1, - 3, - 6, - 9, - 12 and
- 15 dB ref. maxi pressure in the refracted field.

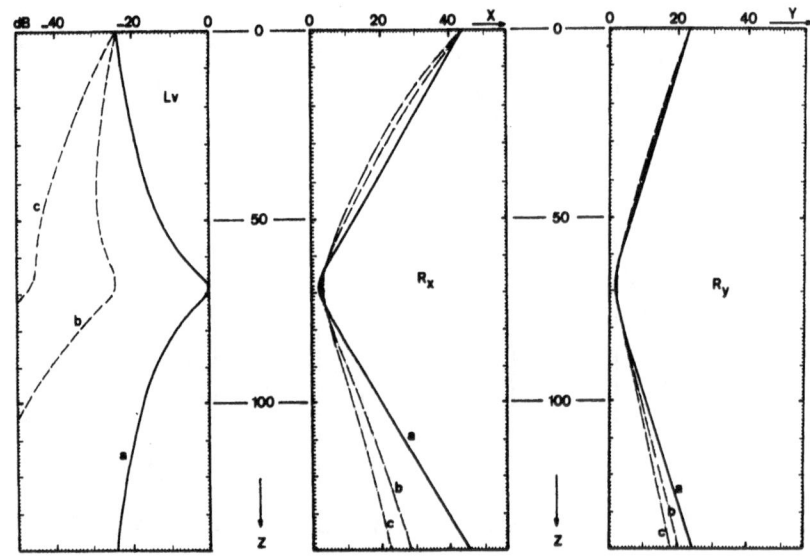

Figure 3-B - Maximal level (Lv) and X-Y resolutions (R_x and R_y)
versus depth :
(a) $\alpha^1 = 0$ dB/λ^1 ; (b) $\alpha^1 = 0,5$ dB/λ^1 ; (c) $\alpha^1 = 1$ dB/λ^1

INTERFACE WATER/STEEL

$c/c^1 = 0.459$; $\rho/\rho^1 = 0.127$; $\theta_0 = 18.94°$
Virtual focuses : $Z_x = 358\,\lambda$; $Z_y = 200\,\lambda$; $\ell_x = \ell_y = 2\,\lambda$

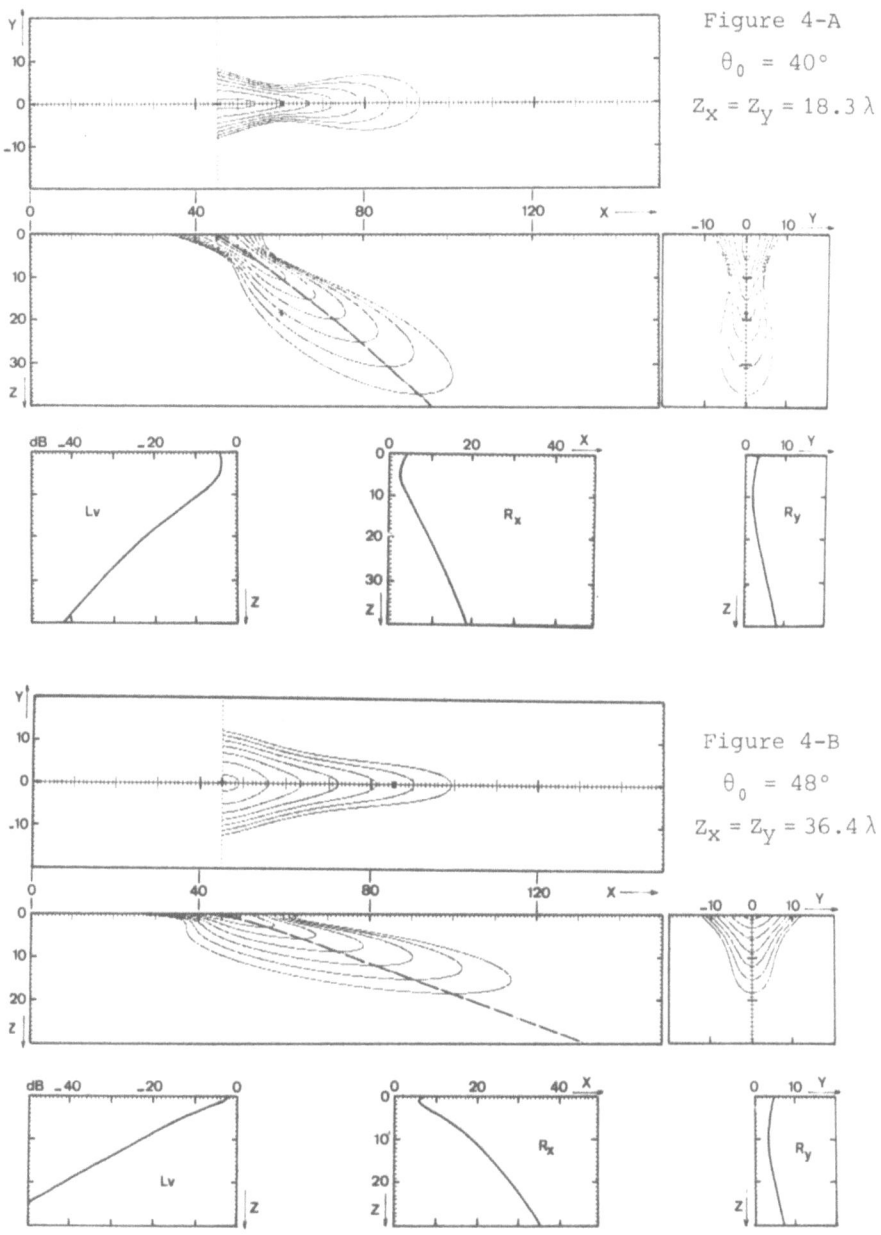

Figure 4-A

$\theta_0 = 40°$

$Z_X = Z_Y = 18.3 \lambda$

Figure 4-B

$\theta_0 = 48°$

$Z_X = Z_Y = 36.4 \lambda$

INTERFACE WATER/SEDIMENT : $c/c^1 = 0.764$; $\rho/\rho^1 = 0.667$; $\alpha^1 = 0.5 \, dB/\lambda^1$

- Transmitter : $11.8 \lambda \times 11.8 \lambda$ located 20λ above interface.
- Equiamplitude lines : $-1, -6, -12, -18, -24, -30$ and $-36 \, dB$ ref. maxi pressure in the refracted field.
- Maximal pressure level (Lv) and X-Y resolutions (R_X and R_Y) versus depth.

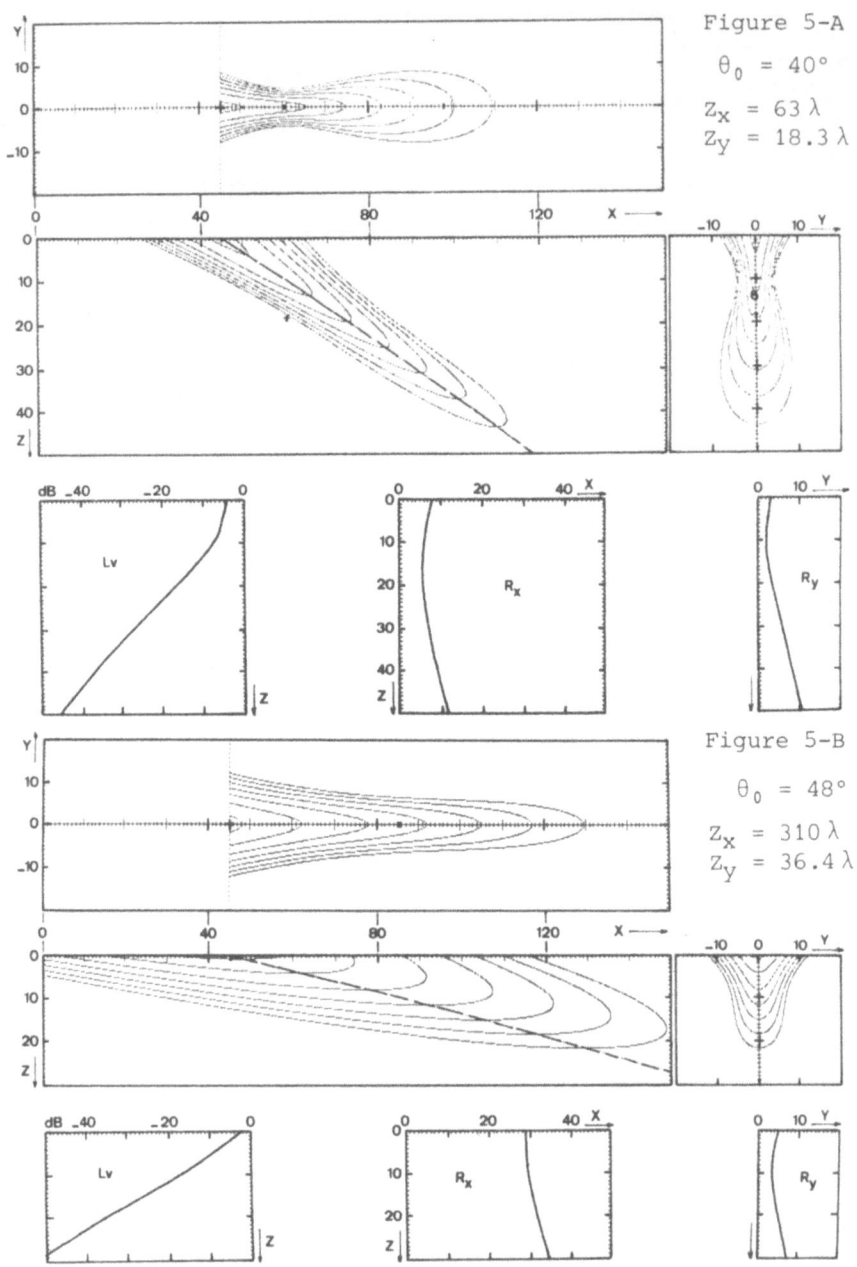

Figure 5-A

$\theta_0 = 40°$

$Z_X = 63\lambda$
$Z_y = 18.3\lambda$

Figure 5-B

$\theta_0 = 48°$

$Z_X = 310\lambda$
$Z_y = 36.4\lambda$

INTERFACE WATER/SEDIMENT : $c/c^1 = 0.764$; $\rho/\rho^1 = 0.667$; $\alpha^1 = 0.5\,dB/\lambda^1$

- Transmitter : $11.8\,\lambda \times 11.8\,\lambda$ located $20\,\lambda$ above interface.

- Equiamplitude lines : -1, -6, -12, -18, -24, -30 and $-36\,dB$
                         ref. maxi pressure in the refracted field.
- Maximal pressure level (Lv) and X-Y resolutions (R_X and R_y) versus depth.

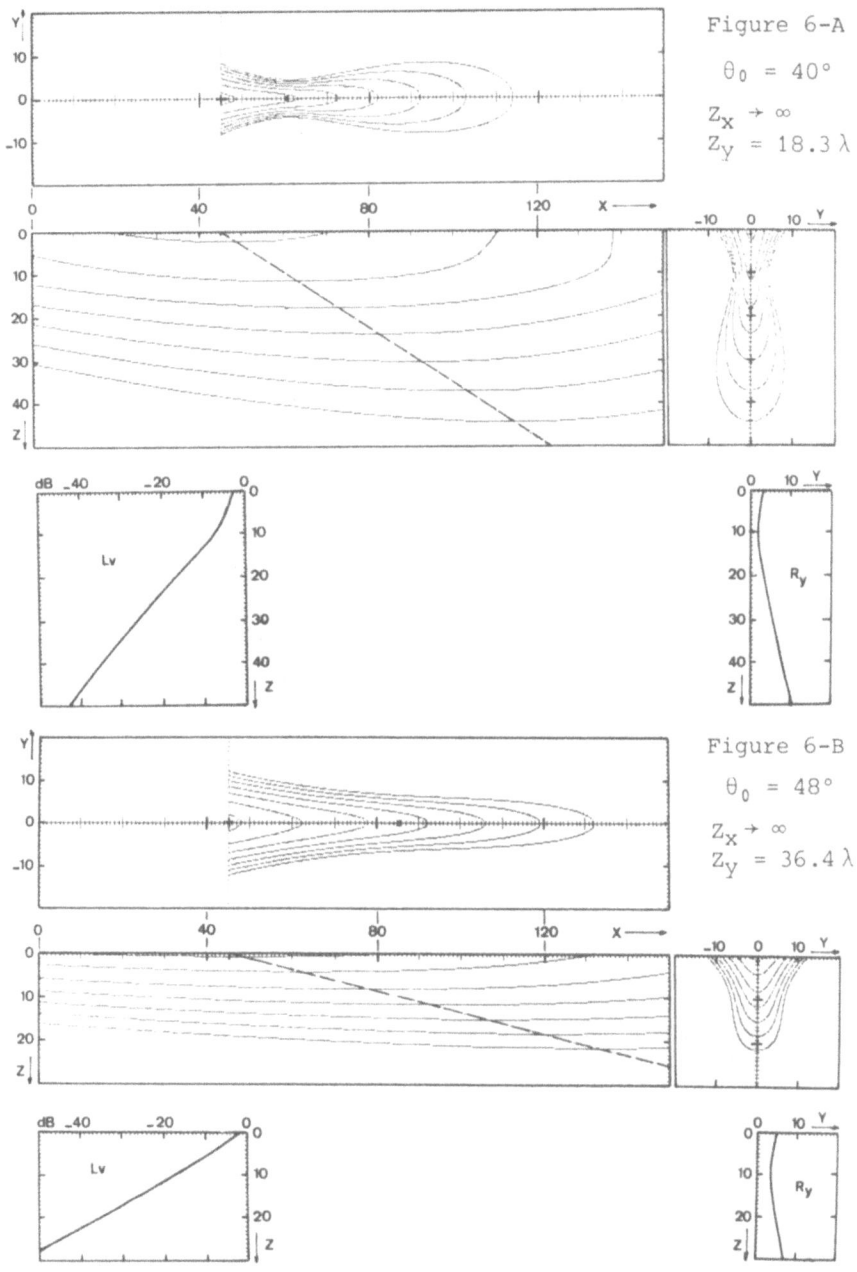

Figure 6-A

$\theta_0 = 40°$

$Z_x \to \infty$

$Z_y = 18.3\,\lambda$

Figure 6-B

$\theta_0 = 48°$

$Z_x \to \infty$

$Z_y = 36.4\,\lambda$

INTERFACE WATER/SEDIMENT : $c/c^1 = 0.764$; $\rho/\rho^1 = 0.667$; $\alpha^1 = 0.5\,dB/\lambda^1$

- Transmitter : $11.8\,\lambda \times 11.8\,\lambda$ located $20\,\lambda$ above interface.

- Equiamplitude lines : -1, -6, -12, -18, -24, -30 and -36 dB
                        ref. maxi pressure in the refracted field.

- Maximal pressure level (Lv) and Y resolution (R_y) versus depth.

So, we retain three choices for the location of F_X :

- a unique virtual focus $Z_X = Z_y$ (Fig. 4) :

- Z_X chosen so that acoustic pressure is maximal at a depth of 20λ
  (Fig. 5); then, $Z_X = 310\ \lambda$ when $\theta_0 = 48°$ ($\ell_X = 18.5\ \lambda$)
  and $Z_X = 63\ \lambda$ when $\theta_0 = 40°$ ($\ell_X = 4.1\ \lambda$)

- in the incident plane (X,Z), the beam is collimated (Fig. 6).

We can notice, from these results, that, in case of absorbing media, the nearer of the interface the virtual point F_X is, the more curved the maxima pressure line is, and the deeper its direction is oriented. On the other hand, maximizing the level at the depth of $20\ \lambda$ leads to a quasi-uniform field in the X direction.

CONCLUSION

We have extended the classical Fresnel approximation to beams propagating in absorbing media and according to an oblique axis, a theory which permits to attain aberrations encountered when focusing through a plane interface. N D E and submarine acoustical examples have been treated for gaussian beams which give the behaviour of any focused beam at least for the main lobe.

REFERENCES

{1} J.W. GOODMANN, "Introduction to Fourier Optics".
{2} K.V. MACKENZIE, "Reflection of Sound from Coastal Bottoms",
        J.A.S.A., 32 (2) pp. 221-231, 1960.
{3} P. ALAIS, "Effets de l'atténuation sur un rayonnement quelconque
        dans un milieu propagatif linéaire absorbant", C.R.Ac.Sc.
        Paris, t.282, Mars 1976, Série A, pp. 547-549.
{4} P. ALAIS and P.Y. HENNION, "Etude par une méthode de Fourier
        de l'interaction non linéaire de deux rayonnements acousti-
        ques dans un fluide absorbant, Cas particulier de l'émission
        paramétrique", Acustica, 43 (1), pp. 1-11, 1980.
{5} M. NOKOONAHAD, E.A. ASH, "Ultrasonic Focusing in Absorptive
        Fluides", Acoustical Imaging, Vol. 12, pp. 47-60, 1983.
{6} R.SAGLIO, A.C. PROT, A.M. TOUFFAIT, "Determination of defects
        characteristics using focused probes", Material Evaluation
        Janvier 1978.

MATERIAL CHARACTERIZATION BY ACOUSTIC LINE-FOCUS BEAM

Jun-ichi Kushibiki, Yasushi Matsumoto,
and Noriyoshi Chubachi

Department of Electrical Engineering
Faculty of Engineering, Tohoku University
Sendai 980, Japan

INTRODUCTION

The line-focus-beam acoustic microscope system has been successfully developed for the material characterization especially in the velocity measurements of leaky waves on the boundary of water/samples through the V(z) curve measurements.[1-4] V(z) curves also contain another information, i.e., attenuation of the relevant leaky waves. The material characterization would be completed by measuring two physical quantities, i.e., attenuation in addition to velocity. The wave attenuation is mainly caused by three important effects: 1) the water loading on samples, 2) the acoustic absorption, and 3) the structural scattering due to surface roughness, grains, pores and boundaries. The effects of the two quantities of leaky waves on V(z) curves have been already clarified by theoretical analysis based on field theory and numerical calculations for the acoustic line-focus beam.[5] It has been also shown that the attenuation could be estimated from the shape of V(z) curves by comparing experiments and theoretical calculations. However, no experimental procedure has been established so far for direct determination of the two quantities of both velocity and attenuation from V(z) curves.

In this paper, we propose a novel method of material characterization of determining both the velocity and the attenuation of leaky waves directly from V(z) curves obtained by the acoustic line-focus beam. In this method, a simple model is developed to represent transducer output V(z), which is approximately formed by a combination of ray and field theories. Experiments are demonstrated for a fused quartz sample using an acoustic line-focus-beam sapphire

lens of 1.0 mm radius at 226.3 MHz. This characterization method
is applied to quantitative estimation for two Mn-Zn ferrite samples
with different grain size, and the results are compared with acoustic
imaging information observed by the reflection type scanning acoustic microscope at a frequency of 440 MHz.

CHARACTERIZATION PRINCIPLE

Figure 1 shows schematically the cross-section geometry of the
acoustic line-focus beam for explaining the construction mechanism
of V(z) curves and for making the material characterization method
of directly determining both velocity and attenuation of leaky waves
on the water/sample boundary from measured V(z) curves. To develop
the direct method, we must necessarily represent the transducer
output V(z) with an approximated description as a function of relative distance z between a lens and a sample. Here, let us consider
a simple case of a single leaky SAW mode existing on the boundary.
This component (#1) contributes to an interference with a directly
reflected component (#0) from the sample in the characterization
region. The transducer output V(z) is expressed as a composition
of these two components. The principle of V(z) curve construction
is conceptually shown in Fig. 2. We now assume that the whole transducer output V(z) can be approximately represented by combining ray
theory for leaky SAW component and field theory for directly
reflected component as follows:

$$V(z)=V_I(z)+V_L(z) \tag{1}$$

where $V_L(z)$ is the characteristic lens response defined as the hypo-
thetical transducer output with respect to z, derived by field
theory, where the response for leaky SAW component (#1) is excluded.
The lens response $V_L(z)$ depends uniquely on the dimensions of the
acoustic line-focus-beam lens and the operating frequency, namely,
acoustic field distribution. The amplitude of $V_L(z)$ is muximum
when the sample is located at the focal point, because the transducer can receive most of acoustic waves reflected from the sample.
As the sample is moved towards the lens along the z-axis, wave components around the beam axis will be dominantly received by the
transducer, so that the amplitude abruptly decreases. The function
$V_I(z)$ is a function derived by ray theory, and it describes mathe-
matically the interference of two components: #0 and #1. A leaky
SAW component would be efficiently excited at the critical angle of
θ_{lsaw} on the water/sample boundary by the acoustic line-focus beam
with wide spatial-frequency components limited by the lens aperture
angle of θ_M. Considering different propagation paths for #0 and
#1 to be detected by the transducer, we can express the $V_I(z)$ as

$$V_I(z)=C \cdot ATT \cdot \exp(j(\xi z+\phi)) \ . \tag{2}$$

The quantity C is the amplitude constant, and ATT stands for the

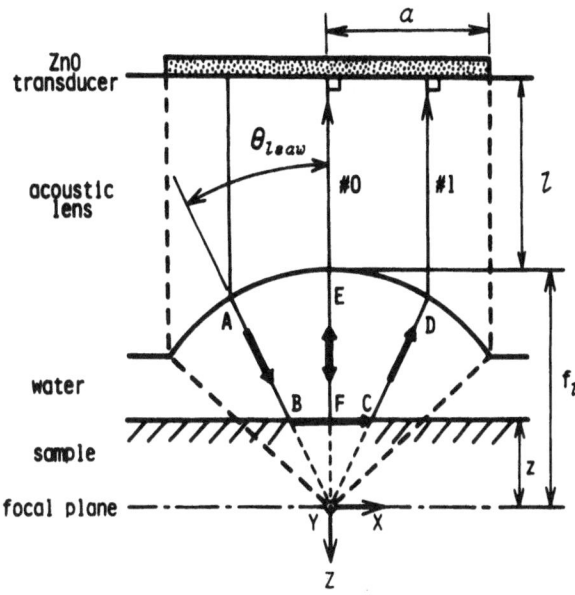

Fig. 1. Cross-section geometry of acoustic line-focus beam for
explaining the construction mechanism of V(z) curves and for
making the material characterization method.

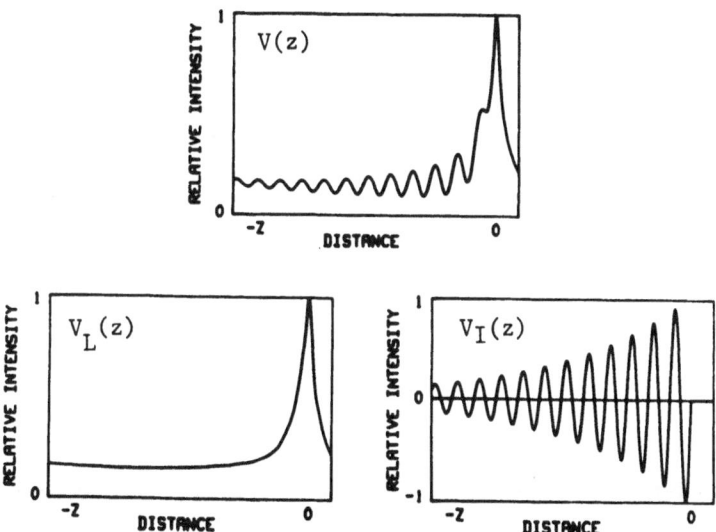

Fig. 2. Decomposition of V(z) curves.

attenuation of the function. In the phase term, ξ is the relative phase change between #0 and #1 per unit translation distance z, and ϕ is the initial phase difference between #0 and #1 at the focal plane. The phase variation of 2π corresponds to the dip interval Δz in V(z) curves. The term of ATT is then given by

$$ATT=\exp(-2\alpha_w t(z)) \cdot \exp(2\gamma z \tan\theta_{lsaw}) \ , \tag{3}$$

where

$$\gamma = 2\pi f \alpha / v_{lsaw} \ , \tag{4}$$

$$\alpha = \alpha_{lsaw} + \alpha_b + \alpha_s \ . \tag{5}$$

The quantity f is the acoustic frequency, α_w is the attenuation coefficient in water of $\alpha_w/f^2 = 25.3 \times 10^{-17}$ s^2/cm at 20°C [6], and $t(z) = \overline{AB}$. The quantities α_{lsaw}, α_b and α_s are the normalized attenuation factors due to the water loading effect, the acoustic bulk absorption effect, and the structural scattering effect, respectively. The quantity v_{lsaw} is the phase velocity for leaky SAW, and it is determined from the dip interval Δz in measured V(z) curves, using the following conventional equation[2]:

$$v_{lsaw} = v_w / (1 - (1 - v_w/2f\Delta z)^2)^{1/2} \ , \tag{6}$$

where v_w is the longitudinal velocity of water of 1483 m/s at 20°C.

Thus, the theory developed above states that extraction of the interference function $V_I(z)$ from experimental V(z) curves makes it possible to determine directly two quantities of velocity and attenuation of leaky SAW for samples, using Eqs.(2)-(6): the velocity is determined from the dip interval of interference and the attenuation from the gradient of the interference amplitude.

The practical procedure in experiments is given as follows:
(a) record of V(z) curve
(b) synthesis of $V_L(z)$ curve from the measured V(z) curve using digital-filtering techniques
(c) extraction of $V_I(z)$ from the subtraction of V(z)$-V_L(z)$
(d) determination of v_{lsaw} from $V_I(z)$
(e) measurement of ATT from $V_I(z)$
(f) determination of γ and α from ATT.

EXPERIMENTS

In order to verify the effectiveness of the proposed method for material characterization described above, experiments are performed for a sample of fused quartz (SiO_2) plate with an optically polished surface, using a line-focus-beam sapphire lens at a frequency of 226.3 MHz. The acoustic lens parameters, namely, curvature radius R, aperture angle θ_M, distance from transducer to lens surface l, and transducer width $2a$, are R=1.0 mm, θ_M=60°, l=12 mm, and $2a$=1.73 mm, respectively. A chalcogenide glass film is deposited on the lens surface as an acoustic antireflection coating with one quarter-wavelength.[7] The acoustic field formed in water by the line-focus-beam lens has linearly focused distribution with -3 dB width of about one acoustic wavelength (8 μm) along the x-axis and about 1 mm along the y-axis shown in Fig. 1, as reported in Ref. 7. The V(z) curves are recorded into a wave-memorizer synchronized with the translation of the sample drived by a stepping-motor along the z-axis with a resolution of 0.1 μm, and then they are processed by a computer in the way described above.

Figure 3 (a) and (b) are the measured V(z) curve and the $V_I(z)$ curve extracted from the V(z) curve, respectively. The velocity of leaky SAW is determined to be 3432 m/s with a dip interval of Δz= 33.1 μm. From the envelope of $V_I(z)$ decreasing at the rate of 103 dB/mm in Fig. 3 (b), we can determine the normalized attenuation factor of α=3.64x10^{-2}, substituting the relevant values into Eqs. (2)-(6). Experimental results are in good agreement with the values calculated from the bulk constants of SiO_2 with differences of about 0.1 % to phase velocity and about 5 % to attenuation factor, as shown in Table 1.

The attenuation measured above can be mostly attributed to the water loading effect. This can be easily understood by the facts: 1) the scattering attenuation factor α_s due to surface roughness, internal defects, etc. is considered to be negligible small, and 2) the bulk absorption attenuation factor α_b mainly caused by the viscous effect is theoretically estimated to be at most 0.1 % of the attenuation factor α_{lsaw} due to the acoustic water loading effect. In this calculation of α_b, the attenuation coefficients, α_l/f^2= 1.74x10^{-18} s^2/cm for longitudinal wave and α_s/f^2=2.08x10^{-18} s^2/cm for shear wave presented in Ref. 8 are employed.

Table 1. Comparison of measured results of velocity
and attenuation for leaky SAW with calculated
results for water/SiO_2 boundary.

Phase velocity v_{lsaw}(m/s)		Normalized attenuation factor α_{lsaw}	
Measured	Calculated	Measured	Calculated
3432	3430	3.64x10^{-2}	3.82x10^{-2}

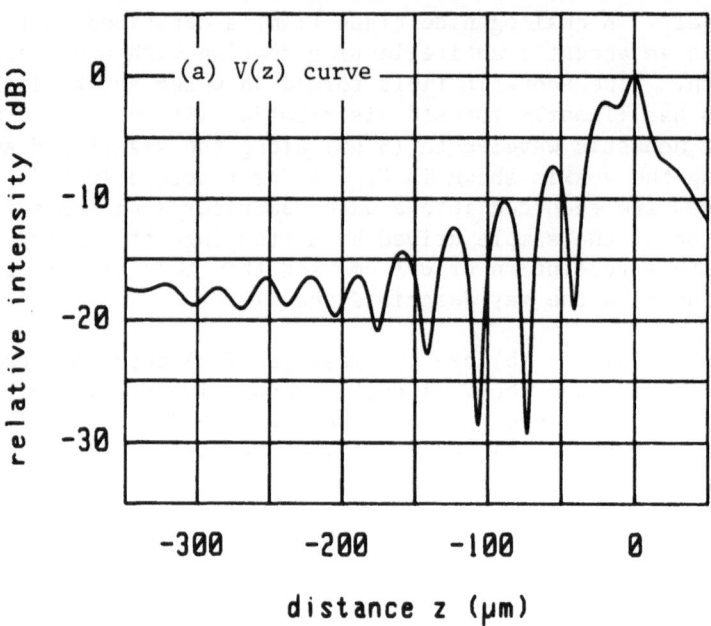

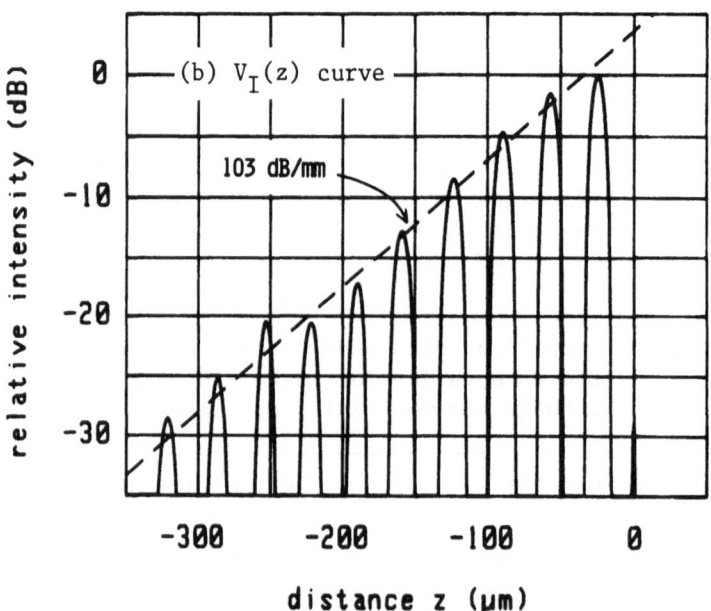

Fig. 3. Measured V(z) curves for water/SiO$_2$ boundary at 226.3 MHz.

COMBINATION WITH ACOUSTICAL IMAGING

An attempt is made to compare the quantitative information of acoustic properties obtained by the method mentioned above using the line-focus beam with the acoustical imaging information obtained by a scanning acoustic microscope using a conventional point-focus beam. Two Mn-Zn ferrite materials with different grain sizes are employed in this study. These ferrites are materials for magnetic heads with average grain sizes of nominal 8 μm (Grain 'S' sample) and 50 μm (Grain 'L' sample), respectively. The surfaces are optically polished, and no significant variations in contrast are observed on the surfaces by an optical microscope.

Figure 4 (a) and (b) show the acoustic images for each sample observed by an acoustic microscope with a point-focus beam operating at a frequency of 440 MHz. This microscope has a resolution of about 2-3 μm. For a Grain 'L' sample, we can clearly recognize individual grains and boundaries with grains of 10-100 μm in size, while for a Grain 'S' sample, we can see smaller grains of several μm, although the image is blurred. It is also estimated that almost the same grains for each sample are distributed evenly in the depth direction beneath the surfaces.

Next, quantitative characterization is performed to these two samples. Figure 5 (a) and (b) show the typical V(z) curves for each sample. The line-focus beam has uniform field distribution of about 1 mm along the line direction in water, much larger than the grain sizes for both samples, so that mean propagation characteristics of leaky SAW can be measured from the V(z) curves. In the V(z) curve for Grain 'L' sample, great attenuation is observed in the amplitude of the interference waveform as compared with the V(z) curve for Grain 'S' sample. Both velocity and attenuation from each V(z) curve are given in Table 2. The attenuation value for Grain 'L' sample is about 60 % larger than that for Grain 'S', corresponding to the grain sizes. As the surfaces are very smooth for both

Table 2. Acoustic properties of leaky SAWs obtained by the acoustic line-focus beam for Mn-Zn ferrite samples shown in Fig. 4.

Sample	Grain size (μm)	Phase velocity v_{lsaw} (m/s)	Normalized attenuation factor α_{lsaw}
Grain 'S'	8	3317	3.68×10^{-2}
Grain 'L'	50	3299	5.87×10^{-2}

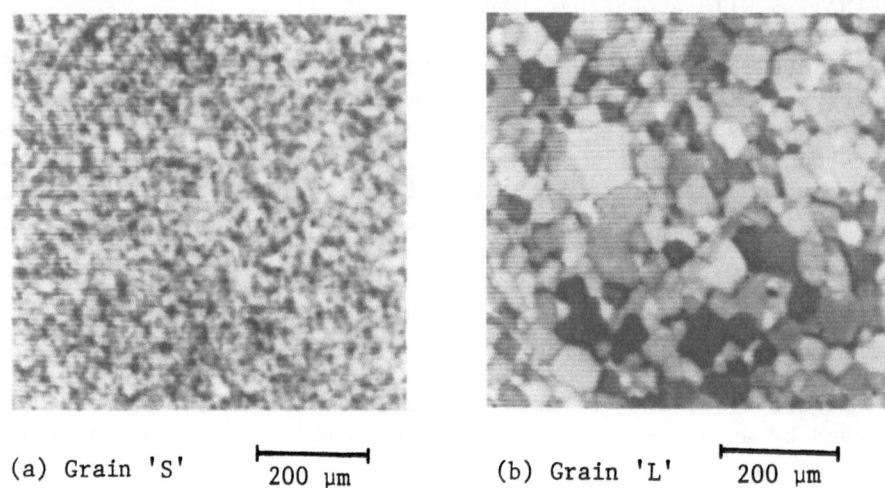

(a) Grain 'S'   ⊢——————⊣
                   200 μm

(b) Grain 'L'   ⊢——————⊣
                   200 μm

Fig. 4. Acoustic images of Mn-Zn ferrite samples observed by
mechanically scanned acoustic microscope on the focal
plane at 440 MHz.

(a) Grain 'S'   (b) Grain 'L'

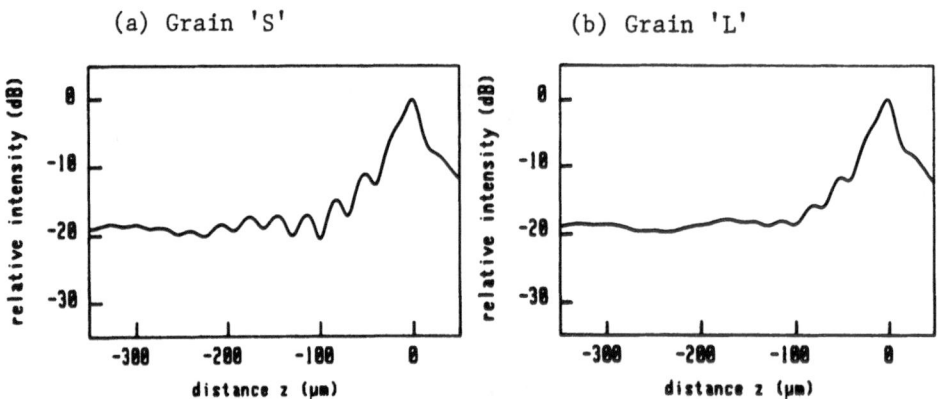

Fig. 5. Measured V(z) curves for Mn-Zn ferrite samples shown
in Fig. 4 at 226.3 MHz.

samples, the scattering occurs beneath the surfaces, especially at
the grain boundaries. The wavelengths of leaky waves for each
sample are calculated to be almost same, namely about 15 μm, using
the values of phase velocity measured at 226.3 MHz. In the case of
Grain 'S' sample, the average grain size is less than the wavelength, where the scattering is in a range of the intermediate
scattering.[9] On the other hand, in the case of Grain 'L' sample,
the average grain size is much larger than the wavelength, so that
reflection and refraction of leaky waves dominantly cause the larger
scattering loss.

CONCLUSION

    This paper has described a novel method for the quantitative
measurement of acoustic properties by means of a line-focus-beam
acoustic microscope by using a simple model derived from a significant combination of ray and field theories. This method enables us
to determine both velocity and attenuation of leaky waves directly
from measured V(z) curves, including the anisotropic properties of
materials. Experiments on the determination of acoustic properties
by this method have been performed for a fused quartz with sufficient
accuracy. Further, an attempt has been demonstrated for two Mn-Zn
ferrite samples with different grain sizes to quantitatively estimate the structural variation of grains by the present characterization method as compared with the acoustic images observed by a
mechanically scanned acoustic microscope. It has been satisfactorily proved that the difference of grain sizes corresponds to that
of attenuation factors of leaky SAWs propagating on the boundary.
The characterization method proposed here is expected to take an
important role for the application of a point-focus-beam acoustic
microscope to be developed widely in the field of material characterization.

ACKNOWLEDGMENTS

    The authors are very grateful to T. Sannomiya for his technical
assistance on experimental instruments, and to K. Horii and
H. Maehara for their helpful discussions and technical assistance
on the experiments. They wish to express their thanks to
M. Sugimura, Matsushita Electric Industrial Co. Ltd., and U. Kato,
Tohoku Metal Industries, Ltd., for the supply of Mn-Zn ferrite samples.
This work was supported in part by the Research Grant-in-Aids from
the Japan Ministry of Education, Science & Culture, and also the
Toray Science and Technology Grants, Japan.

REFERENCES

1. J. Kushibiki, A. Ohkubo, and N. Chubachi, Anisotropy detection
   in sapphire by acoustic microscope using line-focus beam,
   Electron. Lett. 17:534 (1981).
2. J. Kushibiki, A. Ohkubo, and N. Chubachi, Acoustic anisotropy
   detection of materials by acoustic microscope using line-focus
   beam, IEEE Ultrasonics Symp. Proc., pp.552-556 (1981).
3. J. Kushibiki, A. Ohkubo, and N. Chubachi, Material characterization by acoustic microscope with line-focus beam, in: "Acoustical
   Imaging, Vol. 12" E. A. Ash and C. R. Hill, eds.,
   Plenum Press, London (1982).

4. J. Kushibiki, K. Horii, and N. Chubachi, Velocity measurement of multiple leaky waves on germanium by line-focus-beam acoustic microscope using FFT, Electron. Lett. 19:404 (1983).
5. J. Kushibiki, A. Ohkubo, and N. Chubachi, Theoretical analysis of V(z) curves measured by acoustic line-focus beam, IEEE Ultrasonics Symp. Proc., pp.623-628 (1982).
6. J. M. M. Pinkerton, The absorption of ultrasonic waves in liquids and its relation to molecular constitution, Proc. Phys. Soc. B20:129 (1949).
7. J. Kushibiki, A. Ohkubo, and N. Chubachi, Linearly focused acoustic beams for acoustic microscopy, Electron. Lett. 17:520 (1981).
8. B. A. Auld, Elastic properties of solids, in: "Acoustic Fields and Waves in Solids, Vol. I" B. A. Auld, ed., John Wiley & Sons, New York (1973).
9. E. P. Papadakis, Ultrasonic attenuation caused by scattering in polycrystalline media, in: "Physical Acoustics, Vol. IV Part B" W. P. Mason, ed., Academic Press, New York (1968).

ULTRASONIC VELOCITY SPATIAL DISTRIBUTION ANALYSIS OF BIOLOGICAL

MATERIALS WITH THE SCANNING LASER ACOUSTIC MICROSCOPE

P.M. Embree, S.G. Foster, G. Bright and
W.D. O'Brien, Jr.

Department of Electrical Engineering, Bioacoustics
Research Laboratory, University of Illinois, 1406 West
Green Street, Urbana, Illinois 61801   USA

The fundamental examination of biological tissue with ultrasound can lead to important diagnostic capabilities. In order to quantify tissue characteristics with ultrasound, the ultrasonic propagation properties of normal and pathological tissues must be characterized and cataloged. An important ultrasonic property is the speed of sound for characterizing tissue [1]. In this work, the Scanning Laser Acoustic Microscope (SLAM) is used to measure the spatial variation of the speed of sound in tissue, thereby providing a quantitative ultrasonic parameter for tissue characterization.

The SLAM (Sonomicroscope 100, Sonoscan, Inc.) provides three different television type images as shown in Figure 1. A laser scans the lower surface of the coverslip in order to detect mechanical disturbances induced by the 100 MHz ultrasonic energy which has passed through the specimen from below. The optical (laser scan) transmission image (Figure 1a) allows the operator to position the sample in the center of the 2 mm x 3 mm field of view. The acoustic image (Figure 1b) shows the amount of ultrasound energy passing through the sample. This signal is proportional to the envelope of the laser detector output. In this image, dark areas correspond to high areas of ultrasonic attenuation and light areas to low attenuation areas. The third image (Figure 1c), the interference image, is produced by electronically mixing the laser detector output with a 100 MHz reference signal.

The interference image consists of approximately 39 vertical light and dark bands which represent locations of constant phase contours of the ultrasonic wave after it has traversed through the

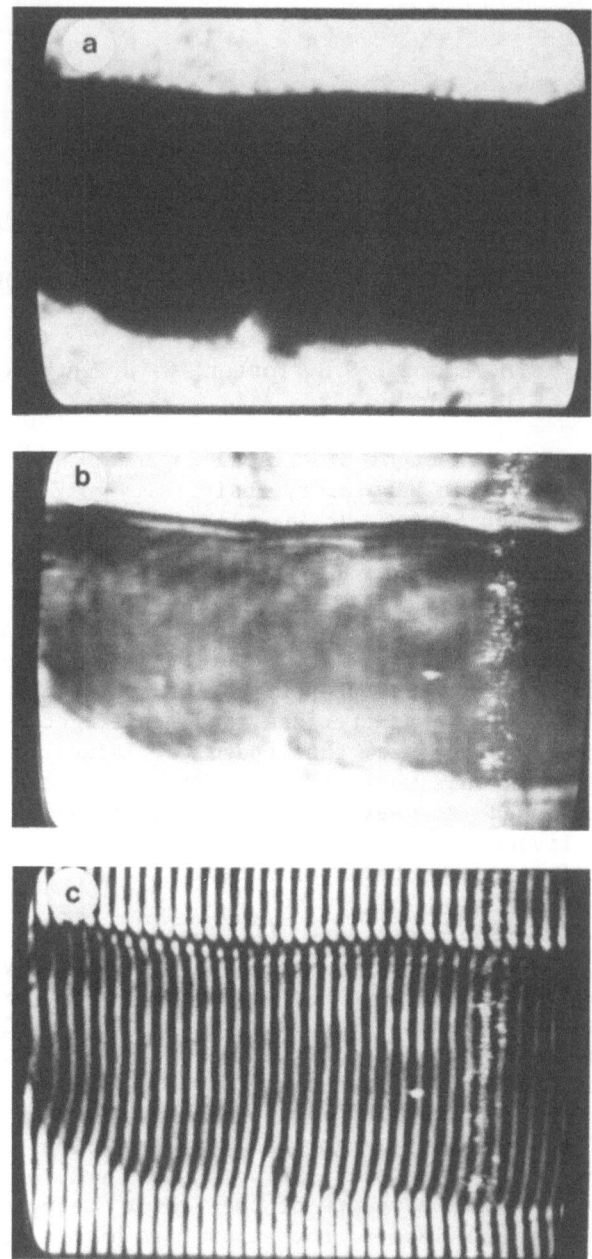

Figure 1. Photographs from the SLAM monitors of a 550 μm thick
           mouse liver specimen. (a) Optial Image, (b) Acoustic
           Image, and (c) Interference Image.

specimen. In saline solution (the normal coupling medium for tissue samples) the interference lines are straight and equally spaced. When a slice of tissue (usually 400 to 800 μm thick) is placed in the saline solution, the interference lines shift to the right at the saline-tissue interface indicating the tissue has a higher speed of sound. In an area of tissue where the thickness is constant, the interference lines appearance is somewhat corrigated, that is, they do not appear straight as in the very homogeneous coupling medium. This could represent a microscopic index of refraction gradient in tissue and may represent a source of ultrasonic back scattering for clinical B-scan imaging systems. To quantify this gradient, the interference lines are subjected to an automated analysis technique which is described herein.

THE DATA ACQUISITION SYSTEM

The interference image produced by the SLAM is digitized by a data acquisition system (Figure 2) that was designed and fabricated at the Bioacoustics Research Laboratory [2]. This system consists of a video amplifier, filter, analog to digital converter and high speed buffer memory which interface the SLAM to the Perkin-Elmer 7/32 32 bit mini-computer. The video signal is monitored on an oscilliscope to assure that the signal is using the full dynamic range of the A/D converter.

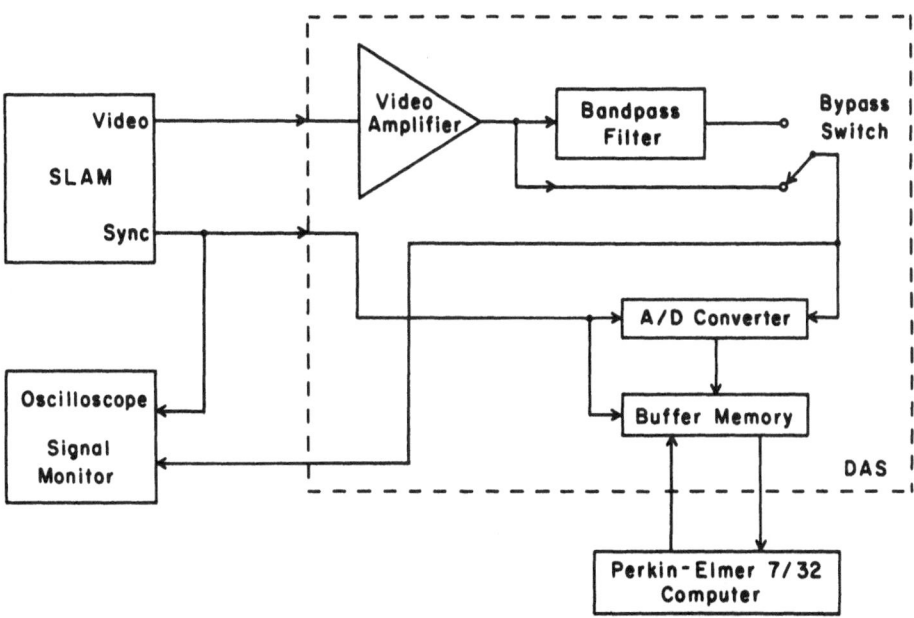

Figure 2. Block diagram of data aquisition system used to digitize the SLAM images.

Since the interference image consists of about 39 lines, the video signal is relatively narrow-band for purposes of extracting the interference images. The video portion duration of a raster line is 52.5 μs so the average time between signal peaks is 1.35 μs or an average frequency of 740 kHz. Thus, the signal to noise ratio of the interference image can be improved significantly by the addition of a bandpass filter. A sixth order Chebyshev filter was designed for this purpose. The passband of this filter is between 700 and 900 kHz with a $\pm$ 0.5 dB ripple. The addition of this filter has allowed the speed of sound in tissue specimens with higher attentuation to be measured more accurately.

One raster line (of which there are a total of 482) at a time is digitized at 28.6 MHz (1514 data points per raster line), and stored in the buffer memory (1024 x 16 bits, 35 ns access time). The data are then transferred into the mini-computer memory. This operation is repeated for all 482 lines. Thus, the interference image consists of 729,748 picture elements.

Several programs were developed to enhance the interference image. One program digitizes the raster line data, averages it n times (selectable between 1-16 times), stores the data on magnetic tape, and outputs a printer (Printonix model 300) representation of the interference image, a typical example of which is shown in Figure 3a. The averaging is accomplished by using a running average algorithm which was developed to compensate for slight and somewhat periodic horizontal jitter of the images which is produced by the SLAM.

It would be desirable to store the whole image in memory and provide for frame averaging but 1.5 megabytes of memory would be required. The present computer system has 300 kilobytes. Thus, the averaging must be performed on a block by block basis. The number of raster lines in the block is the same as the desired number of averages. The block of lines are arranged in order of position in the image. The top line of the block is averaged n-1 times, the next raster line is averaged n-2 times, and so on, until the next to the last line in the block is averaged once. The program then digitizes each line once more, adding these to the previous lines. The top line is now averaged n times and is transfered to magnetic tape. The start of the block now moves down one raster line and continues the averaging process until all 482 raster lines have been stored on magnetic tape.

By increasing n, the quality of the resulting image is improved. Unfortunately, the total digitization time is increased. Approximately 20 minutes are required to digitized the data and store a single image on magnetic tape, when four averages are taken. For eight averages the sample time is increased to 25 minutes.

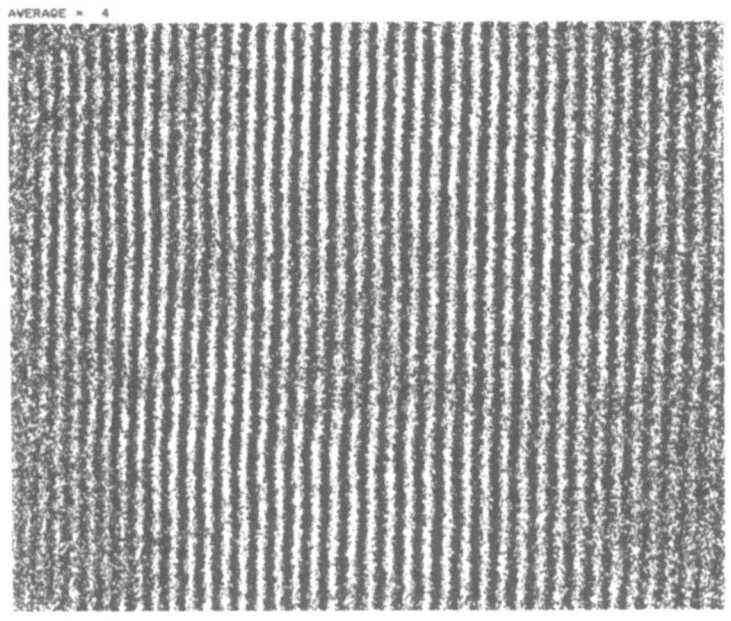

(a)

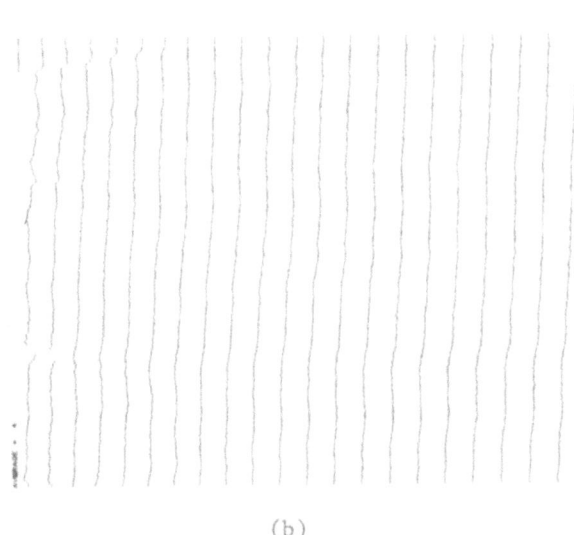

(b)

Figure 3. Enhanced interference images of a blank field in which the medium was water, that is, no biological tissue specimen was present. The image data were averaged 4 times. (a) The upper image is histogram equalized to improve its contrast for the line printer. (b) The lower image has been enhanced by a correlator receiver technique.

The program then enhances the interference image with a correlated receiver algorithm which compares the waveform of each raster line to that of a reference waveform stored in memory, detects a relative maximum in the correlated result and defines the position of the interference line. The reference waveform consists of 17 data points and was obtained from a saline solution image averaged a large number of times. The enhanced interference image consists of interference lines which are each one data point in width. Each raster line now consists of about 39 data points. Figure 3b shows the results of this program. An important characteristic of this image is that it is represented by a more manageable 482 x 39 element array.

CALCULATION OF THE SPEED OF SOUND

The calculation of the specimen's speed of sound, c_x, from the interference line image can be performed by using: [3]

$$c_x = \frac{c_o}{\sin\theta_o} \left\{ \tan^{-1}\left[\frac{1}{(1/\tan\theta_o) - (N\lambda_o/T \sin\theta_o)} \right] \right\} \qquad (1)$$

where c_o is the speed of sound in the reference medium, λ_o is the wavelength of sound in that same medium, θ_o is the angle of the beam from the normal in the reference medium, T is the thickness of the specimen, and N is the normalized lateral fringe shift (N=ab/ac - see Figure 4). Using Snell's Law, θ_o can be determined for the reference medium:

$$\theta_o = \sin^{-1}\left[\frac{c_o}{c_s} \sin\theta_s \right] \qquad (2)$$

where c_s is the speed of sound in the fused silica stage (5968 m/s) and θ_s is the angle at which the generated sound waves travel through the stage with respect to the normal ($\theta_s = 45^o$). In this project, the materials being examined are biological specimens; therefore, saline solution (0.9% NaCl) is the reference medium and is isotonic with the specimens. Saline has a known speed of 1507 m/s at a temperature of 22°C, hence $\theta_o = 10.2^o$.

The previous method employed [1,3] to determine the speed of sound in a specimen was performed manually by assessing the location of lines a, b, and c (see Figure 4) to calculate the normalized lateral fringe shift, N = ab/ac. For this particular method the operator had to determine the best fit of the fringe lines and choose points from which to measure the distances ab and ac. This procedure was quite time consuming and provided only a few velocity values for the specimen. The computer program that has been developed eliminates the need to do this manually,

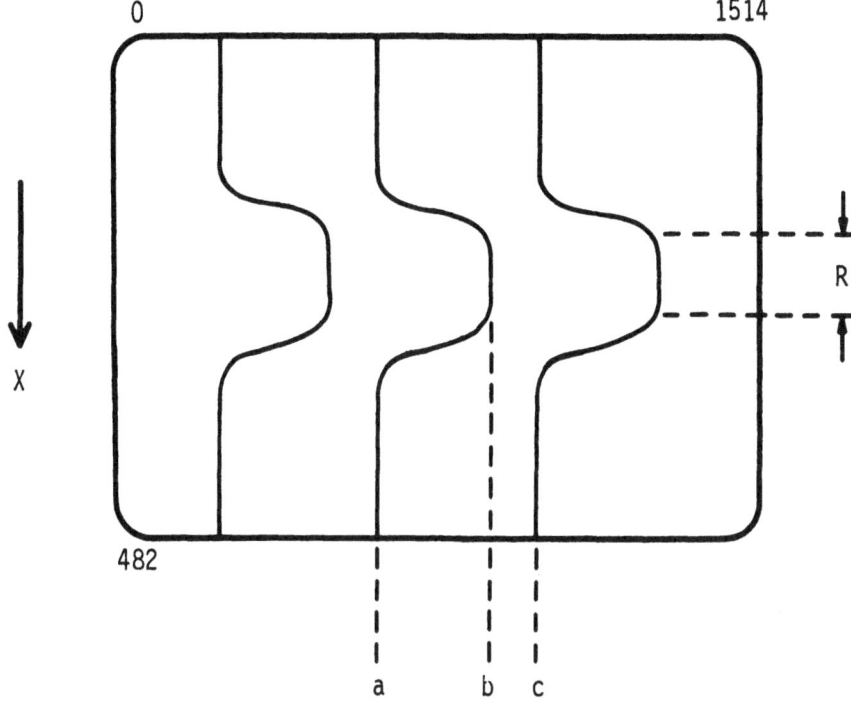

Figure 4. Schematic representation of the fringe shift,
           N = ab/ac.

increases the area of tissue from which the speed of sound can be
calculated, and involves minimal operator interaction.

As can be seen in Figure 3b, the enhanced interference lines
are not always continuous and can have a significant amount of
jitter. In order to automate the speed of sound determination, it
is necessary for an algorithm to identify the discrete
interference lines. Thus, a computer program was written to
follow the interference lines and smooth them by curve fitting.
If a line is not continuous (i.e., a point is missing) then the
program adds a point at a position determined by a linear
interpolation based on the nearest 10 points. The program informs
the user how many points have been added in which line
(interference lines are numbered sequentially from left to right).
The resulting continuous interference lines are then smoothed by
a sliding 17 point quadratic curve fitting algorithm which can be
considered a low pass filtering operation.

Two adjacent interference lines are used to determine the
distance ac provided that the specimen is placed on the microscope
stage within the boundaries x = 51 to x = 432 (see Figure 4). The

array for a single interference line consists of 482 data points. The top and bottom 50 points (100 points total) of each pair of interference lines are used to determine the location of lines a and c. The best fit straight lines for a and c are found by using a least squares linear curve fit. If lines a and c are perfectly vertical in orientation, the slope of these lines will be zero. The y intercept of these lines will vary from 0 to 1514, depending upon the location of the interference lines. With the equations for lines a and c defined, the distance ac is calculated for each point x = 1 to 50 and x = 433 to 482 and the average ac value determined is used for the remainder of the calculations of the speed of sound for that specimen.

The width of region R, (see Figure 4) where the distance ab is calculated, is a variable under operator control. The rationale for this is that not all of the data in the region between raster lines 51 and 432 may be useful because (1) the specimen may not occupy the entire region between raster lines 51 and 432, (2) the specimen may not be uniformly thick over this region because of edge effects and (3) it may be necessary to avoid certain inhomogeneities in the tissues, such as blood vessels. All of these can be assessed by the acoustic microscopist from the images. The horizontal distance ab is calculated R times and N is determined using each of these ab values and the average ac value previously calculated for that area of the specimen. Given the sample thickness, the speed of sound data set (R total values) is determined from equation 1.

Various statistical parameters are calculated from each speed of sound data set, namely, the standard deviation, mean, mode, median, skewness and kurtosis.

The standard deviation is defined as,

$$S = \sqrt{\frac{1}{R} \sum_{j=1}^{R} (x_j - \bar{x})^2} \qquad (3)$$

where $\bar{x}$ is the mean value, R is the number of samples, x_j is the jth speed value along the region R. The standard deviation is a measure of the concentration of values around the mean.

The mode is that value which occurs more frequently than any other value; graphically it appears as the peak of a distribution. Of a set of numbers arranged in order to magnitude, the median is the value of the exact middle position of the set. In this program, to calculate the median, the data set length is always chosen to be odd, thus requiring only one algorithm.

Skewness is an indication of the "sidedness" of a distribution with respect to its mean. By definition,

$$SKEW = \frac{3(\bar{x} - median)}{S} \tag{4}$$

The distribution is said to be skewed right and SKEW will be positive if more values of the data set are to the right of the mean. Similarly if SKEW is negative the distribution is shewed left.

Kurtosis is a measure of the shape of distribution which reveals the degree of peakedness of the distribution as compared to a normal distribution. Kurtosis is defined by a dimensionless 4th moment around the mean or,

$$KURTOSIS = \frac{\frac{1}{R} \sum_{j=1}^{R} (x_j - \bar{x})^4}{S^4} \tag{5}$$

For a normal (Gaussian) distribution KURTOSIS would be zero. A distribution is said to be leptokurtic (spikey, very peaked) if KURTOSIS is greater than zero and platykurtic (flat) if less than zero.

RESULTS AND DISCUSSION

A white, ICR female mouse (Harlan Sprague Darley) approximately 6-7 months old was sacrificed by means of spinal cord dislocation so as to not introduce any drugs into the tissues. The liver was quickly excised (within 5 minutes post mortem) and immediately placed in isotonic saline solution. The sample is then trimmed with a razor blade into a rectangular piece suitable for viewing with the SLAM. The speed of sound can only be determined by this technique if there is no discontinuity of the interference line between the saline and tissue. Thus, the edges of the specimen are bevelled.

The sample is then placed in the center of the microscope slide, on the plastic sheet, and surrounded by a spacer which has a thickness slightly greater than the sample. The spacer prevents the sample from being crushed or distorted by the coverslip, insures a constant distance between the slide and the coverslip, and keeps the coverslip level with respect to the stage. The area surrounding the tissue is filled with saline, which serves as the reference media.

A typical output of a program which calculates all of the previously described speed of sound statistical data is shown in Figure 5 for a 460 µm thick, fresh liver specimen which was averaged 4 times. The output is shown for the 22nd and 23rd interference lines.

The slopes, y intercepts and x intercepts for lines a and c (see Figure 4) of these two interference lines are listed. Figure 5 shows graphically the velocity distribution for interference line 22 between x = 210 and x = 270 (R = 61 data points) along with the minimum speed (1561.3 m/s), the maximum speed (1571.0 m/s), the average (mean) speed (1567.4 m/s) and the standard deviation (2.3 m/s). The distribution is playtkurtic, is skewed to the right of the mean, has a median speed of 1566.5 m/s and has a mode at 1566 m/s at where there are 23 values.

Figure 6 shows the smoothed continuous fringe lines of a 550 µm liver sample. The statistics for each of the 33 line pairs are shown in Table 1. For this set of data, the standard deviation is small and the skew and kurtosis values are close to zero indicating that the liver sample is quite homogeneous. A plot of the average speed of sound, one standard deviation error bounds and the median speed versus distance across the sample is shown in Figure 7. The speed of sound is rather constant and the standard deviation is small as would be expected from a relatively homogeneous specimen.

The speed of sound for several thicknesses of mouse liver and mouse spleen has been determined [4]. In each case 61 data points were used in the center of the specimen. The average speed of sound are 1556.3 m/s for mouse liver and 1565.0 m/s mouse spleen. The average standard deviation is 2.1 m/s for liver and 3.2 m/s for spleen. These results are comparable to those cited in the literature for similar tissues using the SLAM. Frizzell and Gindorf [5] obtained values of 1565 ± 8 m/s for sheep liver and 1567 ± 13 m/s for cat liver.

In conclusion, an automated technique for determining the speed of sound at many points across a sample has been developed. The speed of sound data set which is generated is much larger than could be generated manually, thus enabling one to use quantitative statistical parameters to study the degree of tissue specimen heterogeneity.

```
NUMBER OF AVERAGES = 4
THICKNESS(M) = 0.00046
LINE#    SLOPE     YBAR      XBAR
 22      -0.02     798.88    241.50
LINE#    SLOPE     YBAR      XBAR
 23      -0.02     838.32    241.50
NS = 210   NE = 270   N = 61
AVERAGE FN = 1.19533
NMAX = 1.26496   NMIN = 1.08027
VMIN(M/S) = 1561.348   VMAX(M/S) = 1571.029
AVERAGE SPEED(M/S) = 1567.367
STANDARD DEVIATION(M/S) = 2.313
PLATYKURTIC  -0.55
MEDIAN SPEED(M/S) = 1566.490
SKEWED RIGHT  1.137
MODE OCCURS AT 1566M/S   VALUE OF 23.00
```

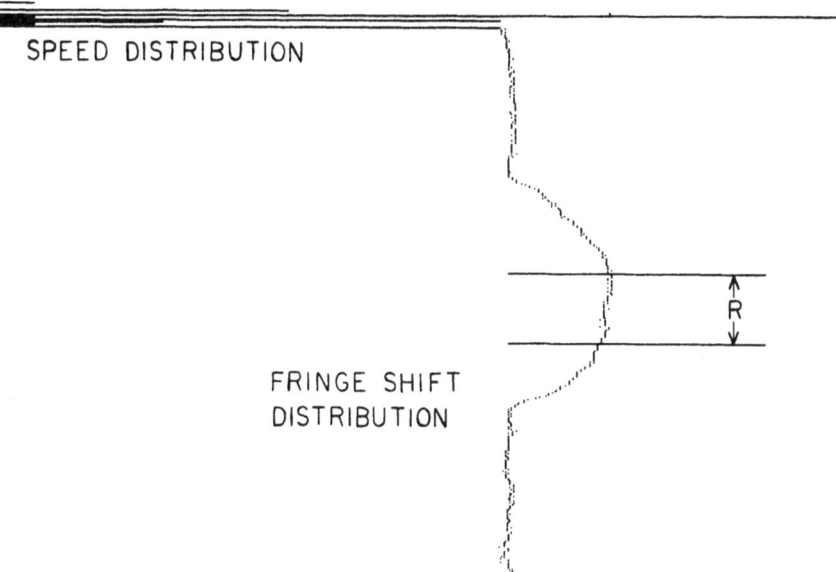

Figure 5. Typical output of the program which calculates the
speed of sound statistical data for one pair of interference lines. (Data for lines 22 and 23 are shown.)

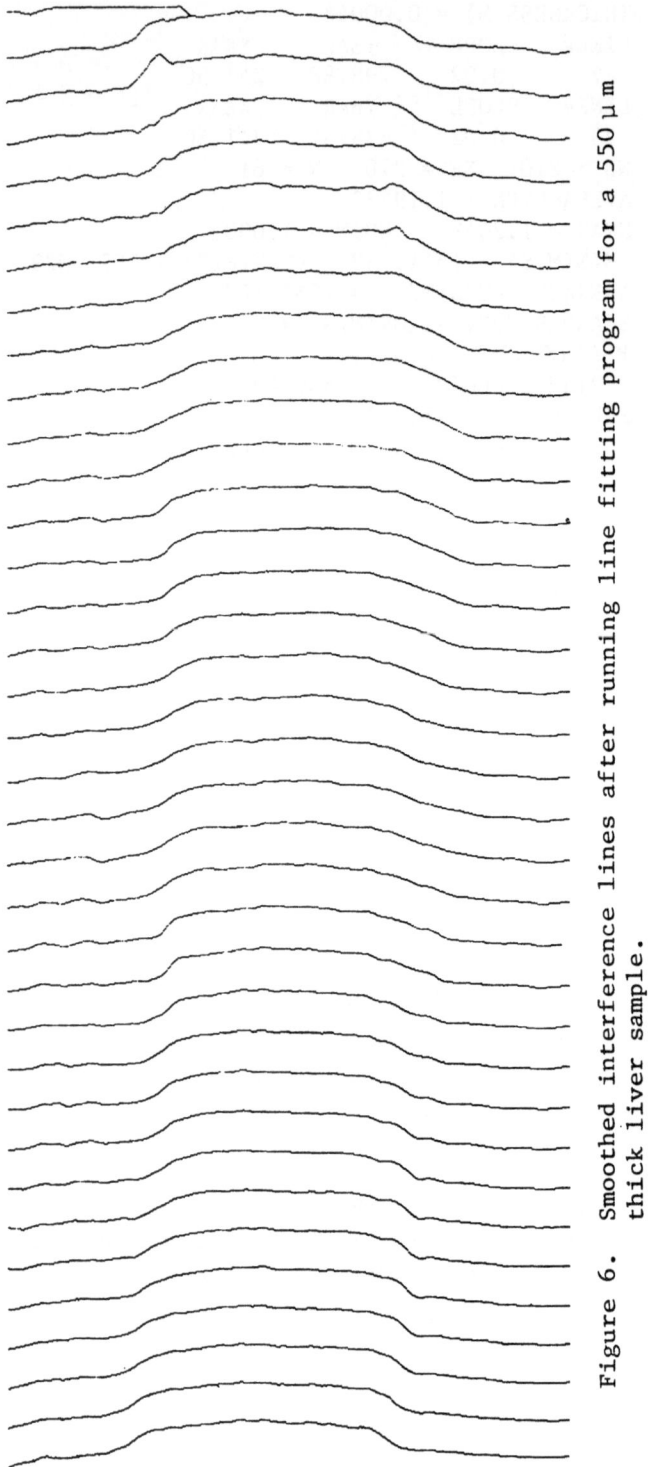

Figure 6. Smoothed interference lines after running line fitting program for a 550 μm thick liver sample.

Table 1. Statistical Speed of Sound Data for 550 μm Thick Liver Sample Averaged Over 121 Points Horizontally. Thirty-three Points Across the Sample are Shown.

NS=180 NE=300 N=121
THICKNESS(M) = 0.00055

LINE #	SLOPE1	YBAR1	SLOPE2	YBAR2	AVE N	NMIN	NMAX	VMIN	VMAX	VAVE	SIGMA	KURTOSIS	VMED	SKEW
2	0.01	127.0	0.01	162.5	0.9453	0.8006	1.0280	1540.30	1550.02	1546.48	2.60	-0.02	1546.82	-0.397
3	0.01	162.5	0.00	198.5	0.9431	0.8368	1.0221	1541.84	1549.77	1546.39	1.91	-0.09	1546.35	0.051
4	0.00	198.5	0.00	235.1	0.9315	0.8554	1.0015	1542.64	1548.88	1545.89	1.56	-0.62	1545.24	1.250
5	0.00	235.1	0.00	271.8	0.9262	0.8391	0.7820	1541.94	1548.05	1545.66	1.61	-0.78	1545.60	0.115
6	0.00	271.8	0.00	308.6	0.9139	0.8438	0.9837	1542.14	1548.12	1545.13	1.52	-0.80	1544.59	1.083
7	0.00	308.6	0.00	345.6	0.9023	0.8444	0.9580	1542.16	1547.02	1544.64	1.39	-0.83	1544.61	0.075
8	0.00	345.6	0.00	383.0	0.8907	0.8375	0.9465	1541.87	1546.53	1543.76	1.04	-0.82	1544.21	-0.189
9	0.00	383.0	0.00	421.1	0.8818	0.8160	0.9175	1540.96	1545.29	1543.76	0.96	-0.03	1544.12	-1.117
10	-0.00	421.1	-0.00	459.4	0.8702	0.8102	0.7361	1540.71	1546.08	1543.26	1.11	0.25	1543.80	-1.448
11	-0.00	459.4	0.00	497.7	0.8635	0.9013	0.7284	1540.33	1545.75	1542.98	1.21	0.63	1543.48	-1.248
12	0.00	497.7	0.00	536.4	0.8656	0.8075	0.9120	1540.59	1545.05	1543.07	1.27	-0.46	1542.82	0.592
13	0.00	536.4	0.00	575.8	0.8744	0.7999	0.9310	1540.27	1545.86	1543.44	1.56	-0.70	1543.60	-0.301
14	0.00	575.8	0.01	614.9	0.8899	0.8028	0.9534	1540.39	1546.82	1544.11	1.79	-0.99	1544.54	-0.717
15	0.01	614.9	0.01	654.3	0.8860	0.8054	0.9465	1540.51	1546.53	1543.94	1.95	-1.06	1544.43	-0.751
16	0.01	654.3	0.00	693.8	0.8771	0.7940	0.9315	1540.02	1545.88	1543.56	1.96	-1.37	1543.75	-0.291
17	0.00	693.8	0.00	732.6	0.8901	0.8254	0.9578	1541.36	1547.01	1544.12	2.02	-1.59	1544.70	-0.870
18	0.00	732.6	0.01	771.9	0.8998	0.8452	0.9534	1542.20	1546.82	1544.53	1.75	-1.66	1544.71	-0.305
19	0.01	771.9	0.00	811.7	0.8949	0.8230	0.9542	1541.26	1546.85	1544.32	1.57	-1.07	1544.58	-0.187
20	0.00	811.7	0.00	851.5	0.8913	0.8295	0.9591	1541.53	1547.07	1544.17	1.44	-0.82	1544.02	0.211
21	0.00	851.5	0.00	891.1	0.8912	0.8476	0.7460	1542.30	1546.51	1544.17	1.34	-0.89	1543.45	1.594
22	0.00	891.1	0.00	931.4	0.8768	0.8212	0.7135	1541.18	1545.11	1543.55	1.02	-0.91	1543.22	0.974
23	0.00	931.4	0.00	972.4	0.8540	0.7738	0.3900	1540.23	1544.11	1542.57	1.09	-1.02	1542.98	-1.110
24	0.00	972.4	0.00	1012.3	0.8511	0.7926	0.0907	1539.96	1544.14	1542.45	1.16	-1.20	1543.03	-1.485
25	0.00	1012.3	0.00	1052.0	0.8702	0.8254	0.7227	1541.35	1545.51	1543.27	1.00	0.21	1543.38	-0.342
26	0.00	1052.0	0.00	1092.1	0.8835	0.8413	0.7235	1542.03	1545.54	1543.83	1.02	-0.82	1544.20	-1.247
27	0.00	1092.1	0.00	1132.5	0.8751	0.8207	0.9153	1541.16	1545.19	1543.47	0.86	-0.14	1543.84	-1.278
28	0.00	1132.5	0.00	1172.4	0.8865	0.8207	0.9165	1541.16	1545.24	1543.96	0.82	0.81	1544.07	-0.384
29	0.00	1172.4	-0.00	1212.7	0.8990	0.8357	0.9343	1541.80	1546.00	1544.50	1.25	-0.71	1544.87	0.893
30	-0.00	1212.7	0.00	1253.2	0.8844	0.8201	0.7691	1541.13	1547.50	1543.88	1.67	-0.43	1543.33	0.979
31	0.00	1253.2	0.00	1293.6	0.8680	0.8102	0.9609	1540.71	1547.14	1543.18	1.78	-0.39	1542.93	0.412
32	0.00	1293.6	0.00	1333.9	0.8767	0.8006	0.9549	1540.30	1546.89	1543.54	1.82	-0.63	1543.56	-0.033
33	0.00	1333.9	0.00	1374.6	0.8815	0.8079	0.9614	1540.61	1547.16	1543.75	1.83	-1.14	1544.08	0.542
34	0.00	1374.6	0.01	1415.0	0.8717	0.7956	0.7520	1540.09	1546.76	1543.33	1.95	-1.13	1543.65	-0.482

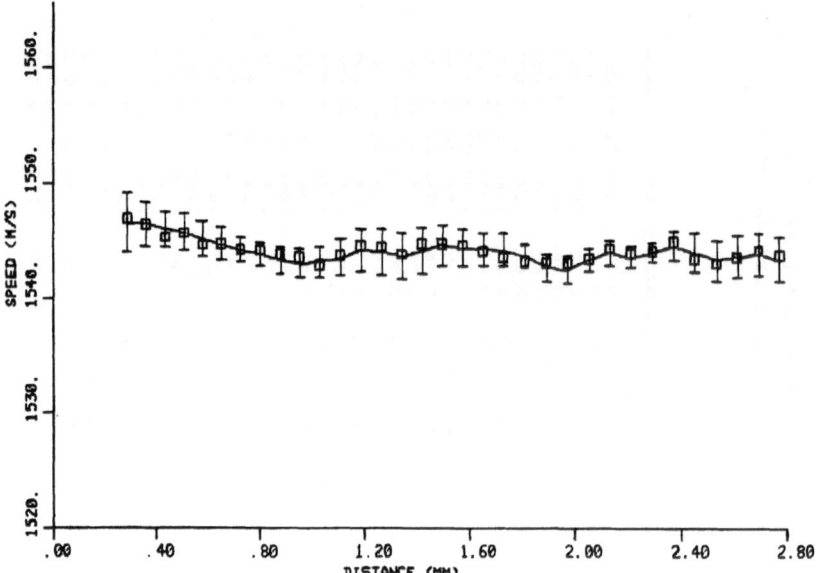

Figure 7.   Speed of sound versus distance acrosse sample.
── average speed.
☐ median speed.   ⊥ ± 1 sigma error bounds.

ACKNOWLEDGEMENTS

The partial support of the National Institutes of Health, National Institute of General Medical Sciences (GM24994) and the National Institutes of Health, National Cancer Institute (CA 36029) are gratefully acknowledged.

REFERENCES

1. O'Brien, W. D., Jr, J. Olerud, K. K. Shung, and J. M. Reid. Quantitative Acoustical Assessment of Wound Maturation with Acoustic Microscopy. J. Acoust. Soc. Amer. 69, 575-579, 1981.
2. Foster, S. An Image Digitizing System for a Scanning Laser Acoustic Microscope. M.S. Thesis in Electrical Engineering, University of Illinois at Urbana-Champaign, 1981.
3. Goss, S. A. and W. D. O'Brien, Jr. Direct Ultrasonic Velocity Measurements of Mammalian Collagen Threads. J. Acoust. Soc. Amer. 65, 507-511, 1979.
4. Bright, G. Quantative Assessment of Material Heterogeneity Using Acoustic Microscopy. M.S. Thesis in Electrical Engineering, University of Illinois at Urbana-Champaign, 1982.
5. Frizzell, L. A. and J. D. Gindorf. Measurement of Ultrasonic Velocity in Several Biological Tissues. Ultrasound in Medicine and Biology, 7, 385-387, 1981.

DIFFRACTION OF ULTRASOUND BY SOFT TISSUES: THE INHOMOGENEOUS CONTINUOUS MODEL

C. M. Sehgal and J. F. Greenleaf

Department of Physiology and Biophysics, Mayo Clinic/
Foundation, Rochester, MN 55905

ABSTRACT

In this paper the theory of wave propagation in continuous-inhomogeneous media is utilized to explain scattering of ultrasound by soft tissues. When a number of scattering volume elements are present, as opposed to one, the scattered energy from a given volume element subsequently encounters other scattering volume elements. As a result of this, the sound energy is diffused instead of being propagated as well-defined waves. This means that the phase relationships are lost and the radiation becomes incoherent somewhat in the fashion described by Foldy[15]. If one were to freeze the ensemble to one instant and observe all the volume elements, each of them would appear as a secondary source emitting radiation in a different phase. Using this concept and assuming that soft tissues do not possess any sharp discontinuities, an equation for attenuation due to scattering is obtained which predicts a near linear frequency dependence for scattering-attenuation. The velocity fluctuations and correlation length for a variety of tissues are estimated from the values of scattering-attenuation using the postulated model. On the basis of these values, ray, Born, and Rytov approximations used to describe sound propagation through soft tissues, are evaluated.

INTRODUCTION

When a beam of sound falls upon a medium, the pressure field associated with it sets small volume elements of the material into periodic oscillations. The material then serves

as a secondary source of sound and radiates energy in different
directions in the form of scattered radiation with wavelength,
λ, equal to the incident sound. The pressure amplitude or intensity, angular distribution and the fine structure of scattered radiation are dictated by the size and shape of scatterers
and their physico-chemical properties with respect to the medium
in which they are embedded. Conversely, from the measurement of
scattering of sound by a material, it should be possible to
estimate characteristic properties of the medium, provided the
theory of interaction of ultrasound with the medium is well
understood.

Such procedures are often used in medical ultrasound to
characterize biological tissues.[1,2] In the past several years,
significant progress has been made in improving the measuring
capabilities of sound-scattering by tissues but relatively
little attention has been devoted to the theory of tissue-sound
interaction. In this paper we specifically address to this
problem. The theory of fluctuations is utilized to derive an
equation that interrelates attenuation (due to scattering) to
the variables that define the intrinsic properties of the
medium. The resulting equation is then tested with the measured
data available in literature.

DISCUSSION ON SCATTERING OF ULTRASOUND

The calculation of scattering can be approached in two
ways: (a) direct calculation of radiation of each scattering
particle and summing all the combinations; and (b) treating
the scattering as a result of statistical fluctuations in composition, concentration and density which in turn results in
the fluctuations in the refractive index of the medium. Both
the approaches have been used in optics and acoustics to study
scattering problems.

(a) The first method was elaborated by Rayleigh[3] and is
suitable for particles much smaller than the wavelength of the
radiation. It also assumes scattering particles to be sufficiently apart so that their relative positions are random, as
in gases and in dilute solutions. In concentrated solutions, the
second assumption does not hold true completely, and there is
structural order (like that in a crystal) such that the energy
scattered from a scatterer of the lattice in different directions is counterbalanced by the energy received by the scatterer from the neighboring scatterers of the lattice, so there
is no net scattering of energy. This in essence means that such
scatterers are exposed to the radiation field of the neighboring
scatterers and the rules of single scattering are no longer

applicable. Twersky[4] proposed a "two phase approximation" to
explain scattering in a dense distribution of scatterers.
According to the approximation some scatterers are treated as
"gas-like" and randomly distributed and the remainder as
"crystal like." Shung et al.[5] recently used Twersky's wave
scattering theory to explain the scattering of ultrasound by
red blood cells. They observed the scattering did not increase
linearly with the hematocrit [HMTC] concentration, as one would
expect on the basis of single scattering.

   (b) The second method of calculating scattering uses the
concept of thermodynamic fluctuations and has been found to be
much more useful for studying light scattering in liquids,
concentrated solutions and systems that are interactive on a
molecular level through weak physical forces. This method is
particularly suited for systems in which direct calculation
of interference from the individual scatterers becomes too
difficult. The concept of fluctuations was first used by
Smoluchowski[6] to explain light scattering in liquids near their
critical temperature. According to him the number of molecules
in a small volume element of a liquid varies from instant to
instant as a consequence of thermal motion. The local fluctuations in density due to a variation in the number of particles give local inhomogeneities in refractive index and thus
scatter light. Smoluchowski's theory was further extended by
Einstein[7] for liquid mixtures to include inhomogeneities in the
refractive index due to fluctuations in concentrations; and by
Liebermann[8] and Chernov[9] to acoustics to explain scattering of
ultrasound due to temperature inhomogeneities in ocean. Later
Chivers[10] lucidly discussed Chernov's theory as applied to biological tissues. Chernov's formalism for hydroacoustic transmission assumes the radiation field at a scattering-volume-
element is not influenced by the radiation scattered by the
neighboring volume elements. In essence this implies single
scattering. On the basis of Chernov's theory[9] as applied to
tissues[10] one may predict a fourth power frequency dependence
for attenuation due to scattering for Rayleigh scatterers; a
quadratic power dependence for directional scatterers
(size $\gg\lambda$); and an intermediate power dependence between 2
and 4 for diffractive scatterers (size $\simeq\lambda$). The experimental
data on ultrasonic absorption and attenuation (sum of energy
loss due to absorption and scattering) in mammalian tissues[11,12]
show the frequency power dependence of attenuation due to scattering to be linear or slightly greater (Fig. 1). The question
that follows is whether the biological tissues that have a
rather complex physical and chemical nature and have spatial
fluctuations (averaged over time) in composition, concentration
and density conform to the laws of single scattering.

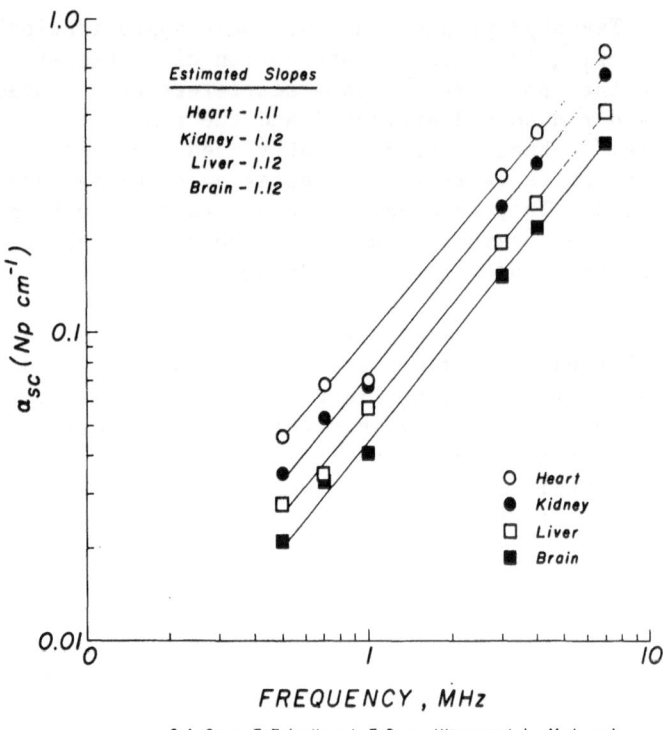

Fig. 1. Frequency-power dependence of attenuation due to
scattering in bovine tissues[11]. (Reproduced with
permission from C. Sehgal et al. Ultrasonic Imaging
(In Press))

There are two ways of estimating multiple scattering
effects that are followed in optics[13]. A simple conclusive test
is to increase the concentration of scatterers in the investigated medium. If the scattered intensity does not increase pro-
portionately, it would indicate multiple scattering. As stated
earlier, Shung et al.[5] have extensively studied the variation of
blood scattering with red blood cell concentration (HMTC). Some
of the representative values obtained from their studies are
listed in Table 1. It is clear from the data that the scattered
intensity does not increase proportionally to the concentration
of HMTC thereby indicating multiple scattering to be significant.

The second test which is more applicable to solid soft
tissues utilizes attenuation as a criterion for estimating
multiple scattering. The intensity of a beam passing through a

Table 1. Backscatter coefficient as a function of Hematocrit
          (HMTC) concentration.

Frequency (MHz)	[HMTC] [a] (%)	Scattering [a] Coefficient x 10^5 cm^{-1}
5	4	4.4
	12	10.7
	22	12.2
8.5	4	2.8
	12	5.8
	24	9.1
15	4	3.4
	12	6.4
	24	7.4

[a] Representative values obtained from Figures 7, 8, and 9
    of Reference 5.

sample is reduced by extinction to e$^{-\tau}$ of its original value.
Here τ represents acoustic or optical length. In optics if
$\tau<0.1$ single scattering is predominent[13]. If $0.1<\tau<0.3$ double
scattering becomes an important factor and if $\tau>0.3$, there is
significant multiple scattering[13]. Ultrasonic attenuation
(amplitude) coefficient for solid soft tissues is ~0.1 Np cm^{-1}
MHz^{-1}.[14] For a path length of 1 cm, the τ value for soft tissues will be 0.2 per MHz. Using the optical criteria in
acoustics, the magnitude of τ in the frequency range 1-10 MHz
would be sufficiently large to consider higher order scattering.
For example, at 5 MHz the value of τ would be ~1. This implies
multiple scattering and it should be accounted for in the scattering theory of soft tissues.

MULTIPLE SCATTERING

     The theory of multiple scattering has been considered with
different degrees of complexities.[10,15,16,17] For our purpose,
we will discuss this problem from a phenomenological point of
view.

     Consider tissue medium to be an assembly of a large number
of scattering volume elements. The scattered energy from a
given volume element would encounter another scattering volume
and subsequently other volume elements in succession. In a
crowded assembly of scatterers, each volume element is exposed

to sound scattered by all other volume elements of the ensemble
in addition to the primary beam which has suffered extinction by
other volume elements. As a result, the phase of the radiation
scattered by a volume element, n, will be influenced by the
scattered waves from the surrounding regions which are themselves corrupted by the scattered energy from the neighboring
volume elements. In principle, one can compute the relationship
among various phases, that is, the phases of scattered waves and
that of the interrogating wave. However, the associated mathematical difficulties can be circumvented by proposing that the
phase, ϕ_n, of the scattered radiation of the volume element

n, is different from the phase, ϕ_n^o, of the incident inter-

rogating beam by an amount ϕ_n^s due to the statistical super-

position of the scattered waves from the surrounding medium
(see Reference 15). ϕ_n may, therefore, be defined as

$$\phi_n = \phi_n^o + \phi_n^s \tag{1}$$

Hereafter, ϕ_n^s will be differentiated from ϕ_n^o by calling the
latter "local phase" as it results due to the interactions
with the surrounding regions.

The scattered intensity $I_{sc}(\theta)$ due to an ensemble of
scattering-volume-elements at the observation point is given by
the equation[18]:

$$I_{sc}(\theta) = \int_v \int_v p_m^*(\phi_m) \, p_n(\phi_n) \, dv_m \, dv_n \tag{2}$$

where subscripts m and n represent two volume elements (Fig.
2), p represents scattered pressure per unit volume; and v
stand for the volume of the scattering element. Assuming
scattered waves to be planar; and the phase difference
$\phi_m^s - \phi_n^s$ between two simultaneous scattering events to be equal
to the product of the distance, q, between the scattering
volume elements and the wave number, k, one obtains by introducing Eq. 1 in Eq. 2[18] the following equation,

$$I_{sc}(\theta) = \int_v \int_v (\cos kq) \, p_m^*(\phi_m^o) \, p_n(\phi_n^o) \, dv_n \, dv_m. \tag{3}$$

Equation 3 is analogous to that obtained by Chernov with a
difference of cos kq term. Following the mathematical analysis

of small perturbations[9,10], the wave equation for a plane
progressive wave in an inhomogeneous media can be solved to
obtain the expressions for the scattered acoustic pressures
$p_n(\phi_n^0)$ and $p_m^*(\phi_m^0)$ in the Fraunhaufer region of the ultrasonic field. Using the expressions for $p_n(\phi_n^0)$ and $p_m^*(\phi_m^0)$ in
Eq. 3 and simplyfing the resulting equation under the
assumptions that the statistical properties of the medium are
spatially homogeneous such that the correlation function
depends only on the <u>relative</u> coordinates between the elements
m and n; and neglecting the density fluctuations it can be
shown that

$$I_{sc}(\theta) = (VA_0^2 k^4 \overline{\mu^2} / \pi R^2 K) \int_0^\infty \psi \cdot q \cdot \cos kq \cdot \sin Kq \, dq, \quad (4)$$

where A_0 = pressure amplitude of the incident plane progressive
         wave;
       V = volume of the tissue being insonified;
       R = distance of the observation point O from the scat-
              tering medium whose center is located at the
              origin of the coordinate system;
       K = $2k \sin \theta/2$;
       $\overline{\mu^2}$ = $(\Delta c)^2/c_0^2$ = velocity fluctuation;
       Δc = deviation of local speed from the average value of c_0;
and ψ = correlation function that relates the properties of the
              medium at one place to those of at a different place.
              In this discussion $\overline{a}$ Markoffian correlation function,
              $\exp(-q/\overline{a})$, is assumed throughout. The quantity $\overline{a}$ is
              the correlation length and it is a measure of the
              distance at which the properties of the two points
              of the medium become uncorrelated. Often $\overline{a}$ is used
              as a measure of mean diameter of the scattering
              volume elements.

Eq. 4 is further developed in the following section to
obtain an equation for attenuation due to scattering.

ATTENUATION DUE TO SCATTERING

By assuming Markoffian correlation function in Eq. 4, the
energy scattered in all the directions except the forward is
summed up and subtracted from the incident energy to estimate
transmitted intensity, I_t. The equation for I_t so obtained
is substituted into the Beer-Lambert equation to obtain an
expression for scattering-attenuation.[18] For a medium
containing identical inhomogeneities the attenuation coefficient

α'_{sc} is given by the equation,

$$\alpha'_{sc} = (4\overline{\mu^2}k^4\overline{a^3})/[(1+k^2\overline{a^2})(1+9k^2\overline{a^2})] \ Np \ cm^{-1} \qquad (5)$$

In the Rayleigh limit ($\overline{ka} \ll 1$) and the geometric limit ($\overline{ka} \gg 1$), Eq. 5 predicts a frequency power dependence of 4 and 0, respectively.

In actuality, tissues possess a spectrum of scattering structures that exhibit regularity over small volumes. This causes the nature of scattering to change from one region to another. For scattering purposes tissues may be regarded to be made up of a large number of regions of volume v_i each of which is defined by a correlation length a_i. If Δv_i represents the volume of the tissue that has a correlation length between a_i and $a_i + \Delta a_i$, the total attennuation coefficient α_{sc} will be given by the equation

$$\alpha_{sc} = \sum_i (\alpha'_i/v_i)\Delta v_i \qquad (6)$$

where α_i is the attenuation coefficient of the volume element v_i. In the limit $\Delta v_i \to dV$, and Eq. 6 can be expressed as an integral which when integrated in the limits of 0 to a_m, where a_m is the maximum value of correlation, yields

$$\alpha_{sc} = (\overline{\mu^2}k/2)[3 \ tan^{-1}ka_m - tan^{-1}3ka_m] \ Np \ cm^{-1} \qquad (7)$$

Terms tan^{-1} can be expanded by Taylor series. For large values of ka_m the terms in the bracket of Eq. 7 show only a weak dependence on k. Consequently α_{sc} approaches a near-linear frequency dependence. As illustrated by Fig. 1, this is the experimental finding. Equation 7 also provides a relationship between the attenuation due to scattering and the intrinsic properties of the tissues $\overline{\mu^2}$ and a_m. In the remaining portion of this section we will estimate the properties of tissues, $\overline{\mu^2}$ and a_m, by using the experimental data available in literature.

Attenuation due to scattering was estimated by subtracting absorption from total attenuation for various mammalian tissues (Table 2). Fig. 2 is a plot of α_{sc} versus

Table 2: Attenuation+, absorption+ and scattering+ data obtained from [11] and [12]

Tissue (Bovine)		Frequency (f) Dependence f in MHz	Frequency (MHz)					
			0.5	0.7	1.0	3	4	7
Brain	α_{att} [11]	0.07 f^1.14	0.032	0.047	0.070	0.24	0.34	0.64
	α_{abs} [11]	0.024 f^1.18	0.011*	0.014	0.029	0.088*	0.123	0.23
	α_{sc}		0.021	0.033	0.041	0.152	0.217	0.41
Heart	α_{att} [11]	0.13 f^1.07	0.060	0.086	0.13	0.41	0.56	1.0
	α_{abs} [11]	0.028 f^1.04	0.014*	0.018	0.033	0.088*	0.118*	0.21
	α_{sc}		0.046	0.068	0.070	0.322	0.442	0.79
Kidney	α_{att} [11]	0.10 f^1.09	0.049	0.070	0.10	0.34	0.47	0.87
	α_{abs} [11]	0.028 f^1.02	0.014*	0.017	0.033	0.086*	0.115*	0.20
	α_{sc}		0.035	0.053	0.067	0.254	0.355	0.67
Liver	α_{att} [11]	0.08 f^1.13	0.038	0.055	0.08	0.29	0.40	0.75
	α_{abs} [11]	0.026 f^1.17	0.010	0.020	0.023	0.094*	0.14	0.24
	α_{sc}		0.028	0.035	0.057	0.196	0.26	0.51
Liver	α_{att} [12]	0.043 f^1.266	0.018*	0.0274*	0.043*	0.173*	0.249*	0.505*
	α_{abs} [11]	0.026 f^1.17	0.010	0.020	0.023	0.094*	0.14	0.24
	α_{sc}		0.0079	--	0.020	0.079	0.109	0.265

*These values were obtained by using the frequency relationship listed in the second column.
+All these coefficients are expressed in units of Np cm^{-1}.

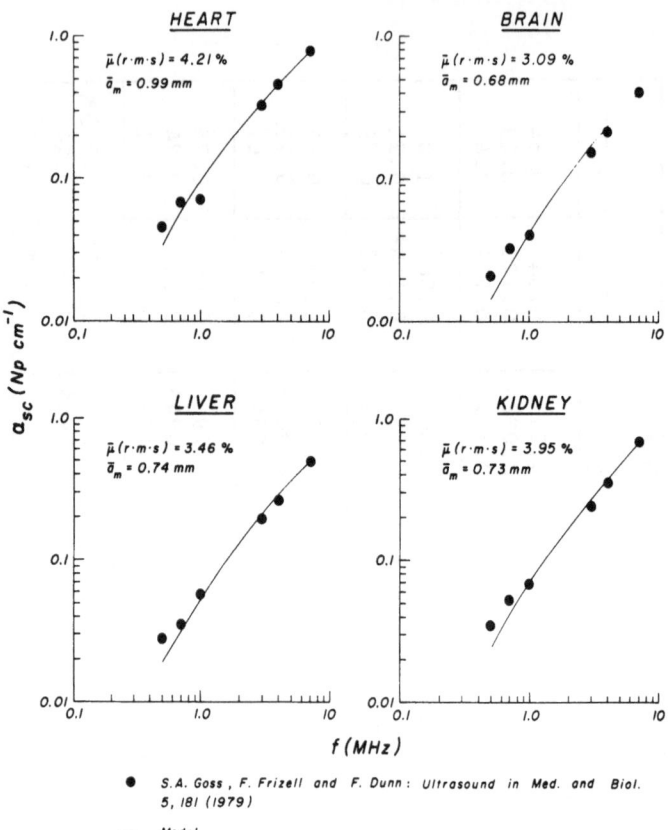

Fig. 2.   Attenuation (scattering) data for various bovine
          tissues (dotted circles) is shown to be fitted by
          Eq. 7 (solid line). The values of α_{sc} were deter-
          mined by subtracting the absorption coefficient
          from the attenuation coefficient. The data were
          obtained from Ref. 11. (Reproduced with permis-
          sion from C. Sehgal et al. Ultrasonic Imaging (In
          Press))

frequency for liver, heart and kidney. It is well demonstrated in the figure that the curves obtained by a least
square fit of scattering-attenuation data to Eq. 7 closely
accounts for the variation in α_{sc} with irradiation frequency
for a number of tissues. The root mean square (r.m.s.) velocity fluctuations and the correlation length as estimated by
fitting scattering data to Eq. 7, are summarized in Table 3.
In normal soft tissues, velocity fluctuations (r.m.s.) range
from 2.8 to 4.2 percent and the correlation length varies from
0 to a maximum of 0.7 to 1 mm. Some earlier rough estimates
of velocity fluctuations have been made. Using the data of

Table 3. Values of $\bar{\mu}$(r.m.s.) and a_m estimated by fitting
          Eq. (7) to the experimental data of Reference 11.

Tissue	$c \times 10^{-5}$ [25] (cm/sec)	$\bar{\mu}$ (r.m.s.) (%)	a_m (mm)	Collagen Concentration (wet weight percent) [21]	
				range	mid point
Heart [11]	1.51	4.21	1.0	0.4-2.6	1.5
Kidney [11]	1.56	3.95	0.73	0.39-1.47	0.93
Brain [11]	1.57	3.09	0.68	0.03-.34	0.19
Liver [11]	1.56	3.46	0.74	0.18-1.1	0.64

Wells,[19] Chiver[10] estimated $\bar{\mu^2}$ to be 7.433 x 10^{-4}. This
corresponds to $\bar{\mu}$(r.m.s.) of 2.72 percent which is in good
agreement with the values predicted on the basis of the present formalism (Table 3).

Fields and Dunn have reported that in collagen fibers,
the static or low frequency Young's modulus of elasticity is
several times ($\sim 10^3$) greater than those of parenchymal tissues.[20] Since the elastic modulus (Y_m) is related to ultrasonic velocity, the presence of collagen in tissues would
introduce a velocity fluctuation. It is, therefore, reasonable
to expect tissues with higher collagen concentration to have
larger $\bar{\mu}$(r.m.s.). A plot of collagen concentration in various
tissues[21] versus the corresponding estimated value of μ(r.m.s.)
shows (Fig. 3) a perfect rank correlation between the two
variables.

APPLICABILITY OF RAY, BORN AND RYTOV APPROXIMATIONS

In this section we will try to evaluate, on the basis of
estimated values of $\bar{a}$ and μ^2, the applicability of ray, Born
and Rytov approximations for the propagation of ultrasound
through soft tissues.

Sound propagates as a ray in a medium with random inhomogeneity if the scale of inhomogeneity, $\bar{a}$, is much greater than
the wavelength, λ, as well as the size of first Frensel zone
in the region of propagation of linear dimension, L,[9] i.e.,

$$\lambda \gg \bar{a} \quad \text{and} \quad \lambda L \gg \bar{a}^2 \qquad (8)$$

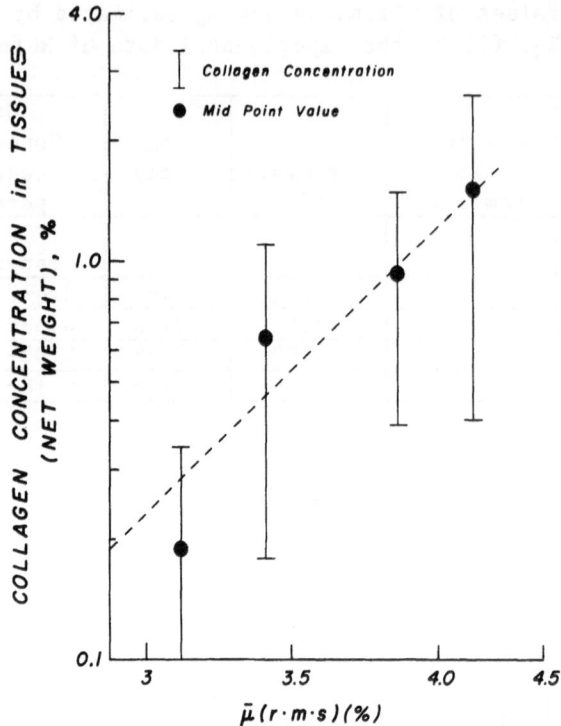

Fig. 3. Correlation between collagen concentration in tissues
         and the parameter $\bar{\mu}$(r.m.s.). The values of collagen
         concentration in tissues were obtained from Ref. 21,
         and the corresponding values of $\bar{\mu}$(r.m.s.) were esti-
         mated by fitting α_{sc} data to Eq. (7). (Reproduced with
         permission from C. Sehgal et al. Ultrasonic Imaging
         (In Press))

        The maximum sizes of inhomogeneities estimated in this
study is of the order of 1 mm. The wavelength in the frequency
range 1 to 10 MHz is 1.54 to 0.15 mm. Since $\lambda \approx a$ the conditions
for ray propagation are not satisfied by soft tissues.

        Next we turn our consideration to Born and Rytov
approximations. When a wave propagates in the tissue medium
with inhomogeneities, the fluctuations in the characteristics

of the wavefield due to superposition of scattered waves and the primary wave must occur. The fluctuations in the properties of the wavefield will be related to the fluctuations of refractive index. The deviation of phase, $\Delta\xi$, from the average value of ξ_0 during the propagation through the tissue-medium of length L is related to velocity (or refractive index) fluctuations by the relationship (Appendix A)

$$\langle(\Delta\xi)^2\rangle = \langle(\xi-\xi_0)^2\rangle = 4\pi^2 \overline{\mu^2} L^2/\lambda^2 \tag{9}$$

For propagation to occur in accordance with the Born approximation, the changes (which in this case is the deviation from the mean value) in amplitude and phase of the incident wave must be small.[23] According to Eq. 9 the deviation in the r.m.s. value of phase grows with the distance traveled by ultrasound. The Born approximation breaks down when phase is in error by, say, $\pi/2$.[24] Velocity fluctuation estimations range from 1% to 5% in tissues (Table 3). Using these values in Eq. 9 it may be concluded that the Born approximation would break down after 5λ and 8λ for 5% and 3% fluctuations, respectively. In a medium of sound speed 1.54 x 10^5 cm sec^{-1} this would correspond to propagation distances of 0.8 cm to 1.23 cm at 1 MHz.

In contrast to the Born approximation, the Rytov approximation requires deviation of phase per wave number to be less than or equal to unity (see Eq. 94 of Ref. 9). According to Eq. 9 $\Delta\xi$(r.m.s.) per wave number is equal to $\overline{\mu}$(r.m.s.) L. Therefore, for Rytov approximation to be applicable $\overline{\mu}$(r.m.s.) L 1. This means that distance over which Rytov approximation is valid is governed by the magnitude of velocity fluctuation only. For $\overline{\mu}$(r.m.s.) of 3 to 5 percent the Rytov approximation will be good for distances 33 cm to 20 cm, respectively.

ACKNOWLEDGMENTS

The authors would like to thank Mr. Chris R. Hansen for providing the computer program and Mr. Steven J. Richardson and Mr. James L. Hanson for the illustrations. We are particularly grateful to Ms. Elaine Quarve for her assistance in preparing this manuscript. This research was supported in part by Grants HL-04664, RR-00007, HV-7-2928, HL-00060, and HL-07000 from the National Institutes of Health and NCI-CB-64041 from the National Cancer Institute.

APPENDIX A

In this section we will derive an expression for phase fluctuations. It is known that

$$\overline{\mu^2} = \frac{\langle (c-c_o)^2 \rangle}{c_o^{\,2}}, \qquad\qquad\qquad A(1)$$

and $\quad \xi = \omega t = \dfrac{\omega L}{c}. \qquad\qquad\qquad\qquad\qquad A(2)$

In the above equations ξ and c represent the phase and velocity of sound in a localized zone, t represents the time at which the sound wave arrives at the point. Eliminating c from Eqs. A(1) and A(2) yields

$$\langle (\Delta\xi)^2 \rangle = \langle (\xi-\xi_o)^2 \rangle = 4\pi^2 \overline{\mu^2} L^2 / \lambda^2. \qquad A(3)$$

REFERENCES

1. M. Linzer and S. J. Norton, Ultrasonic tissue charac-
    terization, Annu Rev Biophys Bioeng 11:303 (1982).
2. R. C. Chivers, Tissue characterization, Ultrasound Med
    Biol 7:1 (1981).
3. Lord Rayleigh, Phil Mag 41:447 (1871).
4. V. Twersky, Multiple scattering of waves in dense
    distributions of large tenuous particles, in "Electronic
    Scattering," M. Kerker, ed., Pergamon Press, (1962);
    C. I. Beard, T. H. Hays, V. Twersky, Scattering by random
    distribution of small scatterers vs. concentration, IEEE
    Trans AP-15 1:99 (1967).
5. K. K. Shung, R. S. Sigelmann, and J. M. Reid, Scattering of
    ultrasound by blood, IEEE Trans in Biomed Eng 6:460
    (1976).
6. M. Smoluchowski, Ann Physik 25:205 (1908); Phil Mag 23:165
    (1912).
7. A. Einstein, Ann Physik 33:1275 (1910).
8. L. Liebermann, The effect of temperature inhomogeneities in
    ocean on propagation of sound, J Acoust Soc Am 23:563
    (1951).

9. L. A. Chernov, in "Wave Propagation in Random Medium," Dover, New York (1960).

10. R. C. Chivers, The scattering of ultrasound by human tissues - some theoretical models, Ultrasound Med Biol 3:1 (1977).

11. S. A. Goss, L. A. Frizzell, and F. Dunn, Ultrasonic absorption and attenuation in mammalian tissues, Ultrasound Med Biol 5:181 (1979).

12. J. D. Pohlhemmer, C. A. Edwards and W. D. O'Brien, Phase insensitive ultrasonic attenuation coefficient determination of fresh bovine liver over an extended frequency range. Med Phys 8:692 (1981).

13. H. C. Van de Hulst, in "Light Scattering by Small Particles," Dover Publication, Inc., New York (1981).

14. S. A. Goss, R. L. Johnston, and F. Dunn, Comprehensive compilation of empirical ultrasonic properties in mammalian tissues. J Acous Soc Am 64:423 (1978).

15. L. L. Foldy, The multiple scattering of waves, Phys Rev 67: 107 (1945).

16. M. Lax, Multiple scattering, Rev Mod Physics 23:287 (1951).

17. V. K. Vardan, V. V. Vardan, Y. H. Pao, Multiple scattering of elastic waves by cylinders of arbitrary cross section .1.SH waves, J Acous Soc Am 63:1310 (1978).

18. C. M. Sehgal and J. F. Greenleaf, Scatter of ultrasound by tissues (submitted for publication).

19. P. N. T. Wells, "Physical Principles of Ultrasonic Diagnosis," Academic Press, New York (1969).

20. S. Fields and F. Dunn, Correlation of echographic visualizability state. J Acous Soc Am 54:809 (1973).

21. J. Pohlhemmer and W. D. O'Brien, Jr., Dependence of the ultrasonic scatter coefficient on collagen concentration in mammalian tissue. J Acous Soc Am 69:283 (1981).

22. M. O'Donnell, J. W. Mimbs, and J. G. Miller, Relationship between collagen and ultrasonic backscatter in myocardial tissue. J Acous Soc Am 62:580 (1981).

23. J. F. Greenleaf, Computerized tomography with ultrasound. Proc IEEE 71:330 (1983).

24. B. S. Robinson, Speckle processing for ultrasonic imaging, Ph.D. Thesis, University of Canterbury, Christchurch, New Zealand, p 10 (1982).

25. P. D. Edmonds and F. Dunn, Introduction: Physical description of ultrasonic fields, in "Methods of Experimental Physics; Ultrasonics," P. D. Edmonds, ed., Academic Press, New York (1981).

TISSUE INFORMATION FROM ULTRASOUND SCATTERING

Sidney Leeman and Joie Pierce Jones

Department of Radiological Sciences
University of California, Irvine
Irvine, CA 92717

INTRODUCTION

Medical ultrasound is now progressing beyond its initial, and successful, phase, wherein its principal purpose was to provide adequately high resolution (in both space and time) images of tissue structure. Increasing emphasis is now being placed on the widely accepted notion that tissue probing with either pulsed or continuous wave ultrasound fields potentially provides more information than is currently displayed in the pulse-echo image. Much justification for this viewpoint (which is commonly stated, without adducing firm evidence) stems from the realization that conventional medical ultrasound images are constructed from a data base which is only a subset of the full data ensemble that is, in fact, measured. Moreover, the measurements, in turn, comprise only a subset of the data that are, in principle, practicably addressed with present day technology. In this sense, ultrasound imaging is probably unique amongst medical imaging modalities. Of course, the mere realization that not all measured data are used for image construction does not, in itself, guarantee that the discarded data contain new, or non-redundant, information. Occasionally, there is also an understandable confusion between the achievement of a "better" (e.g. higher resolution, more displayed dynamic range, etc) image, and the development of techniques that generate perhaps even rather poor images with an intrinsically different (tissue) information content. This last point can be illustrated without attempting to accurately define the concept of "information" displayed: there is an intuitive appreciation of the differences between a high and low resolution ultrasound reflectivity map of tissue (as crudely

displayed in a conventional B-scan) on the one hand, and a
reflectivity map and an attenuation image, on the other. The
former two attempt to portray the same intrinsic tissue
information, but to different degrees of fidelity. The latter
two differ markedly in their "information content" quite
irrespective of the relative resolution, or fidelity, to which
they may individually be obtained.

Projections of the ultimate utility of future (yet to be
discovered!) information extraction techniques have ranged from
the ludicrously optimistic to the crushingly pessimistic. It is
our contention that a realistic assessment may be made by looking
to the physics of the ultrasound/tissue interaction. In this
manner, a better appreciation of the range of information
amenable to probing may be obtained. This is a vast field for
investigation, and we shall heavily circumscribe our treatment.
What cannot be answered here, is the medical utility of the
extracted information: this is a problem that only protracted
clinical trials can resolve.

INFORMATION EXTRACTION WITH ULTRASOUND

The basic medical ultrasound (imaging) "experiment" may be
essentially described as follows: an input ultrasound wave is
generated, which is allowed to penetrate into tissue, where it
interacts. Waves generated by this interaction, including the
modified, emergent incident wave are detected outside the
tissue. The aim is to make some statement about the tissue from
measurements on (a subset of) the emergent waves ("OUT") and a
knowledge of the incident wave ("IN"). Tissue is thus an
"OUT/IN" modifier, about which information is extracted from a
study of "OUT/IN" relationships. This description is virtually
identical to that used to define scattering experiments, and it
is immediately clear that scattering theory will play a
fundamental role in our analysis.

Given the fact that it is a medical ultrasound investigation,
some details can be added to the above description. The input
field is likely to be a longitudinal pressure wave, with a
frequency in the 1-10 MHz range; for most applications, a pulsed
field will be employed--broad band, but with a central frequency
well within the indicated range. In principle, there is no
difficulty in assuming the incident wave to be accurately
known. There can be less certainty about specifying the
(acoustic) nature of tissue. It is clearly an inhomogeneous
medium, varying in both space and time, the latter often
manifesting itself as an approximately cyclic variation of
properties. Tissue is very richly structured, with spatial
correlation lengths spanning a wide spectrum, well beyond the
wavelength scales of the so-called "diagnostic" frequency range.

Ultrasound attenuates fairly rapidly within most soft tissues,
at a rate that is definitely frequency-dependent. Since there
is some experimental evidence to support the idea that shear
waves do not propagate over significant distances within soft
tissues, it is probably reasonable to describe the wave as a
propagating pressure wave. Experimental techniques are not, in
general, sufficiently sophisticated to unambiguously detect
non-linear effects within tissues at the moderate power levels
and frequencies employed for imaging, and there is thus no
compelling reason, at present, to resort to the added complexity
of non-linear theories. Soft tissues certainly scatter
ultrasound waves, and it is known that scatter-generating
structures consist of fluctuations in at least the density,
compressibility and ultrasound absorption parameters. The
relative strength of scattering from these sources, as well as
the overall scattering strength, are not known, but it is
generally assumed (if only for computational convenience) that
the scattering is, in some sense, weak. Soft tissues exhibit a
complicated composition, and a complicated thermodynamics, and,
in some cases at least, are clearly not isotropic media: the
significance of these factors has not been fully assessed in
most appolications and we disregard them, for simplicity, in the
following. Many of the (acoustical) properties of soft tissues
change with disease, and this ultimate diagnostic potential of
ultrasound provides one of the most exciting areas for
investigation.

     It seems clear that there are basically three classes of
information that can be extracted about soft tissues via
ultrasound probing: some statement can be made about tissue
acoustical properties (such as attentuation, impedance, etc.),
or about tissue structure (i.,e., the spatial distribution of
the acoustic parameters), or about tissue dynamics (i.e. the
temporal distribution of the acoustic parameters). In practice,
methods for achieving that end may be conveniently classified
into three types of technique: pulse-echo imaging, quantitative
imaging, and tissue characterization. Before we can examine
these techniques in more detail, we will have to consider more
closely the description of scattering, which is fundamental to
all these approaches.

SCATTERING OF ULTRASOUND IN TISSUES

     Consider an ultrasound wave, ψ, in tissue: in general,
this will be a function of both position and time, but we will,
for convenience, not specifically indicate these variables. A
static tissue (which is the only case we consider here) is
source-free, and ψ will obey:

$$L\psi = 0 \qquad\qquad (1)$$

Here, L is some (linear) operator depending on the tissue, and containing, in general, both spatial and temporal derivatives. Equation (1) is then clearly adequate to describe the propagation of pulses, and is not limited to continuous waves only, as has been, rather surprisingly, stated in more than one treatment. We shall, henceforth, confine ourselves to the study of <u>pulse</u> propagation and scattering. Any realistic wave equation for ultrasound propagation in soft tissues will not be immediately soluble. A common approach is to decompose the linear operator, L, as

$$L \equiv P - R \qquad (2)$$

The decomposition (2) is not unique, but it is usual to ensure that the operator P is independent of (tissue parameter) fluctuations, these being contained within R. We suggest that this decomposition, often done quite unthinkingly, should be performed with care and is, in fact, a crucial step in developing solutions of (1). Indeed (2) moulds the "physical" conceptual basis of the scattering problem. We now depart somewhat from the usual conventions in allowing P to contain also (large-scale) variations in the <u>mean</u> values of the acoustic parameters; R then contains only fluctuations about that mean value. Thus, effects such as refraction, and reflection from large-scale smooth tissue interfaces, are considered to be incorporated in P. These are, in principle, relatively well-understood, and solvable, problems so that we can still maintain that, with the specification of appropriate initial and boundary conditions (and probably with the invocation of realistic approximations) a solution, ψ_0, may be found to the equation,

$$P\psi_0 = 0 \qquad (3)$$

The correct interpretation of Equation (3) cannot be overemphasized. It describes the propagation of the same incident pulse as appropriate to the original problem (1), in an idealized, small-scale fluctuation-free, possibly piece-wise continuous medium. In practice, we will regard the choice of P to have been such that the solution of (3) represents (for a bounded input to the tissue half-space) a possibly distorting, but non-the-less effectively bounded pulse, travelling through the medium; it may depart from its initial path as a result of refraction effects, and "echoes" may be generated at major interfaces. However, the overall picture is that of the propagation of well-defined bounded pulses--there is no scattered "halo" of ultrasound being generated, to any significant extent.

The primary idea underlying the introduction of the operator, P, is that ψ_0 presents, in some sense, an approximate solution to (1): such an expectation cannot be adequately fulfilled unless the values of the (mean) tissue parameters appearing in P are known. We provisionally assume this to be the case, or, at least, that quite reliable estimates may be made. Another important aspect of (3) is that its solution may be written as

$$\psi_0 = P\psi_B \qquad (4)$$

Here, the pulse within the idealized medium described above is expressed as some kernel, P, acting on the known input, i.e., the initial, boundary values of the pulse, denoted here as ψ_B. Given the linearity of the problem, P is determined, in principle, by P: we shall refer to the former as the "propagator."

The full solution to (1) is now written in terms of the approximate solution as

$$\psi = \psi_0 * \psi_s \qquad (5)$$

ψ_s remains to be determined, and * denotes some operator. The so-called Rytov scheme has, as its starting point, the choice that * be scalar multiplication. However, we here choose to pursue a more transparent and intuitively more meaningful pathway, whereby * denotes simple addition. Thus

$$\psi = \psi_0 + \psi_s \qquad (6)$$

Hence, from (1), (2) and (3):

$$P\psi_s = R(\psi_0 + \psi_s)$$

If it be now assumed that

$$|R\psi_s| \ll |R\psi_0| \qquad (7)$$

it follows that, to good approximation,

$$P\psi_s = R\psi_0 \qquad (8)$$

This is a linear, inhomogeneous differential equation for ψ_s, whose solution may be found by standard methods, to ultimately yield

$$\psi = \psi_0 + GR\psi_0 \qquad (9)$$

There are, in fact, a number of subtleties involved in arriving at (9), but we will merely draw attention to a few points. The operator, G, is the "Green function", and is evaluated from P in association with a rather specific boundary condition-- ultimately derived from the so-called Sommerfeld radiation

condition. Essentially, this insists that, in accordance with
our physical intuition, the field scattered from a very small
(compared to wavelength) inhomogeneity, should behave as an
outgoing (possibly angle modulated) spherical wave at very large
distances. Moreover, the ψ_S component of the field will
tend to have a similar behavior at large distances from the
small-scale inhomogeneous region $(R \neq 0)$ because of its
dependence on G, as indicated in (9). Clearly, the ψ_S-
component vanishes if the small-scale tissue fluctuations vanish,
and arises only from regions where ψ_0 is non- zero. These
properties lead us to associate ψ_S with our intuitive
notions of a scattered field: (9) then implies that the actual
field within the tissue may be regarded as a sum of an "incident"
(ψ_0) and "scattered" (ψ_S) field. Note that our treatment
has forced us to the conclusion that small-scale fluctuations in
any acoustic parameter (including the absorption) will give rise
to scattering. The essential distinction between what is a
scattered and incident field originated in our ability to call
on special methods to solve, with high degree of accuracy, a
wave equation such as (3), which depends on slowly varying,
and/or piece-wise, continuity of the medium, and our inability
to attack the problem in general when small-scale fluctuations
are present. This point is clarified when we consider that
pulse transmission across a large, smooth impedance discontinuity
(say) is a relatively straightforward problem compared to the
case when the interface contains significant structure on a
wavelength scale. Or compare the problem of a pulse propagating
in a medium with slowly varying velocity (where ray "optics" may
well be applicable) compared to the case of propagation through
a medium showing random, small fluctuations in velocity.
Equation (7) is a statement that the first Born approximation
applies. Given the observed weakness of signals scattered from
soft tissues, it is likely that this approximation is valid in
most cases. There are certain situations which seem in conflict
with this approximation, e.g., where reverberations are clearly
seen in a B-scan image. However, it is often reasonable to
interpret these as arising from multiple reflections of the
ψ_0 component of (6), so that (7) may hold under rather more
general conditions than is at first apparent. It is certainly
possible to relax the first Born approximation, but at the
expense of a rather more cumbersome description. We shall
maintain the simplicity and, in all likelihood, accuracy of (7).

It is often forgotten that that scattering theories are
exactly that: they deal with the calculation of the scattered
field within the medium. The actual field is then the sum of
this field, and the internal incident field. Thus, a scattering
object which has the dimensions and properties to cast a "shadow"
(region of low acoustic field) when insonicated by a particular
pulse, must show relatively strong forward scatter. This is so

that the scattered and incident waves can interfere to produce relatively <u>weak</u> sonic fields in the shadow zone. Nonetheless, there exist speculations that observed regions of enhanced acoustic intensity in tissues originate from theoretically strong forward <u>scattering</u> into those regions by appropriate tissue structures: these attempts are in clear contradiction with (9), and with the concepts of scattering theory in general.

PROPAGATOR-REFLECTOR CONCEPT

The Green function, G, appearing in (9), describes how the wave from a point source propagates within the fictitious medium described in (3). It is thus easy to appreciate that there is an intimate connection between G and the propagator, P, introduced in the previous section. Indeed, a more careful analysis shows that the Green function, when acting as kernel inside an integral expression, as in (9), is equivalent to the propagator, P, in (4). We are thus led to a very intuitively appealing conceptual framework for the underlying structure of the scattered field, schematically shown in Fig. 1. Consider a

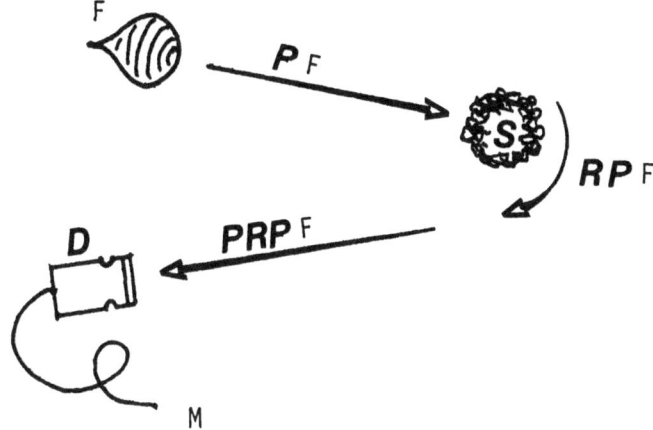

Fig. 1. Highly schematic and symbolic representation of the scattered field. The notation is explained in the text.

known input pulse, F, incident on some tissue-like medium. We are interested in the scattered field from the small region, S, depicted in the figure. The pulse actually arriving at S has the form PF, where P is the "propagator" (integral operator) introduced earlier. The region, S, modifies the pulse in accordance with its scattering properties. We shall investigate this in more detail below, but here describe the modifying effect by the action of an operator, R, which may, appropriately, be referred to as the "reflector." Thus the region S acts as a source of (scattered) sound, described as RPF, and this gives rise to sound waves propagating through the medium. Since this propagation process is described by the propagator, P, the field arriving at a detector may be symbolically written as PRPF. If the detection process is embodied in the operator D, then it clearly follows that the measured signal, M, is given by:

$$M = DPRP\text{F} + \text{N} \tag{10}$$

where N denotes possible (additive) noise. The structure of Eq. (10), which we refer to as the "fundamental imaging equation," is intuitively appealing, even without recourse to scattering theory. However, the more formal considerations leading up to (9) clearly show the close relationship between the two approaches. We emphasize that (10) is written in very symbolic notation, and actually involves integrations over many variables, not explicitly shown. In particular, it should be noted that, despite the suggestive compactness of S in Fig. 1, Eq. (10) actually embodies scattering from throughout the medium of interest.

The propagator-reflector approach for the description of ultrasound scattering from human tissues appears to have been first used by Leeman (8th Acoustical Imaging Symposium) in a derivation of the impediographic equations. Rather similar ideas were, in more clouded fashion, used by Berkhout et al (10th Acoustical Imaging Symposium) in their development of wave extrapolation imaging. The technique has also been used directly by Lefebvre (10th Acoustical Imaging Symposium). We feel that the method is rather more general than the implied dependence on the first Born approximation, since it allows phenomenological and heuristic modifications to both the propagator and reflector operators. Some of these will be equivalent to including, at least partially, higher order corrections. There is also the possibility of incorporating experimental data, e.g., by modelling the propagator to fit findings in pulse transmission experiments. This last, somewhat surprising, contention may be illustrated in a particularly telling example. Given the absence of small scale fluctuations (i.e. scattering!) in the description (3), it seems apparent that, in a lossy medium, the propagator should depend on the absorption characteristics of the medium. However, it does not seem at all unreasonable (in fact, it seems rather more appropriate) to model P, in (10), on the measured attenuation properties of the medium, thus including some aspects of higher Born approximations in the treatment.

It may be observed that the propagator, as introduced in (10), is closely associated with the propagation of the coherent wave (as described, for example, in Ishimaru, 1978), with the

reflector being associated with the incoherent wave. We will
not develop this idea further here, but rely on the relatively
simple connection with formal scattering theory outlined before.

It is of some interest to examine more carefully the
structure of the reflector operator. Given the purely
compressional nature of the linear acoustic waves considered in
medical ultrasound, it is clear that the small-scale
fluctuations giving rise to scattering are those in the density,
compressibility, and absorption of the medium. Absorption
processes in tissues are not well understood, and it is likely
that more than one parameter will be required for their adequate
description. A small density inhomogeneity presents as a dipole
scatterer, a small compressibility inclusion as a monopole
scatterer, but scattering by absorption fluctuations is poorly
investigated and understood. The density and compressibility
components of the "reflector" appear as well-known local
differential operators, but the absorption component cannot be
definitively written down. Some idea of the complicated
modelling needed to arrive at an adequate propagator/reflector
description may be obtained from the excellent paper by Mueller
(1980). A somewhat physical motivation for the reflector
operator structure is given by Jones and Leeman (1982). It is
important to appreciate that scattering into different
directions reveals different physical properties of the
scattering structure. For example, consider a small tissue
inhomogeneity, comprised of local fluctuations in both the
density and compressibility of the medium. The back-scattering
depends on the characteristic impedance variation of the
inhomogeneity, the forward scattering depends on its velocity
variation, whereas the right angle scattering depends on the
compressibility variation alone.

QUANTITATIVE IMAGING

Quantitative imaging is the attempt to produce a quantitative
mapping of either a single ultrasound/tissue interaction
parameter, or, at least, a well-defined, invariant combination
of such parameters. Conventional pulse-echo B-scan images,
despite their great medical importance, are not quantitative
images for two reasons. Firstly, image densities depend rather
markedly on system settings and choice of transducer. Secondly,
even if such system artefacts could be allowed for, the resultant
image would essentially be a reflectivity map of tissue:
unfortunately, reflectivity is a complicated function of both
tissue interaction parameters as well as tissue morphology. The
quantitative image in principle, will not suffer from these
defects: it would be independent of system settings and
scanning procedure, and the image density will relate directly
to an intrinsic tissue property, uncomplicated by geometrical

considerations. In this sense, the quantitative ultrasound image
presents a new dimension of tissue information. Unfortunately,
quantitative imaging demands elaborate data acquisition
arrangements, as well as extensive, computer-based, manipulation
of the acquired data.

   Two major approaches to quantitative imaging are being
developed. Transmission imaging, as computerized ultrasound
tomography, has been utilized to generate both attenuation and
velocity maps of tissues. The difficulties associated with such
methods are well-known, and will not be belabored here. We
merely point out that, as seen from the formalism developed in
the previous section, the technique measures a combination of
both propagator and reflector, the latter by virtue also of the
forward scattering that is monitored. A fundamental, and not
easily circumvented, difficulty is the importance of a relatively
simple propagator structure (viz. "straight-line propagation").

   A second approach is the attempt to generate quantitative
scatter images--i.e., the attempt to display, essentially, the
reflector, R, structure. The proliferation of techniques may be
conveniently classified by their data acquisition methods.
Diffraction tomography, and its many variants (including
filtered back-propagation) is characterized by the measurement
of the scattering into all angles. Impediography, including its
variant, reflectivity tomography, is characterized by the
measurement of the back-scattered signal only. Wave
extrapolation imaging utilizes what may be termed retro-
scattering--i.e., scattering into a relatively small cone about
the strictly back-scattered direction. This classification by
data acquisition method is, in fact, dictated by the physics of
the ultrasound/tissue interaction: each class of technique will
be expected to measure different parameters or combinations of
parameters, of the interaction. On the other hand, the variants
of each technique are merely different approaches towards mapping
the same parameter (combination).

   Quantitative scatter imaging methods are nothing but
realizations of special solutions to the inverse scattering
problem. The first step in solving such a problem is the
specification of a wave equation (i.e. a physical model),
solution of the direct problem, and only then is the uncovered
relationship between scattering parameters and measured
quantities inverted. The specification of the physical model is
a crucial step in the procedure, since only those interaction
parameters which are incorporated into the original wave equation
can be imaged. However, tissue may not conform to the postulated
model in all respects, and the measured data may well not be
totally compatible with the original wave equation. Under these
circumstances, the inversion scheme, which would be near-exact

only under the hypothesis that the postulated physical model
applies, produces an image which is, in some sense, incorrect,
and, thus, hardly quantitative.

Much emphasis has been placed on constructing inversion
schemes to underpin the quantitative scatter image, but little
thought has been given to the limitations of the approach. We
may identify three such problem areas:

(i)   Resolution: the attained resolution of the final image
is the problem that has received the most attention. This is
ultimately a data acquisition and computing problem. In this
sense, we may relate resolution primarily to the "data model."

(ii)   Distortion: the final image may show geometrical
distortions of the tissue structures. We include in this
category also amplitude distortions, viz. the displayed value
of the interaction parameter is distorted from its true value
(in an unknown way). We have merely to consider the effects
of refraction and simple attenuation to appreciate that
distortion effects ultimatey depend on the correctness of the
propagator model.

(iii) Fuzziness: an image displaying the desired interaction
parameters (combination) may be corrupted in an unpredictable
way by some other interaction parameter(s), not included in
the original physical model. Since we are dealing here
primarily with quantitative scatter imaging (although our
general considerations apply also to transmission imaging) it
is clear that fuzziness stems from an incorrect reflector
model. Note that a fuzzy image may appear quite sharp--i.e.
be apparently highly resolved!

Much attention has focused on the resolution problem; the
distortion problem has been addressed mainly within the context
of transmission imaging; while fuzziness has been virtually
unappreciated as a problem. We emphasize that the three
problems are, in fact, inter-related. Consider a conventional
impediography inversion procedure. The three steps involved
here are measurement of back-scattered signals, deconvolution to
obtain the impulse response, and inversion via an "impediography
equation" which relates the impedance profile to the impulse
response. The actual form of the impediography equation depends
on the physical model assumed. Consider a constant velocity
model in which frequency-independent attenuation (propagator)
and impedance fluctuations only (reflector) are assumed. The
final impedogram will be fuzzy if it turns out that back-
scattering from absorption fluctuations (not included in the
inversion scheme) are significant. If the attenuation is, in
fact, frequency dependent, then the ensuing pulse distortions

will ensure an incorrect impulse response, which will, in turn,
ensure a resolution loss in the displayed impedogram. However,
we will nonetheless indicate the following scheme even bearing
in mind that there may be some cross-links between the columns:

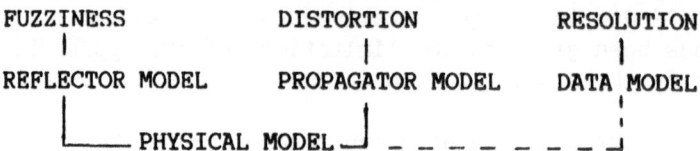

It should be clear that physical modelling underpins the entire
framework of quantitative imaging.

An interesting example is afforded by diffraction tomography.
Initially, the technique was based on the lossless Helmholtz
equation (velocity fluctuations only). Given the measurement of
angle scattering on which the technique is based, it is clear
that the original technique would give rise to a distorted and
fuzzy image. However, a much more detailed analysis by Mueller
(1980), based on quite a general visco- elastic tissue model,
has revealed that diffraction tomography, while not being fuzzy,
displays an image of a rather unexpected combination of
interaction parameters. An appreciation of the fuzziness
concept runs deep in Mueller's work, even though the idea is not
formulated as above, or analyzed as a general problem of
quantitative imaging.

Another interesting, and poorly addressed, problem
associated with quantitative imaging is one we dub "freedom
counting." The idea is easily explained by a simple example.
Consider a two-dimensional velocity distribution, $c(x,y)$. The
velocity field shows two degrees of (spatial) freedom, x and y.
It is clear that in order to reconstruct the velocity map,
measurements with two degrees of freedom will also be needed.
For example, in (two-dimensional) transmission tomography a
time-of-flight projection (one degree of freedom) must be
measured at all angles (second degree of freedom) in order to
achieve a reconstruction; in (two-dimensional) diffraction
tomography, the field scattered at all angles (one degree of
freedom) must be measured for all angles of incidence (second
degree of freedom). In order to measure both the attenuation,
$\alpha(x,y)$, and velocity distribution, $c(x,y)$, of a given (two-
dimensional) structure, it is clear that two sets of independent
data, each with two degrees of freedom, will be required. The
simple expedient of freedom counting often allows a preliminary
judgment as to whether an elaborate, and usually highly
mathematical, scheme can, in fact, reconstruct all the postulated
interaction parameters, within the context of the suggested data
set.

TISSUE CHARACTERIZATION

We will make only brief comments, since a more detailed assessment has been made by Leeman and Jones (1982). Tissue characterization methods fall into three classes (see Leeman and Jones (1982)), which we here analyze in terms of the propagator/reflector concept:

i)   Parameter estimation: attempts to find features of the propagator, e.g. mean attenuation.

ii) Structure characterization: attempts to identify characteristic features of the reflector, e.g., correlation length of fluctuations.

iii) Dynamic characterization: attempts to characterize the time dependence of either the propagator or reflector, or both.

If we recall again the fundamental imaging equation, (10), it is clear that the measured data set, M, depends not only on tissue properties (P and R), but also on system properties (D and F). It is for this reason that meaningful parameter and structure characterization methods are so difficult in practice. In fact, many tissue characterization methods are pre-doomed to failure because they are essentially "M-characterization," rather than tissue-characterization methods. The one possible exception to this is dynamic M-characterization. Since D and F can easily be arranged to be fixed (on a physiological timescale), it is clear that a characterization of the time dependence of M is, in some sense, equivalent to a dynamic characterization method (i.e. measuring time-dependence of P and R).

REFERENCES

Ishimaru, A., 1978: "Wave Propagation and Scattering in Random Media," Academic Press, N.Y.

Jones, J.P. and Leeman, S., 1982: Int. Workshop on Physics and Engineering in Medical Imaging, IEEE Computer Soc. Press (Cat. No. 82CHI751-7), 247-258.

Leeman, S. and Jones, J.P., 1982: Proc. 1st Int. Symp. on Med. Imaging and Image Interpretation, ISMII '82, IEEE Computer Soc. Press (Cat. No. 82CH1804-4), 179-184.

Mueller, R.K., 1980: Ultrasonic Imaging, 2, 213-222.

STABLE IMPEDANCE PROFILE RECONSTRUCTION

BY SPIKE TRAIN APPROXIMATION

Horst Schwetlick* and Wolfgang Sachse

Department of Theoretical and Applied Mechanics
Cornell University, Ithaca, New York - 14853 U S A

*Institut für Elektronik
Technische Universität Berlin, 1000 Berlin F R G

Abstract

This paper describes a procedure for reconstructing the impedence profile of a one-dimensional medium from ultrasonic pulse-echo data. The procedure is based on an extension of the method of characteristics in downward continuation.   Results of ultrasonic experiments are presented which make use of excitation wavelets possessing their maximum energy in the first half cycle and which are approximated by a series of spike pulses. A stable reconstruction of impedance profiles corresponding to plexiglas plates in water is demonstrated.

Introduction

The problem of reconstructing the one-dimensionnal acoustic impedance profile from wave propagation data arises in many fields of engineering and physics, including seismology, ultrasonic materials characterization, medical imaging, and ocean acoustics, among others. Here, we consider a procedure for analyzing the incident and received ultrasonic signals of waves propagating through an inhomogeneous medium which has stratifications perpendicular to the direction of wave propagation.   The stratifications are treated as a spatially varying acoustic impedance, ρc , where ρ is the density of the medium and c is its wavespeed. The aim of the measurements then is to determine the acoustic impedance along the direction of wave propagation from the measured incident and reflected waveforms. Up to now, several approaches have been developed for processing the ultrasonic signals. Among the simplest of these is the solution obtained by the Born approximation which can be shown to correspond to the impediography equation /1/. Other methods, in which the multiple reflections of ultrasonic waves from the impedance discontinuities in the medium are considered, include the Gelfand-Levitan method /2/ and that developed by Goupillaud /3/. The characteristics of these and other methods are summarized in a review article /4/.

In the method to be described, the reconstruction procedure is restricted to waves propagating in a one-dimensional medium for which the sound field is that of a plane wave such as in the far-field of a piston radiator. For that case, the equation of motion can be written as:

$$v_{tt} = A^{-1}(y) \{A(y)v_y\}_y \qquad\qquad (1)$$

where $v = v(t,y)$ represents the sound field particle displacements which are a function of time t and travel time, y. The latter corresponds to the distance in the direction of wave propagation. The function $A(y)$ is the acoustic impedance which is to be determined. The sound pressure, $f(t)$, and particle velocity, $g(t)$, and the assumption that the system is quiescent for time less than zero are introduced as boundary conditions in the y - t plane to the system, that is:

$$v_y(t,0) = f(t)/A(0) \qquad\qquad (a)$$

$$v_t(t,0) = g(t) \qquad\qquad (b) \ (2)$$

$$v(t,y) = 0 \ \text{ for } \ t < y , t < 0 \qquad\qquad (c)$$

As will be shown, $f(t)$ and $g(t)$ can be obtained from the reflection and excitation signals. From a numerical solution of the wave equation, the acoustic impedance is found by downward continuation for the wavefield along the characteristic lines /5/.

Similar methods have also been recently described by Driessel and Symes /6/ and Bube and Burridge /7/. The method of characteristics has the advantage of being the basis for a fast algorithm, requiring of order N^2 multiplications and additions, and it is capable of reconstructing the impedance profile without first deconvolving the wave signals. Numerical simulations with broadband excitation and reflection pulses of minimum delay type have shown /8/ that this procedure yields a stable solution for the impedance profile even when the data is highly noise perturbed. In all practical measurement situations, however, a transducer is used to launch and to detect the ultrasonic signals of the medium. While the principles of operation of a transducer may be based on various mechanical, electromagnetic or thermal phenomena, they invariably possess a response sensitivity which is limited to a particular band of frequencies. With band-limited data, the result obtained from the reconstruction procedure is non-unique, thus, infinitely many impedance profiles can match the same reflection and incident waveforms. This leads to instabilities in the numerical process, and additional a priori information is required to achieve a stable solution. With the procedure for approximating the incident wavelet as outlined in this paper, a stable solution can be achieved.

Reconstruction From Band-Limited Data

It is well known from deconvolution theory that a transfer function, $w(t)$, possesses only a stable and causal inverse when $w(t)$ is strictly a minimum delay-wavelet (see e.g. Robinson /9/). The method of characteristics implies a deconvolution procedure and it is therefore expected to be stable for an incident pulse which possesses the minimum delay property. This has been demonstrated by numerical experiments of a number of reconstruction simulations. As mentioned earlier, a strictly minimum delay signal is generally not obtained with normal piezoelectric transducers as used in conventional ultrasonic systems. For this reason, the inclusion of a priori information is needed to stabilize the reconstruction procedure. This can be achieved in two ways. In one, the reconstruction procedure itself is made more robust, and in the other, the measured signals are deconvolved prior to processing them, so that they are converted to minimum delay type.

In one deconvolution procedure, use is made of a priori information, in which the spectrum of the reflection data is parameterized so that it can be extrapolated to obtain the missing spectral components with a prediction filter. This method, based on the theory of maximum entropy spectral estimation, has been developed by Scheuer /10/. A similar method for stabilizing the reconstruction algorithm has recently been described by Santosa and Symes /11/. Their procedure is to select from the infinitely many impedance profiles which are consistent with the measured data, that which has the minimum L_1 - norm of the reflection coefficients A/A'. The further development of these methods in practical situations is still a topic of investigation.

The approach considered here is based on the minimum delay concept mentioned above, that is, the excitation signal is replaced by a signal of minimum delay which contains the same energy as the incident wave signal. From analysis of the equivalent circuit of a piezoelectric transducer, Cook /13/ and Redwood /14/ showed that the transducer response can be decomposed into a spike-train, $p(t)$, and a smoothing function, $s(t)$, that is

$$w(t) = p(t) * s(t) \qquad (3)$$

where $*$ denotes the convolution operator. When the transducer is used as a receiver, the corresponding spike-train has the minimum delay property.

Symes /12/ has shown that small perturbations in the incident pulse, $f(t)$, lead only to small perturbations in the obtained impedance profile. Numerical experiments have confirmed that smoothed reflection data give a stable solution of a smoothed impedance profile when the incident pulse is approximated by a delta function. Thus, the data entering the reconstruction algorithm are

$$f(t) = \delta(t) \tag{4}$$

$$g(t) = -\delta(t) + r(t) * s(t) \tag{5}$$

where $s(t)$ is the unipolar smoothing function. A spike-train approximation of these data is obtained by convolving Eqns. (5) by $p(t)$, the spike-train function. The latter is obtained by approximating the ultrasonic signal detected from a step reflector by a series of equivalent energy delta functions. Then,

$$f(t) = p(t)$$
$$\tag{6}$$
$$g(t) = p(t) + r(t) * s(t) * p(t)$$

The convolution $r*s*p = d(t)$ corresponds to the measured reflection data. Since $p(t)$ is a minimum delay spike-train, we obtain by this approximation, a stable reconstruction. The problem of decomposing the incident wavelet into a smoothing function and a spike train is achieved by an optimization procedure using the reconstruction algorithm itself. Since $w(t)$ is the reflection data of a single reference reflection, which is known, individual spikes comprising the spike-train can be adjusted in time and amplitude until an optimum reconstruction of the reference impedance profile is obtained. The resulting profile is a smoothed version of the actual one.

Thus, beginning with a set of measured reflection data, $d(t)$, and the reference reflection data from a known step reflector, $w(t)$, the entire reconstruction algorithm procedes as follows:

1. A spike train $p(t)$ representing $w(t)$ is generated

2. The boundary conditions from $p(t)$ and $d(t)$ are constructed as follows:

$$f(t) = p(t)$$

$$g(t) = -p(t) + d(t)$$

3. Using the method of characteristics, the impedance profile, A(y) , is reconstructed.

The Acoustical Experiment

In the acoustical experiment, only longitudinal waves were desired, thus the experiment was carried out in water in an immersion tank. In order to utilize the minimum delay spike train approximation, a sharp and discontinuous excitation pulse is needed. As this is difficult to attain with a conventional ultrasonic transducer, a spark source operating in distilled water was used. A typical pulse generated by this source is shown in Fig. 1(a). The reflecting layers were plexiglas plates of various thicknesses, oriented parallel to the surface of the water. In order to approximate a plane wave propagation condition, the layers were located as far from the source as possible, subject to the requirement of an adequate signal-to-noise waveform and the absence of unwanted reflections from the walls of the tank. In the experiments, this was about 15 cm.

The reflection signal was detected by a point transducer, 0.5 mm in diameter, amplified 40 dB, and digitized by a 8-bit waveform recorder to be transferred for subsequent processing to a laboratory minicomputer. The reflected signals from the plexiglass layer specimens were gated and a trend in the signals which was the result of the electromagnetic interference from the spark was minimized by a linear trend removal algorithm. The spike-train approximation used in these experiments was constructed from the reflection signal obtained from the first water-plexiglas interface and is shown in Fig. 1(b). It is likely that the approximated incident signal includes some of the effects of a non plane-wave propagating condition.

The signal corresponding to the reflection from a 12 mm thick plexiglas plate is shown Fig. 2(a) and the result of applying the reconstruction algorithm is shown in Fig. 2(b). Also shown in this figure is the expected impedance profile. A second reconstruction example is shown in Fig. 3. Here the test specimen consisted of a 2.19 mm plexiglas plate directly on top of the 12 mm thick plexiglas layer in the immersion tank such that a thin layer of water was between them. In this case, it is difficult to discern from the reflection data the location of the plate boundaries. The result of the reconstruction procedure is shown in Fig. 3(b). The small gap between the two plates is evident in the reconstructed profile. This result also shows the smoothing of the impedance profile which occurs when the spike train approximation is used to represent the incident wavelet.

It is expected that in processing the signals reflected from many layers, the result of the reconstruction becomes increasingly imprecise because the amplitude of the reflected wave signal decreases. This is a

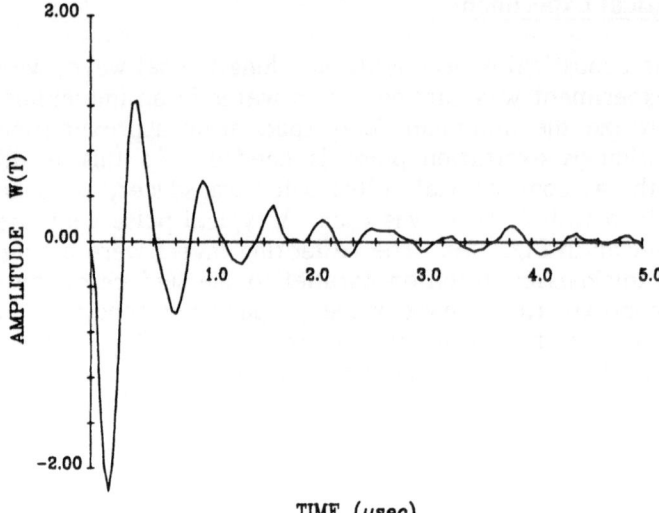

Figure I(a) - Incident wavelet corresponding to the spark source signal
reflected from the plexiglas reference reflector.

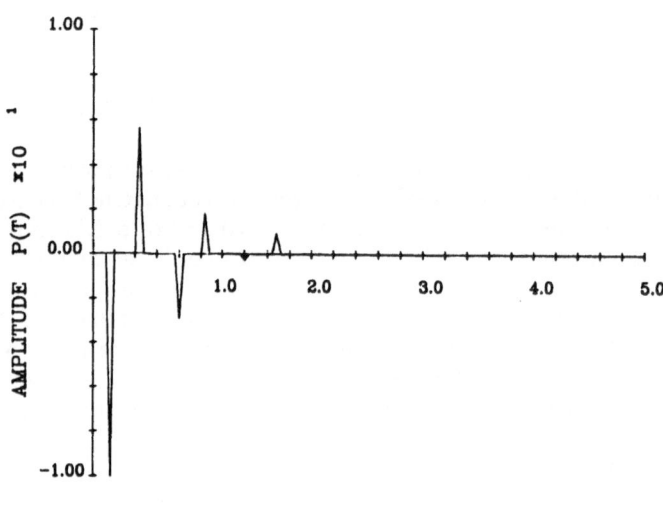

Figure I(b) - Spike-train approximation to the signal in (a).

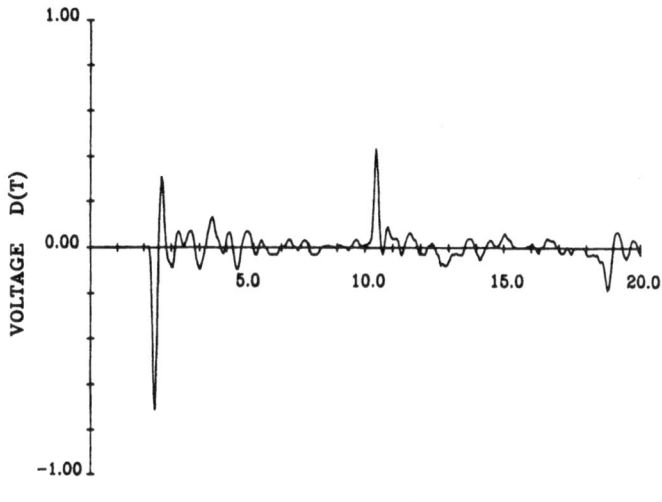

Figure 2(a) - Reflected signal from a 12 mm thick plexiglas plate.

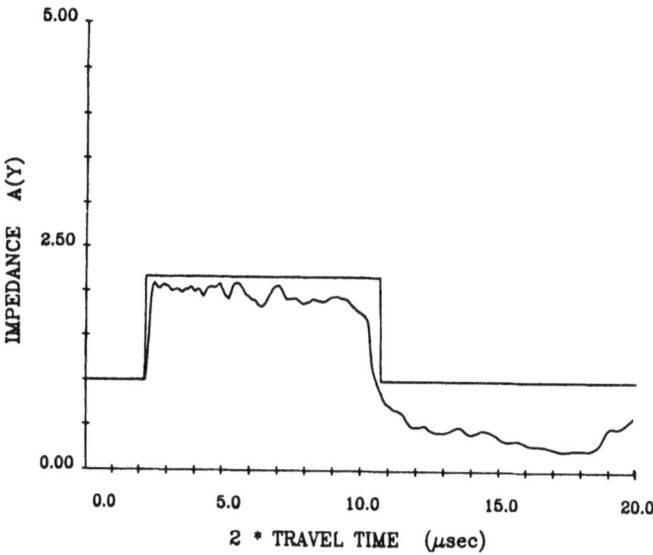

Figure 2(b) - Actual and reconstructed acoustic impedance profiles.

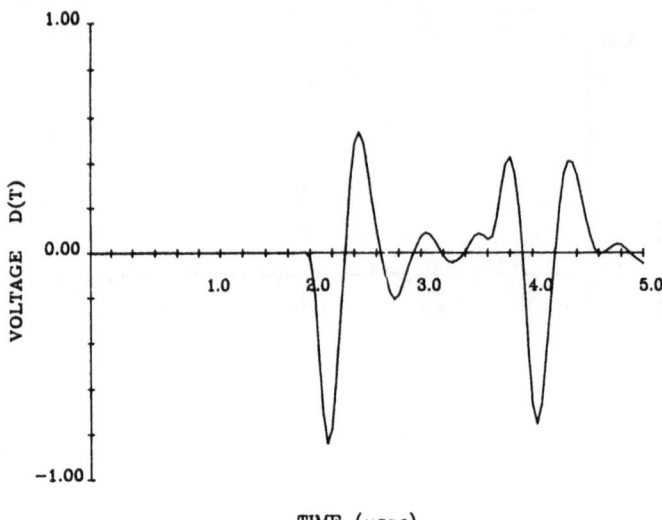

Figure 3(a) - Reflected signal from two plexiglas layers,
2.2 and 12 mm thick.

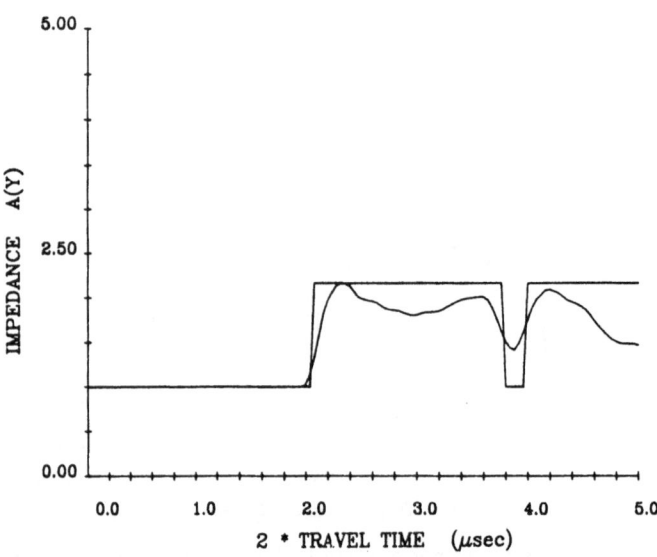

Figure 3(b) - Actual and reconstructed acoustic impedance profiles.

result of ultrasonic damping effects of the medium and the propagation of waves which are non-planar across the interfaces of the plate layers. A limit is also reached when the signal-to-noise ratio of the detected signal reaches the resolution of the waveform digitizer. The results of Figs. 2(b) and 3(b) show that for a limited time interval it is possible to recover information on the acoustic impedance. For longer times, the result obtained from the reconstruction becomes unstable. It is likely that improved results would be attainable if a more reproducible minimum wavelet generating source were available than that obtained from the spark source. The use of a shaped ultrasonic pulse generator is suggested but this awaits investigation.

Conclusion

It has been shown that actual ultrasonic waveforms obtained from plexiglas layered specimens can be processed to obtain the impedance profile of the layering. Key to the procedure is the representation of the excitation wavelet obtained in an experiment with a spike series approximation which is of minimum delay type and the utilization of the method of characteristics impedance profile reconstruction algorithm. The spike-train approximation is a series of delta functions representing the original excitation wavelet and having an energy equivalent to it. By this procedure, the method of characteristics becomes stable and useful to the inversion of actual ultrasonic waveforms.

Acknowledgement

The authors appreciate the discussions they have had with F. Santosa and W. Symes. This work was supported by the Materials Science Center at Cornell University.

References

1.   J.P. Jones, "Ultrasonic Impediography and its Application to Tissue Characterisation", Recent Advances in Ultrasound in Biomedicine, D.N. White, Ed., Research Study Press, Forest Grove, IL (1977).

2.   F. Santosa, "Numerical Scheme for the Inversion of Acoustical Impedance Profile based on the Gelfand-Levitan Method", Geophys. J. Royal Astr. Soc., 70, 229-243 (1982).

3.   J. Ware and K. Aki, "Continuous and Discrete Inverse Scattering Problems in a Stratified Medium", J. Acoust. Soc. Am., 45, 211-221 (1969).

4.   R. Burridge, "The Gelfand-Levitan, the Marchenko, and the Gopinath-Sondhi Integral equation of Inverse Scattering Theory, Regarded in the Context of Inverse Impulse-Response Problems", Wave Motion, 4, 305-323 (1980).

5.   F. Santosa and H. Schwetlick, "The Inversion of Acoustical Impedance Profile by the Method of Characteristics", Wave Motion, 4, 99-110 (1982).

6.   K.R. Driessel and W.W. Symes, "Fast and Accurate Algorithms for the Seismic Inverse Problem in One Space Dimension", AMOCO Production Co. Technical Report, Tulsa, OK (1981).

7.   K. Bube and B. Burridge, "The One-Dimensional Inverse Problem of Reflection Seismology", To appear in SIAM Review (1983).

8.   H. Schwetlick, "Inverse Methods in the Reconstruction of Acoustical Impedance Profile", J. Acoust. Soc. Am., 73, 1179-1186 (1983).

9.   E.A. Robinson and M.T. Silvia, "Digital Foundations of Time Series Analysis", Vol. 1. Holden-Day, San Francisco, CA (1979).

10.   T.E. Scheuer, "The Recovery of Subsurface Reflectivity and Impedance Structure from Reflection Seismograms", M.S. Dissertation, University of British Columbia, Vancouver, B.C., (1982).

11.   F. Santosa and W.W. Symes, "Inversion of Impedance Profile from Band-limited Data", 1983 Proceedings of the International Geoscience and Remote Sensing Symposium (IGARSS), IEEE Cat. No. 83CH1837-4, Inst. Elect. Electr. Eng., New York, NY (1983), pp. 6.1 - 6.7.

12.   W.W. Symes, "On the Map between Coefficients and Solution of Wave Evolutions", Cornell-ONR/SRO Report No. 6, Cornell University, Ithaca, NY (1983).

13.   E.G. Cook, "Transient and Steady-State Response of Ultrasonic Piezoelectric Transducers", IRE Intl. Conv. Rec., Part 9, 61-65 (1956).

14.   M. Redwood, "A Study of Waveforms in the Generation and Detection of Short Ultrasonic Pulses", Applied Materials Research, 2, 76-84 (1963).

GOLAY CODES FOR SIMULTANEOUS

MULTI-MODE ULTRASONIC IMAGING

B. B. Lee and E. S. Furgason

School of Electrical Engineering,
Purdue University, West Lafayette, Indiana 47907.

Abstract

In previous studies, a multi-mode phased array system was described which simultaneously transmits a set of uncorrelated pseudo-random Golay codes and then, upon reception, isolates the signals and their associated modes using correlation receivers. In this study the performance of the same system is evaluated under the presence of moving targets and clutter through the use of a generalized ambiguity function and a system signal-to-noise ratio formula. Results indicate that, although the special Golay code properties of self-noise and cross-correlation cancellation are diminished as the velocity of a target increases, the degradation of system performance will be minimal for all but the fastest moving targets in medical imaging applications. For most practical imaging situations the Golay code simultaneous multi-mode transmission system will still provide an improvement in speed and/or signal-to-noise ratio over conventional sequential multi-mode systems.

Introduction

One of the most important developments in medical ultrasonic imaging has been the development of rapid scanning techniques for creating two dimensional images of moving targets within the body. Systems which use rapid scanning have allowed medical specialists to make "real time" B-scan images of fetuses and fast moving organs such as the heart. Such techniques require sequential multiple-mode operation in which the modes correspond to different scan directions. These different scan directions can correspond to different transducer elements in a linear array, different rotation positions in a mechanical scanner, or different steering angles in an electronically steered phased array. A thorough review of these techniques can be found in a recent article by Ramm and Smith[1].

257

The basic advantage of a linear or phased array is that no physical movement of the ultrasonic transducer is needed to create an ultrasonic image since all the steering of scan directions is done electronically. Avoiding mechanical movement of the transducer has made it possible to create much higher resolution ultrasonic images[2].

Unfortunately, all of the present scanning techniques require sequential excitation of the various modes such that the system control must wait until all the detectable echoes from a mode have been received before switching to the next mode. If each sequential mode requires the same amount of time, T, and M modes are required to complete an image, the sequential system requires an amount of time MT to complete an image. If all M modes can be completed simultaneously over a period of time less than MT, an improvement in system speed will be realized. Ideally such a system will increase the system operation speed by a factor of M. However in practice, cross-talk interference between the modes will occur which will require extra time to reduce, and which must be minimized by proper beam forming and optimum detection methods.

From a different standpoint, if an improvement in system speed is not required, an improvement in system signal-to-noise ratio can be realized by transmitting M modes simultaneously and time averaging over the time MT.

In ongoing work we have investigated a system which simultaneously transmits M random[3] or pseudo-random signals[4] and then isolates these signals upon reception using correlation receivers. A simplified example of a simple two direction simultaneous phased array system is shown in Fig. 1. The number of

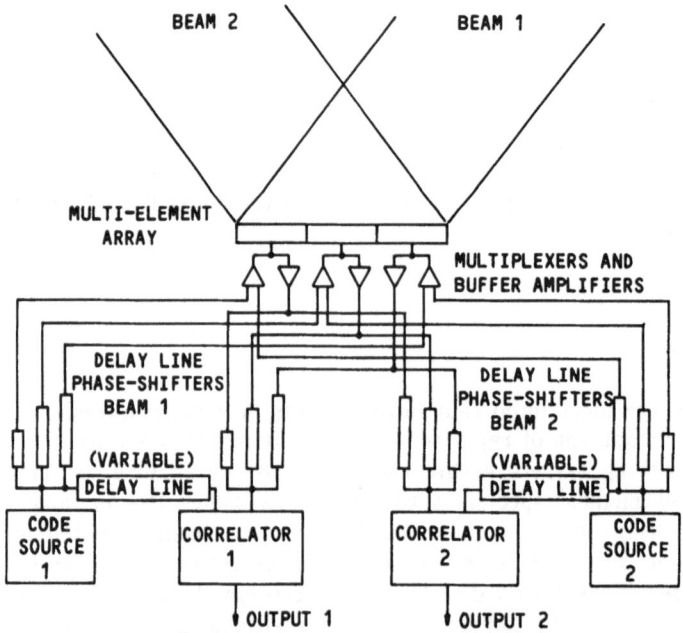

Fig. 1   A two-mode simultaneous transmission phased array.

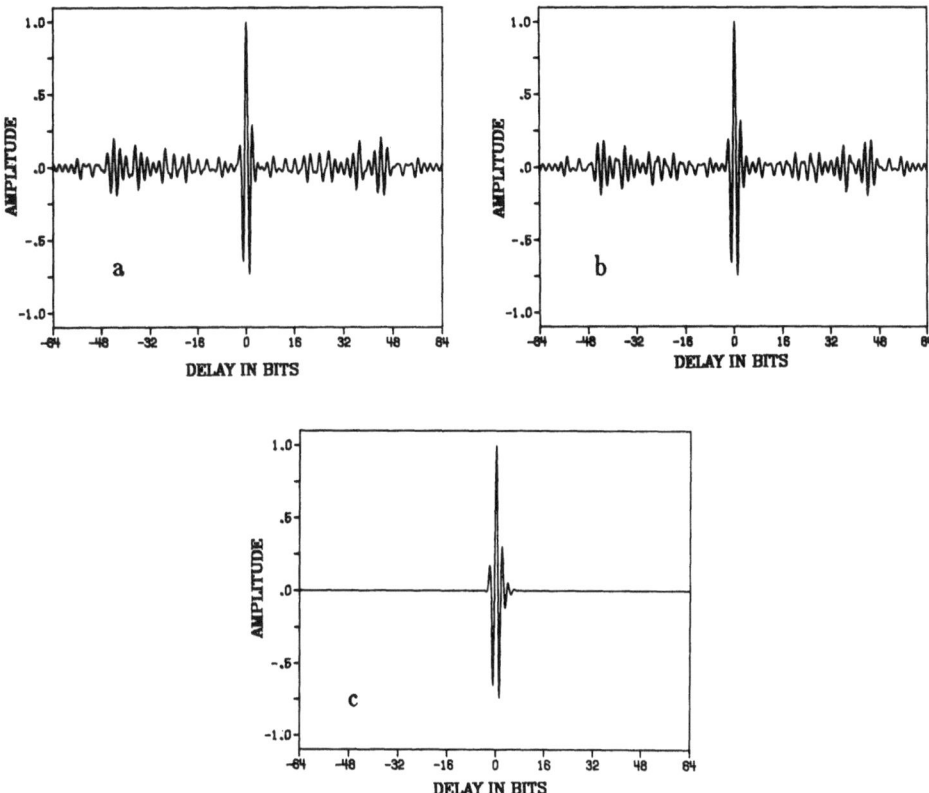

Fig. 2   Simulated correlation outputs for a. a 64 bit Golay code,
b. the complementary code to a., c. normalized sum of a. and b.

scan directions can be extended beyond the two shown by adding more signal
sources and delay lines. Preliminary studies of similar systems have also been
made by Tournois[5] and Miwa et. al.[6]. Such a system retains the longitudinal
resolution of conventional pulse-echo systems since the full bandwidth of the
transducer is utilized for all modes, and minimizes the mean squared interference
error between modes. This simultaneous transmission system has been shown to
produce an increase in system speed over sequential systems which is somewhat
less than M depending upon a number of factors including the cross-talk between
modes, the number of modes M, and the clutter-to-noise ratio which is present.
The cross-talk between modes depends on the the amount of cross-correlation
between the transmit codes and the amount of physical overlap of the mode
beam patterns.

In an investigation of single-mode operation of correlation systems which
image stationary targets we have shown that special binary pseudo-random codes
called Golay codes provide optimum performance under all input signal-to-noise
ratio situations[8]. Golay codes are paired complementary codes, described by
Golay[9], which exhibit the remarkable property that when the correlation functions of two complementary codes are added, the range sidelobes cancel, as
shown in Fig. 2. The correlation system shown in Fig. 3 transmits the paired

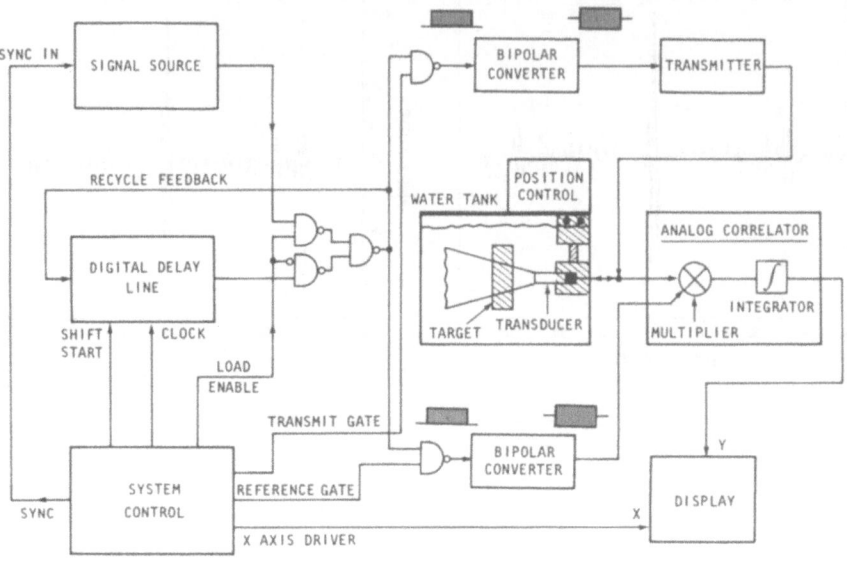

Fig. 3   Single-mode digital Golay code correlation system.

Golay codes in consecutive transmit bursts and then effectively sums the the correlation functions in the integrator of the correlator[7]. A prototype system has been demonstrated which cancels the self-noise to a level of -35 dB compared to the correlation peak[7], Fig. 4.

Certain special pairs of complementary Golay code pairs have also been shown to have the unique property of zero cross-correlation, for all relative delays, in two transmit bursts[10]. The zero cross-correlation property remains invariant when synthesis techniques are applied to generate long Golay codes from shorter Golay codes with the zero cross-correlation property. A set of four short two bit basis Golay code pairs has been proffered[7] which can be used to

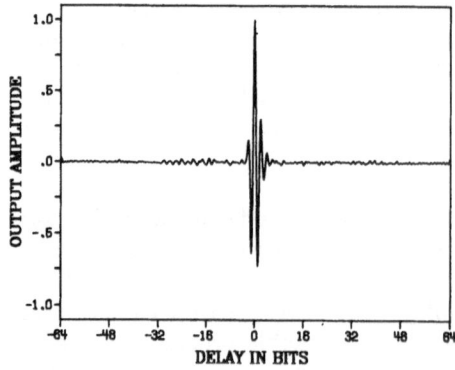

Fig. 4   Measured digital Golay code system output.

generate larger sets of longer codes with low cross-correlation levels and zero cross-correlation between each code pair and one other pair in the code set. The cross-correlation of those pairs with non-zero cross-correlation was also found to decrease as the square root of the code length, n, in the same manner as the self-noise and cross-correlation of random noise signals[3]. Since the cross-correlation cancellation requires two burst correlation the maximum possible speed improvement of a Golay code simultaneous transmission system is M/2[7].

The previous study[7] of simultaneous transmission using Golay codes did not investigate the effects of moving targets on either the autocorrelation or the cross-correlation functions of Golay codes. Since the self-noise and cross-correlation cancellation require the summation of two autocorrelation or two cross-correlation functions, which correspond to transmit bursts separated by the transmit repetition period, movement of the target between transmit bursts will produce correlation functions which will not align as required for peak superposition and self-noise and cross-correlation cancellation. Takeuchi[11] studied a phase-modulated system using Golay codes and estimated that Doppler shifts would cause self-noise which follows a curve similar to $\cos(1/2\theta)$, where θ is the phase difference between received complementary codes - due to Doppler shift.

In the following paper a more analytical study of the effects of moving targets on self-noise cancellation, peak superposition, and cross-correlation cancellation is made through the use of a generalized ambiguity function proposed by Kelly and Wishner[12]. The results are used to determine the target velocity limits of a Golay code correlation system. The results of this study are then included in an output signal-to-noise formula for a simultaneous multi-mode system under the presence of background clutter and moving targets. This signal-to-noise ratio formula is then used to determine a speed improvement formula which describes the speed improvement available in simultaneous transmission systems over conventional sequential systems.

Generalized Ambiguity Function

In order to describe the effects of moving targets on the output of a correlation system using a wideband transmit signal, Kelly and Wishner[12] have developed a generalized ambiguity function. The generalized ambiguity function can be represented in a number of equivalent integral forms. Assuming, for simplicity, that the source/receiver is stationary and that the target has zero acceleration, one convenient form of the generalized ambiguity function is

$$G(\tau,\nu) = \int_{-\infty}^{\infty} A(\omega)\, A^*[\omega\nu]\, e^{-j2\pi\omega\nu\tau}d\omega, \tag{1}$$

where τ is the time delay difference between the received signal and the reference signal, $A(\omega)$ is the complex frequency spectrum of the transmit signal, and $\nu \simeq 1 - 2v/c$; where v is the target velocity, and c is the velocity of sound. Note that for a given ν, $G(\tau,\nu)$ represents the output of a correlation system for a moving target. A simple substitution of τ/ν for τ shows that for a fixed ν, the integral can be viewed as the inverse Fourier transform of $A(\omega)A^*(\omega\nu)$, which maps into the τ domain scaled by the factor ν. The variable ν is normally very close to unity since c is very large compared to v for practical imaging situations. Thus $G(\tau/\nu,\nu) \simeq G(\tau,\nu)$ and the scale factor on τ can be ignored in the inverse Fourier transform calculation for simulations.

Computer simulations of the generalized ambiguity function were generated in the following manner. The transmit waveform was first bandlimited in the frequency domain by truncating the spectrum at the first null, which corresponds to the code clock rate. This spectrum was weighted using a Hamming window to reduce truncation sidelobes. An inverse digital Fourier transform was then calculated in which each discrete frequency component was shifted an amount in frequency corresponding to the velocity of the target. This produced a compressed

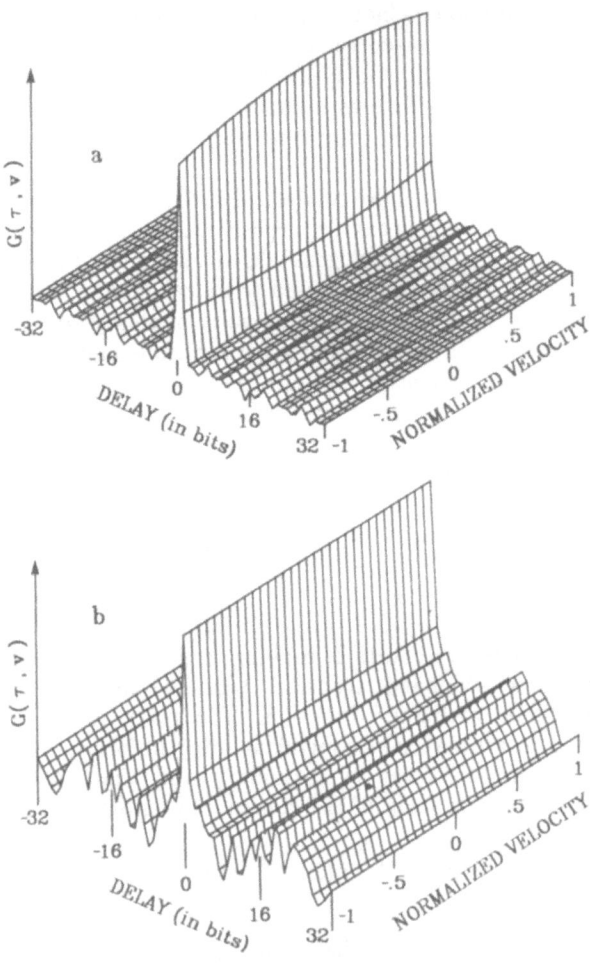

Fig. 5   Generalized ambiguity function for
a. two 32 bit Golay code pairs.
b. one 31 bit m-sequence.

time waveform corresponding to the received Doppler shifted waveform. The
Fourier transform of the Doppler shifted wave form produces $A[\omega\nu]$. The generalized ambiguity function over τ, for a constant ν, is then the inverse Fourier
transform of the product of $A(\omega)$ and $A^*[\omega\nu]$.

The value of v was varied and correlation functions over τ were calculated
to generate the generalized ambiguity function over τ and v. The generalized
ambiguity functions for a 31 bit m-sequence and for two 32 bit complementary
Golay codes are shown in Fig. 5. The velocity axis is normalized to the velocity
which results in a Doppler shift frequency for the upper frequency of the transmit
code, which is one-half the transmit repetition frequency. Doppler shift frequencies below this frequency will then fulfill the Nyquist sampling criterion. We
shall call this velocity the maximum unambiguous velocity. In the m-sequence
ambiguity function the self-noise changes in form but remains essentially constant in power for different velocities. The Golay code ambiguity function, however has zero self-noise for zero velocity and the self-noise increases with increasing velocity to approximately the same amplitude as the m-sequence self-noise.
Significant self-noise cancellation of greater than 20 dB occurs for values up to
30% of the maximum unambiguous velocity.

As a comparison, the generalized ambiguity functions for several other
waveforms were determined. A single pulse produces a single narrow ridge, Fig.
6. This ridge represents an ideal imaging function since target velocity does not
effect the amplitude or the resolution of the correlation output. If two pulse time
averaging is employed by correlating two one bit bursts, the ambiguity function
changes to the sum of two slanted ridges, as shown in Fig. 7. This two burst
averaging is equivalent to the Golay code addition without the presence of self-
noise. The triangular correlation function can be seen to decrease in amplitude
and widen with increasing velocity. Thus the benefits of time averaging are also
degraded when applied to moving targets.

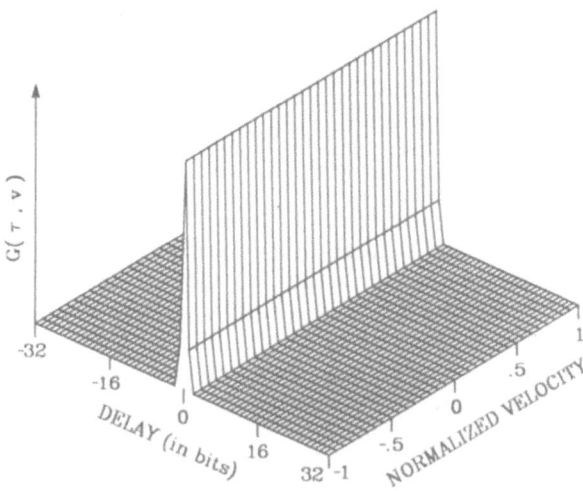

Fig. 6   Generalized ambiguity function for a single pulse of width δ.

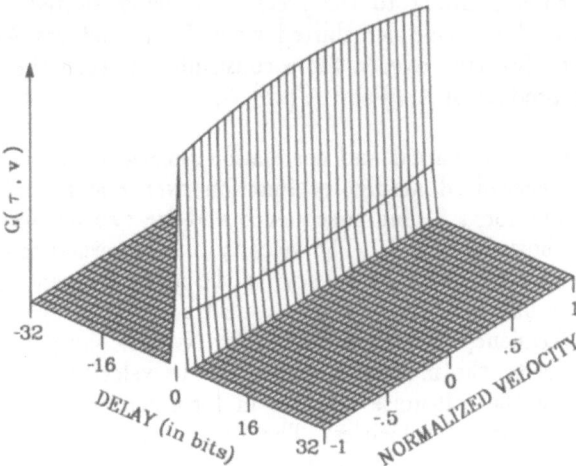

Fig. 7   Generalized ambiguity function for two single pulses of width δ separated by one repetition period.

Single-Mode Signal-to-Noise Ratio

In an evaluation of correlation flaw detection systems,[13] a signal-to-noise ratio formula was presented which included the effects of self-noise, clutter, and background receiver noise. This formula was derived assuming a very simple clutter situation — a uniform distribution of randomly distributed clutter targets each with the same relative orientation and backscattering cross-section. It was also assumed that all the noise sources are uncorrelated so that all their powers add. The total signal-to-noise ratio at the output of the correlator can then be represented as

$$\text{SNR} = \frac{P}{C(1 + \frac{2r}{N}) + \eta(\frac{b}{nN}) + P(\frac{b}{nN})}, \tag{2}$$

where P is the peak signal power due to a desired target, C is the average clutter interference power that would be seen by an ideal pulse-echo system, r is a constant which accounts for the variability of the self-noise levels of different pseudo-random codes (measurements in a sample of large grain austenetic stainless steel indicated that a practical value of r is approximately $1/2$ [13]), η is the average background receiver noise, n is the number of bits in the transmit burst, N is the number of transmit bursts correlated, and $b = 1/(\delta B)$; where δ is the pulsewidth of a single binary bit of code and B is the half power bandwidth of the received signal. The term $2r/N$ is included to account for the self-noise contribution to the clutter power. The total signal-to-noise ratio for a conventional pulse-echo system is described by equation (2) with $r = 0$ and $b/(nN) = 1$.

The signal-to-noise ratio formula, equation (2), was derived for flaw detection applications and it was assumed that the desired target and the clutter targets were all stationary with respect to the transducer. If the effects of movement are included the signal-to-noise ratio can be modified so that

$$SNR = \frac{P\,S(v_t)}{C[Q(v_c) + (\frac{2r}{N})R(v_c)] + \eta(\frac{b}{nN}) + P(\frac{b}{nN})R(v_t)} ,\qquad (3)$$

where $S(v_t)$ is the peak power variation of the desired signal with target velocity, $Q(v_c)$ is the variation in the average power of the desired signal with clutter velocity, and $R(v_t)$ is the variation of the average self-noise power with target velocity.

The velocity dependent terms, $S(v)$, $Q(v)$ and $R(v)$ can be determined using the previously described generalized ambiguity function. The peak and average power of the correlation function of the desired signal vary with velocity if two transmit burst averaging is used, regardless of the form of the transmit burst. The generalized ambiguity function for two transmit burst averaging using a single pulse in each burst, Fig. 7, describes the effects of velocity variations on the peak without any effects of self-noise present. The peak and average power variations with velocity, $S(v)$ and $Q(v)$ are shown in Fig. 8.

The variable $R(v)$ is code dependent and can be determined by subtracting the generalized ambiguity function of Fig. 7 from the generalized ambiguity function of the given code. This isolates the self-noise variation with velocity. The variable $R(v)$ can then be calculated by determining the average self-noise power variation with velocity and dividing this by the average self-noise power in the

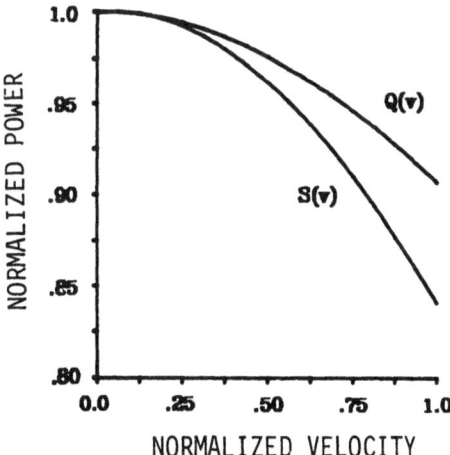

Fig. 8   Normalized peak power variation with velocity, $S(v)$, and normalized average power variation with velocity, $Q(v)$, for two transmit burst averaging.

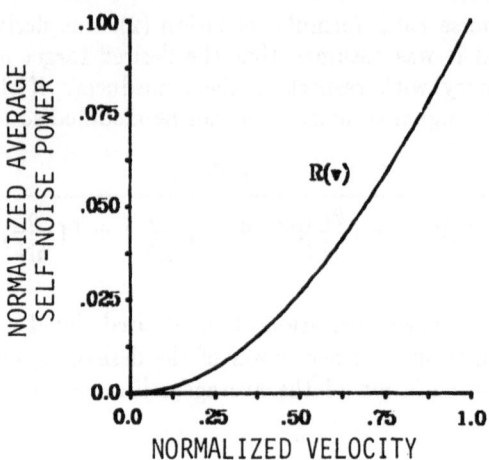

Fig. 9   Normalized average self-noise power variation with velocity, R(v), of
two complementary 32 bit Golay codes.

correlation function for a single burst of code. Using this process on the Golay
code ambiguity function results in an average self-noise power variation with
velocity as shown in Fig. 9. Without averaging the self-noise power of a pseudo-
random code does not vary with velocity as can easily be seen in Fig. 5.

The signal-to-noise ratio formula can now be used to determine the
optimum type of system to use for given clutter level and target velocity,
whether Golay code correlation or conventional pulse-echo. Similar assumptions
and comparisons to those made in references [3] and [13] can be made for different
target and clutter velocities and clutter and noise levels. However, the goal of
this paper is to discuss a simultaneous transmission system, so a further discussion of a single-mode system will not be undertaken.

Generalized Cross-Ambiguity Function

A generalized cross-ambiguity function can be generated in the same
manner as the previous generalized ambiguity function. The cross-ambiguity
function for two typical zero cross-correlation 32 bit Golay code pairs is shown in
Fig. 10. The average normalized power of the cross-ambiguity function versus
target velocity was found to be nearly identical to the average self-noise power
variation with target velocity, R(v). The average cross-correlation power was
normalized to the average power in the cross-correlation of two Golay codes.
This agreement in variation with target velocity is not a surprising result since
self-noise and cross-correlation noise both arise from finite time cross-correlation
of uncorrelated signals and the processes of self-noise cancellation and cross-
correlation cancellation are essentially the same — they both require the summation of two signals, from two consecutive transmit bursts, which are opposite in
sign.

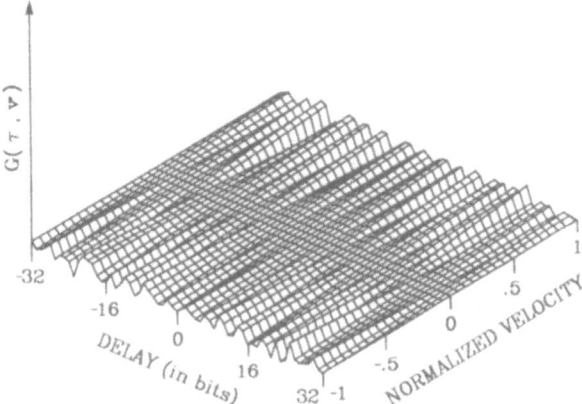

Fig. 10 Generalized cross-ambiguity function for two 32 bit complementary Golay code pairs.

Multi-Mode Signal-to-Noise Ratio

An extension of the single-mode signal-to-noise.ratio formula can be made to include the presence of extra noise sources from each of the interfering simultaneous modes, in a manner similar to that presented in reference[3] for a random signal system.

If the noise signals from the interfering modes are considered to be uncorrelated with respect to each other and with respect to the other noise signals present in the single-mode SNR formula, the noise signals can all be summed. Assuming that the cross-correlation between the transmit signals varies in power with transmit burst length and receiver bandwidth in the same manner as the cross-correlation with background receiver noise (This has been verified experimentally for random signals[3]) the signal-to noise ratio formula for a'multi-mode system is

$$SNR = \frac{P_i\, S(v_t)}{C_i\, Q(v_c) + \eta(\frac{b}{nN}) + \sum_{j=1}^{M} (\mu_j^i)^2 (\frac{b}{nN})[R_j(v_t)\, P_j + R_j(v_c)C_j\,(\frac{2rn}{b})]}, \quad (4)$$

where the subscript i denotes the desired mode, the subscript j denotes interfering signals, and the variable μ_j^i is the cross-talk overlap of the jth beam mode on the ith beam mode. This cross-talk coefficient is squared since the array can steer on transmission and reception.

In medical ultrasonic imaging applications, a wide range of target velocities will be encountered, depending upon the organ or body section under investigation. The highest velocities of about 120 cm/sec occur in the cardio-vascular system in the aorta[14]. Maximum heart wall velocities on the order of 20 cm/sec are reached during the cardiac cycle[15]. In other imaging situations the velocities encountered are much lower, for example, the velocity of the foetal chest wall is only about 4 cm/sec [16]. In typical imaging applications the maximum scan distance is about 15 cm and the corresponding maximum repetition rate is 10 KHz, if the velocity of the tissue under study is approximately the velocity of water, 1500 m/sec. The maximum unambiguous velocity which can be imaged by a single-mode conventional pulse echo system is then about 75 cm/sec. A single-mode Golay code system can then image a target which moves at less than 25 cm/sec with more than 20 dB of self-noise cancellation and even for velocities up to 75 cm/sec the cancellation is still 10 dB. The Golay code system can thus image all but the fastest moving targets in the body with minimal degradation in signal-to-noise ratio.

Assuming that the target and clutter velocities are less than 30% of the maximum unambiguous velocity, then $S(v_t)$ and $Q(v_c)$ are approximately unity, and $R_j(v_t)$ and $R_j(v_c)$ are approximately zero for $i = j$, and one other mode. If $P_j = P$, $C_j = C$, $R_i(v) = 1$ for all but the desired mode and one other mode, and $\mu_j^1 = \mu$, for all $M-1$ interfering modes, and letting $r = 1/2$, then equation (4) reduces to

$$ SNR = \frac{P}{C + \eta(\frac{b}{nN}) + \mu^2(M-2)[P(\frac{b}{nN}) + C(\frac{1}{N})]} . \tag{5}$$

A comparison to conventional sequentially excited arrays can now be made using this simplified formula.

For comparison with the simultaneous transmission SNR, the SNR of a sequentially excited phased or linear array is then

$$ SNR = \frac{P}{C + \eta}. \tag{6}$$

Note that in a phased array each simultaneous mode would be connected to each array element so that the power for each mode might have to be reduced in order that the peak average power is not exceeded for any array element. In this case the above comparison SNR would not be valid for the phased array system. We will assume, however, that the sequential phased array is not average power limited and that the average power limit is not exceeded in the simultaneous phased array. This SNR comparison formula will always be valid for a linear array since in a linear array each different code source would be connected to a different array element so that the same amount of power can be transmitted into each array element for both simultaneous and sequential excitation.

If the SNR of the simultaneous array is to be greater than the sequential array system we must have

$$ N > (\frac{b}{n}) + \mu^2(M-2)[\frac{P}{\eta}(\frac{b}{n}) + \frac{C}{\eta}]. \tag{7}$$

Assuming n is large (long transmit bursts) so that the terms with n in the denominator can be ignored, the restriction for greater SNR becomes

$$N > \mu^2(M-2)[\frac{C}{\eta}]. \tag{8}$$

The maximum speed improvement of the simultaneous system over the sequential system is the ratio of the number of modes M to the number of transmit bursts N required by a simultaneous system to reach the same SNR, so that the maximum possible speed improvement is

$$\frac{M}{N} = \frac{M}{Int[\mu^2(M-2)(\frac{C}{\eta})]}, \tag{9}$$

where Int chooses the smallest integer greater than or equal to the argument.

These equations indicate that whether a simultaneous transmission system is better than a sequential system depends on the type of application. The equations show that a simultaneous transmission system will provide the most speed improvement in low clutter-to-noise ratio situations. This is not surprising since correlation systems have been shown to be optimal for noise dominated situations.

Note that the equations of speed improvement have no meaning for a two-mode simultaneous transmission system. If two modes are desired, as in the system of Fig. 1 and if the proper zero interference codes are chosen, a Golay code simultaneous transmission system will operate at the same speed as a sequential two-mode pulse-echo system, with the increased signal-to-noise ratio provided by correlation, even in the worst case of completely overlapping beams. This two-mode system would have the same signal-to-noise ratio enhancement as the single-mode system, with the same good performance in the presence of clutter and interfering targets.

Conclusions

We have determined the limits of self-noise and cross-correlation cancellation of a Golay code system in the presence of moving targets under two burst correlation. Assuming that these limits are not exceeded, a general signal-to-noise ratio formula for a multi-mode Golay code system, under the presence of clutter and moving targets, was simplified, and then used to determine the speed improvement available with a simultaneous transmission system over conventional sequential systems. It was also assumed for simplification that all the interfering modes were identical in power and cross-talk overlap, and that certain noise terms could be made negligible by using long transmit bursts. Under the constraint of identical signal-to-noise ratio for both sequential and simultaneous systems, a formula for calculating the expected speed improvement of the simultaneous Golay code system over conventional sequential systems was determined for the linear and phased arrays. The amount of speed improvement was shown to be inversely proportional to the cross-talk coefficient, the number of interfering modes minus two, and the clutter-to-noise ratio.

In most practical imaging situations a Golay code simultaneous transmission system will provide improved performance in clutter and interfering targets compared to previous simultaneous transmission systems and will still provide a speed improvement over sequential multi-mode systems. The amount of speed improvement would be a factor of at most M/2 for all number of modes, M, if the amount of beam overlap is small. If two zero cross-correlation Golay codes are used in a two-mode system, the speed of operation is independent of the beam overlap, and even total overlap of the two beams is possible.

References

1. O. T. Von Ramm, and S. W. Smith, "Beam Steering with Linear Arrays," IEEE Trans. on Biomed. Eng., Vol. BME-30, No. 8, August 1983, pp. 438-452.

2. F. L. Thurstone and O. T. Von Ramm, "A New Ultrasound Imaging Technique Employing Two Dimensional Electronic Beam Steering," Acoustical Holography, Vol. 5, P. S. Green, Editor, Plenum Press, pp. 249-259, 1974.

3. B. B. Lee and E. S. Furgason, "The Use of Noise Signals for Multi-Mode Operation of Phased Arrays," Journal of the Acoustical Society of America, Vol. 68, No. 1, pp. 320-328, July, 1980.

4. B. B. Lee and E. S. Furgason, "Pseudo-Random Codes for Multi-Mode Operation of Phased Arrays," 1980 IEEE Ultrasonics Symposium Proceedings, Boston, MA, pp. 941-944, November 5-7, 1980.

5. P. Tournois, "Acoustical Imaging Via Coherent Reception of Spatially Coloured Transmission," 1980 IEEE Ultrasonics Symposium Proceedings, Boston, MA, pp. 747-750, November 5-7, 1980.

6. H. Miwa, H. Hayashi, T. Simura, and K. Murakami, "Simultaneous Multifrequency Ultrasonography the Principle and Technology," 1981 IEEE Ultrasonics Symposium Proceedings, Vol. 2, pp. 655-659, October 14-16, 1981.

7. B. B. Lee and E. S. Furgason, "Golay Codes for Simultaneous Multi-Mode Operation in Phased Arrays," 1982 IEEE Ultrasonics Symposium Proceedings, San Diego, CA, 1982.

8.   B. B. Lee and E. S. Furgason, "Digital High-Speed Golay Code Flaw Detection System," 1981 IEEE Ultrasonics Symposium Proceedings, Chicago, IL, pp. 888-891, October 14-16, 1981.

9.   M. J. E. Golay, "Complementary Series," IRE Trans. on Info. Theory, Vol. IT-7, No. 4, pp. 82-47, April, 1961.

10. C. -C. Tseng and C. L. Liu, "Complementary Sets of Sequences," IEEE Trans. on Infor. Theo., Vol. IT-18, No. 5, September 1972, pp. 644-652.

11. Y. Takeuchi, "An Investigation of a Spread Energy Method for Medical Ultrasound Systems, Part Two: Proposed System and Possible Problems," Ultrasonics, Vol. 17, No. 5, pp. 219-224, 1979.

12. E. J. Kelly and R. P. Wishner, "Matched-Filter Theory for High-Velocity Accelerating Targets," IEEE Trans. on Military Electronics, Vol. MIL-9, No. 1, pp. 55-69, January, 1965.

13. B. B. Lee and E. S. Furgason, "An Evaluation of Ultrasound NDE Correlation Flaw Detection Systems," IEEE Trans. Sonics and Ultrasonics, Vol. SU-29, No. 6, pp. 359-369, 1982.

14. R. S. Renemen and M. P. Spencer, "Local Doppler Audio Spectra in Normal and Stenosed Carotid Arteries," Ultrasound in Med. and Biol., Vol. 5, No. 1, pp. 1-11, January, 1979.

15. J. A. Ambrose, B. D. King, L. E. Teichholz, D. T. LeBlanc, M. Schwinger, and J. H. Stein, "Early Diastolic Motion of the Posterior Aortic Root as an Index of Left Ventricular Filling," J. Clin. Ultrasound, Vol. 11, No. 7, pp. 357-364, September, 1983.

16. L. W. Korba, R. S. C. Cobbold, and A. J. Cousin, "An Ultrasonic Imaging and Differential Measurement System for the Study of Fetal Respiratory Movements," Ultrasound in Med. and Biol., Vol. 5, No. 2, pp. 139-148, 1979.

SPECTRAL EQUALISATION: A SIGNAL PROCESSING TECHNIQUE TO INCREASE

PULSE-ECHO SIGNAL BANDWIDTH

L.F. van der Wal*, A.C. Dijkhuizen* and G.L. Peels**

* Institute of Applied Physics TNO-TH,
  P.O. Box 155, 2600 AD Delft, The Netherlands
**University of Technology
  Laboratory of acoustics and seismics
  P.O. Box 5046, 2600 GA Delft, The Netherlands

ABSTRACT

The spectral bandwidth of received pulse-echo signals often is far
less than the bandwidth originally generated by the transmitter
electronics. This decrease in bandwidth may e.g. be caused by
attenuation and/or transmitter/receiver characteristics. Processing
techniques, such as inverse filtering, aim to restore the original
signal bandwidth in order to improve temporal resolution of the
pulse-echo data.
Spectral equalisation is a relatively unknown processing technique
with similar objectives. It aims at equally distributing the signal
energy within the available spectral bandwidth, i.e. to flatten the
frequency spectrum.
In this paper we will discuss two types of application:

  - Spectral equalisation on broad-band attenuated signals.
    This type of application may successfully be applied in
    seismic data processing.
  - Spectral equalisation on sequentially transmitted narrow-band
    signals covering different bandwidths.
    Presently this technique is being used in an acoustic
    sub-bottom profiler to improve depth-resolution.

Processing application, parameters and results of spectral
equalisation will be discussed on the basis of computer modelled
(synthetic) data.

1. INTRODUCTION

In most fields within the echo-acoustic discipline, frequency-
dependent attenuation is a well-known phenomenon [1,2,3].
In some cases, e.g. in tissue characterisation (medical echography
[4]) and vertical seismic profiling (seismic exploration [5]),
accurate determination of attenuation characteristics is pursued in
itself.
Most frequently however, attenuation is looked upon as a perturbation of the recorded data, since it decreases spectral bandwidth and
limits penetration depth.
Since attenuation both depends on frequency and travel time, its
effects are difficult to compensate. An additional problem forms
the dependency on medium characteristics, which are far from always
accurately known. Finally, although this assumption is often made,
attenuation is not a linear function of frequency in most cases,
and it does not show zero-phase characteristics. Thus phase
distortion and dispersion effects may occur, interfering with phase
sensitive signal processing techniques, e.g. synthetic focussing
and inverse filtering.

In summary, accurate compensation of attenuation effects is time
consuming and not easily performed on-line. Therefore, in most real-
time systems, attenuation is corrected as a function of travel time
only, using time-variant gain settings (TVG), based on the
estimated attenuation of the centre frequency only.
Time-variant amplification may be successfully applied in media,
where attenuation varies little with frequency, e.g. in underwater
acoustics.
In those fields where data processing is performed off-line, e.g.
seismic migration and synthetic focussing, attenuation effects may
be corrected more accurately, i.e. as a function of both travel
time and frequency. However, since time-variant frequency filtering
is a mere deterministic process, it is often interfered with by
medium characteristics. This quite often results in recursive
interactive processing, which is not only very delicate but time-
consuming and expensive as well.

The spectral equalisation process, discussed in this paper, differs
from the above mentioned processing techniques in that it is a data
adaptive method, which requires little or no prior knowledge on the
medium under investigation. In the authors opinion it may prove an
acceptable mean between simple time-variant amplification and
extensive digital processing.

2. SPECTRAL EQUALISATION

Spectral equalisation aims at restoring the originally transmitted
frequency spectrum within the recorded data. Spectral equalisation
is not a processing technique which removes all attenuation

effects however: the main objective of the process is to remove the
relative differences in the energy spectrum, caused by attenuation
and/or geometrical spreading.
Thus the frequency dependent attenuation effects are removed first.
Attenuation effects with respect to travel time can be compensated
for after spectral equalisation using a simple TVG-technique.

As mentioned in the introduction, the spectral equalisation process
is a data adaptive method, applicable in many situations where
little a priori knowledge of the medium is available.
The process itself is probably best explained from an example.
Figure 1a shows a broad-band echo-registration, with a flat zero-
phase frequency spectrum, shown in figure 2a. Before spectral
equalisation is applied, the broad-band signal is decomposed in a
given number (3 in our example) of narrow-band signals, using
cos^2-windows in the frequency domain (figures 1b, c and d).
From figures 2b, c and d it can be seen that the cos^2-windows have
identical bandwidths and add up to the original frequency spectrum
exactly.

However, if the transmitted signal passes through an attenuating
medium, the received frequency spectrum will no longer be flat, nor
zero-phase. And consequently the narrow-band signals will no longer
have equal signal energy.

Now the objective of spectral equalisation is to restore the
original shape and bandwidth of the transmitted spectrum by
removing these mutual differences in narrow-band signal energy,
i.e. to flatten (or rather equalise) the received frequency
spectrum.
Since we want to be able to discriminate in time (or depth),
spectral equalisation is always performed in the time-domain. With
respect to the processing itself, we may follow a number of
different approaches. A rather simple, but fast method of spectral
equalisation is based on the narrow-band signal envelope, and will
be discussed in section 3.
Presently a more sophisticated spectral equalisation technique,
especially with respect to signal to noise ratio, is being applied
in an acoustic sub-bottom profiler system, which uses a parametric
sound source [6].
Due to the low efficiency of the parametric effect the profiler
system can only generate secondary signals with a maximum bandwidth
of 5 kHz. Although generation of secundary signals with a larger
bandwidth is possible, the resulting signal to noise ratio in those
cases is unacceptable.
To improve the depth-resolution of the system, three difference
frequencies (10, 15 and 20 kHz) are therefore generated
sequentially. The received echoes are A/D-converted and stored in a
digital memory, after which in-phase addition of the data yields
the desired broad-band response.

DEPTH (M)

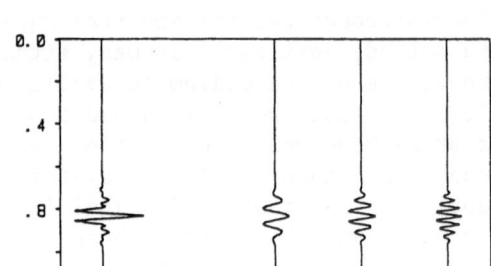

Figure 1: A broad-band echo-registration is decomposed in a given
number of narrow-band signals before spectral
equalisation is applied.

FREQUENCY (kHz)

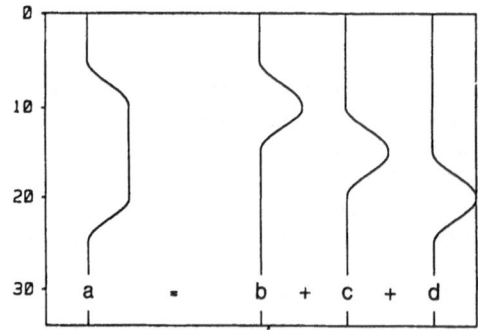

Figure 2: Decomposition in narrow-band signals is achieved, using
$\cos^2$-windows in the frequency domain.

However, since the source strength of the secondary signals depends on (frequency)^2, the individual narrow-band signals show large differences in amplitude and signal to noise ratio [6,7]. Spectral equalisation is necessarily applied to achieve both the desired shape of the frequency spectrum and an acceptable signal to noise ratio. This rather special type of spectral equalisation will be discussed in section 4, on the basis of a large number or examples.

Note that in the sub-bottom profiler discussed above, the decomposition in narrow-band signals is achieved in hardware, using a special data acquisition method, and not in software during signal processing, as mentioned earlier. The similarity of both methods is obvious, since figure 1 may be read from left to right, as well as from right to left.

3. SPECTRAL EQUALISATION BASED ON SIGNAL ENVELOPE

In this section we will discuss a rather straightforward type of spectral equalisation based on signal envelope information only. As mentioned in the abstract, this type of processing can be applied to seismic data. An example with respect to a common-shot gather will be given at the end of this section.

Figure 3a shows a noise-free broad-band registration, which contains two distinct echoes. The result of narrow-band decomposition is shown in figures 3b, c and d. Regarding the first echo, the narrow band signals show an amplitude ratio of 0.5 : 0.2 : 1.0*. Regarding the second echo, the amplitude ratio of the narrow-band signals has changed due to frequency-dependent attenuation: the difference between the high- and low-band has become smaller, the difference between the low- and centre-band is increased. Relative differences in signal energy can now be removed on the basis of signal envelope. Denoting the narrow-band envelopes as ENV1(t), ENV2(t) and ENV3(t) respectively, the mean envelope is defined as:

$$ENV(t) = \frac{1}{3} [ENV1(t) + ENV2(t) + ENV3(t)], \qquad (1)$$

as shown in figure 3e.

---

* Although these differences in signal amplitude seem very unrealistic, they were chosen here to correspond with the examples shown in section 4, where the sequentially generated narrow-band signals show similar differences in source strength.

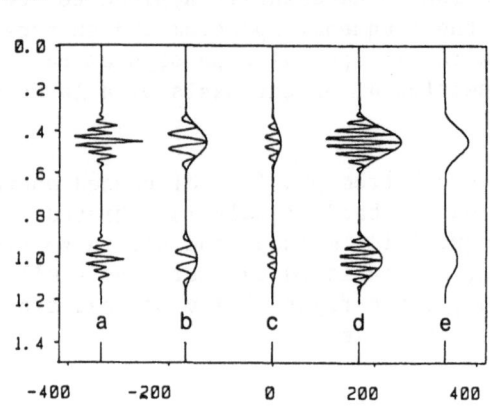

Figure 3: Decomposition of the broad-band registration (a) shows
          large differences in narrow-band signal amplitude as
          well as frequency-dependent attenuation effects (b,c,d).
          Spectral equalisation is based on mean signal envelope
          (e).

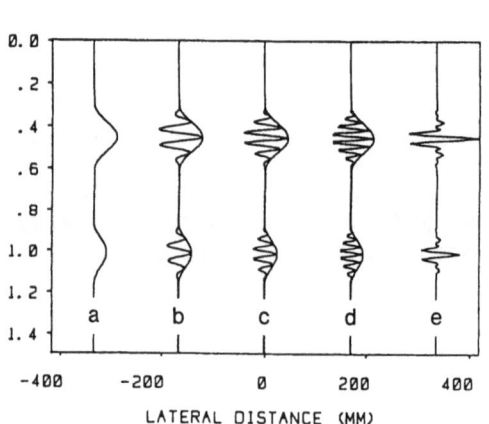

Figure 4: Spectral equalisation of the narrow-band signals, shown
          in figure 3, is based on mean signal envelope (a).
          Processed narrow-band signals show equal envelopes
          (b,c,d), while depth-resolution is improved after
          addition (e).

Now, spectral equalisation is achieved by applying the following apodisation functions to the r.f.-narrow-band signals:

$$\text{low band} \quad : W1(t) = ENV(t)/ENV1(t),$$
$$\text{centre band}: W2(t) = ENV(t)/ENV2(t), \qquad (2)$$
$$\text{high band} \quad : W3(t) = ENV(t)/ENV3(t).$$

Results of the process are shown in figure 4. Figure 4a again shows the mean signal envelope, figures 4b, c and d show the processed narrow-band signals (equal envelopes!) and figure 4e finally shows the broad-band result after addition. The increased depth-resolution can be clearly seen in both echoes.

Remarks:

1. The described method of spectral equalisation does not affect the signal to noise ratio, since both noise and reflection data are processed in the same way.
2. Spectral equalisation does not compensate for depth-dependent attenuation effects (Note the difference in amplitude between the first and second echo in figure 3e).
3. The method is strictly data adaptive, since no a priori knowledge on medium or attenuation characteristics is used. Hence, phase distortion due to attenuation, is not corrected for by spectral equalisation.
4. The data adaptive nature of the method, allows a highly selective suppression of band-limited interference (e.g. ground-roll in seismic exploration; see example).

Example:

In figure 5a a seismic common-shot gather is shown, showing an excessive amount of ground-roll. Using spectral equalisation based on signal envelope, this band-limited interference can be adequately suppressed as shown in figure 5b.

4. SPECTRAL EQUALISATION BASED ON SPECTRAL ENERGY

As mentioned in section 2, spectral equalisation in this section will be discussed on the basis of an acoustic sub-bottom profiler system, which uses a parametric sound source [6]. The system sequentially generates three difference frequencies, covering three adjacent frequency bands, i.e. 7.5 - 12.5 kHz, 12.5 - 17.5* kHz and 17.5 - 22.5 kHz. Now spectral equalisation becomes essential to

---

* Actually the system generates a 5 kHz (2.5 - 7.5 kHz) difference frequency. However, since the 5 kHz band shows a very low signal to noise ratio under sailing conditions, the 15 kHz band, which is generated as a first uneven harmonic, is used during data processing (see reference [6]).

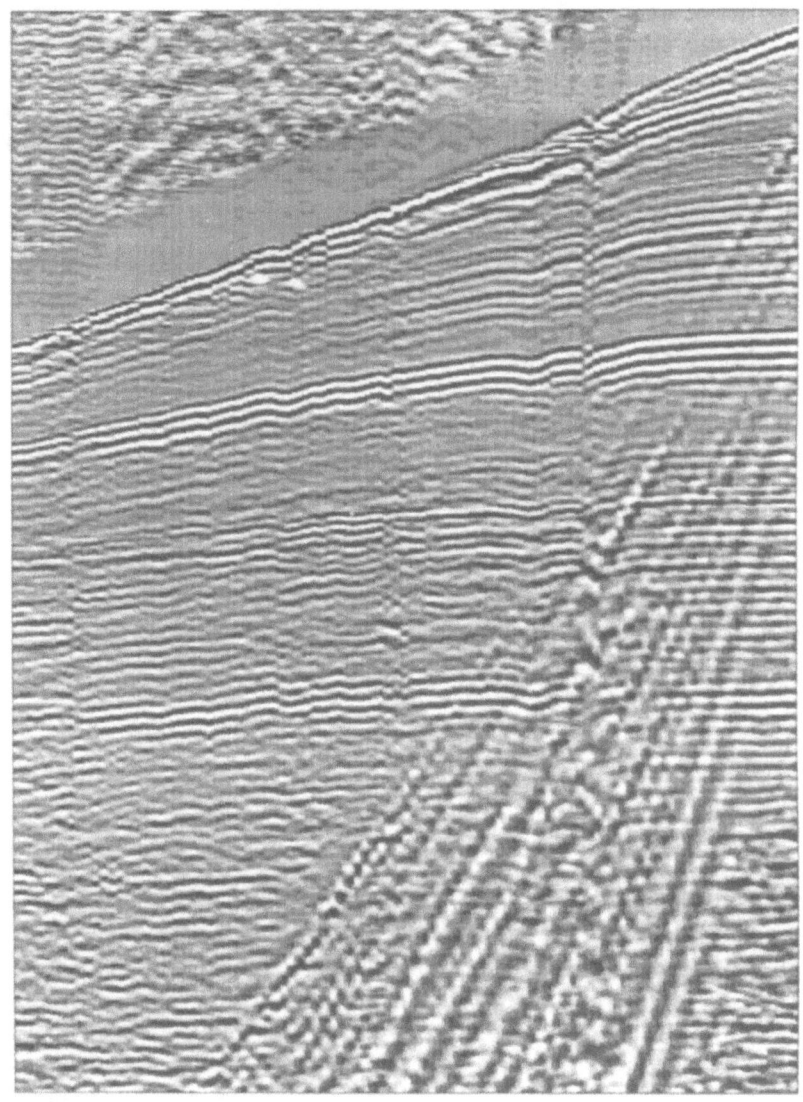

Figure 5a: Seismic common–shot data showing an excessive amount of
          band–limited ground–roll in the lower right corner
          (courtesy of the Institute of Groundwater Survey TNO,
          Delft, The Netherlands).

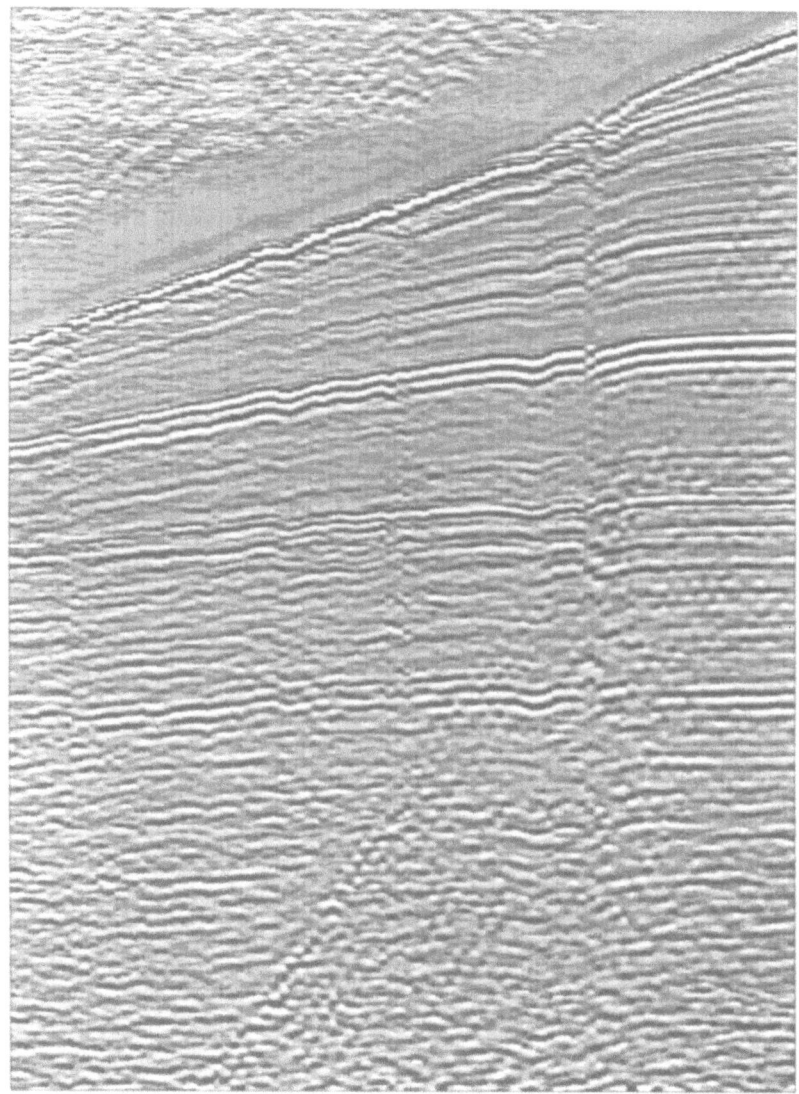

Figure 5b: Data from figure 5a shown after spectral equalisation
based on mean signal envelope. Note the selective
suppression of ground-roll interference (courtesy of the
Institute of Groundwater Survey TNO, Delft,
The Netherlands).

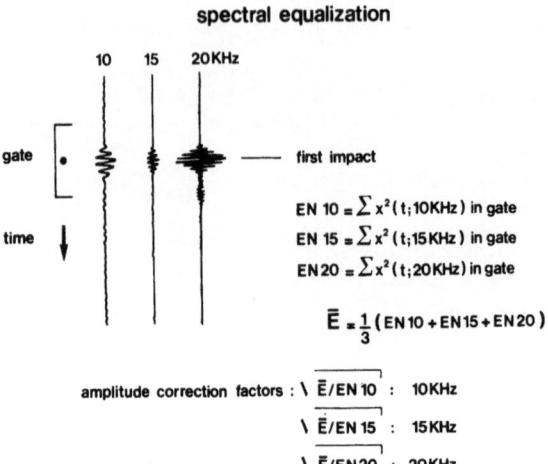

Figure 6: Schematic illustration showing the basic principle of
spectral equalisation based on spectral energy content.

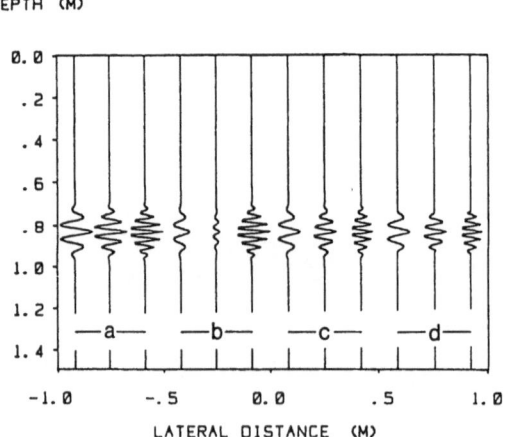

Figure 7: Simulated narrow-band sub-bottom reflections, both ideal
(a) and with realistic differences in source strength
(b). Results of spectral equalisation using a 16-sample
window and 0% (c) and 20% (d) energy levels.

eliminate differences in source strength, geometrical spreading and sub-bottom attenuation.

The process will be explained from figure 6. Starting at the top of the region of interest, we apply a time window of a given length to the digitized 10, 15 and 20 kHz registrations. The energy within these windows is denoted as EN10, EN15 and EN20 respectively. With the mean energy, E = 1/3 (EN10 + EN15 + EN20), the following correction factors can be applied to the data sample in the middle of each window:

$$10 \text{ kHz}: \sqrt{E/EN10},$$

$$15 \text{ kHz}: \sqrt{E/EN15}, \qquad\qquad (3)$$

$$20 \text{ kHz}: \sqrt{E/EN20}.$$

With respect to the first sub-bottom reflection, these apodisation factors may be regarded as a correction of differences in transmitted acoustic energy and differential losses (absorption, spherical spreading) within the water layer above. For all subsequent data samples - the window is shifted from sample to sample along the time-axis - the apodisation expresses a compensation for relative differences in attenuation.

Under sailing conditions the registrations of the sub-bottom profiler show a relatively high noise level. The described method of spectral equalisation offers an attractive opportunity to increase the signal to noise ratio, since we can set the correction factor to zero, if the energy in the window is less than a given percentage of the total energy in the registration. If so, we assume the window does not contain a sub-bottom reflection, only noise.
In the following we will discuss parameter selection with respect to windowlength and energy level to obtain a minimum loss of reflection data and a maximum suppression of noise. Results will be presented on the basis of computer-generated data, for good flexibility only.

4.1. Results on synthetic data

In figure 7a ideal registrations are shown, i.e. three noise-free records, which have equal signal energy. In figure 7b we have simulated registrations with a realistic difference in source strength (ratio 0.5 : 0.2 : 1.0). Results shown in figures 7c and 7d are obtained from figure 7b, using spectral equalisation with a 16-sample window (ca. one quarter of the narrow-band impulse response) and an energy level of 0 and 20% respectively.

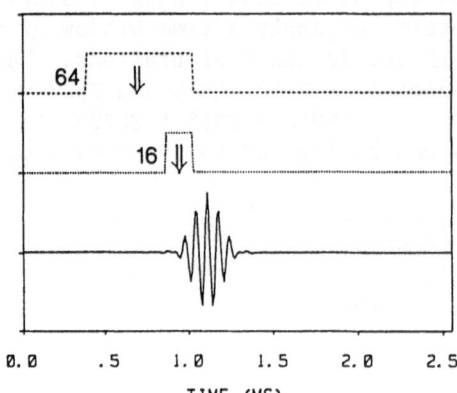

Figure 8: Schematic illustration of the low amplitude distortion
effect. Using a small time window in combination with a
high energy level, low amplitude values will be set to
zero.

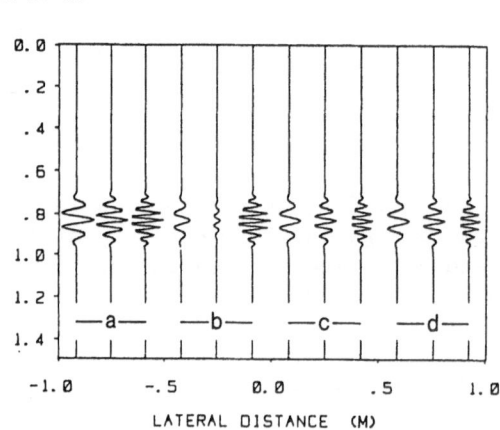

Figure 9: Simulated narrow-band sub-bottom reflections, both ideal
(a) and with realistic differences in source strength
(b). Results of spectral equalisation using a 64-sample
window and 0% (c) and 20% (d) energy levels.

We see that, when no energy level is used, the equalisation process corrects the registrations perfectly well. However, when a 20% level is applied, there is some distortion at the extremities of the impacts. The reason for this is explained in figure 8.
The 16-sample window has to be "pushed" into the impact for more than half of its length, before the energy in it exceeds 20% of the total energy in the registration. Correction factors with value 0 are therefore applied to small amplitudes both at the beginning and the end of the impacts. This effect may not so easily occur if we use a larger window, because then the center of the window, where the correction factor is applied, is located further from the actual impact.

Results with a 64-sample window and energy levels of 0 and 20% respectively are shown in figure 9. Note the distortion at the extremities has now disappeared (figure 9d).

To create a more realistic situation we add band-filtered white noise to the registrations. Although the spectral density of the noise is the same in each registration, the signal to noise ratio will vary due to the simulated differences in source strength. Thus the best signal to noise ratio is obtained in the 20 kHz registration.

Spectral equalisation is carried out using both a 16- and a 64-sample window. Energy levels were chosen 0, 1 and 5% in both cases. Since the registrations have a total length of 1024 samples and only contain a single sub-bottom reflection, the 1% energy level should give good results when using the 16-sample window. The higher 5% energy level is probably the best choice for the larger 64-sample window.
Processing results are shown in figure 10 for the 16-sample window and in figure 11 for the 64-sample window. As expected both figures show there is no adequate noise suppression if we do not apply an energy level. Figure 11b even shows that, using the larger 64-sample window, the noise in the 15 kHz registration is somewhat amplified, while the noise in the 20 kHz registration is slightly attenuated. This may be explained from the fact that, because of the larger windowlength, correction factors with values larger (15 kHz) and smaller (20 kHz) than 1 are applied to noise samples located just in front and after a sub-bottom reflection. The width of this "noise-area" is mainly determined by the windowlength, not by the applied energy level.

Using the 16-sample window, an adequate suppression of the noise is obtained with an energy level of 1% (figure 10c). However, due to the very low signal to noise ratio in the 15 kHz registration, here some of the noise will not be suppressed. Increasing the energy level to 5% (figure 10d) removes the noise completely, but the extremities of the sub-bottom reflections become distorted. Using

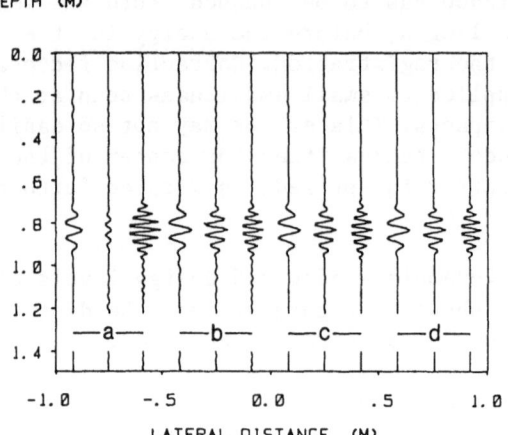

Figure 10: Simulated narrow-band sub-bottom reflections with
band-limited noise (a). Results of spectral equalisation
using a 16-sample window and 0% (b), 1% (c) and 5% (d)
energy levels.

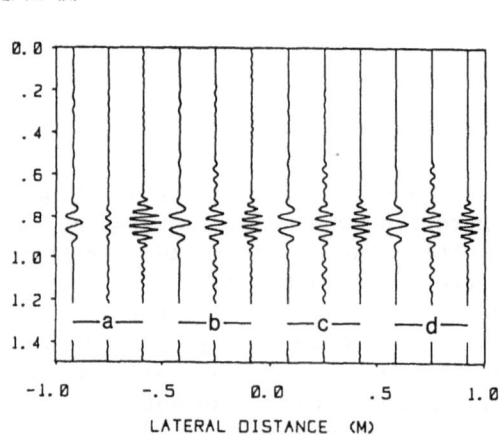

Figure 11: Simulated narrow-band sub-bottom reflections with
band-limited noise (a). Results of spectral equalisation
using a 64-sample window and 0% (b), 1% (c) and 5% (d)
energy levels.

the 64-sample window, we can see that the 1% energy level is too
low (figure 11c). Even with an energy level of 5% a distinct
"noise-area" can still be seen in both the 15 kHz and 20 kHz
registrations (figure 11d).

Summarizing the results of this section:

1. Best equalisation results are obtained with small time windows
   (i.e. small with respect to the narrow-band impulse response).
   Large windows will give rise to so-called "noise-areas" around
   distinct sub-bottom reflections.
2. The choice of the energy level should be based on the signal to
   noise ratio of the recorded data. Since high energy levels tend
   to distort the received waveforms, the level is best chosen as
   low as possible.
3. Spectral equalisation should not be applied to registrations
   with a very poor signal to noise ratio. Due to the data adaptive
   nature of process, no discrimination is made between strong
   noise contributions and indistinct reflection data.

4.2. Practical aspects

In practical situations the sub-bottom will not form a perfectly
horizontal layer, due to either bottom-topography or "heave"-
movements of the ship.
Figure 12 gives a somewhat exaggerated example of a non-horizontal
sub-bottom registration. As shown in figure 13, spectral
equalisation of these registrations results in largely distorted
reflections, because we apply large correction factors to noise
samples. For this reason, the received echoes have to be locally
stratified, before the spectral equalisation process can be
applied. In the profiler system this stratification is based on the
h.f.-bottom trigger, which is derived from the 200 kHz primary
frequency.
Figure 14a shows a locally stratified sub-bottom with two distinct
reflections. As mentioned in section 3 (see figures 3 and 4) the
amplitude variations in the first bottom reflection are caused by
differences in source strength and geometrical spreading.
Variations in succeeding reflections are also affected by frequency-
dependent attenuation.
Results of spectral equalisation (16-sample window; energy levels
of 0, 1 and 5% respectively) are shown in figures 14b, c and d. All
figures show the lateral equalisation of signal energy, as
expected. Best results, with respect to signal enhancement, are
obtained using the 1% energy level. Although there is some noise
left in the 15 kHz registration, the low-level signal amplitudes
are not distorted.

Note that, since spectral equalisation does not correct attenuation
effects which depend on travel time, the vertical variation in echo-

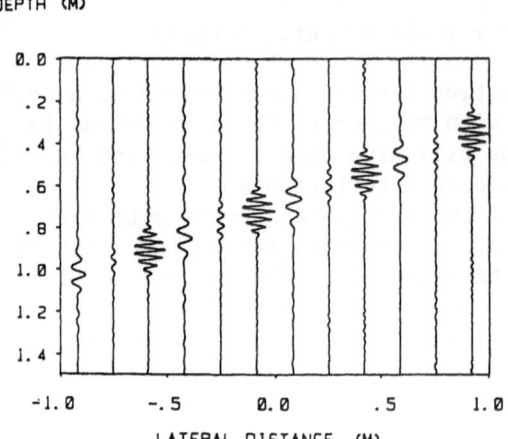

DEPTH (M)

Figure 12: Simulated example of a possible non-horizontal
sub-bottom registration.

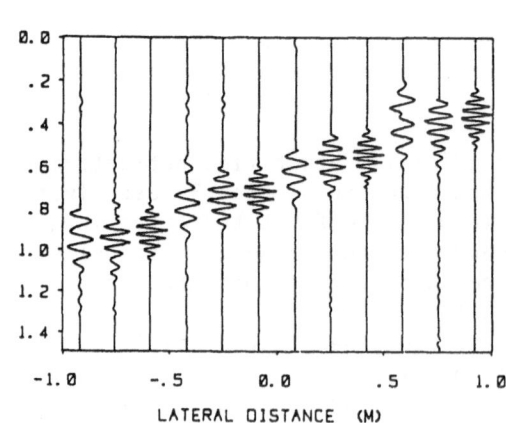

DEPTH (M)

Figure 13: Results of spectral equalisation of the data shown in
figure 12, using a 16-sample window and an energy level
of 1%.

strength is maintained. These time-dependent effects can be easily
corrected for after spectral equalisation, as will be shown in the
following paragraph.

4.3. Spectral equalisation in combination with time-variant
     amplification.

Since spectral equalisation has removed all frequency-dependent
attenuation effects, the compensation for time-dependent effects,
using time-variant amplification, is no longer approximate, but
theoretically correct.
In addition time-variant amplification will not result in an unacceptable enhancement of noise contributions, since spectral
equalisation offers a good opportunity for noise suppression.

Results of time-variant amplification after spectral equalisation
are shown in figure 15. Figure 15a shows ideal noise-free
sub-bottom responses (no differences in source-strength, no
frequency-dependent attenuation). Figure 15b illustrates the
actually recorded data, previously shown in figure 14a. The result
of spectral equalisation is shown in figure 15c. Finally figure 15d
shows the processed data after time-variant amplification, based on
the (estimated) depth-dependent attenuation of the 15 kHz centre
frequency only.

From figure 15d we may conclude that:

a. Time-variant amplification accurately corrects depth-dependent
   attenuation effects, after spectral equalisation.

b. The enhancement of noise-contributions, due to time-variant
   amplification, is quite acceptable.

CONCLUSIONS

In this paper the authors have described a relatively unknown data
processing technique, which is refered to as spectral equalisation.
A number of processing applications, with respect to the correction
of frequency-dependent attenuation, have been discussed on the
basis of computer modelled data. Processing results on real data
(seismic exploration) are also shown.
Since spectral equalisation is a data adaptive method, satisfactory
results are obtained, with little or no a priori knowledge on
medium of attenuation characteristics.
Equalisation based on spectral energy, as discussed in section 4,
offers a good opportunity to increase the signal to noise ratio of
the recorded data.

Both frequency- and time-dependent attenuation effects can be
compensated for using spectral equalisation in combination with a
simple TVG-technique.

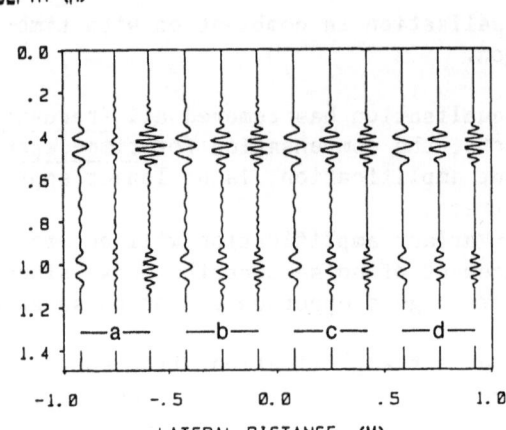

Figure 14: Simulated narrow-band sub-bottom reflections, showing
realistic differences in source strength as well as
frequency-dependent attenuation effects (a). Results of
spectral equalisation using a 16-sample window and 0%
(b), 1% (c) and 5% (d) energy levels.

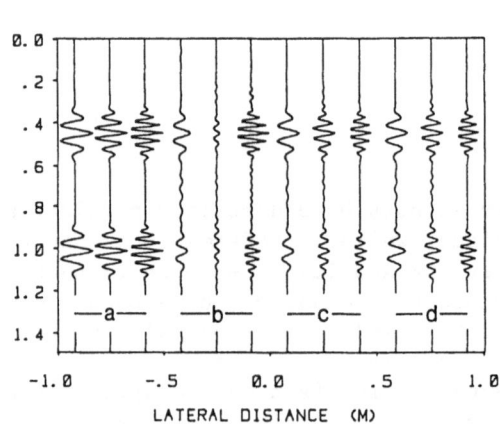

Figure 15: Simulated narrow-band sub-bottom reflections, both ideal
(a) and realistic (b). Results of spectral equalisation
using a 16-sample window and an energy level of 1% (c)
and of subsequent time-variant amplification (d).

In the authors opinion the on-line application of spectral equalisation may become feasible in the near future, using modern digital hardware and microprocessor controlled data processing.

REFERENCES

1. C.R. Hill: "Interactions of Ultrasound with Tissues", Ultrasound in Medicine, M. de Vlieger et al. (editors), pp. 14 - 20, Excerpta Medica, Amsterdam, 1974
2. J.E. White: "Seismic Waves: Radiation, Transmission and Attenuation", McGraw-Hill Inc., New York, 1965.
3. C.S. Clay and H. Medwin: "Acoustical Oceanography: Principles and Applications", John Wiley & Sons, New York, 1977.
4. "Ultrasonic Tissue Characterization: Clinical Achievements and Technological Potentials", J.M. Thijssen (editor), Stafleu's Scientific Publishing Company, Alphen a/d Rijn - Brussels, 1980.
5. B.A. Hardage: "Vertical Seismic Profiling; Part A: Principles", Geophysical Press, London - Amsterdam, 1983.
6. L.F. van der Wal, D.Ph. Schmidt and A.J. Berkhout: "Acoustic Determination of Sub-Bottom Density Profiles using a Parametric Sound Source", Acoustical Imaging, Vol. 12, pp. 721 - 732, Plenum Press, New York - London, 1982.
7. L.F. van der Wal, D.Ph Schmidt and A.J. Berkhout: "In situ Density Measurements using a Broad-Band Parametric Sound Source", Acoustical Imaging, Vol. 13, Plenum Press, New York - London, 1983.

THE MODIFIED DISCRETE WIGNER DISTRIBUTION

AND ITS APPLICATION TO ACOUSTIC WELL LOGGING[*]

Dwight D. Day[**]
Texas Instruments
P.O. Box 405
Lewisville, TX 75067

Rao Yarlagadda
School of Electrical and
  Computer Engineering
Oklahoma State University
Stillwater, Ok 74078

ABSTRACT

The full wave acoustic well log gives a large amount of information about the surrounding formation, which is encoded into a complex wavetrain and is not easily interpretable. This paper presents a technique using the Wigner Distribution to recover some of the information encoded in these wavetrains.

I. INTRODUCTION

Some of the first acoustic well logs were done in the 1950's by Vogel and Summers and Broding. The tool (commonly called sonde) used for these logs consisted of one transmitter and one or two receivers. At that time, only a rudimentary knowledge of the acoustics of the borehole were known; however it was known that the compressional wave in the formation would be the first to effect the received signal. Using this fact, the logs were

---

[*]This research has been supported by a consortium of companies. The member companies are AMOCO Production Company, ARCO Oil and Gas Company, EXXON Production Research Company, Gearhart Industries, Inc., MOBIL Research and Development Inc., Seismograph Service Coroporation, SOHIO Petroleum Company, and TEXACO Corporation.

[**]Formerly with Oklahoma State University.

developed to measure the compressional wave travel time through
the formation. The present day sonde has two transmitters and two
receivers, all to obtain a more accurate measure of compressional
travel times.

It is apparent that compressional travel time is only a part
of the information that is available from the acoustic log.
Observing the traces produced by a receiver, several different
occurrences can be seen in the trace. These include several wave
types including compressional, shear, Stonley waves, etc. What
caused all these phenomena, and how they could be used has been
the subject of a good deal of research. Our problem here is to
determine the wave arrival times of compressional and shear waves.
This is not that easy since the compressional and shear wavelets
may overlap, covering up the beginning of the shear wave. Our
approach in solving this problem is to use nonstationary spectral
analysis of the wave-trains, and then identify the wave arrivals
from this analysis.

II. ACOUSTIC LOGGING

To produce a full-wave acoustic log, first a sonde, suspended
by a steel cable containing several conductors, is lowered into
the borehole. As the sonde is drawn back up the borehole the
transmitters emit bursts of acoustic waves at regular intervals of
depth. These waves travel out through the borehole fluid to the
formation as compressional waves. As the waves strike the formation, they cause several different types of waves to occur in the
formation. These different waves will then move through the
formation exciting compressional waves in the borehole fluid,
which will carry the impulses back to the receivers on the sonde.
The acoustic impulses are then converted to electrical signals at
the receivers. The electrical signals are transmitted to the
surface and recorded. Identification of the different wave types
from these recorded signals is the major objective of this research.

A. Compressional Wave

One of the uses of the acoustic log is for to measure the
slowness of the formation compressional wave. This is accomplished by assuming that the fastest moving wave is the formation
compressional wave, and that the first signal to appear is the received waveform will be the result of the formation compressional
wave. The arrival of the compressional wave can be estimated as
the time when the received signal exceeds some threshold value.
In this way, the time required for the compressional wave to
travel from the transmitter to the receiver can be measured. Once

the travel time is determined, some compensation must be made for
the time required for the waves to travel from the transmitter to
the formation and then back to the receiver. This compensated
time is now divided by a compensated sonde length to give compressional slowness (the inverse of velocity). The compensated sonde
length will be slightly shorter than the actual transmitter receiver spacing; this is due to the refraction angle of the compressional wave as it enters the formation.

The compensation factors are hard to determine and can cause
considerable error. To avoid these compensation errors, two
receiver tools were developed. By using two receivers, the compensation for borehole travel is unnecessary. The slowness of the
compressional wave through formation is simply the difference of
arrival times at the receivers divided by the receiver spacing.
It has been recognized that even this tool is prone to error. If
the tool is tilted in the hole, errors can develop. To correct
for this error, a two transmitter, two receiver tool was developed. In our analysis we will use the single transmitter, dual
receiver tool (See Figure 1). All this development of acoustic
sondes has been done so as to assure a good measurement of compressional slowness or velocity in the formation. But what information can be gained from knowing the compressional velocity?

A relationship between the compressional velocity and the
formation porosity was proposed by M. R. J. Wyllie [1]. This
relationship called Wyllie's "time average formula" is as follows.

$$\frac{1}{V} = \frac{\phi}{V_f} + \frac{(1-\phi)}{V_{ma}} \tag{1}$$

where:

ϕ is the fractional porosity of rock,
V is the compressional formation velocity,
V_f is the velocity of the pore space fluid, and
V_{ma} is the velocity of the rock matrix.

This equation is now rewritten in terms of slowness (Δt) and then
solved for porosity   to give

$$\phi = \left[\frac{\Delta t - \Delta t_{ma}}{\Delta t_f - \Delta t_{ma}} \right] \tag{2}$$

It can be seen from this equation that an "a priori" knowledge or
at least an estimate of the formations make up is required before
(1) or (2) can be applied. The value Δt_f does not vary greatly
for most borehole fluids, with the exception of gases, and can be

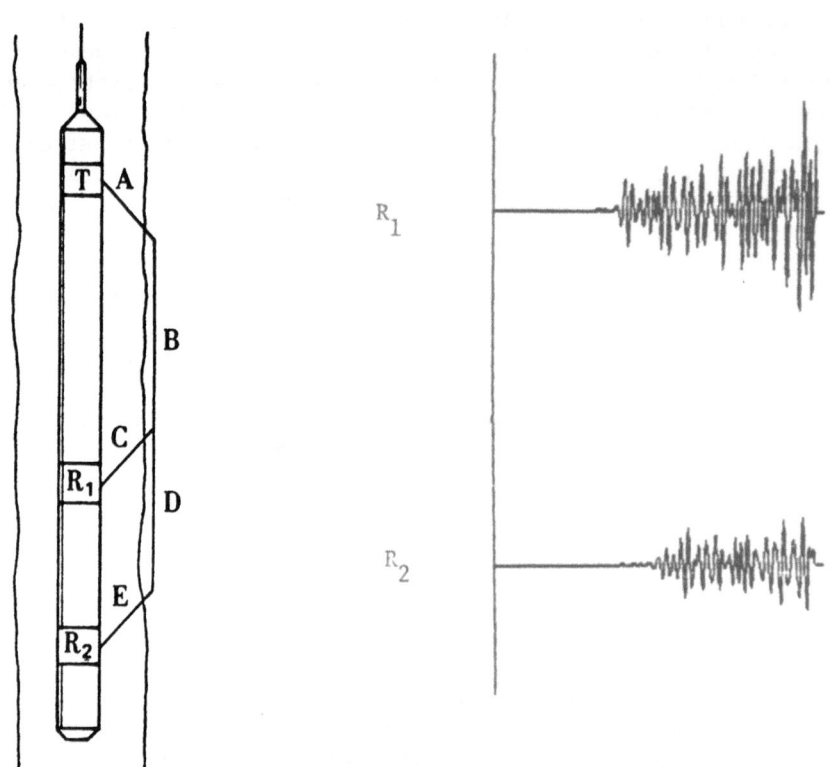

Figure 1.   (1 transmitter, 2 receiver) with
            representative waveforms.

assumed to be approximately 189 μ sec per foot. The value of Δt for the rock matrix can change drastically depending on the lithology of the formation about the borehole. Given some knowledge of the area and formation in question, an estimate of matrix velocity can usually be made. These estimates usually range from 167 μ seconds per foot for some shales and 43.5 μ seconds per foot for dolomite. Recent studies, show that a more accurate form of this equation would be [2]

$$V = (1-\phi)^m \quad V_{ma} + \phi V_f$$

This equation, known as the "Raymer-Hunt-Gardner" equation, uses a value of m=2 for sandstones and 2.0 to 2.2 for carbonates. This equation agrees well with the Wyllie equation for porosity in the range of 0.25 to 0.30.

B. Shear Wave

As was stated in the introduction, there is more information in the wave train than simply compressional velocities. With the introduction of digital wave form recorders, it is now possible to more fully analyze the wave train from the acoustic log and recover more of the information encoded there in.

One wave, other than the compressional, that is of interest is the shear wave. This wave has not been used in the past, since it is slower than the compressional wave and its arrival is usually covered by the trailing end of the compressional wavelet. This means that more complex methods will be needed to identify the shear wave, especially its arrival time from the received signal.

In 1963, G. R. Pickett [3] showed, from laboratory measurements made on core samples, some of the properties of the shear wave. First, Pickett showed that the shear wave velocity is more sensitive to porosity changes than the compressional wave. Also, the shear wave velocity obeys similar linear laws in its relationship to porosity as does the compressional wave. This means that, if an accurate measure of shear wave velocity could be found, the calculation of porosity using the shear wave would be less susceptible to errors. Also shown in Pickett's paper is a relationship of lithology to the ratio of compressional velocity to shear velocity. The ratio DTR is more commonly written as the slowness of the shear wave divided by the compressional slowness, and falls into three major groups shown in Table 1. The values in Table 1 are only close fits to the Pickett data [3].

Table 1.

DTR Fitting Pickett's Data

Lithology	DTR
Sandstone	1.58-1.78
Dolomite	1.8
Limestone	1.9

Prior to Pickett's **work** it was observed that a reduction in acoustic amplitudes was usually a good indicator of formation fracturing. Pickett's analysis of shear wave amplitudes showed that they are more sensitive to fracturing than the compressional waves and shear wave amplitudes would be very useful. This obviously depends upon a consistent procedure for identifying the shear wave from the wave train.

C. Wave Characteristics

Before we discuss the techniques to identify shear wave arrivals from the acoustic wave-train, we will first discuss some of the properties of compressional and shear waves.

The actual path of travel for the compressional wave is rarely as clean as that shown in Figure 1. When the wave strikes the formation, part of the acoustic energy is radiated back into the borehole all the way down. The energy radiated back into the borehole is then reflected by the tool and reenters the formation to proceed down with the rest of the compressional wave. These multiple paths make the compressional wave very reverberant, which causes the spectrum of the compressional wave to have sharp peaks at certain frequencies.

For the case of the shear wave, a similar reaction occurs, with waves being generated in the fluid by the shear waves in the formation. These generated waves reflect off the tool and return to the formation causing new shear and compressional waves to be generated. This implies that the shear wave is also reverberant and has large spectral peaks. However, since the velocity of the shear wave is lower than that of the compressional wave, the spectral peaks of the shear wave will be at different frequencies. In fact, it has been shown both analytically and experimentally that there is a frequency separation between the shear and compressional waves [4, 5].

D. Guided Fluid Waves

The remainder of the wave train is primarily made up of guided fluid waves. These waves include Stonely, reflected conical, and pseudo-Rayleigh waves. The exact usefulness of these waves is not apparent; however, there is speculation about using the Stonely wave for detection of fractures. The basic concept is that the energy of the Stonely wave will be dissipated into a fractured or permeable formation more than into a solid formation, meaning a large drop in Stonely wave amplitudes. To identify the wave arrival times, we will make use of nonstationary signal analysis. A useful tool is Wigner Distribution [6].

III. THE MODIFIED WIGNER DISTRIBUTION

In this section, an alternate definition of the Wigner Distribution is proposed, which allows for more efficient computation. This new definition, referred to as the Modified Wigner Distribution, maintains many of the properties of the classic Wigner Distribution. Some of the basic properties are listed and the derivation of a fast algorithm is included.

A. Origin and Definition

The Wigner Distribution (WD) can be credited to Eugene Wigner from his work in quantum mechanics [11]. It was defined for a continuous signal $f(t)$ in the form

$$W_f(t,w) = \int_{-\infty}^{\infty} f(t+\tau/2) \ f^*(t-\tau/2) \ e^{-jw\tau} \ d\tau \qquad (3)$$

where (*) corresponds to complex conjugation and w is the usual radian frequency symbol. Recently, the WD has been found to have applications in signal processing for time spectral analysis. Towards this, a definition was proposed for discrete signals by Claasen and Mecklenbrauker [7, 8,] and is given below

$$\widetilde{W}_f(n,\theta) = \sum_{k=-\infty}^{\infty} f(n+k) \ f^*(n-k) \ e^{-j2k\theta} \qquad (4)$$

An alternate definition to that of (4) is given below, which has some advantages that will be apparent later. The alternate Wigner Distribution

$$W_f(n,\theta) = \sum_{k=-\infty}^{\infty} f(n+k+1) \quad f^*(n-k) \quad e^{-j2(k+1/2)\theta} \tag{5}$$

which will be referred hereafter as the Modified Auto Wigner Distribution (MAWD).

The Wigner Distribution can be defined for two different signals also. The Modified Cross Wigner Distribution (MCWD), is defined by

$$W_{fg}(n,\theta) = \sum_{k=-\infty}^{\infty} f(n+k+1) \quad g^*(n-k) \quad e^{-j2(k+1/2)\theta} \tag{6}$$

B. Mathematical Properties

In the following some of the interesting properties of the MAWD are listed. Proofs are omitted for brevity.

1. Inverse Operation. The inverse of (5) is

$$f(n+r+1) \quad f^*(n-r) = \frac{1}{\pi} \int_{-\pi/2}^{\pi/2} W_f(n,\theta) e^{j2(r+\frac{1}{2})\theta} \, d\theta \tag{7}$$

The problem with this result is that f(n) cannot be found directly. Rather, some value must be assumed or known for the first non-zero point f(a). Unlike the definition for the Wigner distribution, where at least the magnitude of f(a) can be found, there is no way to recover anything more than the array f(n+r+1) ($f^*(n-r)$).

2. Real Valued. The MAWD is always real. That is,

$$W_f(n,\theta) = (W_f(n,\theta))^* \tag{8}$$

3. Symmetric and Periodic Spectral Components. The symmetry of the MAWD in the frequency domain can be characterized as

$$W_f(n,\theta) = W_{f^*}(n,-\theta)$$

The periodicty of the MAWD in the frequency domain is characterized by

$$(-1)^{\ell} \; W_f(n,\theta) = W_f(n,\theta+\ell\pi) \tag{9}$$

where ℓ is an integer.

It should be noted that the magnitude of the MAWD is periodic with period πf_s (f_s assumed to be 1 Hz in previous development) and not $2\pi f_s$ as is the case in the Fourier transform of discrete signals. This implies that an analog signal must be sampled at twice the Nyquist rate to avoid aliasing.

4. Time Limited Signals have Time Limited MAWD's. Let $f(n)$ be a time limited signal with

$$f(n) = 0 \qquad \text{for } n < n_1 \text{ and } n_2 < n \text{ with } n_1 < n_2.$$

Then the MAWD of $f(n)$ is also time limited, with

$$W_f(n,\theta) = 0 \qquad \text{for } n < n_1 \text{ and } n_2 \geq n. \tag{10}$$

5. Relationship to the Fourier Transform of $f(n)$. The relationship between the MAWD and the Fourier transform of $f(n)$ is important. First, the Fourier transform of $f(n)$ is

$$F[f(n)] = F(\theta) = \sum_{\ell=-\infty}^{\infty} f(\ell) \; e^{-j\ell\theta}, \tag{11}$$

and the inverse

$$F^{-1}[F(\theta)] = f(n) = 1/2\pi \int_{-\pi}^{\pi} F(\theta) \; e^{-jn\theta} \, d\theta \tag{12}$$

The MAWD of $F(\theta)$ is defined as

$$W_F(\theta,n) = 1/2\pi \int_{-\pi}^{\pi} F(\theta+\xi) \; F^*(\theta-\xi) \; e^{j2(n+1/2)\xi} \, d\xi \tag{13}$$

From this, we can state the interesting property that

$$W_F(\theta,n) = W_f(n,\theta) \tag{14}$$

Next we will consider the inverse for the MAWD of $F(\theta)$. The inverse is

$$(1/2)[F(\theta+\psi)\ F^*(\theta-\psi) - F(\theta+\psi+\pi)\ F^*(\theta-\psi-\pi)]$$

$$= \sum_{k=-\infty}^{\infty} W_F(\theta,k)\ e^{-j2(k+1/2)\psi} \equiv A \tag{15}$$

Noting (14), we have also

$$A = \sum_{k=-\infty}^{\infty} W_f(k,\theta)\ e^{-j2(k+1/2)\psi}$$

6. The Effect of Windowing on the MAWD. If the function $f(n)$ is not time limited, it would be impossible to calculate $W_f(n,\theta)$, as it contains an infinite summation. To estimate the MAWD at some time, we will need to window the function $f(n)$. The question is what effect will this have on our estimate of $W_f(n,\theta)$.

Let $h(n)=f(n)\ g(n)$, where $g(n)$ is a time limited window function. It can be shown that

$$W_h(n,\theta) = 1/\pi \int_{-\pi/2}^{\pi/2} W_f(n,\xi) \cdot W_g(n,\theta-\xi)d\xi \tag{16}$$

or

$$W_h(n,\theta) = W_f(n,\theta) * W_g(n,\theta) \tag{17}$$

The effect of windowing on the MAWD is very similar to that found in short time Fourier analysis, except here the window effects only the frequency resolution. The time resolution of the MAWD is un-affected by the window.

C. Some Properties of the Modified Cross Wigner Distribution

In the following some of the interesting properties of the

Modified Cross Wigner Distribution (MCWD) (see equation (6)) are listed. Again, the proofs are omitted for brevity.

 1. Inverse Operation. The inverse operation of the MCWD is

$$
f(n+r+1) \; g^*(n-r) = 1/\pi \int_{-\pi/2}^{\pi/2} W_{f,g}(n,\theta) \; e^{j2(r+1/2)\theta} d\theta \qquad (18)
$$

 2. Periodic Spectrum. The MCWD is periodic with respect to its frequency variable. This can be expressed as

$$
(-1)^{\ell} \; W_{f,g}(n,\theta) = W_{f,g}(n, \theta + \ell\pi) \qquad (19)
$$

 3. Relationship to the Fourier Spectrum. As in the case of the MAWD, the MCWD can be shown to have a close relationship to the Fourier transforms of the corresponding signals. The definition of the MCWD for the $F(\theta)$ and $G(\theta)$ is given by

$$
W_{F,G}(\theta,n) = (1/2\pi) \int_{-\pi}^{\pi} F(\theta+\xi) \; G^*(\theta-\xi) \; e^{j2(n+1/2)\xi} \; d\xi \qquad (20)
$$

where $F(\theta)$ and $G(\theta)$ are the Fourier transforms of functions $f(n)$ and $g(n)$.

Based on (20), it can be shown that

$$
W_{F,G}(\theta,n) = W_{f,g}(n,\theta) \qquad (21)
$$

Now we will consider the inverse of $W_{F,G}(\theta,k)$. The inverse is given by

$$
1/2[F(\theta+\psi) \; G^*(\theta-\psi) - F(\theta+\psi+\pi) \; G^*(\theta-\psi-\pi)]
$$

$$
= \sum_{k=-\infty}^{\infty} W_{f,g}(\theta, k) \; e^{-j2(k+1/2)\psi} \equiv E. \qquad (22)
$$

By using (22), E can also be expressed by

$$E \quad = \sum_{k=-\infty}^{\infty} W_{f,g}(k,\theta) \; e^{-j2(k+1/2)\psi} \tag{23}$$

An interesting result follows from (23) when $\psi=0$, and is

$$F(\theta) \; G^*(\theta) \; = \; \sum_{k=-\infty}^{\infty} W_{f,g}(k,\theta)$$

where we have assumed that $F(\theta)$ and $G(\theta)$ are band limited to $\pm [\pi/2]$ f_s.

4. The Effect of Windowing on the MCWD. It is impossible to numerically calculate the MCWD for non time limited signals as the MCWD contains an infinite summation. To estimate the MCWD at some time n, it will be necessary to window the sequence about the time n. The question is how the windowing will effect our estimate of the MCWD?

Assume we have two functions, $f(n)$ and $g(n)$, which are windowed by two finite windowing functions, $w_1(n)$ and $w_2(n)$, respectively. Let $h(n) = f(n) \; w_1(n)$ and $e(n) = g(n) \; w_2(n)$ then the MCWD of $h(n)$ and $e(n)$ can be expressed as

$$W_{h,e}(h,\theta) = 1/\pi \int_{-\pi}^{\pi} W_{f,g}(n,\xi) \; W_{w_1,w_2}(n, \theta - \xi) \; d\xi \tag{24}$$

The effects of windowing on the MAWD is very similar to those found in Short-Time Fourier Analysis (STFA). One important point is that the time resolution of the MAWD is not effected by the window length, which is not the case in STFA. To achieve good frequency resolution we need a good window function. To demonstrate the differences between windows, plots were generated for the MAWD for two windows used in STFA. These plots are shown in Figures 2, and 3.

Figure 2 is the MAWD for the center point of a rectangular window. Note, the MAWD does not decrease very fast for this window. The rectangular window can thus cause extreme blurring of the MAWD. This can be alleviated by using windows, such as Hamming, Kaiser windows, etc. Figure 3 is the MAWD for a Kaiser window [9]. The Kaiser window can be adjusted by changing an input parameter "Beta".

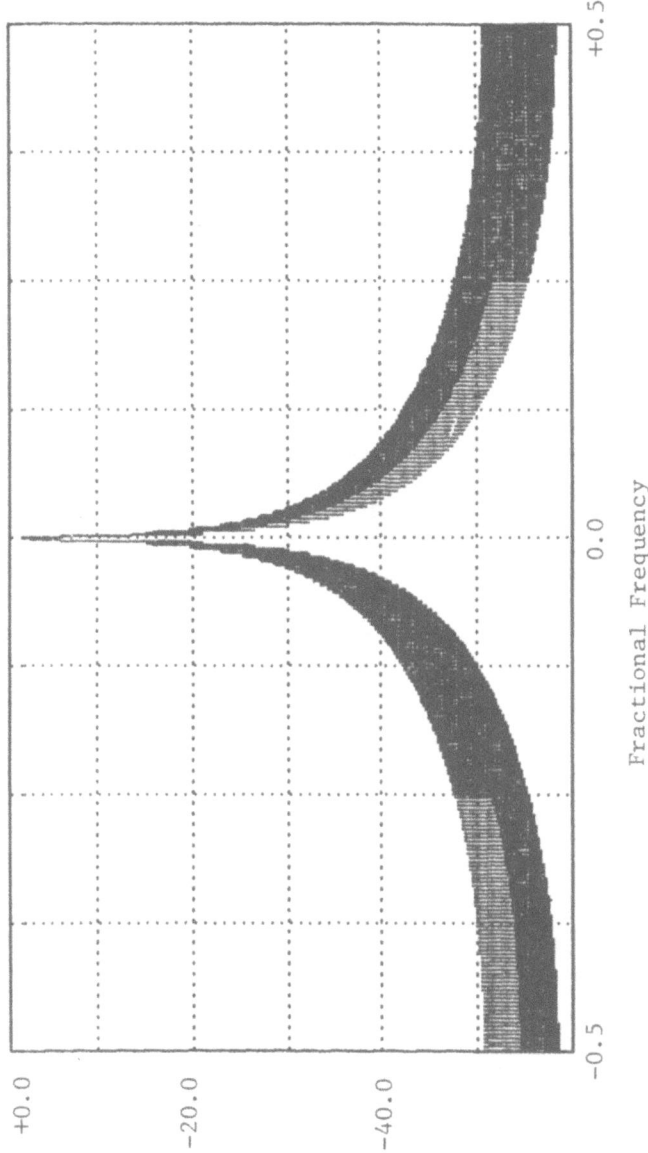

Figure 2.   Rectangular Window

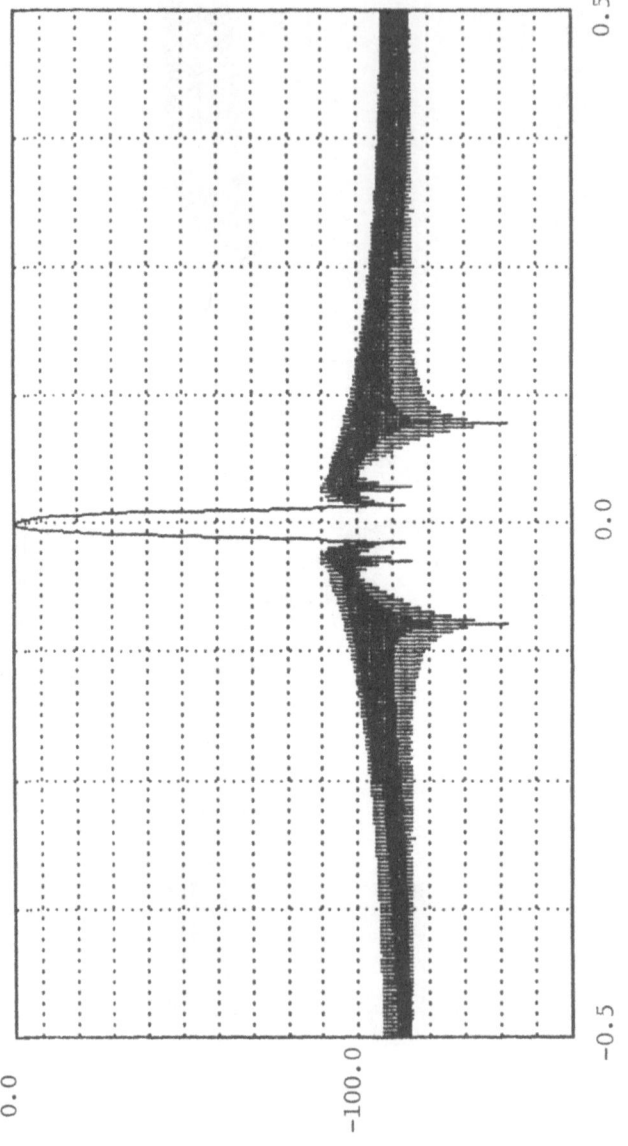

Figure 3. Kaiser Window (Beta = 5.5)

5. Computation of the MAWD and MCWD. Algorithms for computing, both the MAWD and MCWD, using the FFT are now derived.

a. For the MAWD. The following operations are performed on the MAWD

$$W_f(n,m) = \sum_{k=-N}^{N-1} f(n+k+1) \; f^*(n+k) \quad e^{-j2(k+\frac{1}{2})(m+\frac{1}{2})\xi}$$

$$= e^{-jm\zeta} \; e^{-j(1/2)\xi} \quad f(n+k+1) \; f^*(n-k) \; e^{-jk\xi} \; e^{-j2km\xi}$$

Using k=ι−N, we have

$$W_f(n,m) = e^{-jm\xi} \; e^{-j(\frac{1}{2})\xi} \sum_{\iota=0}^{2N-1} \; f(n+\iota-N+1)$$

$$f^*(n-\iota+N) \; e^{-j\iota\xi} \; e^{jN\xi} \; e^{-j2\iota m\xi} \; e^{j2Nm\xi}$$

$$= e^{-j(m+1/2)\xi} \; e^{j2(m+1/2)N\xi} \sum_{\iota=0}^{2N-1} [f(n+\iota-N+1) \; f^*(n-\iota+n) \quad e^{-j\iota\xi}]$$

$$e^{-j2\,\iota\,m\xi}$$

$$W_f(n,m) = e^{-j(m+1/2)\,\pi/2N} \quad (-1)^{m+1/2}$$

$$\sum_{\iota=0}^{2N-1} [f(n+\iota-N+1) \; f^*(n-\iota+N) \; e^{-j\iota\,\pi/2N}] \quad e^{-j\,m\iota2\pi/2N} \quad (25)$$

where we have used $\zeta = \pi 2N$. The summation over ι is a DFT of length 2N, which may be a power of 2. If it is a power of 2, we can use the FFT to compute it.

b. For the MCWD. When the same operations and substitutions are applied to the MCWD, as were applied to the MAWD in the last section, we have

$$W_{f,g}(n,m) = e^{-j(m+1/2)\pi/2N}\ (-1)^{m+1/2}$$

$$\sum_{\iota=0}^{2N-1} [f(n+\iota-N+1)\ g^*(n-\iota+N)\ e^{-j\iota\pi/2n}\ e^{-j\iota m\ 2\pi/2N}$$

The summation over   is a DFT of length 2N, which may be a power of two. If it is a power of two, we can again use the FFT to compute it.

Alternatively, the definition of the MAWD as given by Claasen and Mecklenbrauker (see (4)) can also be implemented in a similar fashion to that of the MAWD. Equation (4) is expressed below

$$W_f(n,\theta) = \sum_{k=-N}^{N} f(n+k)\ f^*(n-k)\ e^{-j2k\theta}$$

where we have assumed that the signal has been windowed.

Using k= -i-N, we have

$$W_f(n,\theta) = \sum_{\iota=0}^{2N} f(n+\iota-N)\ f^*(n-\iota+N)\ e^{-j2(\iota-N)\theta}$$

Substituting $\theta=m\zeta$, with $\zeta=\pi/2N+1$, we have

$$W_f(n,m) = \sum_{\iota=0}^{2N} f(n+\iota-N)\ f^*(n-\iota+N)\ e^{-j2\iota m\xi}\ e^{+j2Nm\xi}$$

$$= e^{+j(2N+1)m\ \pi/(2N+1)}\ e^{-jm\xi}\sum_{\iota=0}^{2N} f(n+\iota-N)\ f^*(n-\iota+n)$$

$$e^{-j2\iota m\xi}$$

The summation is now a DFT of length 2N+1. The length of the DFT is odd, and therefore makes it less efficient to compute. It is clear that the Modified Wigner Distribution is more efficient.

IV. APPLICATION OF THE MAWD AND MCWD TO THE ACOUSTIC LOG

In this section, we discuss how the MAWD and the MCWD can be used to analyze the acoustic log.

A. Introduction

It is possible that the MAWD and MCWD could be used to re-cover a large amount of information from the traces received from the acoustic log. However, the identification of the shear wave and its arrival time will be our first objective.

It is known that the shear wave travel time will approxi-mately range from $\sqrt{2}$ to 2.2 times the compressional travel time. As was discussed earlier, compressional travel time can be found using rather simple techniques. Once compressional travel time has been determined, we will then have a range of time in which we will look for the shear wave to arrive. This can greatly reduce the amount of computation required.

The determination of the shear wave is now dependent upon determination of compressional travel times. Compressional wave travel times are usually determined by thresholding the received signal. Thresholding is a dependable technique, provided, no noise is in the signal. If noise is present, erroneous data will result. Noise is usually not a problem with the acoustic log.

B. The MAWD

Using the definition of the MAWD developed in section III.B, the MAWD was applied to two real traces. The traces and the re-sulting Wigner distributions are shown in Figures 4 and 5. The Wigner distributions are displayed as perspective plots with time and frequency as indicated. The acoustic traces that are used to derive MAWD's are displayed directly above them.

The first trace has a transmitter receiver spacing of 10 ft. and the second has a spacing of 12 ft. In each of these plots estimated arrival times are indicated by arrows, where arrow 1 corresponds to the compressional wave arrival and arrow 2 corre-sponds to the shear wave arrival. It is clear from Figure 4 that the shear wave arrival time cannot be marked as clearly as in Figure 5. This is due to the interaction between the shear and compression wavelets, which did not separate clearly in the case of the shorter spacing.

C. The MCWD

When using the MCWD to analyze acoustic traces, we will be using traces taken from two receivers similar to those shown in Figure 1. The following procedure was used in applying the MCWD.

1. Compressional travel times were found for each trace.

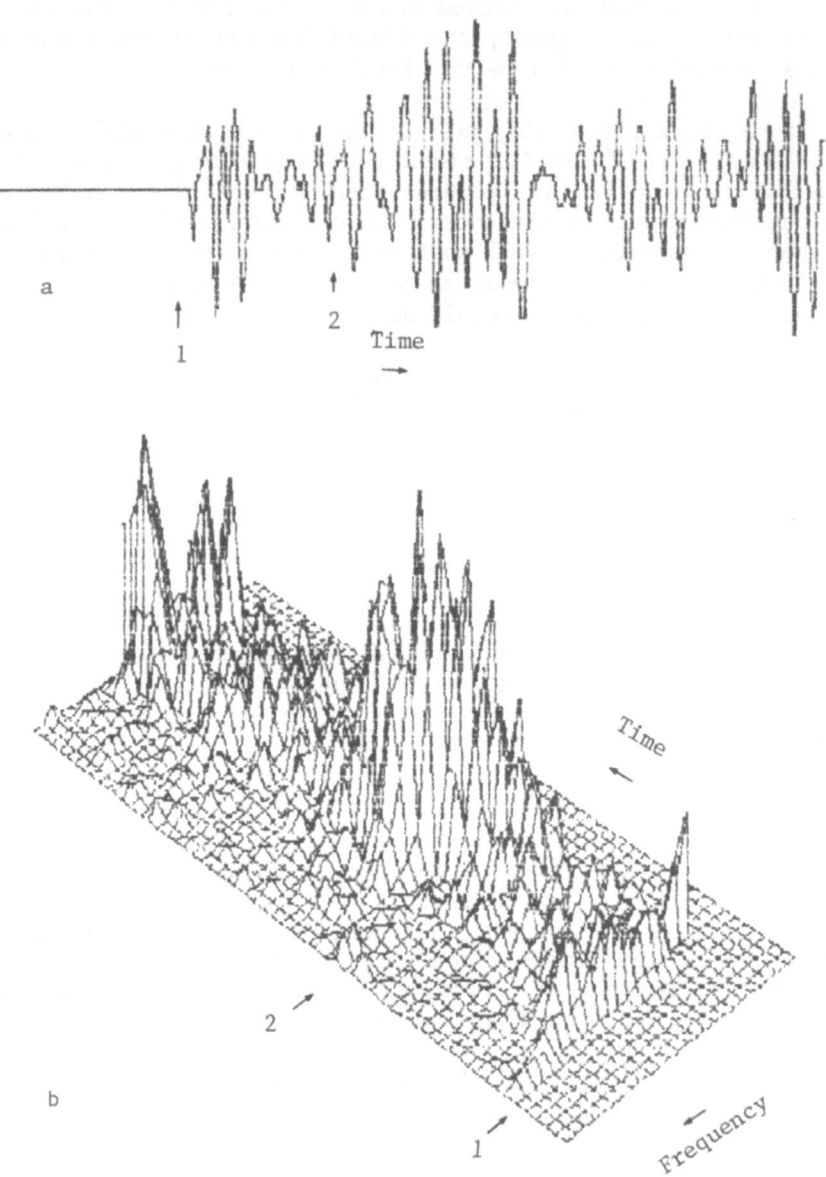

Figure 4.   a.   Acoustic trace (10 ft. transmitter-
                 receiver spacing),
            b.   Corresponding MAWD perspective plot.

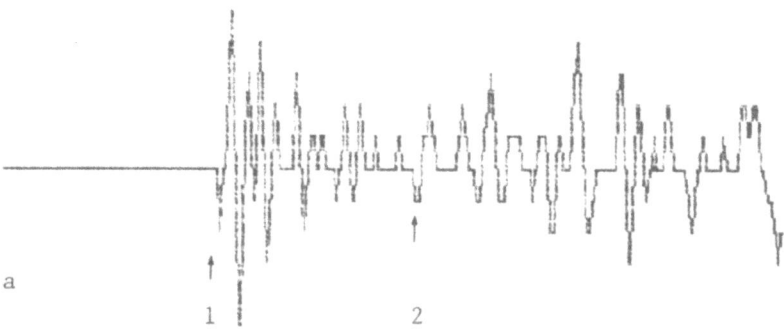

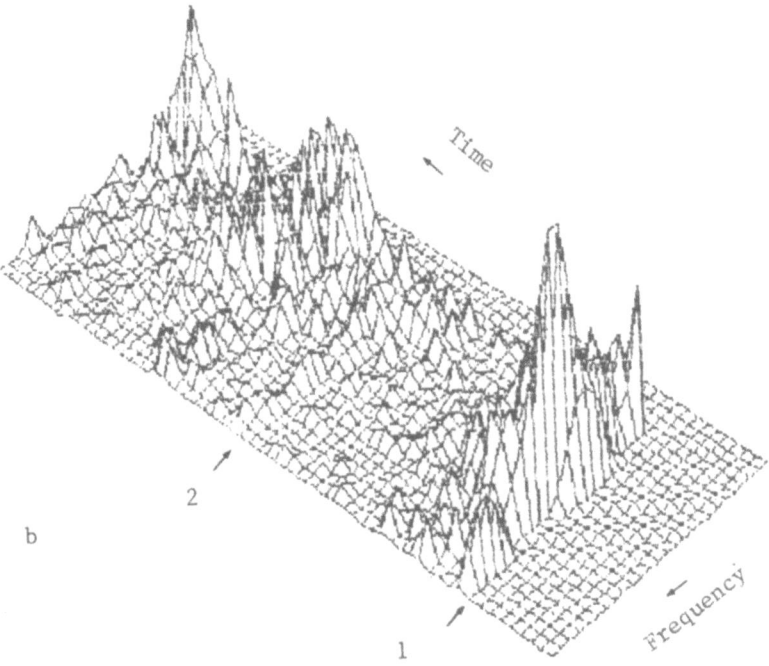

Figure 5. a. Acoustic trace (12 ft. transmitter-
                receiver spacing),
            b. Corresponding MAWD perspective plot.

2. Estimates of the times when the shear wave should arrive are
    calculated for both traces using their compressional ar-
    rivals.
3. A window function is chosen that takes in all of the shear
    wave arrivals estimates for the near receiver.
4. This window is now applied to both traces and the MCWD is
    computed at the center point of the window. The phase of the
    MCWD for a certain range of frequencies is then recorded.
5. The trace from the far receiver is now advanced one point in
    time.
6. Steps 4 and 5 are now repeated, until the far trace has been
    advanced such that the center point of the window is beyond
    the traces maximum shear wave arrival estimate.

The interpretation of the recorded phase is as follows. We
know that the shear wave in the far trace will appear later than
the shear waves in the near trace and ideally we might consider
this to be a time delay of the shear waves spectral components.
Now, if we were to advance the far trace by a time equal to this
time delay and then calculate the MCWD, the result should be the
same as the MAWD, at least for the spectral components of the
shear wave. We know that the MAWD is always real having a phase
shift of zero. Therefore, we need only find an advancement time
such that we have approximately zero phase shift through out our
frequency band. The advancement time satisfying the above requirement will equal the delay or travel time of the shear wave
between our two receivers.

This type of analysis was applied to the two acoustic traces
in Figures 4 and 5. Obviusly, this cannot be illustrated effectively by a perspective plot. However, this can be best illustrated using color density plots. From the color plots, a consistent band of zero phase shift can be identified between 5.5 to 11
KHz for this particular case. The frequency range of the compressional wave tends to be higher than this range. Using this band
as the approximate advancement or delay time and a two foot receiver spacing, we calculate a shear wave velocity of 9300 ft/sec.
This is in close agreement with the shear wave velocity of 9508
ft/sec. found using the core analysis.

V. CONCLUSIONS

In this paper, we have shown how the MAWD and the MCWD can be
used in the analysis of full-wave acoustic well logs. Of the two
processes, the MAWD unlocked more of the information encoded in
the acoustic traces. One problem is the amount of calculations
involved in computing the MAWD. To properly find shear wave
travel times, we need to calculate the MAWD for approximately 100
to 150 points. This is about equivalent to 100 to 150 FFT's, and

that is only for one trace. If we are to borehole compensate, we would be using at least two traces which doubles our computations. On the other hand the MCWD uses the two traces in its original computation and the result tends to be already compensated. Also, we need to find only the MCWD for 100 to 150 points, corresponding to 100 to 150 FFT's. The MCWD is the best approach, if only the shear wave arrival times are required. The MAWD and MCWD have been applied to actual well log data and the perspetive plots indicate that the methods do indeed mark the arrival times of compressional and shear waves.

VI. REFERENCES

[1] Wyllie M.R.J., Gardner G.H.F. and Gregory A.R., "Some Phenomena Pertinent to Velocity Logging," Journal of Petroleum Technology, pp. 629-636, July 1961.   •

[2] Leslie H.D. and Mons F., "Sonic Waveform Analysis: Applications," Society of Professional Well Log Analysts, 23rd Annual Logging Symposium, Paper GG, July 1982.

[3] Pickett G.R., "Acoustic Character Logs and Their Applications in Formation Evaluation," Journal of Petroleum Technology, pp. 659-667, June 1863.

[4] Paillet F.L., "Predicting the Frequency Content of Acoustic Waveforms obtained in Boreholes," SPWLA, 23rd Annual Logging Symposium Paper SS, June 1981.

[5] Scarascia S., Columbi B. and Cassinis R., "Some Experiments on Transverse Waves," Geophysical Prospecting, Vol. 24, pp. 249-568.

[6] Wigner E., "On the Quantum Correction for Thermodynamic Equilibrium," Physics Review, Vol. 40, pp. 749-759, 1932.

[7] Claasen T.A.C.M. and Mecklenbrauker S.F.G, "The Wigner Distribution - A Tool for Time-Frequency Signal Analysis Part II: Discrete Time Signals, "Philips Journal of Research, Vol. 35, No. 4/5, pp. 276-300, 1980.

[8] Claasen T.A.C.M. and Mecklenbrauker W.F.G., "Time-Frequency Signal Analysis by Means of the Wigner Distribution," Proc. ICASSP 81, IEEE Int. Conf. on Acoustics, Speech, and Signal Processing, pp. 69-72, 1981.

[9] Kuo F. F. and Kaiser J.F., "System Analysis by Digital Computer," John Wiley and Sons Inc., New York, New York, 1966.

TWO DIMENSIONAL STATISTICAL ANALYSIS OF SYNTHETICALLY FOCUSED IMAGES

P.R. Mesdag and A.J. Berkhout

Delft University of Technology
Lab. of Acoustics and Seismics
P.O. Box 5046, 2600 GA Delft, The Netherlands

Introduction

In previous reports our group has described a novel approach to acoustic imaging (Berkhout, 1982; Berkhout et al, 1982; Mesdag et al, 1982; de Vries et al, 1983). The heart of this approach is formed by a computerized wave field extrapolation technique. By means of this technique the distortions due to the propagation of the acoustic wave are removed from the measured data. For medical purposes the acoustic wave field is scanned by a linear array operated in backscatter mode.

In earlier work, reported at this convention (Mesdag et al, 1982), the authors described the basic limitations of an acoustic data acquisition system. These limitations are envisaged most clearly in respect to the spatial and temporal Fourier spectrum of the produced images. In their report the authors concluded that their method of analysis based on the two dimensional Fourier spectrum is less successful if the complexity of the interrogated media increases. In the final comments the use of a two dimensional auto correlation function was suggested as a way around this problem.

In this paper above suggestion will be substantiated and the limits of both Fourier (or: diffraction) analysis and auto correlation analysis will be discussed.

In order to give some information on the preprocessing involved, we will start with a brief résumé of our approach to acoustic image reconstruction.

The preprocessing

Basically, our approach to acoustic imaging consists of four
separate steps. In this paper only part of the final analysis step
will be treated in detail. In order to understand the format of the
data to be analysed, the treatment of the data prior to analysis
must at least be mentioned here.

Step 1: Measurement and storage of the r.f. data.
By means of a linear array, pulsed, backscatter technique the broad
band r.f. data for a two dimensional image are gathered and stored
in the memory of a digital system (Ridder et al, 1981). The
pressure field at the surface of the medium is now sampled both
axially and laterally, typically forming approximately a 1 Mbyte
pool of data.
This first measurement step can also be simulated by computer,
which was actually done for most of the data used in the first part
of our study. In earlier studies the validity of the simulations
was tested and reported on (a.o. Mesdag et al, 1982).

Step 2: Synthetic focusing (image reconstruction).
As a second step all data are treated by a synthetic focusing
procedure (Berkhout et al, 1982). A computerized inverse wave field
extrapolation algorithm removes the distorting instrumentation and
propagation effects from our measured data. Inverse wave field
extrapolation can be seen as a "mapping" algorithm, transforming
the dimensions of the measured data (x-t) into the dimensions of
the image (x-z), by aid of wave theory.
Synthetic focusing of correctly measured data produces the best
image one can possibly obtain within the physical limitations of
the acoustic measurement system (see step 1).

Step 3: Selection of an area of interest.
After synthetic focusing we select an area of interest from the
total image for local analysis. Due to the high resolution and the
large dynamic range of the image the selected sample volume is
affected only minimally by information coming from other parts of
the image.

Step 4: Image analysis.
This is the subject of our paper.

Two dimensional Fourier analysis

In this paper one of the tools used for image analysis is the two
dimensional Fourier transform. The two dimensional Fourier spectrum
of an image is calculated by performing a Fourier transform along
the two spatial coordinates of the image

$$ r(x,z) \xrightarrow{\text{2D FFT}} \tilde{r}(k_x, k_z). \tag{1} $$

If, e.g. the image contained one spatial deltapulse the entire
Fourier space would show a white spectrum

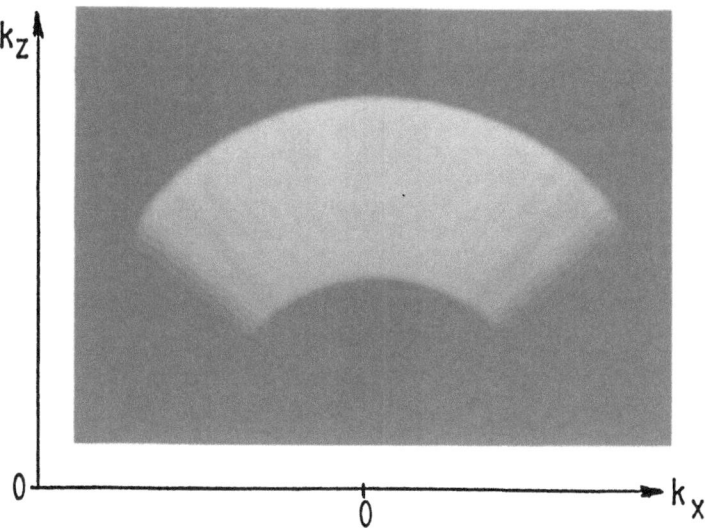

Fig. 1: Two dimensional Fourier spectrum of one point
scatterer, as measured by a linear array with elements
which emit a white spectrum and have an omnidirectional
sensitivity.

$$|\tilde{r}(k_x, k_z)| = 1.0 \quad \text{for all } k_x \text{ and } k_z. \tag{2}$$

In any measurement system the aperture is limited and the emitted
pulse is band limited. Hence we are not able to reconstruct a
spatial deltapulse, even though the point scatterer may have been
infinitely small. In the spatial frequency domain this means that
equation 2 does not hold for all k_x and k_z. Typically the Fourier
space would be filled as in figure 1 (see also Mesdag et al, 1982).

By performing a two dimensional Fourier transform the wave field is
split up into its plane wave components.
This means that we have simulated far field conditions, as every
complex point in Fourier space represents the amplitude and phase
of a plane wave of which the frequency is

$$f = \frac{c}{2\pi} k = \frac{c}{2\pi} \sqrt{k_x^2 + k_z^2}, \tag{3}$$

where c is the sound propagation velocity of the medium. The angle
of incidence of the plane wave is given by

$$\alpha = \tan^{-1}(k_x/k_z) = \sin^{-1}(k_x/k). \tag{4}$$

In this two dimensional Fourier space angle dependent single
frequency backscatter measurements as described by e.g. Nicholas
(1981), and single angle swept frequency backscatter measurements
are illustrated as circles with $|k|$ = konstant, and radii with

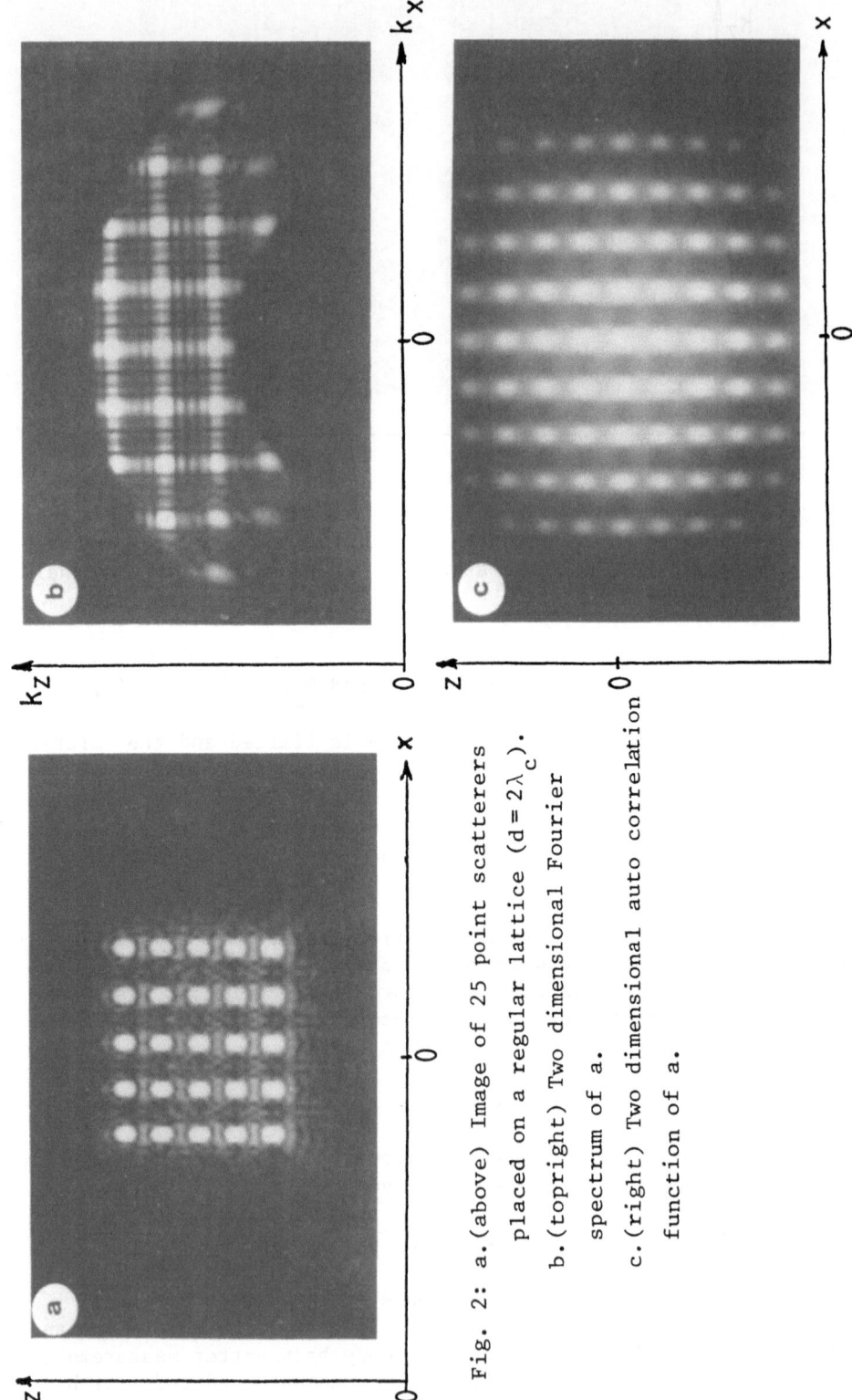

Fig. 2: a.(above) Image of 25 point scatterers
placed on a regular lattice $(d = 2\lambda_c)$.

b.(topright) Two dimensional Fourier
spectrum of a.

c.(right) Two dimensional auto correlation
function of a.

α = konstant respectively. Therefore, $|\tilde{r}(k_x,k_z)|$ may also be referred to as the two dimensional <u>diffraction pattern</u> of the sample volume we are looking at.

Two dimensional auto correlation analysis

The second analysis method discussed in this report is the two dimensional auto correlation analysis. In principle the auto correlation function is closely related to the Fourier spectrum as

$$R(x,z) = r(x,z) * r(-x,-z), \qquad (5)$$

where $R(x,z)$ is the two dimensional auto correlation, and where * denotes a two dimensional convolution in x and in z.
In the spatial frequency domain equation 5 becomes

$$\tilde{R}(k_x,k_z) = \tilde{r}(k_x,k_z)\tilde{r}^*(k_x,k_z) = |\tilde{r}(k_x,k_z)|^2, \qquad (6)$$

where the superscript * denotes that the conjugate of $\tilde{r}$ is meant. Unfortunately the analysis procedure is slightly more complicated, as in statistical media the auto correlation function of the r.f. data does not show the expected characteristic distances.

The reason for this is simple to see if we imagine an image with only two scatterer pairs. The respective distances between the pairs being Δx and $\Delta x + \frac{1}{2}\lambda_c$. Here λ_c is the wavelength of the central frequency of the r.f. signal. The sidelobes of the autocorrelation function of one pair will now destructively interfere with the sidelobes of the other pair.

If we use the envelope detected r.f. signal as input for the auto correlation routine, then all sidelobes will add coherently and the characteristic distances can be found.

A disadvantage of this approach is that the resolution of the auto correlation function will be worse than that of the image, as we shall see later on.

An array of point scatterers

Two examples will be given of arrays built up of point scatterers. In the first example the scatterers are placed exactly according to a two dimensional lattice. The image of the lattice, its two dimensional spectrum and auto correlation function are shown in figure 2.

The characteristic lattice distance $d = 2\lambda_c$ and the fractional bandwidth of the transducer F.B.W. = 70%.
All figures are shown linearly on an 8 bit display, theoretically giving them a dynamic range of 48 dB.
Notice how in this example the Bragg diffraction conditions are satisfied, giving a regular interval in the Fourier spectrum.

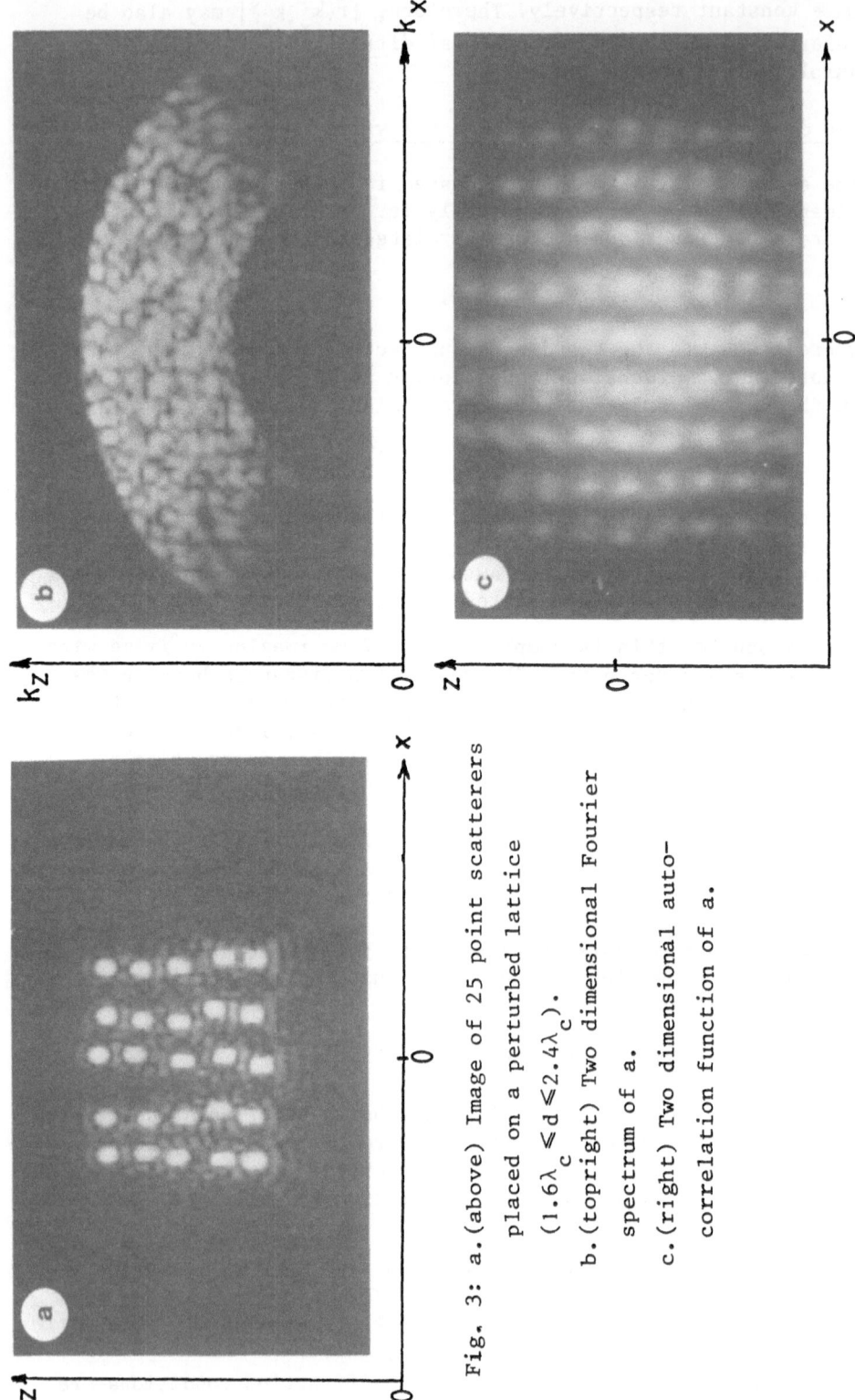

Fig. 3: a.(above) Image of 25 point scatterers
         placed on a perturbed lattice
         ($1.6\lambda_c \leq d \leq 2.4\lambda_c$).
      b.(topright) Two dimensional Fourier
         spectrum of a.
      c.(right) Two dimensional auto-
         correlation function of a.

In the second example of figure 3 a small perturbation has been added to the regular lattice $1.6\lambda_c \leqslant d \leqslant 2.4\lambda_c$. Now the Bragg diffraction conditions are not satisfied any more, rendering detailed analysis of diffraction patterns useless.

Directional scatterers

As the size of the scatterers increases the backscattered energy will become dependent on the angle of the transducer in relation to the scatterer. In previous reports care was always taken to evade this problem, so only omnidirectional pointscatterers were considered, of which the dimensions were much smaller than the wavelength of the highest frequency of the acoustic pulse.

In this section the effect of a finite scatterer size on the acoustical backscatter behaviour will be studied. Instead of inserting an acoustical deltapulse in the model the pulse was transformed into a boxcar function in a direction at an angle $\phi = 45^\circ$ to the x-axis of the image:

$$r(x,z) = \sum_{i=0}^{M} r_o \, \delta(x-x_o-\xi_i)\delta(z-z_o-m\xi_i), \qquad (7)$$

where $m = \tan\phi = 1$ and where $M\Delta\xi /\cos\phi$ is the length of the boxcar function. No changes were observed in the acoustical data for scatterer sizes smaller than $0.5\lambda_c$. Above $0.5\lambda_c$ the point spread function (P.S.F.) of the scatterer changes, but also the Fourier spectrum deformed appreciably. Note that this value of $0.5\lambda_c$

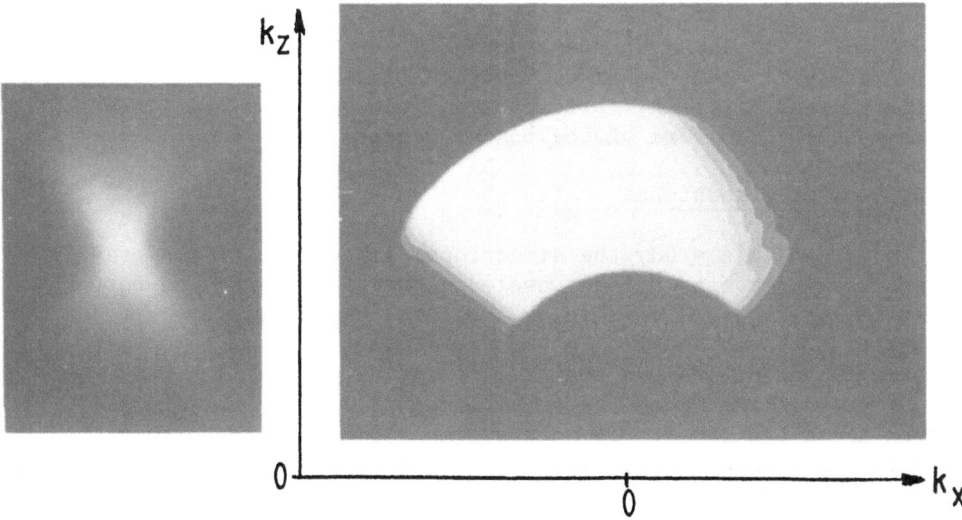

Fig. 4: Point spread function and two dimensional Fourier spectrum of a slanting scatterer with a length of $0.5\lambda_c$.

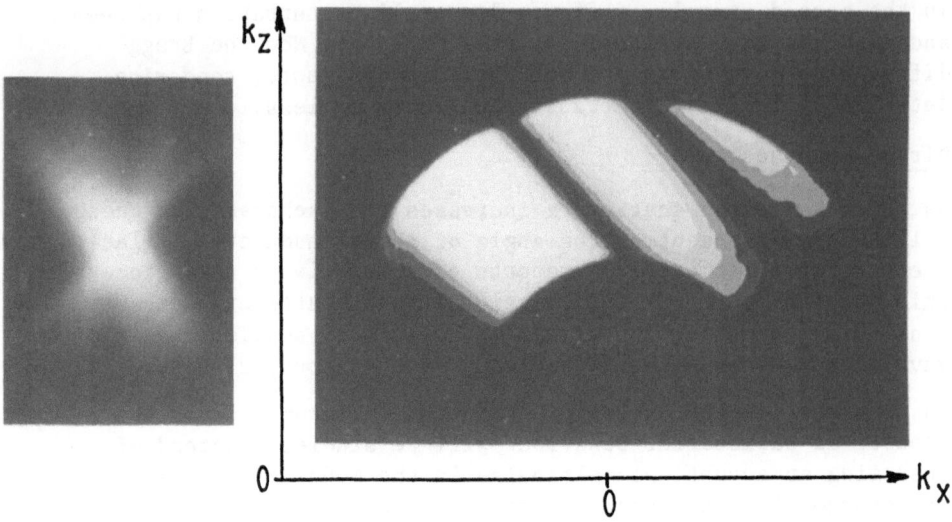

Fig. 5: Point spread function and two dimensional Fourier
        spectrum of a slanting scatterer with a length of
        λ_c.

coincides with the theoretical resolution (-6dB) of this particular
imaging system (Berkhout et al, 1982).
Figures 4 and 5 show the P.S.F's and the Fourier spectra for a
scatterer length of $0.5\lambda_c$ and λ_c respectively.

Equation 7 may be rewritten as a convolution of a spatial deltapulse with a slanting boxcar function. Therefore the response of
the imaging system to this scatterer may also be rewritten as a
convolution of the systemresponse (P.S.F. of a point scatterer)
with the same slanted boxcar function. The Fourier spectra of
figure 4 and 5 are multiplications of the spectrum of figure 1 with
the Fourier transform of the boxcar (viz. a sinc function).

An array of directional scatterers

To complete this study the directional scatterers of the previous
section with a size of one wavelength were placed on the perturbed
lattice of figure 3. Figure 6 shows the resulting image, its
Fourier spectrum and its auto correlation function. As opposed to
earlier conclusions concerning diffraction pattern analysis, the
Fourier spectrum shows the directionality of the scatterers.
Even after the density of scatterers is increased such that the
image shows only speckle and the auto correlation function shows no
detail, the Fourier spectrum keeps the directional information
(figure 7).

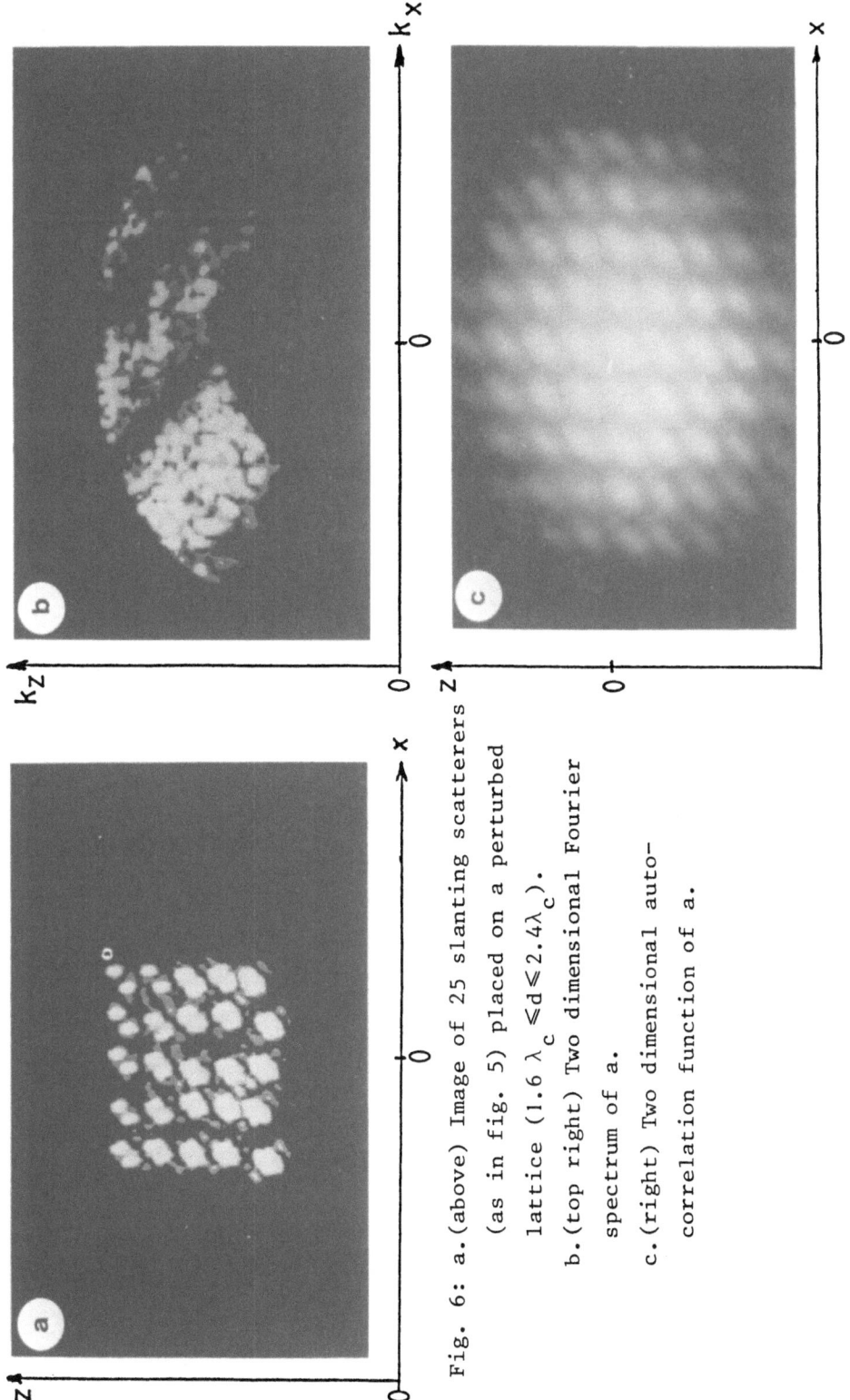

Fig. 6: a.(above) Image of 25 slanting scatterers
(as in fig. 5) placed on a perturbed
lattice $(1.6\lambda_c \leq d \leq 2.4\lambda_c)$.
b.(top right) Two dimensional Fourier
spectrum of a.
c.(right) Two dimensional auto-
correlation function of a.

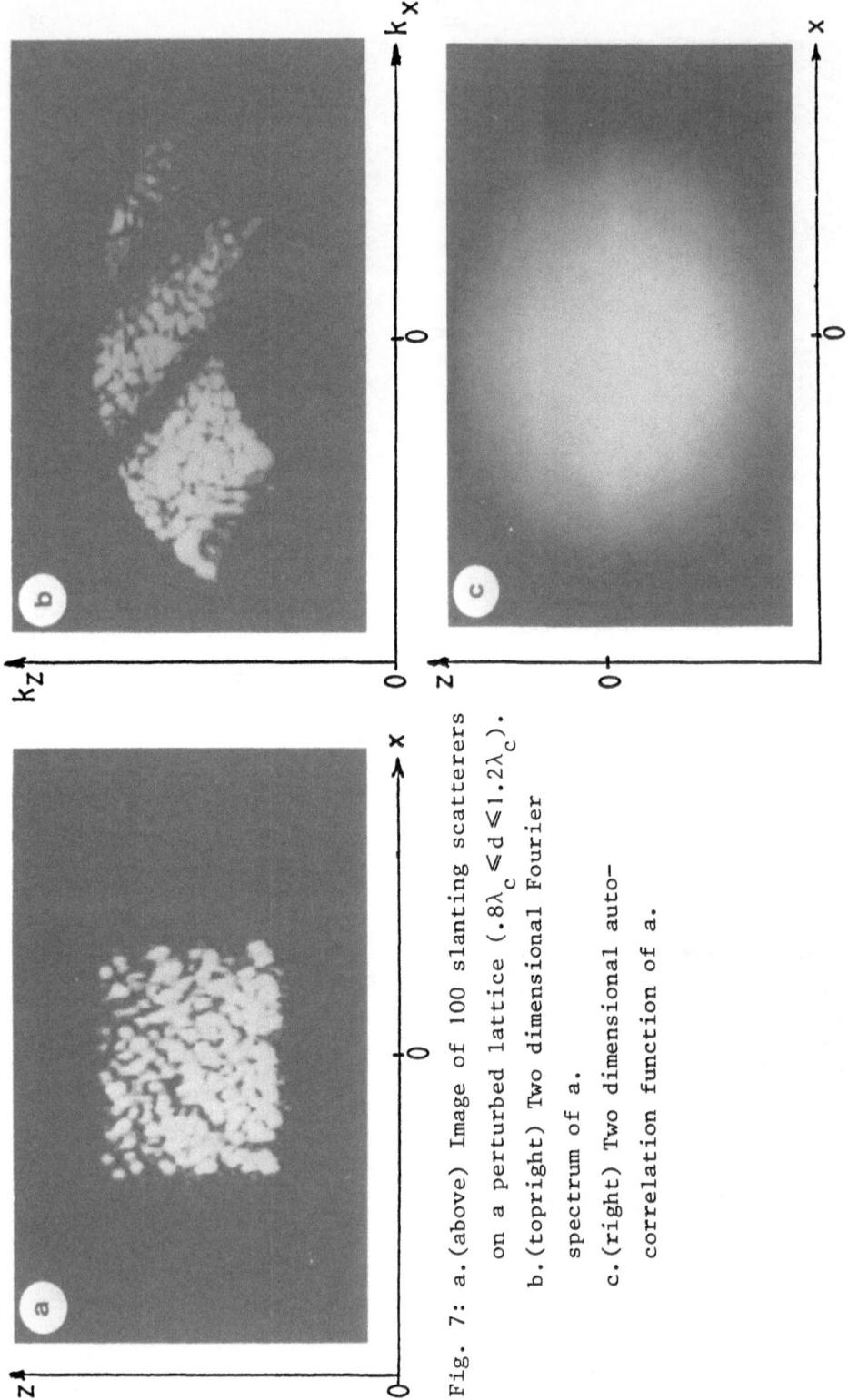

Fig. 7: a.(above) Image of 100 slanting scatterers

on a perturbed lattice ($.8\lambda_c \leqslant d \leqslant 1.2\lambda_c$).

b.(topright) Two dimensional Fourier

spectrum of a.

c.(right) Two dimensional auto-

correlation function of a.

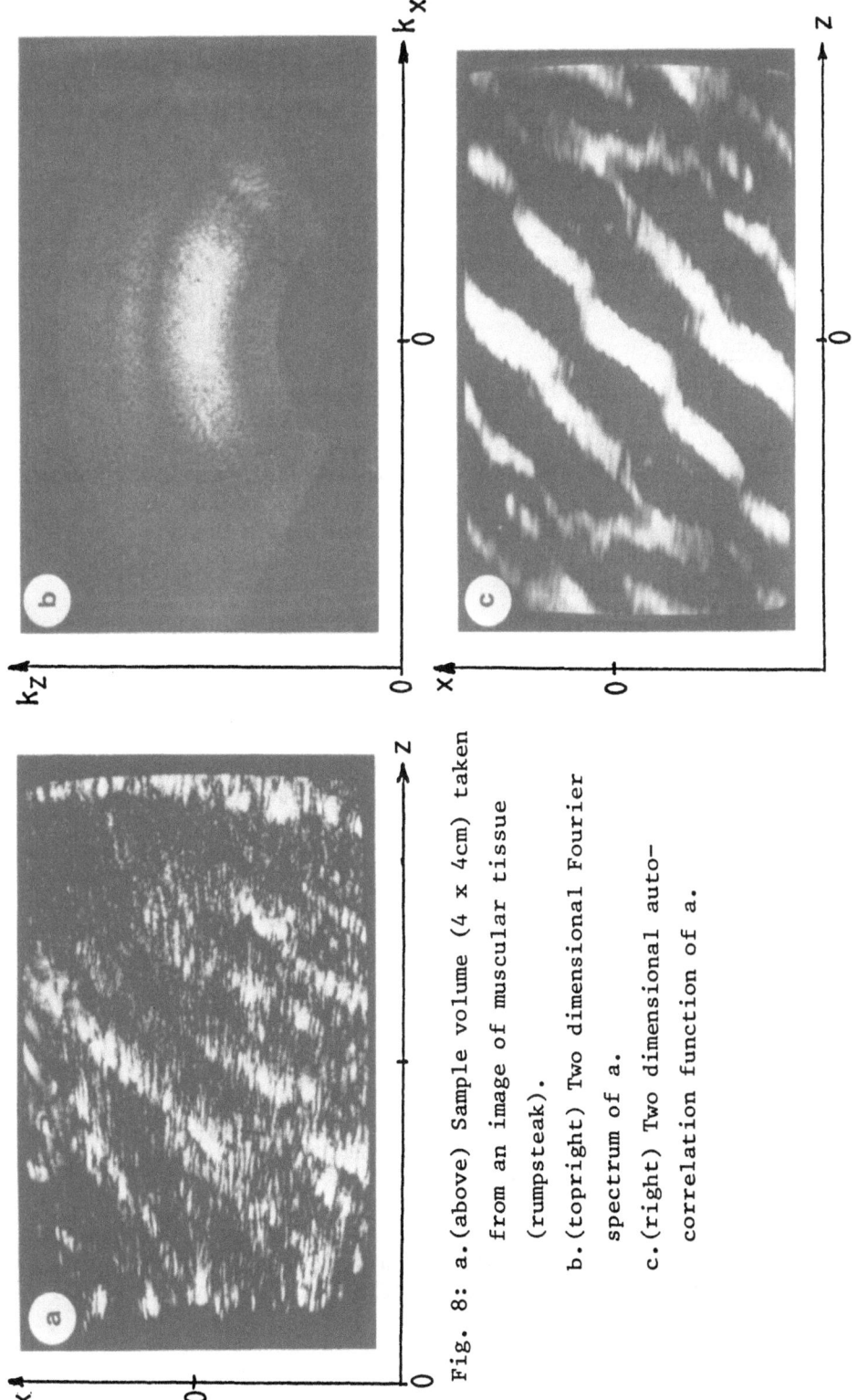

Fig. 8: a.(above) Sample volume (4 x 4cm) taken
from an image of muscular tissue
(rumpsteak).

b.(topright) Two dimensional Fourier
spectrum of a.

c.(right) Two dimensional auto-
correlation function of a.

Tissue characterization

Finally a practical illustration is given in figure 8. Here part of an image is shown of bovine muscular tissue. The size of the sample volume is approximately 4x4cm. The image is built up of mainly point scatterers as can be observed in the Fourier domain.

The auto correlation function shows two features, which might be characteristic for muscular tissue. Firstly there is a large scale periodicity present in the image, of which the dimension is about 1cm and secondly we see a small scale anisotropy denoted by the asymmetry of the centre of the auto correlation function.

Conclusions

In this paper once more the idea has been substantiated that, when studying the fine structure of a medium by means of acoustical backscatter techniques, there exists a theoretical limit to the amount of acoustical information which can be extracted without a priori knowledge of that medium. If great care is taken not to loose any information during data acquisition and image reconstruction the theoretical limit of approximately half a wavelength can be reached.

Though Fourier transformation and auto correlation of an image do not add any information to that already present in the image, some features are more readily extracted by one analysis technique than the other:
- The image itself is pre-eminently suited for the study of singular features (e.g. contours, reflectivity strength).
- The Fourier space shows any angular preference of local reflectivity parameters.
- The auto correlation function extracts any reflectivity features which repeat themselves regularly.

References

Berkhout, A.J., 1982, Seismic Migration -imaging of acoustic energy by wave field extrapolation, Elsevier, Amsterdam-Oxford-New York.

Berkhout, A.J., Ridder, J., Graaff. M.P. de, 1982, New possibilities in data measurement, signal processing and information extraction; philosophy and results, Acoustical Imaging, Vol. 12, pp. 269 - 290.

Mesdag, P.R., Vries, D. de, Berkhout, A.J., 1982, An approach to tissue characterization based on wave theory using a new velocity analysis technique, Acoustical Imaging, Vol. 12, pp. 479 - 491.

Nicholas, D., Merton, J., Hill, C.R., 1981, The application of
    diffraction analysis to liver and thyroid disease, Acoustical
    Imaging Vol. 11.

Ridder, J., Berkhout, A.J., Wal., L.F. van der, 1981, Acoustic
    imaging by wave field extrapolation, Part 2: Practical aspects,
    Acoustical Imaging, Vol. 10, pp. 541 - 567.

Vries, D. de, Berkhout, A.J., 1983, A note on the effect of
    velocity errors in computarized focusing techniques, Journal of
    Acoust. Soc. Am, 74(1), pp 353 - 356.

Acknowledgement:

These investigations were supported in part by the Netherlands
Foundation for Technical Research (S.T.W.), Technical Science
Branch of the Netherlands Organization for the Advancement of Pure
Research (Z.W.O.).

ACOUSTICAL IMAGING TECHNIQUES APPLIED TO GENERAL TRANSDUCER DESIGN

G.H. Harrison, E. Balcer-Kubiczek, and D. McCulloch

Division of Radiation Research
University of Maryland School of Medicine
Baltimore, MD 21201

INTRODUCTION

The Problem

The increasing applications of ultrasound in medical diagnosis have been accompanied by corresponding advances in transducer development, as evidenced by the proceedings of this series of symposia. Our research group is investigating ultrasonic bioeffects, an area largely innocent of specialized transducer technology. Some studies employ typical diagnostic ultrasound exposure systems and exposure conditions to simulate clinical insonation in tissue, at the expense of exposure quantification. Other studies are aimed at establishing dose-response relationships and thresholds for ultrasonically-induced effects; in some cases, these experiments are well characterized in terms of exposure conditions, but with few exceptions, the target areas are insonated with very non-uniform beams.

A typical exposure arrangement calls for cells suspended in a transmission vessel to be insonated using a uniformly-excited disk transducer; the ultrasound beam passes through the vessel with minimal interaction with the vessel walls (Chapman et al.,1979). An intensity profile which might be measured across the target area is shown in Fig. 1. This profile is highly non-uniform; although commonly quoted, peak or spatial average intensity values are poor descriptors for such a beam.

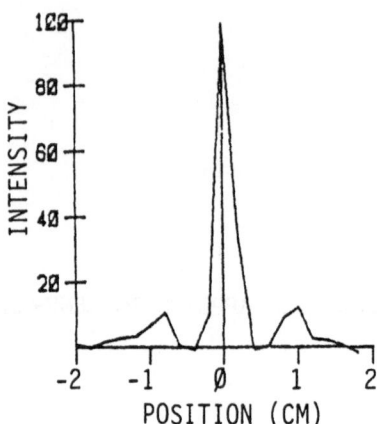

Fig. 1. Observed intensity profile in the far-field of a 2 cm
disk transducer operating at 1.7 MHz in water.

The Objective

Specialized systems involving standing wave chambers and
frequency mixing (Christman, 1981),the concentration of irradiation
samples within a small central area (Li et al., 1977), and multi-
electrode (Kossoff, 1971), or multi-transducer systems have
exhibited uniform insonation. We have chosen the objective of
developing single-electrode disk transducers producing a
travelling-wave, single-frequency CW beam with maximized intensity
uniformity within a given radius, and minimal intensity outside
that radius, and with this optimization occurring along an appreciable distance in the beam propogation axis. If such a transducer
could be produced, it would have wide applicability in bioeffects
studies, and would also be useful in ultrasonic physiotherapy and
thermotherapy for tumor treatment. Additionally, such a field, if
made uniform over a sufficient volume, might be useful in
diagnostic imaging as a constant-amplitude reference beam.

Approaches

Our approach to modeling new transducers was: 1) to investigate the effect of shaded transducer electrodes and other simple
alterations of uniformly excited disk transducers (such alterations
have been investigated previously in the context of focussing or

side-lobe reduction, but not for uniform insonation); and 2)   to
perform inverse-source calculations to determine source transducer
velocity distributions corresponding to optimized intensity
distributions at a given object plane.

Any useful attempt at transducer modeling must be experimentally
verified since we sought to produce real transducers for bioeffects
studies. Our first attempt at producing transducers with the
continuously varying velocity distributions required by shading or
inverse-source solutions consisted of simply etching electrode
patterns on transducer disks; the damping effects of transducer
mounting and edge effects (mechanical and electrical) near
electrode boundaries on otherwise discontinuous amplitudes were
then to be manipulated emprically to approximate the required
continuous distributions. As another approach to producing continuously varying amplitude distributions, opto-acoustic
transducers (OATs) (Elliott et al., 1978) were investigated and
fabricated. Although the acoustic output of an OAT may be limited,
the OAT should serve as a useful prototype design aid for our
transducers. Although the response time and phase properties of
the OAT constitute serious limitations in imaging applications,
these limitations should not degrade the performance required for
producing uniform CW fields.

METHODS

Experimental Methods and Materials

All transducers in this study started as 3.8 cm diam. PZT-4
discs resonant at 1.7 MHz, poled, and electroded with silver
(Vernitron Piezoelectric Corp.). In the case of etched electrode
transducers, the electrode pattern was taped onto the transducer
prior to etching away the electrode with 2M $Fe(NO_3)_3$.   All
transducers were air-backed, supported in their mounts by a thin
styrofoam ring and a bead of silicone rubber sealant around the
circumference, in order to minimize edge damping.

To construct an OAT, one electrode was removed from a
transducer and a Cd/ZnS layer typically 20 μm thick was
silk-screened on the PZT surface prior to sintering according to
the procedure described by Elliott (1978). The thickness uniformity
was approximately 6%. Then, silver or aluminum grids were
evaporated onto the photoconductive layer. Grid spacing was of the
order of an ultrasonic wavelength, and the grid thickness was such
to produce approximately 50% light transmission. Transducers were
then repoled and mounted for evaluation.

A water-filled scanning tank lined with Wallgone[R] Fe_2O_3-
loaded rubber absorber (Consumer Usage Laboratories,Inc.) was used

for experimental measurements.   Computer-controlled data
acquisition was performed with a Dapco model NP10-3 hydrophone
(active element dia. 0.6 mm).   The resolution selected for the
lateral scans presented in this paper was 0.2 cm.

Theoretical Methods

Referring to Fig. 2, we sought to relate a circularly symmetric
source velocity distribution $V(r)$ to the ultrasonic amplitude
distribution $F(y,z)$ on an object plane at z. The determination of
the intensity I from V was accomplished by numeral integration
described below, whereas we have identified the determination of V
from I as an inverse scattering problem with a useful analytical
solution reported in the RESULTS section. The intensity I on the
object plane is

$$I(y,z) = \langle \mathrm{Re}^2 \; F(\overline{y},z) \rangle = \tfrac{1}{2} \langle \overline{F}\overline{F}^* \rangle \qquad (1)$$

where the brackets denote time averaging.

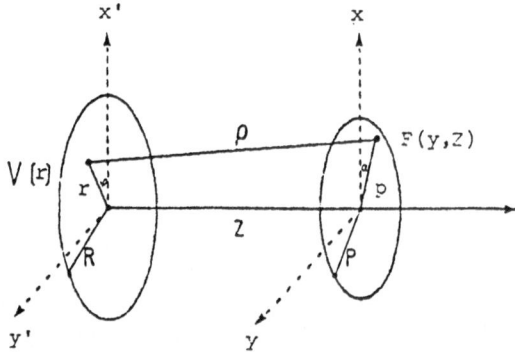

Fig. 2. Coordinate systems for relating source velocity $V(r)$ to
the acoustic field $F(y,z)$ on the object plane.

The acoustic field from a circularly symmetric source, $V(r)$, is
given by the superposition integral:

$$F(y,z) = C \; e^{i\omega t} \int_0^R \int_0^\pi V(r) \left[\frac{e^{ik\rho(r,\varphi)}}{\rho(r,\varphi)} \; d\varphi \right] r \; dr \qquad (2)$$

where $\rho^2 = y^2 + z^2 + r^2 - 2ry \cos\varphi$. We have dropped the
variable x because in the case of a circularly symmetric source at

a fixed value of z=Z, we have at our disposal the adjustments of the x' and y' directions such that x = 0.

The 2-dimensional integral in eq. (2) was approximated by a product rule using a 16-point Gauss-Legendre formula in the r variable and a 48-point trapezoid rule in the φ variable. This numerical technique yielded an efficient computer algorithm for acoustic intensity calculations from a variety of theoretical transducers described by different source distributions, V(r).

In our initial modeling efforts, we obtained the intensity patterns for several distribution functions V(r), which permitted construction of simple practical transducers in emulation of the theoretical V(r).

It has been noted previously that the radiation fields in the source and object planes are related by a Fourier transform. The general case is intractable, but strict Fourier inversion of eq. (2) can be performed assuming the validity of the Fresnel approximation in far field, i.e., z⩾15 cm for a 1 MHz transducer. It is in this case possible to modify eq.(2) in rectangular coordinates to obtain the following:

$$\widetilde{F}(x,y) = c_f \int_{-\infty}^{\infty} \int_{-\infty}^{\infty} \widetilde{V}(x',y') \exp\left\{-\frac{ik}{z}(xx' + yy')\right\} dx' \, dy' \qquad (3)$$

where:

$$\widetilde{F}(x,y) = F(x,y) \exp\left\{-\frac{ik}{2z}(x^2+y^2)\right\} \qquad (4a)$$

and

$$\widetilde{V}(x',y') = V(x',y') \exp\left\{\frac{ik}{2z}(x'^2 + y'^2)\right\} \qquad (4b)$$

we then can Fourier invert (3) and obtain (5):

$$\widetilde{V}(x',y') = c_t \int_{-\infty}^{\infty} \int_{-\infty}^{\infty} \widetilde{F}(x,y) \exp\left\{\frac{ik}{z}(xx' + yy')\right\} dx \, dy \qquad (5)$$

For clarity, we combined all the constants into proportionality factors c_f and c_t. We note that in what follows the z variable has a rigorous meaning of a fixed parameter.

The purely mathematical result given by eqs. (3-5) has been established previously. For practical applications, it is important to note that the experimentally measured quatratic quantity is I, defined in eq. (1).

Thus it follows that several different forms for F(x,y) can give the same I(x,y) since the phase information is lost in the products. We used this observation and selected the functional form for $\widetilde{F}$ instead of F since they differ only by a phase factor. The first step of our modeling technique was the analytical integration of eq.(5), at a fixed z plane and assumed form for $\widetilde{F}$. The source distribution was obtained from $\widetilde{V}$ eq. (4b). To test our procedure, we substituted the derived source distribution V into eq. 2, and generated back the field intensity F(y,z). Since the source distribution extends beyond the geometric source size, the integration eq.(2) incurs a truncation error. The agreement between assumed theoretically I distribution and that obtained by re-integration was satisfactorily in agreement with what we might expect by imposing the final limits on the region of integration.

RESULTS

How to Use an Ordinary Disk Transducer

As an initial example and test of our approach, a uniformly-excited transducer was modeled and experimentally characterized. We found that the optimum location for insonation according to our criteria was the region, between the last central-axis minimum and last maximum in intensity, where the side-lobe peaks matched the central peak. These regions are shown in Figs. 3 and 4a for theoretical and experimental results, respectively. Also shown in Fig. 4b is the experimental far-field profile--a nearly Gaussian distribution, which is commonly selected for bioeffect studies. In further demonstration of the inadequacy of this distribution, note that even if we choose a small target area with radius 1 cm, over 60% of the area is exposed to intensities less than 25% of the peak intensity, as shown in Fig. 3b. Comparison of the distributions shows the advantage gained from using the merged side-lobes along with the central peak, at least according to criteria for uniform insonation. However, the experimental data shown in Fig. 4a demonstrate poorer uniformity than predicted by theory.

The Annular Transducer

In order to evaluate the fringing effect at electrode edges on the surface of transducers, an annular transducer was experimentally evaluated. The transducer was fully electroded on the back surface, while the annular electrode on the radiating side

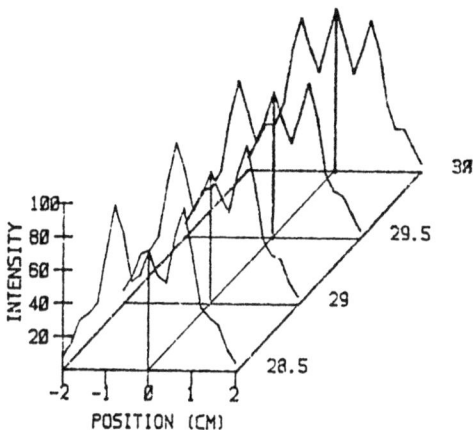

Fig. 3   Calculated intensity distribution produced by a 1.9-
cm-radius disk transducer uniformly excited at 1.7 MHz, at
the optimum location for uniform insonation.

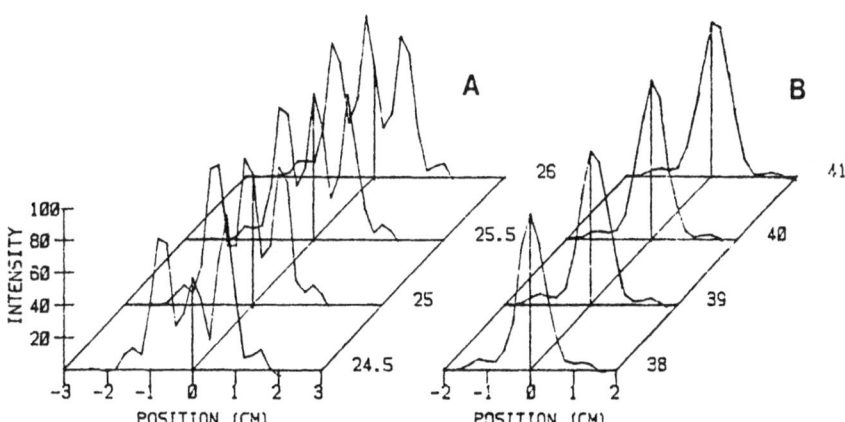

Fig. 4 a)   Experimentally   observed   intensity   distribution
corresponding to Fig. 3; b) Same as a), except in the far
field.

had an inner radius of 1.3 cm, and an outer radius corresponding to the transducer radius of 1.9 cm. The observed intensity distribution is shown in Fig. 5.

Resultant intensity distributions were compared with the computed distributions of 3 theoretical transducer velocity distributions shown in Fig. 6. Figure 6a represents an annular electrode without any damping or fringing.

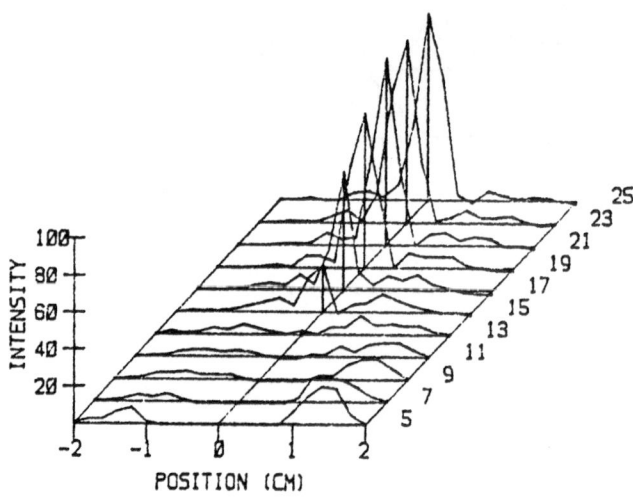

Fig. 5 Experimentally observed intensity distribution produced by a 1.9-cm-radius annular transducer with inner radius of electrode = 1.3 cm, frequency 1.7 MHz.

Figure 6b represents an example of a general polynomial:

$$V(r) = ar^4 + br^2 + c \tag{6}$$

Zero velocity at the outer edge, which could occur in some transducer design requirements, was satisfied by the set of co-efficients:

$$b = \frac{2}{R^2}\left[(1-c) + \sqrt{1-c}\right] \qquad a = \frac{b^2}{4(c-1)} \quad < 0 \tag{7}$$

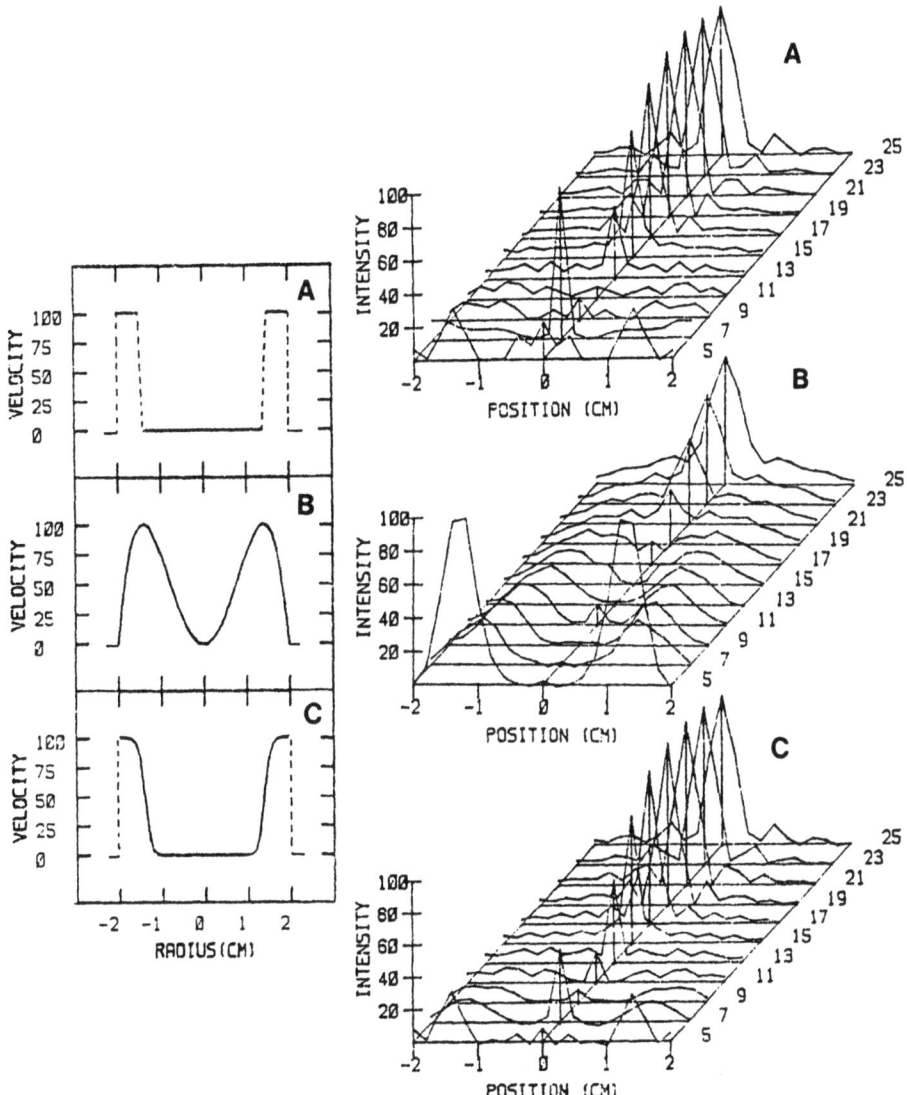

Fig. 6   Three annular velocity distributions as described in the
text, along with corresponding calculated intensity
distributions, frequency = 1.7 MHz.

where c determines the off-center velocity weighting with the maximum occurring at $r = \sqrt{-\dfrac{b}{2a}}$. The closest approximation to the annular transducer is obtained with c = 0

$$V(r') = -4r'^4 + 4r'^2 \quad \text{where } r' = r/R \tag{8}$$

Figure 6c depicts V(r) corresponding to the Woods-Saxon nuclear potential, chosen as a reasonable representation of an exponentially damped annulus. The resulting intensity distributions for these source distributions are also shown in Fig. 6. Examination of these results reveals that except for the polynomial case, the distributions exhibit the expected pattern for annular transducers, that is, diminished near-field intensities and a sharp central-axis beam beyond. It appears that the "Woods-Saxon" model may best reproduce the experimental annulus transducer, since the near-field dimunition most closely resembles that of the experimental field.

The "Fermi-Dirac" Transducer

Returning to the objective of uniform insonation from shaded transducers, it was found theoretically that a uniform velocity distribution with exponential edge shading could produce good intensity uniformity on certain object planes. A convenient expression to parameterize the source velocity was the Fermi-Dirac distribution:

$$V(r) = \left\{ 1 + \exp\left(\frac{r - r_o}{a}\right) \right\}^{-1} \tag{9}$$

The velocity distribution for such a transducer operated at 1.7 MHz, $a_0 = 0.125$ and $r_0 = 1.4$ cm is shown in Fig. 7.

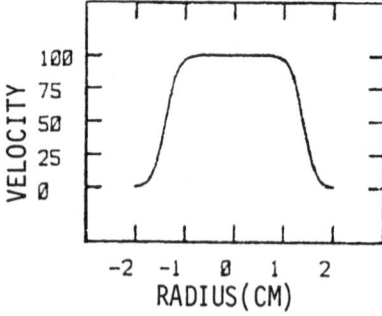

Fig. 7 The velocity distribution of a "Fermi-Dirac" transducer operated at 1.7 MHz with $a_0 = 0.125$ and $r_0 = 1.4$ cm.

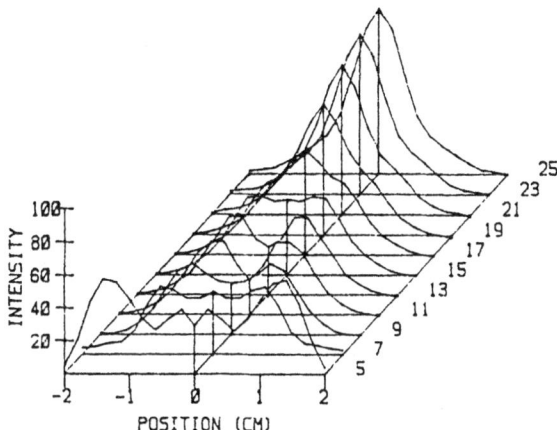

Fig. 8 Calculated intensity distribution produced by the "Fermi-Dirac" velocity distribution shown in Fig. 7.

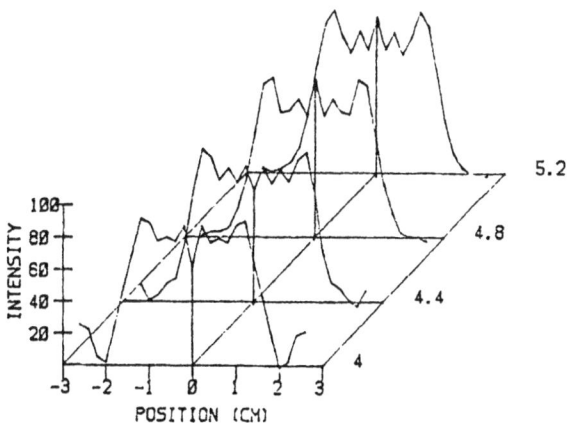

Fig. 9 Optimum insonation pattern calculated for a "Fermi-Dirac" transducer with a_0 = 0.075, r_0 = 1.8 cm, and frequency = 1 MHz.

Good intensity uniformity according to our criteria was
obtained theoretically at depths of 5 and 13 cm (Fig. 8).
These patterns further improved with source distributions decaying
faster near the edge; for a_0 = 0.075 and r_0 = 1.8 cm at 1 MHz, the
intensity distribution between 4 and 5.2 cm in depth is shown
in Fig. 9. These intensity profiles exhibit good uniformity
over an appreciable area, and rapid roll-off beyond, as was
desired. To achieve this result experimentally, several electrode
geometries were investigated. First, in analogy with the results
for annular transducers, we reasoned that the use of front and rear
circular electrodes with different diameters might result in the
exponential type of velocity decay we sought. Accordingly, the
velocity potential shown in Fig. 7 was emulated by a radiating
electrode diameter of 1.2 cm and a rear electrode diameter of
1.6 cm. This attempt resulted in a poor intensity pattern at all
depths, with side-lobes sharply separated from the central axis
peak. A successful electrode pattern was an asteroidal, 9-point
pattern scaled from that proposed by Kikuchi (1978) to achieve
side-lobe reduction. Our experimental results indicated large variations of intensity along the beam axis, but also relatively broad
regions of uniform insonation at certain depths (Fig. 10). This
easily-constructed transducer type may be a practical solution of the
problem we posed.

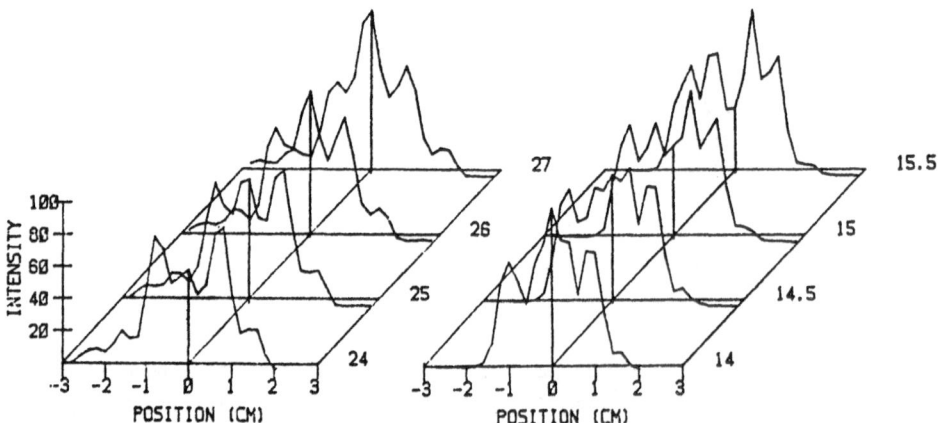

Fig. 10 Experimentally observed insonation patterns for a
1.9-cm-radius transducer with a 9-point asteroidal
electrode pattern similar to that proposed by Kikuchi
(1978).

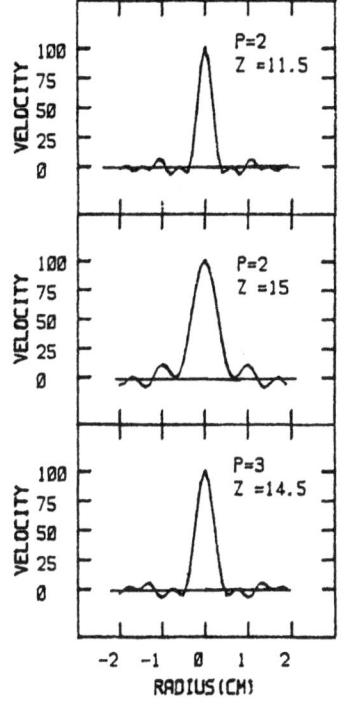

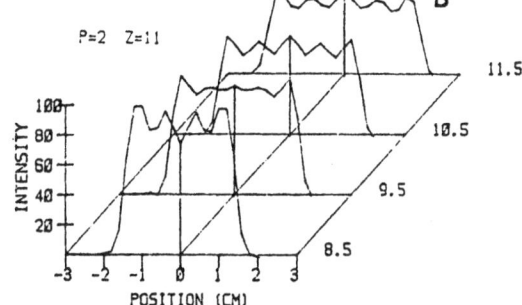

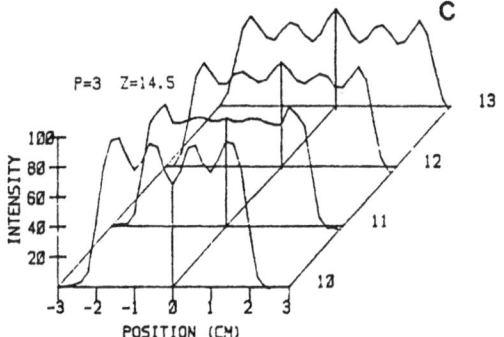

Fig. 11 Velocity distributions from inverse scattering solutions for uniform insonation, with Z and P as shown in the figures and defined in the text.

Fig. 12 Calculated intensity distributions produced by transducers with velocity distributions specified by inverse scattering solutions truncated at radius = 2 cm.

Inverse Scattering Solution

Using the notation defined above in METHODS, the following problem was solved analytically for the far-field: on an object plane at z = Z, if

$$\tilde{F}(x,y) = circ(p) = \begin{cases} 1 \text{ for } p \leqslant P \\ 0 \text{ otherwise} \end{cases} \tag{10}$$

then on the source plane, apart from constant factors,

$$V(r) = \left(\frac{k}{2\pi}\right)^2 \frac{2P}{r} \, e^{-\frac{ik}{2z}(2z^2 + r^2)} \, J_1\left(\frac{k}{z} \, rP\right) \tag{11}$$

where J_1 is the first-order Bessel function of the first kind, order one. This source function, truncated beyond r = 2 cm, is shown in Fig. 11 for various values of parameters P and z=Z. It is evident that this V(r) is rapidly varying and requires phasing, and that the pattern depends critically upon Z and P. Several examples of this V(r), truncated beyond r = 2 cm, were fed back into eq. (2) to calculate the intensity distributions shown in Fig. 12.

Figure 12a shows that uniform intensities over large areas are obtained over a wide range of object plane depths, but that the intensities decrease rapidly with depth. Figures 12b, 12c detail selected volumes of uniform insonation for P = 2 and 3 cm.

CONCLUSIONS AND FUTURE WORK

We have experimentally shown that simple shading techniques can be employed to produce improved intensity uniformity for ultrasonic bioeffects studies. Analytical solutions of the inverse scattering problem for uniform insonation have been obtained, and very uniform fields over large areas of selected object planes were calculated; however, the needed transducer velocity distributions may be difficult to achieve in practice. These results suggest the direction for future work: 1) to investigate the limits and sensitivities of the relatively simple shading techniques described here; 2) to pursue studies with the OAT prototypes we have con-structed; 3) to seek the inverse scattering solutions for more realistic object plane distributions in order to find simpler source distributions; and 4) to seek source solutions corresponding to assumed insonation pattern over a given volume instead of over an object plane.

ACKNOWLEDGEMENT

This research was supported by National Science Foundation Grant ECS-821366.

REFERENCES

Chapman, I.V., MacNally, N.A., and Tucker, S., 1979, Ultrasound-induced charges in rates of influx and efflux of potassium ions in rat thymocytes in vitro, Ultrasound Med. Biol. 6:47.

Christman, C.L., 1981, Average dose absorbed by biological specimens in a diffuse ultrasonic exposure field, J. Acous. Soc. Am., 70:946.

Elliot, S., 1978, Ph.D. Thesis, Department of Electrical Engineering, University of California of Santa Barbara.

Elliot, S., Domarkas, V., and Wade, G., 1978, Frequency characteristics of opto-acoustic transducers, IEEE Trans. Sonics and Ultrasonics SU-25:346.

Kikuchi, Y., 1978, Transducers for ultrasonic systems.Chapter 6, in: "Ultrasound: Its Applications in Medicine and Biology", F.J. Fry, ed., Elsevier, Amsterdam.

Kossoff, G., 1971, A transducer with uniform intensity distribution, Ultrasonics 9:196.

Li, G.C., Hahn, G.M., and Tolmach, L.J., 1977, Cellular inactivation by ultrasound, Nature 267:163.

FRESNEL ZONE PLATE AND
FRESNEL PHASE PLATE PATTERNS FOR ACOUSTIC TRANSDUCERS

M. Mortezaie and G. Wade

Department of Electrical & Computer Engineering
University of California
Santa Barbara, CA 93106

ABSTRACT

An ultrasonic transducer with an electrode pattern in the shape of a Fresnel zone plate (FZP) is useful in acoustic imaging. However a major problem in focusing with an FZP is that the final image will have a large zero-order component and this will restrict the dynamic range of the system. One way to solve this problem is to use a Fresnel phase plate pattern (FPP).

We employ a linear spatial system model to determine the transducer aberrations caused by wave generation and propagation inside the tranducer. The focal-plane intensity distribution is calculated for both the FZP and FPP, assuming similar parameters. The low-pass filtering effect of aberrations in the transducers using the FZP and the FPP are studied in the spatial-frequency domain and are compared. The analysis predicts that, due to these aberrations, the FPP can give better resolution than the FZP by a factor of almost two. Experimental results that support the linear model are presented.

INTRODUCTION

Ultrasonic piezoelectric transducers with an electrode pattern in the form of a Fresnel zone plate (FZP) on the air side and a full planar electrode on the water side have been used in many acoustical imaging systems to focus the acoustic beam [1-4]. The goal is to obtain an acoustic pressure pattern in the shape of an FZP. However, a major problem in FZP focusing is that the final image will have a large undiffracted (zero spatial frequency) component. This restricts the dynamic range of the detector and increases the signal to noise ratio [5]. One way to solve this problem is to use Fresnel phase plate (FPP) patterns [6,7].

In this paper we employ a linear system model of the transducer to predict the focal-plane intensity distribution of the FZP and the FPP transducers. The transfer function of the system, relating the electrode pattern to the focal-plane acoustic distribution, takes into account the mechanical aberration caused by wave generation and propagation inside the transducer and the diffraction of waves from the front surface of the transducer to the focal plane.

To simplify the analysis a one-dimensional variation of the electrode pattern in the x-direction is assumed. A transducer with such a variation would focus the acoustic energy to an x-directed line at the appropriate position in the z-direction and thus would be analogous to a cylindrical lens. Mechanical aberrations cause the focal-plane distribution in either an FZP or an FPP transducer to be broader than that predicted by diffraction from an FZP or an FPP acoustic pattern. Comparing the focal-plane acoustic distribution for the FZP pattern with that for the FPP pattern shows that the resolution, taken to be the focal-spot width at 40% of the maximum intensity, is improved by almost a factor of two in the FPP case.

SPATIAL-SYSTEM MODEL OF THE TRANSDUCER

The transducer shown in Fig. 1(a) is modeled as the linear spatial system in the block diagram of Fig. 1(b). The transfer function can be represented as the product of two factors. The first is denoted H_A and relates the acoustical field distribution in front of the transducer to the electrode pattern. The second is denoted H_D and represents the effect of wave propagation from the transducer surface (z=0) to the FZP focal plane (z=f). H_A is called the aberration-limited transfer function because it represents the effect of mechanical wave-spreading in the transducer [8]. This is analogous to aberrations in optical imaging systems. H_A is a complex quantity whose magnitude is given in Fig. 2 as a function of the normalized spatial frequency, $\nu_x = \lambda f_x$. f_x is the spatial frequency and λ is the wavelength in water. To obtain the result in Fig. 2, only compressional waves have been taken into account. The contribution due to shear waves has been neglected because they are weakly coupled in this case. The piezoelectric plate we considered was of PZT-5A material and was resonant at 3.4 MHz when in contact with water.

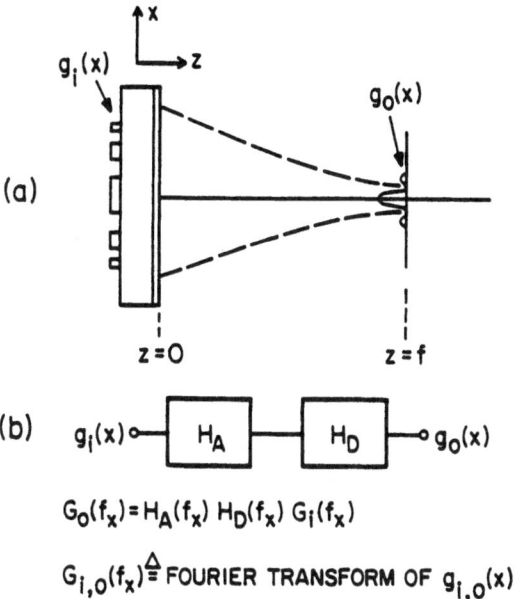

Fig. 1 (a) Transducer system (b) Linear system model of the transducer.

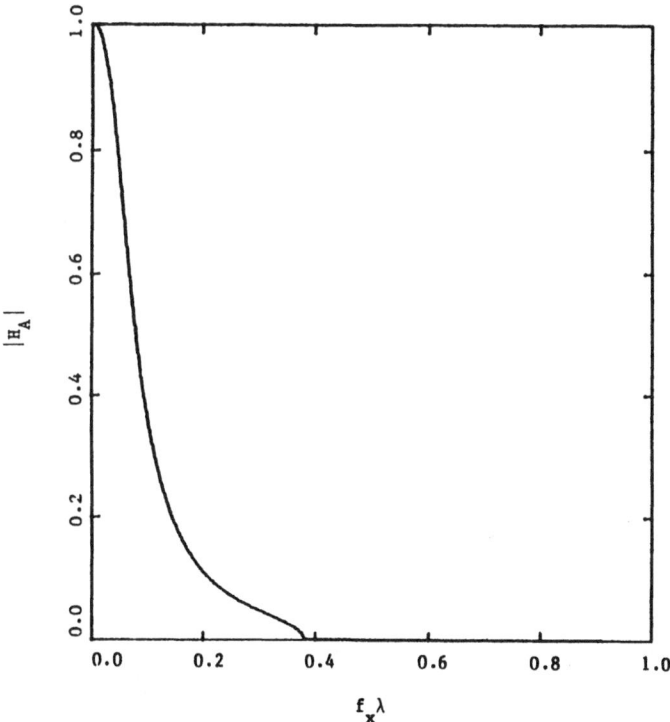

Fig. 2 Magnitude of the aberration-limited transfer function versus normalized spatial frequency.

The normalized cutoff spatial frequency is given by the ratio of the phase velocity v in water to the phase velocity v_o in the piezoelectric material, as follows:

$$\upsilon_{xc} = \lambda f_{xc} = \frac{v}{v_o} = \frac{v}{(C_{11}^E/\rho)^{\frac{1}{2}}}$$

where ρ is the density of the piezoelectric material, and C_{11}^E is one of its elastic constants.

H_D characterizes the effect of wave propagation from the z=o to the z=f plane. Using the planar spectrum approach [9], H_D may be represented by:

$$H_D(f_x) = \text{rect} \ (\frac{\lambda f x}{2}) \ \exp[j(2\pi f/\lambda)(1-f_x^2\lambda^2)^{\frac{1}{2}}]$$

where rect $(\cdot)$ is the rectangle function, equal to unity when the absolute value of its argument is less than or equal to $\frac{1}{2}$, and zero otherwise. The transfer function of the transducer is given by:

$$H(f_x) = H_A(f_x) \ H_D(f_x)$$

To study the resolution of a transducer, its focal-plane distribution may be analyzed. The size of the generated focal spot is a figure of merit for resolution. Smaller focal spots correspond to better resolution.

The focal-plane distribution $g_o(x)$ can be found by the following:

$$G_o(f_x) = H_A(f_x) \ H_D(f_x) \ G_i(f_x)$$

$$g_o(x) = \int_{-\infty}^{\infty} G_o(f_x) \ \exp(j2\pi f_x x) \ df_x$$

where $G_o(f_x)$ and $G_i(f_x)$, are the spatial spectrum of $g_o(x)$ and the electrode pattern, $g_i(x)$, respectively.

FRESNEL ZONE PLATE PATTERN

A transducer with a Fresnel zone plate pattern of focal length f = 4.0 cm was examined theoretically using the transducer model [10]. A PZT 5-A ceramic polarized in the thickness direction was assumed for the transducer. The thickness of the transducer was assumed to be the half the compressional wavelength in the ceramic at the resonant frequency 3.4 MHz. The transducer was assumed to be air-backed and transmitting its energy into water. The side in contact with the water was assumed to be covered by a full electrode. An electrode pattern in the shape of a one-dimensional FZP pattern was assumed to be placed on the air side. The tranducer was considered to be driven at the imaging frequency of 3.4 MHz.

Fig. 3 shows the calculated variation of focal-plane intensities as a function of the normalized lateral distance x/λ, for a 2, 4, and 6-zone transducer. The dashed curves represent the aberration-free case, i.e. the case in which $H_A(f_x) = 1$ and are simply the Fresnel diffraction patterns for an FZP. It can be seen that the resolution increases as the number of zones increases. The solid curves represent focal-plane intensities

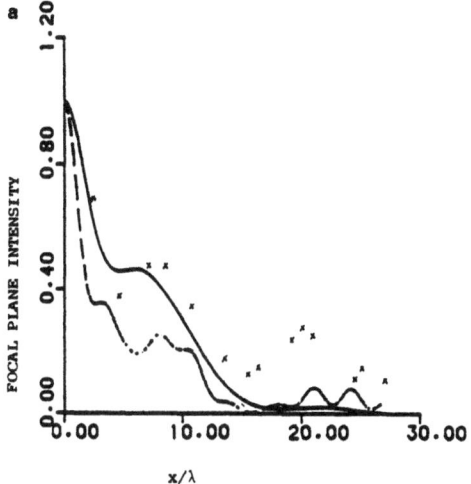

Fig. 3   Focal plane intensities of an FZP transducer for:
(a) 2 zones (b) 4 zones (c) 6 zones.
Dashed curves represent the ideal diffraction-limited case, while the solid curves correspond to the inclusion of aberrations. Crosses show the result of the experimental measurements. (continued)

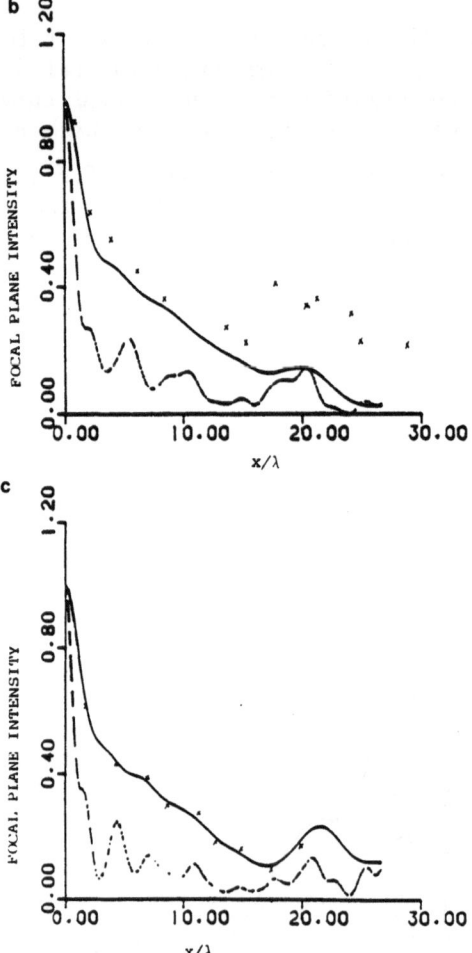

Fig. 3 continued

when aberrations are taken into account. Attenuation of the high spatial frequencies inside the transducer is apparent from the cutoff of the transfer function shown in Fig. 2. Because of this attenuation, the width of the focal spots are increased over what they would otherwise be. This results in poorer resolution as compared to the dashed curve in Fig. 3. Since spatial frequencies above v_{xc} are absent, increasing the number of zones beyond a certain limit does not improve the resolution. The optimum value of N may be found from properties of the transducer material and the propagation medium as well as specifications of f and λ for the zone plate.

In the experiment a PZT-5A square transducer was employed. Parameters of the transducer were the same as in the theoretical analysis. The crosses in Figs. 3(a)-(c) represent the results of experimental measurements taken with a one-dimensional FZP electroded transducer for the same number of zones.

In spite of a few sources of error, the experimental results are close to the theoretical predictions.

TRANSDUCERS WITH FRESNEL PHASE PLATE PATTERNS

In the previous section the focusing and resolving power of transducers with Fresnel zone plate electrode patterns have been analyzed. Many imaging systems [1-6] utilize FZP transducers because of the ease of production. However a major problem with FZP focusing is that the final image will have a large undiffracted (zero spatial frequency) component. This restricts the dynamic range of the detector and increases the signal-to-noise ratio [5]. One way to solve this problem is to use Fresnel phase plate (FPP) patterns [6,7]. In practice the FPP transducers are realized by a three-step poling process [6]. In this process, a high electric field is applied across the ceramic while it is immersed for a minute in peanut oil at a temperature of about 130°C. All the dipole molecules in the material then become aligned in the direction of the applied field. Mechanical deformation in the material depends on the polarization of the applied field.

The first of the three steps is to obtain a uniformly-poled transducer with a direction of polarization as shown in Fig. 4(a). The second step consists of providing the transducer with an electrode in the shape of an FZP pattern on one side and a full electrode on the other side as shown in Fig. 4(b). The transducer is then poled using this electrode configuration by supplying the transducer with a voltage in the opposite direction from that of the original poling voltage. In the third step, a full electrode replaces the FZP electrode.

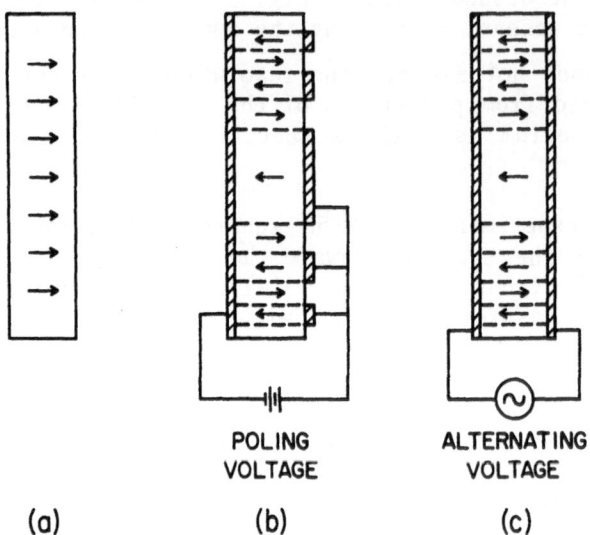

POLING   ALTERNATING
VOLTAGE   VOLTAGE

(a)   (b)   (c)

Fig. 4   Three-step poling process to produce an FPP trans-
         ducer.

An alternating voltage applied to the transducer excites
adjacent zones with opposite deformations and produces an effect
that resembles that of an FPP electrode. The focal plane distribution of the FPP transducer, shown in Fig. 5, is found by utilizing the spatial model of the transducer. The specifications f =
4.0 cm and 3.4 MHz frequency are the same as for the FZP case. In
the passband of the transducer transfer function the FPP has more
spectral content than does the FZP. This leads to a more efficient reproduction of the acoustic FPP pattern compared to that for
an FZP pattern. The resolution can be expected to be better in
FPP transducers than in FZP transducers. A camparison of the
focal-plane acoustic intensity distribution of Fig. 3 for the FZP
case with that of Fig. 5 for the FPP case shows the resolution is
better by almost a factor of two in the FPP case.

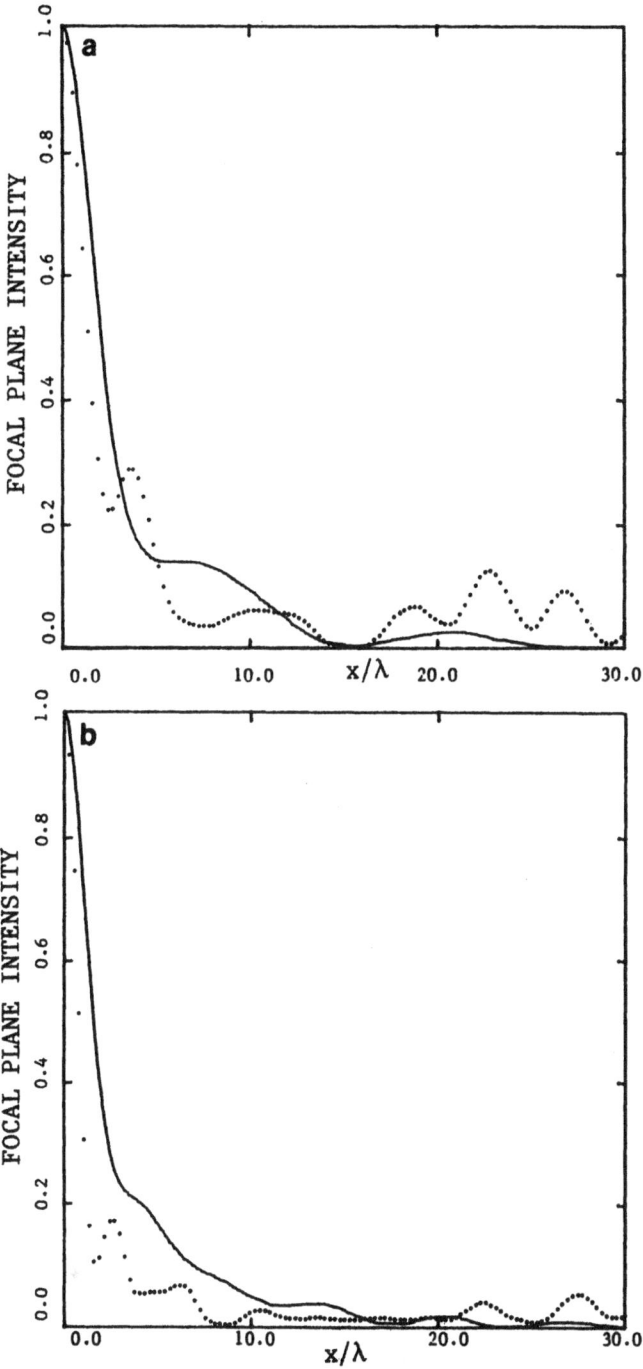

Figure 5   Focal plane intensities of an FPP transducer for:
(a) 2 zones   (b) 4 zones   (c) 6 zones. Dashed
curves represent the ideal diffraction-limited system.   Solid curves correspond to the inclusion of
aberration.(continued)

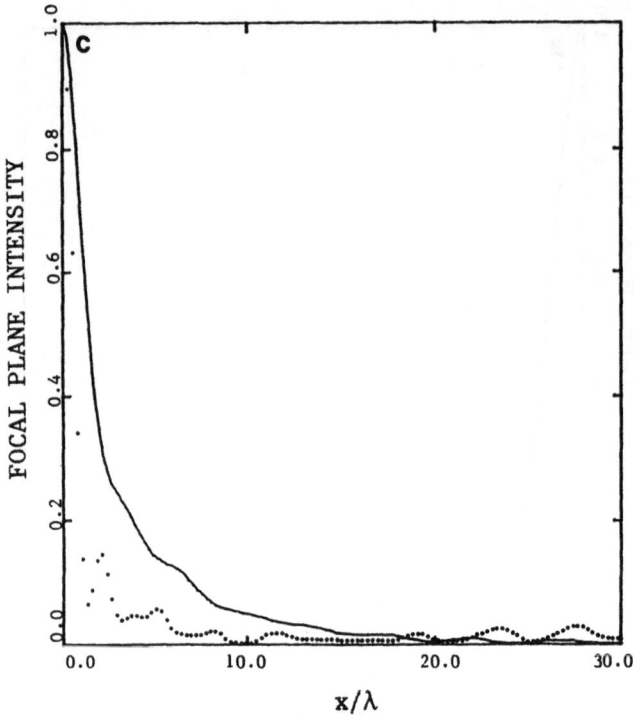

Figure 5 (continued)

CONCLUSIONS

FZP and FPP electrode patterns for transducers have been compared by utilizing a spatial model. The focal-plane intensity
distribution has been calculated and it has been shown that the
resolution in each case (FZP or FPP) is poorer, because of aberrations, than it would be for simple diffraction from an FZP or
FPP acoustical pattern. A theoretical comparison of the FZP case
with that of the FPP case shows that resolution is better by
almost a factor of two in the FPP case.

ACKNOWLEDGEMENTS

We would like to thank Michael L. Galindo for typing this
manuscript. We are also appreciative of support for this work by
the National Science Foundation under Grant NSF-ECS-79-18779.

REFERENCES

[1] S.A. Farnow and B.A. Auld, "Acoustic Fresnel Zone Plate Transducers," Appl. Phys. Lett., Vol. 25, pp. 681-682, 1974.

[2] K. Wang, V. Burns, G. Wade and S. Elliot, "Opto-Acoustic Transducers for Potentially Sensitive Ultrasonic Imaging," Opt. Eng., Vol. 16, No. 5, pp. 432-439, September/October 1977.

[3] K. Wang and G. Wade, "A Scanning Focused-Beam System for Real-Time Diagnostic Imaging," Acoustical Holography, Vol. 6, Ed. N. Booth, Plenum Press, New York, pp. 213-228, 1975.

[4] D.C. Webb, "Scanned Acoustic Imaging at Microwave Frequencies," Ph.D. Dissertation, Stanford University, 1971.

[5] D.J. Stigliani, R. Mittra and R.G. Semonin, "Resolving Power of a Zone Plate," Journal for the Optical Society of America, Vol. 57, Number 5, 1967.

[6] S.A. Farnow and B.A. Auld, "An Acoustic Phase Plate Imaging Device," Acoutical Holography Vol. 6, Ed. N. Booth, Plenum Press, New York, 1975.

[7] R.W. Wood, Physical Optics, pp. 37-39, (Dover, New York, 1975).

[8] B. Noorbehesht, G. Flesher and G. Wade, "Spatial Response of Arbitrarily Electroded Piezoelectric Plates by Plane Wave Decomposition," Ultrasonic Imaging, Vol. 12, pp. 102-121, 1980.

[9] J.W. Goodman, Introduction to Fourier Optics, McGraw-Hill, New York, 1976, pp. 54.

[10] B. Noorbehesht, M. Mortezaie, G. Wade and C. Schueler, "Effect of Mechanical Aberrations on the Resoluton of Fresnel Zone Plate Transducers," Acoustical Imaging, Vol. 11, J. Powers, Ed., Plenum Press, New York, 1981.

[11] B. Noorbehesht, M. Mortezaie and G. Wade, "Resolution of Fresnel Phase Plate Transducers," Proceeding of the Eleventh IEEE Ultrasonics Symposium, October, 1981.

A NEW TECHNIQUE FOR REALIZING ANNULAR ARRAYS

OR COMPLEX SHAPED TRANSDUCERS

P. Alias, P. Challande and C. Kammoun

Institut de Mécanique Théorique et Appliquée
Université Pierre et Marie Curie, Paris, France

B. Nouailhas and F. Pons

Direction Etudes et Recherches, Electricité de France
Saint-Denis, France

INTRODUCTION

We propose here a new technique for making arrays of transducers
with arbitrary shapes and specially annular arrays. It is well known
that the simplest solution which consists in photoetching different
electrodes on a unique piezoelectric plate does not give satisfying
results. The main reason is that exciting localized stresses in a
solid plate creates a lateral propagation in terms of Lamb modes
which perturbs the emitted ultrasonic field, mainly in creating
secondary lobes of high level. The mechanical separation of the
transducers obtained by cutting the plate according to the contours
of the different transducers is not often a satisfying solution.
First, it may be a difficult and expensive technology if the contours
are not straight. But, in general, unpleasant characteristics remain
in the behaviour of each isolated transducer. If the width of the
transducer is larger than the thickness, the electromechanical
behaviour exhibits different modes which may be strongly excited
simultaneously with a transient signal in such a way that the
emitted field has much larger duration and directivity strongly
different from the piston-like one. In the simplest kind of array,
i.e. a linear array of identical transducers, a satisfying solution consists in cutting the piezoelectric plate with a pitch
smaller than the thickness, each transducer resulting from the
electrical association of adjacent blocks of ceramic. In that case,
it has been shown {1-4} that each mechanically isolated block exhibits a first resonant mode at a frequency which is distinct enough
from the other harmonics, so that even a wide band transient may
excit the fundamental mode only. Unhappily, this solution cannot be

extended to other arrays. Even for annular arrays, machining circular slots remain difficult and expensive. Moreover, the radial
width of the different transducers is very often preferably not
constant : this is the case of Fresnel law annular arrays which
exhibit transducers of equal areas with an emission or a reception
separated by equal delays in a focusing technique. A subdivision
in elementary rings of width over thickness ratio (W/T) less than
one, if technologically possible, would lead to different W/T ratios
for the different transducers. We propose here a technique for
making such arrays, which is relatively easy to achieve and seems
to overcome most of the problems presented not only by Fresnel
annular arrays, but by any array of arbitrary shape. Results concerning Fresnel circular and elliptic arrays adapted to medical or
N D E applications will be presented.

THE TECHNOLOGY

The very simple basic idea is to extend the 1-D solution exposed earlier for making linear arrays to a 2-D solution consisting
in separating by straight slots in the ceramic plates identical
blocks of square (or diamond) area with a W/T ratio less than one,
so that each individual block must exhibit the same interesting
electromechanical behaviour. If the characteristic retained pitch W
is also smaller than the characteristic length of the transducers
of the array (for example, the width of the rings in a Fresnel
annular array), a mosaic electrical association of these elementary
cells may approach correctly this desired shape. In fact, this association is automatically done in the retained technique for achieving the array (Fig. 1). The metallized ceramic plate is photoet-
ched on one side according to the desired geometry of the array.
The electrical connections are done and this side is moulded to
obtain the backing (or the front adapting plate). Then, the ceramic
plate is cut at about ninety per cent of the thickness according
to two arrays of slots of regular pitch W. The slots are then filled
with an insulating material of low acoustical impedance, such as
polyurethan, and the recreated surface is then metallized to constitute the ground electrode. The front plate (or the backing) is
then mouldered to achieve the array.

A theoretical approach of the electromechanical behaviour of
this isolated ceramic block is relatively easy through finite element method, but taking in account the mechanical coupling of a cell
with the adjacent ones through the constitutive material of the
array and radiative coupling in the propagating medium and in the
backing is a much heavier task, which is being achieved in the mean-
time in parallel with our experimental approach. Anyway, the theoretical or experimental acquisition of the radiation of a unique
excited cell inside the array is not sufficient to achieve the
computation of a real array : the contours of the electrodes deli-
miting the different transducers are disposed in an aleatory way

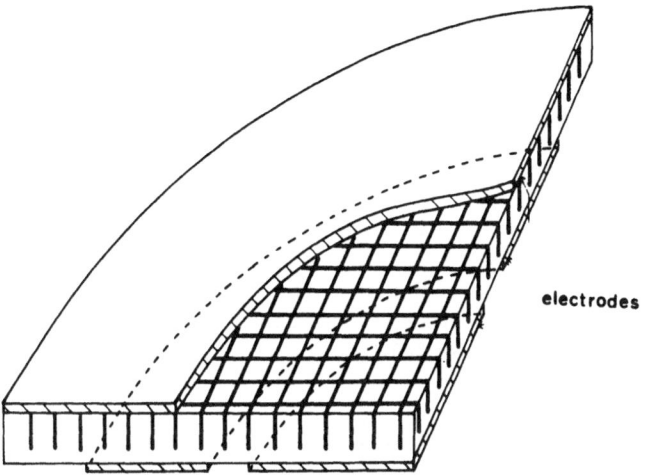

Figure 1 - Exploded view of the array in the retained technology.

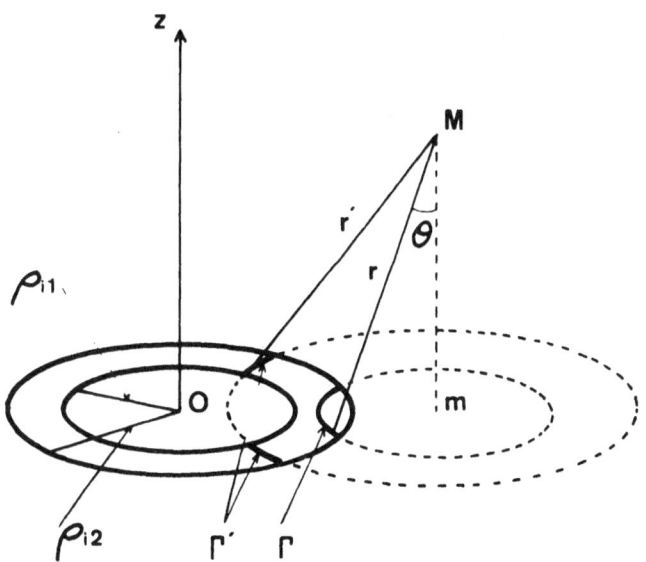

Figure 2 - Emission of the ith ring of an annular array.

relatively to the mesh of slots, so that many cells are just partially metallized. So, in a first experimental approach, rather
than studying the elementary problem of the radiation emitted by
one cell, we have preferred to compare the results obtained from
a Fresnel annular array achieved by this technique to the theoretical results obtained with a simple model when the different transducers have a piston-like oscillation.

THE FRESNEL ANNULAR ARRAY

If we retain in our theoretical approach {5-6} the same piston-
like oscillation $V_0(t)$ for each ring electronically delayed at
emission by τ_{ie} for the ith ring, the velocity potential $\phi_i(M,t)$
induced at any point M in front of the transducer is, according to
the second Rayleigh Sommerfeld formula, (Fig. 2) {7} :

$$\varphi_i(M,t) = \int \frac{V_0(t+\tau_{iE} - r/c)}{2\pi r} \, dS_i = V_0(t) * h_{iE}(M,t) \quad (1)$$

$$h_{iE}(M,t) = \int \frac{\delta(t+\tau_{iE} - r/c)}{2\pi r} \, dS_i = \frac{c \, L_i(r)}{2\pi r \, \sin\theta} , \quad (2)$$

where $r = c(t + \tau_i)$ and L_i is the length of the intersection Γ_i of
the sphere of radius r centered at M with the surface of the ring,
Fig. 2.

For the whole array (N rings) :

$$\varphi(M,t) = V_0(t) * h_E(M,t) , \quad h_E = \sum_{i=1}^{i=N} h_{iE} . \quad (3)$$

In fact, we shall measure the electrical response $U_R(M,t)$ from the
emission electrical excitation $U_E(t)$ coming from a small target
localized in M, and we may write, using time domain convolution
products :

$$U_R = U_E * E_E * h_E * T * h_R * E_R , \quad (4)$$

where E_E and E_R are the electromechanical impulse responses at
emission and at reception, T(t) the quasi-omnidirectional target
retrodiffusion response, and $h_R(M,t)$ the diffraction impulse response at reception which is identical to h_E except that the delays
τ_{IR} used for receiving may be different. It results from (4) that
the echographic response U_R is governed by :

$$U_R(M,t) = S(t) * g(M,t) \quad (5)$$

where :

$$S(t) = U_E * E_E * T * E_R \tag{6}$$

is a spatially invariant signal determined by the nature of the transducer and of the target, and :

$$g(M,t) = h_E(M,t) * h_R(M,t) \tag{7}$$

is the emission reception impulse response associated to the diffraction effects. For a ring of internal and external radiuses ρ_{i1} and ρ_{i2}, and a point A on the axis at the distance z from the transducer, the impulse response h_{iE} reduces to the rectangle function, according to, in the Fresnel approximation :

$$h_{iE}(A,t) = Rect\left[(t + \tau_{iE} - \rho_{io}^2/2cz - z/c)/\Delta\tau_i\right]$$
$$\Delta\tau_i = \frac{\rho_{i2}^2 - \rho_{i1}^2}{2cz} \quad , \quad \rho_{io}^2 = \frac{1}{2}(\rho_{i1}^2 + \rho_{i2}^2) . \tag{8}$$

If the array is a Fresnel array, $\Delta\tau_i$ is the same for the N rings and the ρ_{10}^2 are in an arithmetical progression, so that, for the electronic focus $F_E(Z_E)$, at emission, for which $\tau_{iE} = \rho_{10}^2/2CZ_E$ the diffraction transfer function h_E is just :

$$h_E(F_E,t) = N \, Rect\left[(t - z_E/c)/\Delta\tau\right] . \tag{9}$$

In our experimental case, we have chosen a Fresnel array of eight rings with an outer diameter of 16 mm, for which an electronic focusing at 50 mm leads to a value of $\Delta\tau$ = 37 ns , which remains small enough in front of the mean period of oscillation of the transducer ($\cong$ 400 ns). It is so justified to admit that, for an electronic adjustment of emission and reception focusing of the delays τ_{iE} and τ_{iR} onto the same focus F, the response $U_R(t)$ from a small target localized in F may be identified to the signal function S(t) defined in (6), according to :

$$U_R(F,t) = C S(t - 2z_F/c) . \tag{10}$$

This is this response (Fig. 3) that we have used in our model to compute $U_E(M,t)$.

Figure 3 shows the theoretical and experimental results obtained for an emission focused at 50 mm when moving radially the target at different depths for which the reception focusing is adjusted, simulating a real echographic situation with a dynamic focusing receiving technique. One may check that the confrontation is remarkable from 25 mm to 50 mm, and that a good lateral resolution is so obtained on an appreciable depth of field. These results, combined with the fact that the signal $U_R(F,t)$ shows on a logarithmic scale a first and regular decay must reinforce the idea that this technology

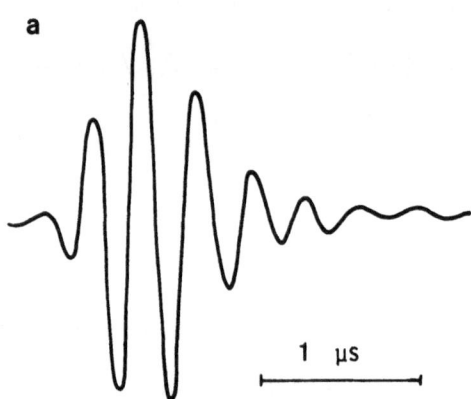

Figure 3 - a) Echographic signal from a small target
at focus electronically adjusted both
at emission and at reception.
b,c,d,e) Theoretical (—) and experimental
(···) echographic response in amplitude
obtained when moving radially the target
at constant depths (25,30,40 and 50 mm).
The electronic focusing is adjusted at
50 mm at emission and at the depth of the
target when receiving.
f) Theoretical (—) and experimental (···)
resolution at 6 and 20 dB versus depth.

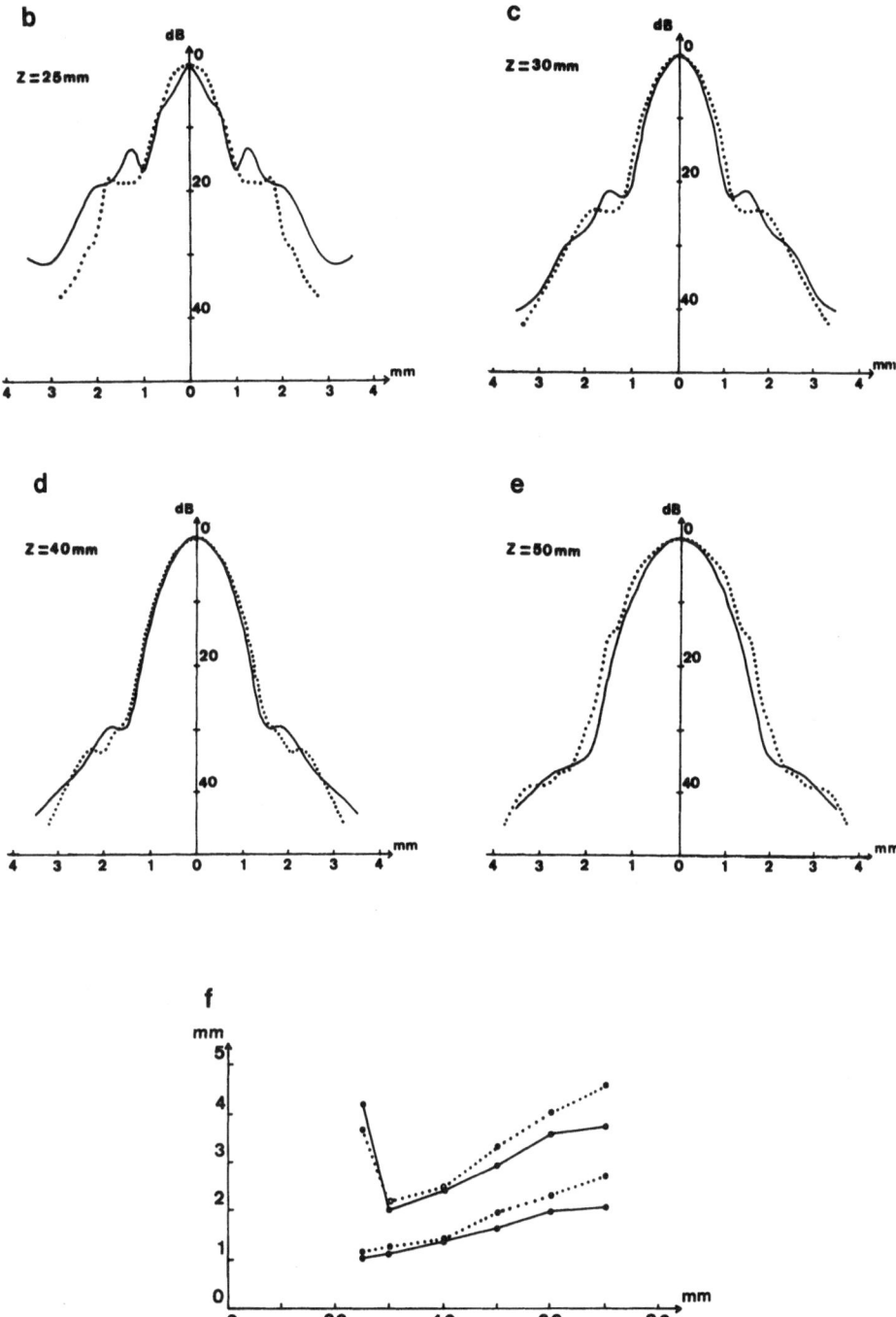

permits to build various arrays with transducers of identical piston-like electromechanical oscillation.

ELLIPTIC FRESNEL ARRAYS

An essential advantage of this technology is to permit the realization of arbitrarily shaped arrays. A first application of this
possibility is the elliptic Fresnel array which permits to focus at
two different depths in two orthogonal planes. Such arrays permit
to obtain a correct focusing behind a plane interface separating
two media of different velocities C_1 and C_2 with an oblique mean
incidence. The figure 4 shows how the virtual focuses F_x and F_y
(for the x and y directions respectively) must be located for obtaining a correct focusing in F, with the relations {8} :

$$\frac{\sin \theta_1}{C_1} = \frac{\sin \theta_2}{C_2} \quad , \quad \frac{OF_y}{OF} = \frac{\cos \theta_1}{\cos \theta_2} \quad , \quad \frac{OF_x}{OF} = \frac{\cos^3 \theta_1}{\cos^3 \theta_2} \quad . \quad (11)$$

This situation is very often met in Non Destructive Evaluation of
defects oriented obliquely or normal to the surface of the investigated material. In the case of the interface water/steel, an elementary calculation shows that, for a mean incidence of 19° in water
$C_1 = 1475$ m/s), the whole transmitted beam is constituted by trans-
versal waves ($C_2 = 3200$ m/s) with a mean incidence of 45°, the longitudinal waves being evanescent ($C_2 = 5900$ m/s). Transducers with
such a fixed double-focused beam are already used {9} and are obtained with an adequate toroidal lens moulded in front of a ceramic
plate. We propose here a different solution with the elliptic Fresnel
array which permits to adjust the focusing depths electronically.
Obviously, the designing of the array and the choice of the Fresnel
laws for the elliptic electrodes dimensions in the ξ and η directions:

$$\rho_{i\xi}^2 = \alpha_\xi n_i \quad , \quad \rho_{i\eta}^2 = \alpha_\eta n_i \quad (12)$$

where the n_i are in an arithmetical progression, imposes a cons-
tant ratio of the focusing depths $CF_x/CF_y = \alpha_\xi/\alpha_\eta$, even if an electronic adjustment of delays used at emission or at reception permits
to vary both CF_x and CF_y. In consequence, a real time electronic dynamic
focusing is possible only for situations where the distance OC is
very small and the α_ξ and α_η must be chosen in the ratio :
$\alpha_\xi/\alpha_\eta = \cos^2 \theta_1/\cos^2 \theta_2$. In other cases, a perfect compensation is met
at one depth only, but electronic adjustments of focusing permits
to achieve at least a better depth of field than fixed focus transducers.

An experimental evaluation of such an elliptic array has been
done with a test block of classical configuration in NDT (Fig. 4).
Five holes, one millimeter in diameter, with a flat bottom, have
been pierced in a face oriented at 45 degrees, i.e. just normally
to the mean beam of shear waves. We have achieved an elliptic array

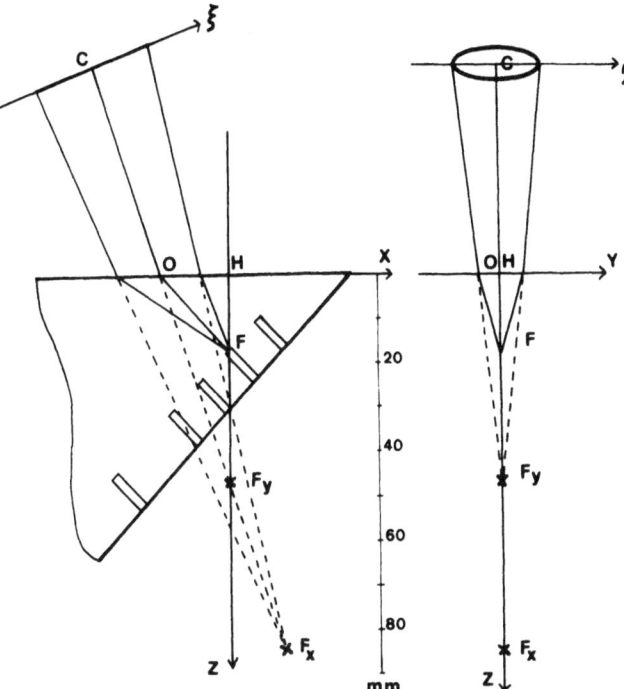

Figure 4 - The elliptic array focusing onto the second hole of
a classical N D E steel target.

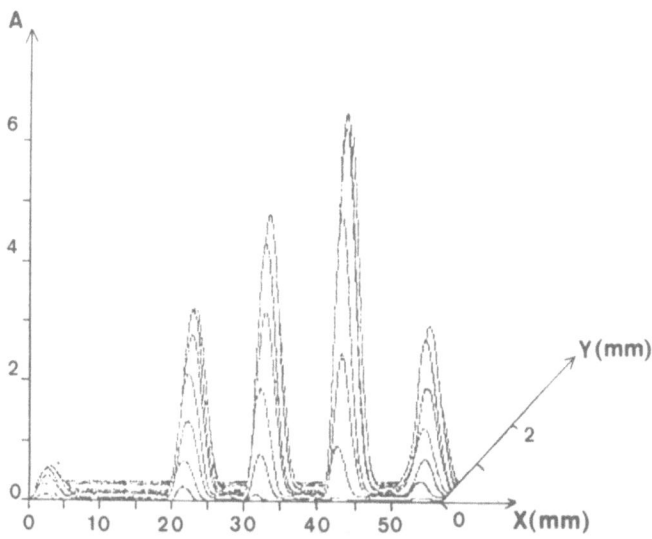

Figure 5 - View in perspective of the echographic response obtained
from the holes when scanning the array in the x direction
for various values of y.

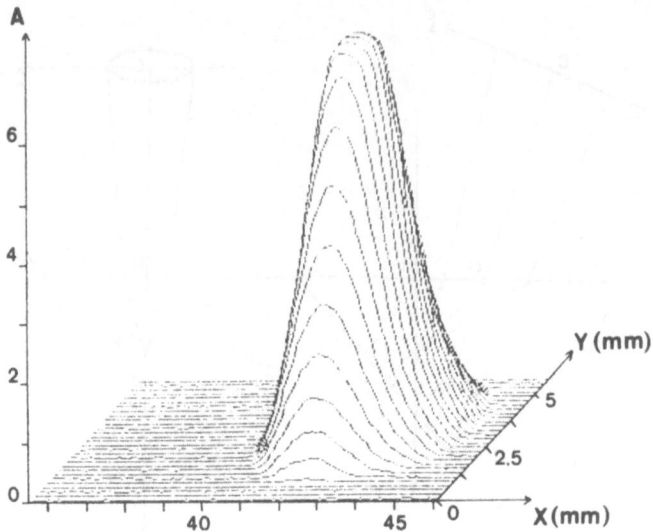

Figure 6 - Enlarged view of the results obtained in the figure 5 for the second hole.

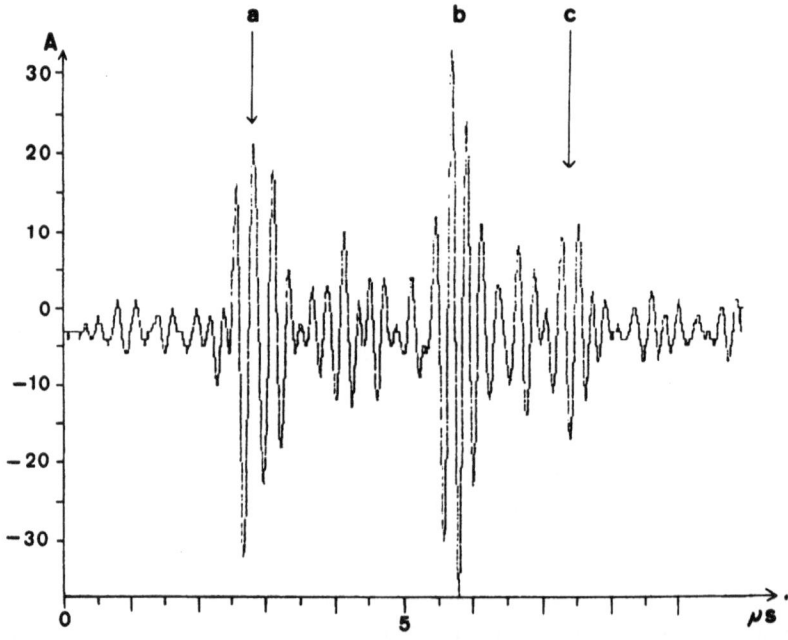

Figure 7 - Echographic signal obtained from a crack in a block of steel coated with 10 mm - austenitic steel.

by photoetching eight elliptic rings in a Fresnel progression in a ceramic plate of 30 mm in diameter with the ratio $(\alpha_\xi/\alpha_\eta)^{1/2} = \rho_\xi/\rho_\eta$ = 1.18. The centered frequency of the obtained echographic signal is nearly 2.4 MHz which corresponds to a wavelength λ = 400 μ in water and 900 μ in steel. Such an array, when disposed at 50 mm above the interface with a 19 degree incidence (Fig. 4), may correctly focus at a depth of 16 mm in the steel, which corresponds to the second of the five holes. With an automatic data acquisition system {10}, we have registered the amplitude of the signal reflected by each one of the five holes for displacements of the transducers in the x direction for different values of y, i.e. scanning the mean beam in vertical planes in the vicinity of the vertical planes of the hole axes (Fig. 5). The figure 6 shows, with an enlarged scale, the same results obtained for the second hole for which the electronic focusing has been adjusted. A lateral resolution at 6 dB, of about 3 mm, both for the x and y directions, is obtained which corresponds nearly to the dimensions of the caustics focused virtually in F_X and F_y and gives a good agreement with the theory for an harmonic beam {11}. An interesting result is that the degradation of the resolution in the x and y resolution is weaker than the predicted one for an harmonic gaussian beam {8}, which could be a consequence of the wide band transient signal used for achieving a good longitudinal resolution. This last characteristic may be evaluated from the figure 7, which shows the signal obtained from a crack located in a block of steel coated with a 10 mm thick austenitic steel, representative of a piece of a nuclear reactor vessel. The crack is located just under the coating oriented normally to the interface and may be seen with a beam of transverse waves at 45 degrees incidence. The edge diffraction effect {11-12} gives two different echos for the upper limit (b) and the lower edge (c) of the crack which is nearly 4 mm deep. The first echo (a) comes from retrodiffusion at the interface water/austenitic steel.

CONCLUSION

We have brought experimental evidence of the interest of achieving complex shaped arrays with a new technology which permits to define various transducers by photoetching one metallized face of a unique ceramic plate which is sawn on the other side by regularly spaced straight slots in two orthogonal (or oblique) directions : this permits to prevent lateral Lamb mode propagation in the array. A good agreement has been obtained with a piston-like oscillation model applied to an annular array typical of medical or N D T applications. Experimental evidence of the advantages of this technique has been given for compensating aberrations encountered in a classical N D T configuration.

We must thank R. LALIMAN and C. SASSIER for their efficient technical cooperation.

REFERENCES

{1} J. SATO, M. KAWABUCHI and A. FUKUMOTO - "Dependence of the electromechanical coupling coefficient on the width-to-thickness ratio of plank-shaped piezoelectric transducers used for electronically scanned ultrasound diagnostic systems", J.A.S.A. 66, 6 (1979).

{2} G.S. KINO and C.S. DESILETS - "Design of slotted transducer arrays with matched backings", Ultrasonic Imaging, 1, pp.189-209 (1979).

{3} B. DELANNOY, C. BRUNEEL, H. LASOTA and G. GHAZALEH - "Theoretical and experimental study of the Lamb wave eigenmodes of vibration in terms of the transducer thickness to width ratio", Jl of Appl. Phys. 52, p. 7433 (1981).

{4} M. Th. LARMANDE, P. ALAIS - "A theoretical study of the transient behaviour of ultrasonic transducers in linear arrays", Acoustical Imaging, 12, pp. 361-370 (1983).

{5} M. FINK - "La focalisation de Fresnel en échographie ultrasonore" Thèse de Doctorat d'Etat, Université P. et M. Curie, Paris (1978).

{6} M. ARDITI, W.B. TAYLOR, F.S. FOSTER and J.W. HUNT - "An annular array system for high resolution breast echography", Ultrasonic Imaging, 4, n° 1, pp. 1 - 31.

{7} J.W. GOODMANN - " Introduction to Fourier Optics".

{8} P. ALAIS, P. CERVENKA - "Fresnel approximation in the non paraxial case and in absorbing media", Acoustical Imaging, 13 (1983).

{9} R. SAGLIO, A.C. PROT, A.M. TOUFFAIT - "Determination of defects characteristics using focused probes", Material Evaluation, (Jan. 1978).

{10} F. PONS, B. GEORGEL, G. JOSSINET, A. LEBRUN - "Response to the evolution of the N D T : the laboratory for studying measurement systems", Proc. of the 6th Int. Conf. on N D E in the Nuclear Industry, Zürich, (Nov. 1983).

{11} W.E. GARDNER and J.A. HUDSON - "Ultrasonic inspection of thick section pressure vessel by the time of flight diffraction method" Proc. of the 5th Int. Conf. on N D E in the Nuclear Industry, San Diego, May 10-13, (1982).

{12} F.L. BECKER - "Near surface crack detection in nuclear pressure vessels", Proc. of the 5th Int. Conf. on N D E in the Nuclear Industry, San Diego, May 10-13, (1982).

A HYBRID TRAPPED ENERGY MODE TRANSDUCER ARRAY

D.V. Shick, H.F. Tiersten, J.F. McDonald, and P.K. Das

Rensselaer Polytechnic Institute
Troy, New York 12181

ABSTRACT

Spatial confinement of the mode shape for thickness-extension-
al (TE) vibrations of the trapped energy mode is influenced by the
form of the dispersion curve for the material in between the electroded elements. In previous treatments the authors have studied
trapping in plates of uniform thickness. In this paper the authors
examine hybrid devices employing a small amount of plate notching.
Provided these notches are not too deep their effect can enhance
the trapping resulting from electrode shorting and mass loading.
Special problems arise, however, with deep notches due to lateral
vibrations within the elements themselves.

INTRODUCTION

Trapped energy mode transducer arrays can be constructed by
simply depositing electrodes of arbitrary shape on a piezoelectric
plate of uniform thickness [1, 2, 3, 4, 5, 6, 7, 8, 9, 10, 11].
Alteration of the dispersion curves for thickness extensional (TE)
vibrations by the mass loading and electrical shorting of the electrodes creates a range of frequencies where the trapping occurs in
these structures. In all cases, however, the spatial confinement
of the mode shape is influenced primarily by the material of the
plate in between the electroded elements. The degree of this confinement can be estimated from the dispersion curve for the unelectroded region. If this curve indicates a purely imaginary wave
number at a given frequency for all dominantly coupled waves then
trapping will occur. The larger the magnitude of the imaginary
wave number the stronger will be the spatial confinement. This
magnitude, however, depends on the plate material (as well as

certain geometric considerations). The search for improved plate
materials is a complex one since a full electromechanical characterization is required for each one in order to obtain its dispersion
curves [15]. Many materials of interest have not yet been thor-
oughly characterized.

Prior discussion of energy trapping has emphasized the monolithic structure for a plate of uniform thickness. While this is
still of great interest, it is worthwhile to examine a hybrid
device which employs notches in the unelectroded region of the
plate in addition to the mass loading and electrical shorting
effects in the electroded region. One way of visualizing the improvement through notching is through the expansion in scale of
the dispersion curves for the unelectroded region which results
from the diminished thickness of the plate. This scale-up of the
dispersion curve can potentially increase the spatial attenuation
resulting from coupling to a branch of the curve corresponding to
an imaginary wave number. Nevertheless, care must be taken when
using notching to insure that the conditions for trapping are still
satisfied.

DISPERSION CURVES FOR ASYMMETRICAL STRUCTURES

Attention is focused in this paper on the case of an asymmetrically notched structure. The geometry for examining this problem is shown in Figure 1. The reason for selection of the asymmetrical notch is dictated by a desire to have the device radiate
into the surrounding medium through the flat face of the plate.
For a symmetrically notched plate or an asymmetrically notched
plate which radiates from the cut face, vibrations in the elements
can radiate laterally into the notches and thence into the surrounding fluid. This has been shown in [12] to degrade the radiation lobe or acceptance angle of an element in the array. For this
reason we consider the plate to be notched on the back (nonradi-
ating) side of the plate while the front (radiating) side is kept
flat. The condition for nonradiation from the notched surface of
the plate can be met be air backing.

It is desirable to have the effective radiating area of a
single element in the array be as small as possible so that the
radiation lobe or acceptance angle will be as large as possible.
Hence the mode shape along the flat part of the plate in front of
the electroded region is of interest. Optimization of this mode
shape is the goal of this research, but a first step in this process (as in prior treatments by the authors) is the development of
an understanding of the relevant dispersion curves.

Attention is restricted to the case of a lightly notched plate.
The difficulty associated with very deep notching is that significant lateral vibrations can arise in the electroded regions. The

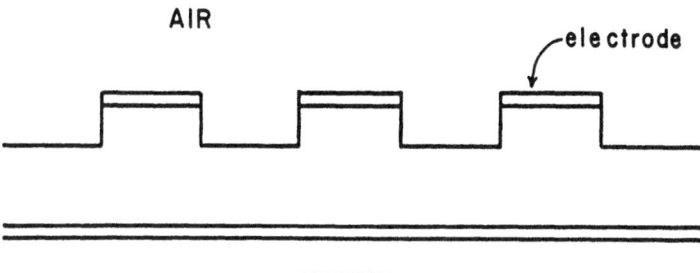

Figure 1   Geometry for the asymmetrically notched
            piezoelectric plate.

satisfaction of the boundary conditions on the free plate faces
created by the notches then demands the involvement of a large
number branches in this region. In fact, the lowest order thickness longitudinal mode possesses a branch which extends all the
way to $\omega=0$. By notching the plate only slightly we can minimize
leakage through coupling to these lowest branches.

Figure 2 shows a complete set of dispersion curves for the
piezoelectric ceramic PZT-7A poled in the thickness direction for
the unelectroded and electroded infinite plates of equal thickness
which have been derived for modes which are symmetric in the thickness dimension. These curves are adequate to describe thickness
extensional and thickness shear behavior of a symmetrical monolithic structure or symmetrically notched structure shown in Figure
3. Such structures would normally be excited in symmetric modes
only, and one can safely ignore the antisymmetric modes.

However, for the asymmetric one-sided notched structure shown
in Figure 1 both symmetric and antisymmetric modes will be excited.
These modes include a flexural motion along the plate augmenting
the thickness extension and shear described in Figure 2. This
creates some additional coupling which can further diminish the
quality of trapping especially if the notch is quite deep. Of
course, it is still the thickness extensional mode which is of
primary interest for transduction in array applications, and it is
desired that this constitute the dominant mode in the final solution.

Figure 4 sketches the first few antisymmetric modes (shown
dashed in the drawing) superimposed on the previous dispersion
curves for the fully electroded plate. The critical frequencies

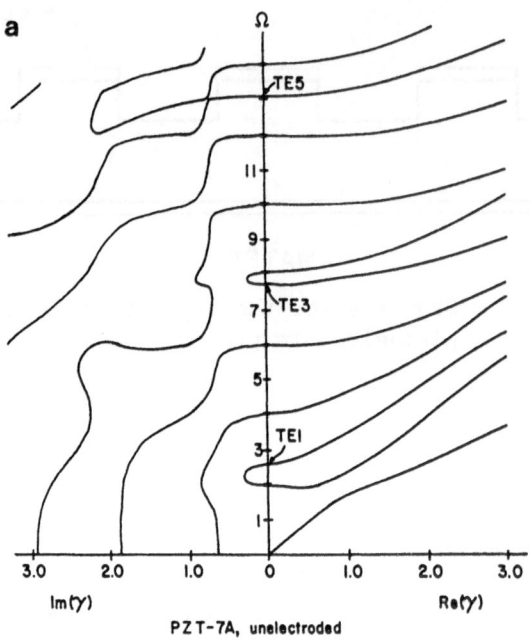

PZT-7A, unelectroded

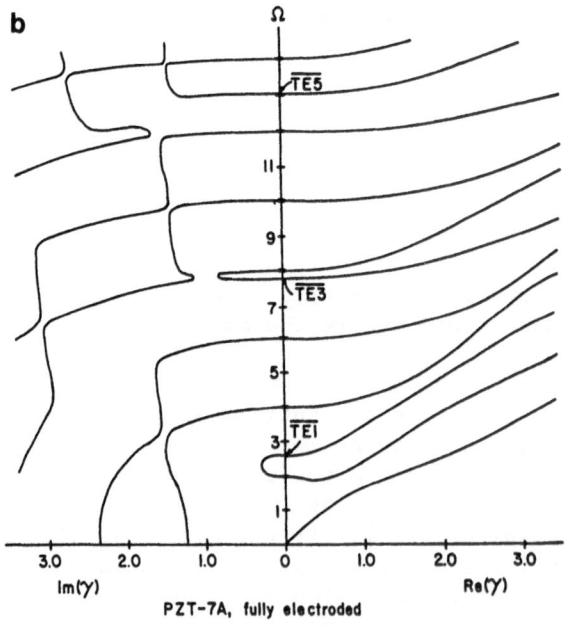

PZT-7A, fully electroded

Figure 2   Dispersion curves for <u>symmetrically</u> excited
modes for (a) unelectroded PZT-7A plate and,
(b) electroded PZT-7A plate.

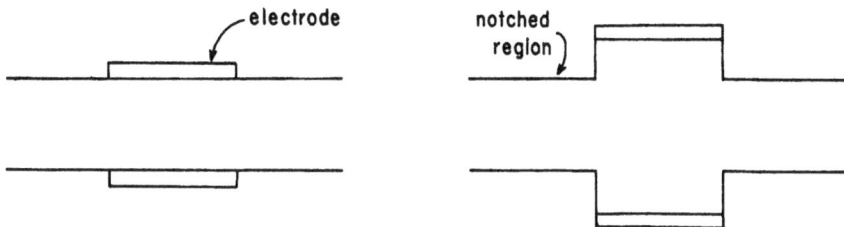

Figure 3   Monolithic and symmetrically notched plates.

for these antisymmetric branches (with $\xi=0$ or zero longitudinal wavenumber) is obtained by an odd-even index interchange [16] with the corresponding symmetric formulas where

$$\omega_n^{ATS} = \frac{n\pi}{2h''} \left(\frac{c_{55}}{\rho}\right)^{1/2} \qquad n = 1, 3, 5, \ldots$$

$$\omega_m^{ATE} = \frac{m\pi}{2h''} \left(\frac{\overline{c}_{33}}{\rho}\right)^{1/2} \qquad m = 0, 2, 4, \ldots$$

in the unelectroded notched plate and

$$\overline{\omega}_n^{ATS} = \frac{n\pi}{2h} \left(\frac{c_{55}}{\rho}\right)^{1/2} (1-R) \qquad n = 1, 3, 5, \ldots$$

$$\overline{\omega}_m^{ATE} = \overline{\eta}_m \left(\frac{\overline{c}_{33}}{\rho}\right)^{1/2} \qquad m = 0, 2, 4, \ldots$$

in the electroded region and where R and $\overline{\eta}_m$ are as defined in Ref. 11, and h" denotes the value of the plate thickness of the unelectroded region. Here ATS and ATE refer to the antisymmetric thickness shear and extension modes. These formulas for the $\xi=0$ intercepts can be used conveniently to locate the antisymmetric branches of the dispersion curve without explicitly including them in subsequent drawings. This is helpful because there are now four sets of curves for the symmetric and antisymmetric case in both the unelectroded and unelectroded regions, and it becomes difficult to visualize them all in the same drawing.

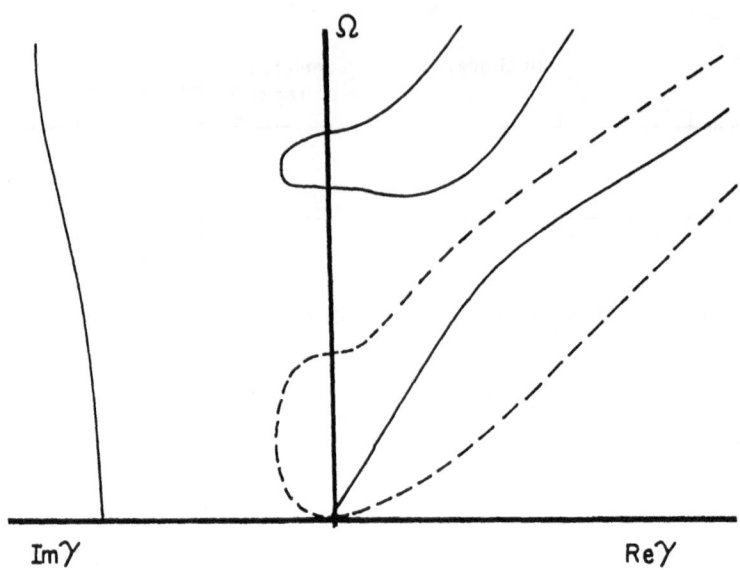

Figure 4   Sketch of first few antisymmetric modes (shown
           dashed) together with lowest symmetric thickness
           modes for the electroded PZT-7A plate.

DISPERSION CURVES FOR SLIGHT NOTCHING

     The dispersion curves in Figure 2 are shown normalized with
respect to the first thickness shear frequency. In other words
$\Omega = \omega/\omega_1^{ATS}$ and $\gamma = 2h\xi/\pi$ where Ω and γ are the normalized frequency and wave number, and $\omega_1^{ATS} = (\pi/2h) (C_{55}/\rho)^{1/2}$. When the
thickness, h, of the plate decreases, the unnormalized dispersion
curves scale up or expand.

     In Figure 5 are shown the dispersion curves relevant to a
notched plate. For the purposes of comparison to a known reference,
the dispersion curve for the electroded region has remained on a
normalized scale, while the unelectroded curve has been expanded by
the ratio h/h".

     It can be seen that expansion of the scale of the dispersion
curve for the notched region can introduce somewhat more attenuation in between the elements along the plate. There will also be
somewhat more available bandwidth. Utilization of this bandwidth
is enhanced by the lowering of the first thickness extensional
frequency of the electroded elements by the mass loading and electrical shorting effect. Further bandwidth enhancement can be
achieved with the use of a tuning inductor [3].

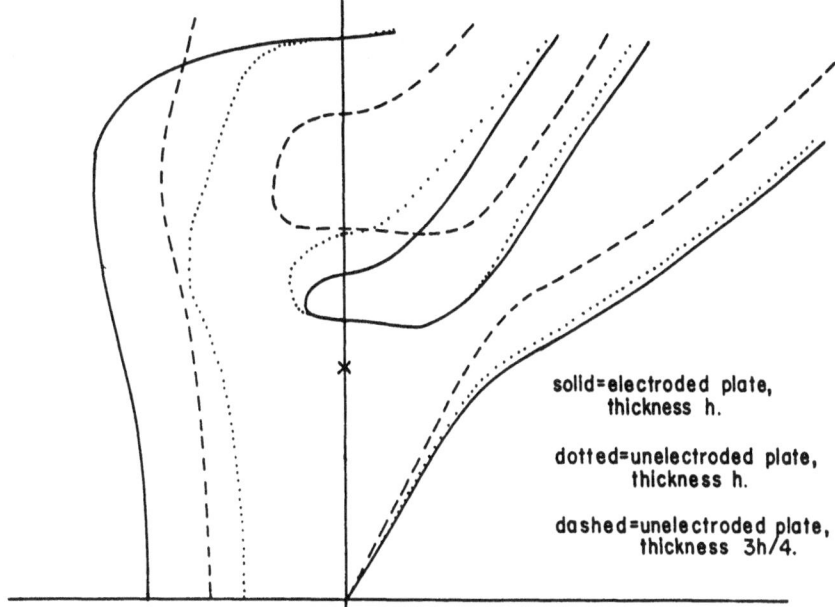

solid=electroded plate,
thickness h.

dotted=unelectroded plate,
thickness h.

dashed=unelectroded plate,
thickness 3h/4.

Figure 5   Superposition of notches (dashed) dispersion
           curves upon electroded (solid) dispersion curves
           for PZT-7A. Note location of antisymmetric
           vibration frequency (cross) for the notched
           unelectroded region.

    Employment of a slightly deeper notch can produce a set of
dispersion curves as shown in Figure 6. In this case energy
trapping can involve the complex branch of the dispersion curve
for the notched region of the plate in conjunction with the fundamental thickness extensional mode in the electroded region. A
loss of trapping can be anticipated where the complex branch joins
the real branch for the unelectroded region. Here again use of a
tuning inductor can be used to widen the bandwidth for trapping.

    Deeper notching than shown in Figure 6 can produce stronger
interactions with the antisymmetric lowest order mode of the unelectroded region of the plate. This occurs as the scale up of
the unelectroded, notched dispersion curve places the lowest antisymmetric mode in the vicinity of the fundamental thickness extensional mode of the electroded region. At this point significant
loss of trapping can occur to flexural motions along the whole
length of the plate.

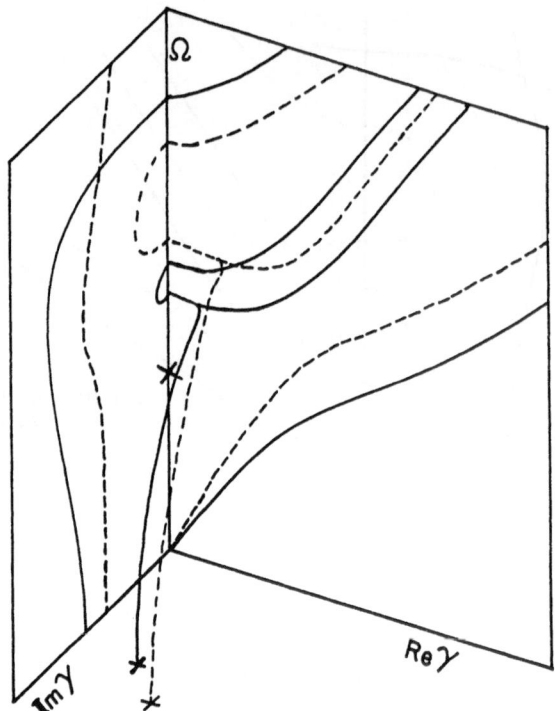

Figure 6   Curves similar to figure 5 but with deeper notch.

NOTCHED TRAPPING WITH OVERTONES

    Much deeper notching can place the branch for the unelectroded,
notched fundamental thickness extensional mode in the vicinity of a
thickness extensional overtone for the electroded region. In this
solution the discontinuity caused by the notch would excite numer-
ous lower order modes creating greater chances for coupling to propagating modes on the plate.

    Another scheme to employ overtones would be to select a mater-
ial such as Lithium Tantalate [11]. For this material the third
thickness extensional overtone lies above the third thickness shear
overtone and hence is suitable for energy trapping. Furthermore,
the magnitude of the dispersion curve for imaginary wave numbers
in the vicinity of this overtone for the unelectroded plate is
about three times larger than achievable with PZT-7A.

    By slightly notching the Lithium Tantalate plate the set of
symmetric mode dispersion curves shown in Figure 7 applies. Again,
provided the notching is slight the attenuation provided by the
notched region will scale up. However, for trapping to occur care

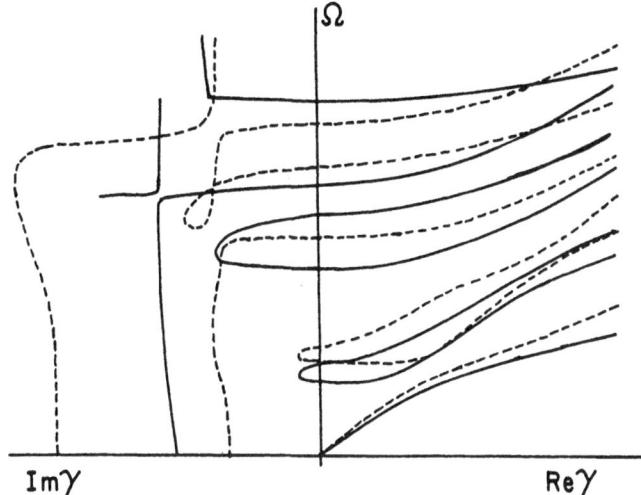

Figure 7 Third overtone of Lithium Tantalate superposing
notched (dashed) dispersion curves upon electroded
(solid) dispersion curves.

must be taken to insure that the next lowest antisymmetric mode
does not scale up to the vicinity of the overtone of interest for
the electroded elements. Therefore, use of a tuning inductor can
still be advantageous in lowering the dispersion curves for the
electroded region.

CONCLUSIONS

    Use of plate notching to enhance the basic energy trapping
due to mass loading and electrical shorting has been examined from
the point of view of the dispersion curves which apply. Due to
the preferred asymmetric notching structure employed care must be
exercised not to notch the plate so deeply that antisymmetric
modes are excited. Some limited enhancement of the interelement
isolation is therefore possible by notching (on the order of 25%).
However, to achieve maximum operating bandwidth the use of a tuning
inductor can be advantageous. Other alterations of the notched
regions are possible by changing materials constants (C_{55}, $\overline{C}_{33}$ or
ρ for example) but similar considerations dictated by the dispersion curves must still apply.

REFERENCES

[1] H.F. Tiersten, J.F. McDonald and P.K. Das, "Monolithic Mosaic
     Transducer Utilizing Trapped Energy Modes", Appl. Phys.
     Lett. 29 (12), 761-763, (1976).

[2]   H.F. Tiersten, J.F. McDonald and P.K. Das, "Two-Dimensional
       Monolithic Transducer Array", in Proc. 1977 IEEE Ultra-
       sonics Symposium, 1977, pp. 408-412.

[3]   H.F. Tiersten, J.F. McDonald, M.F. Tse and P.K. Das, "Mono-
       lithic Mosaic Transducer Utilizing Trapped Energy Modes",
       in Acoustical Holography, edited by L. Kessler, (Plenum,
       New York 1977), Vol. 7.

[4]   H.F. Tiersten, B.K. Sinha, J.F. McDonald and P.K. Das, "On
       the Influence of a Tuning Inductor on the Bandwidth of
       Extensional Trapped Energy Mode Transducers", Proc. 1978
       IEEE Ultrasonics Symposium 1978, pp. 163-166.

[5]   P.K. Das, G.A. White, B.K. Sinha, C. Lanzl, H.F. Tiersten and
       J.F. McDonald, "Ultrasonic Imaging Using Monolithic Mosaic
       Transducer Utilizing Trapped Energy Modes", in Acoustical
       Imaging, edited by A.F. Metherell (Plenum, New York, 1978)
       Vol. 8, pp. 119-135.

[6]   H.F. Tiersten and B.K. Sinha, "An Analysis of Extensional
       Modes in High Coupling Trapped Energy Resonators", Proc.
       1978 Ultrasonics Symposium, 1978, pp. 167-171.

[7]   P.K. Das, S. Talley, H.F. Tiersten, and J.F. McDonald,
       "Increased Bandwidth and Mode Shapes of the Thickness-
       Extensional Trapped Energy Mode Transducer Array", Proc.
       1979 Ultrasonics Symposium, 1979, pp. 148-152.

[8]   H.F. Tiersten and B.K. Sinha, "Mode Coupling in Thickness
       Extensional Trapped Energy Resonators", Proc. 1979 IEEE
       Ultrasonics Symposium, 1979, pp. 142-147.

[9]   D.V. Shick, H.F. Tiersten and B.K. Sinha, "Forced Thickness
       Extensional Trapped Energy Vibrations in Piezoelectric
       Plates", Proc. 1981 IEEE Ultrasonics Symposium, 1981, pp.
       452-457.

[10]   P.K. Das, S. Talley, R. Kraft, H.F. Tiersten and J.F. McDonald
       "Ultrasonics Imaging Using Trapped Energy Mode Fresnel Lens
       Transducers", in Acoustical Imaging, edited by K.Y. Wang,
       (Plenum Press, New York, 1980) Vol. 9, pp. 75-92.

[11]   D.V. Shick, H.F. Tiersten, R. Kraft, J.F. McDonald and
       P.K. Das, "Enhanced Trapped Energy Mode Array Transducer
       Using Thickness Overtones", in Acoustical Imaging, edited
       by E.A. Ash and C.R. Hill, (Plenum Press, New York, 1982)
       Vol. 12, pp. 351-360.

[12]   S.W. Smith, O.T. VonRamm, M.E. Haran, and F.L. Thurstone,

"Angular Response of Piezoelectric Elements in Phased Array Ultrasound Scanners", IEEE Trans. on Sonics and Ultrasonics, Vol. 50-26 (#3) May 1979, pp. 185-191.

[13] M.G. Maginness, J.D. Plummer and J.D. Meindl, in Acoustic Holography, edited by P. Green, (Plenum Press, New York, 1973) Vol. 5, p. 619.

[14] H. Watanabe, K. Nakamura, H. Shimizu, "A New Type of Energy Trapping Caused by Contributions from Complex Brances of the Dispersion Curves", 1980 IEEE Ultrasonics Symposium 1980, pp. 825-828.

[15] H. Jaffe and D.A. Berlincourt, "Piezoelectric Transducer Materials", Proc. IEEE 53, 1372, 1965.

[16] H.F. Tiersten, "Linear Piezoelectric Plate Vibrations, Plenum Press, 1969.

[17] M.Th. Larmande and P. Alais, "A Theoretical Study of the Transient Behavior of Ultrasonic Transducer in Linear Arrays", in Acoustical Imaging, edited by E.A. Ash and C.R. Hill, (Plenum Press, New York, 1982) Vol. 12, pp. 361-370.

ANNULAR SURFACE WAVE TRANSDUCER FOR MEDICAL IMAGING

Ron Vogel and Greg Pollari

College of Engineering
Information Division
University of Iowa
Iowa City, Iowa 52242

INTRODUCTION

Rayleigh to compressional wave conversion has been described by Toda and Murata [1] for plane waves and by Farnell et. al. [2] for radial waves. The phenomena results in compressional plane waves in a fluid, which propagate away from the solid surface containing the Rayleigh waves at an angle equal to the inverse cosine of the ratio of the velocity in fluid to the velocity on the surface. With cylindrically symmetrical Rayleigh waves, the compressional waves in the fluid form a solid hollow cone. The convergence of these waves forms a line of focus in the fluid, the extent of which is determined by the radius of existence of the Rayleigh waves on the solid interface. This result has unique applications in acoustic microscopy [3,4].

Beyond the extent of the region of focus, significant acoustic energy still exists both on and off axis [3]. It is this energy which makes the annular Rayleigh to compressional wave converter or "planar acoustic microscope lens" a candidate for medical imaging transduction.

This report will describe the pattern of the acoustic field beyond the region of focus. It will give theoretical and experimental results on the field pattern, compare transduction efficiency with thickness transducers, and compare pulse shapes with thickness transducers. In addition, the pattern of the unloaded Rayleigh wave will be shown as it was measured with a laser probe of the surface.

THEORETICAL DEVELOPMENT

The configuration of the planar annular acoustic antenna is shown in Figure 1.

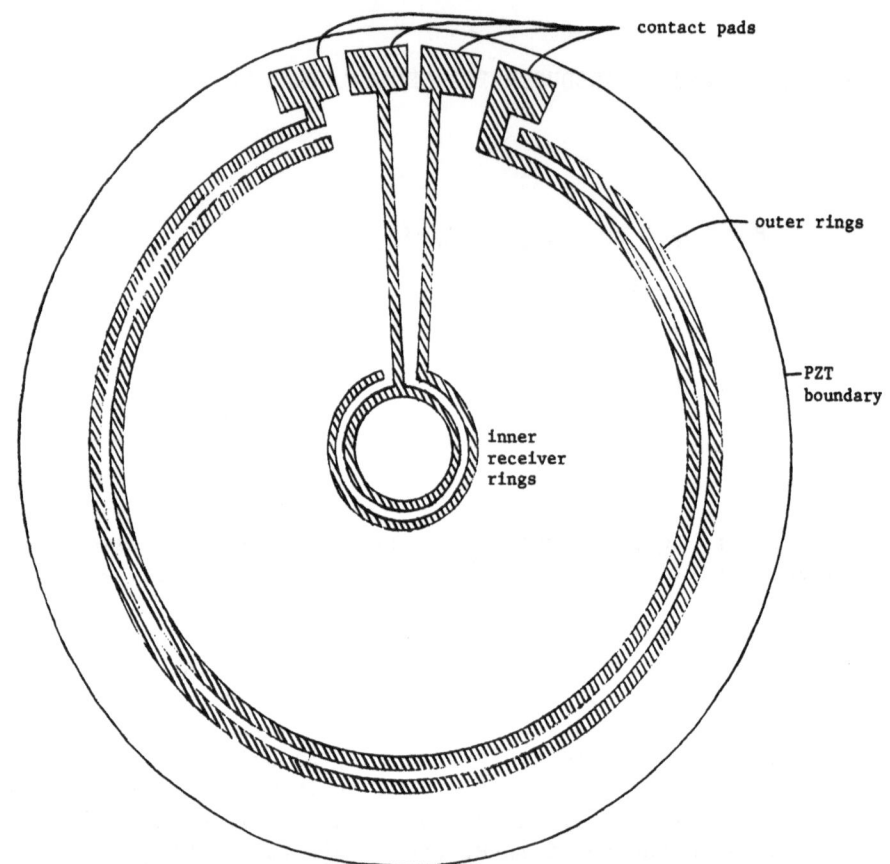

Figure 1. Geometry of Annular Surface Wave Transducer.

The analysis of annular Rayleigh waves on an infinite interface between a Z-cut hexagonal piezoelectric and a vacuum was analyzed by Day and Koerber [5]. The form of the particle displacement in the z direction on the surface, at z = 0, is

$$U_z = H(k'r')e^{j\omega_o t} \tag{1}$$

where ω_o is center angular frequency, t is time, $k' = \frac{2\pi}{\lambda'}$, λ' is the wavelength on the surface, $j = \sqrt{-1}$, and H is a Hankel

function. A Hankel function of the first kind will represent
inward traveling waves and a Hankel function of the second kind
will represent outward traveling waves.

    To match this waveform to the two ring source shown in
Figure 1, an inward and outward traveling wave will be chosen to
match the waveform under the ring electrodes. The outward
traveling wave is then absorbed by damping material at the edge of
the piezoelectric. The inward traveling wave undergoes
modification at the center which resembles a reflection or a
diffraction, the strength of which depends on the precision of
convergence and the type of material. The disturbance which then
travels outward is reduced by an amount which depends on the
previously mentioned factors and on the surface loading
condition. The strength of this outward traveling wave to be used
in the analysis will be found from measurements made under typical
loading conditions. The form of the waves near the rings is shown
in Figure 2.

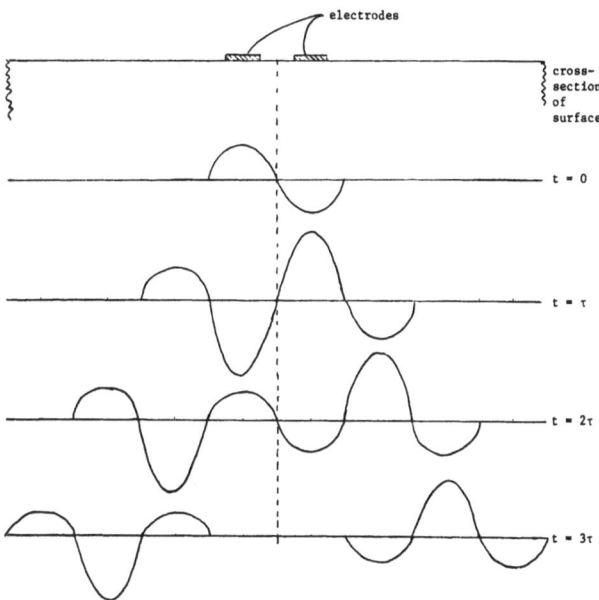

Figure 2.   Waveforms Near Rings Resulting from One Cycle
            Excitation.

    Further details of the radial dependence of the surface wave
on the transducer face while loaded with fluid were determined by
the following method. Radial dependence was first determined on

an unloaded transducer with a laser probe [6]. Figure 3 shows the
results of this probe.

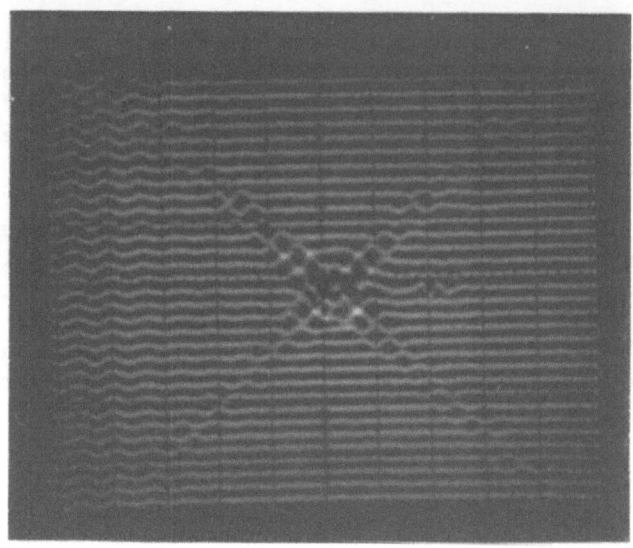

Figure 3. Laser Probe Image of Annular Surface Wave.

        In this picture, the horizontal axis is time and the vertical
axis is distance along a line passing through the convergent point
of the surface waves. The displacement of a trace from its
equilibrium position is proportional to surface particle
velocity. The pattern corresponds to an annular wave moving
toward the center, converging, and then propagating outward. The
waveform results from an electrical impulse applied to a four ring
circular transducer. The wave can be seen to increase towards the
center, and at the center the display is limited by the imaging
technique. The true amplitude is very large. To compare this
with a loaded surface, use of the inner set of rings shown in
Figure 1 was used to measure the amplitude at that radius. This
measurement was made on the loaded and unload surface. These two
measurements allow a calculation of the attenuation due to fluid
loading. Furthermore, by measuring the out-going wave at the
inner rings for the loaded and unloaded surface, the loss at the
center can be assessed. Calculations using these measurements
lead to a loaded surface wave of the following form.

1. Inward traveling waves are

$$U_z^1 = A_1 e^{-\alpha(R-r')} \, H_o^{(1)}(k'r') \, e^{j\omega_o t}$$

    starting at R, the outer edge of the rings.

2. Outward traveling waves are

$$U_z^2 = A_2 e^{-\alpha(r'-R+d)} \; H_0^{(2)}(k'r')e^{j\omega_0 t}$$

starting at R-d, the inner edge of the rings, and

$$U_z^3 = A_3 e^{-\alpha r'} H_0^{(2)}(k'r')e^{j\omega_0 t} \; u(t - \frac{R}{c})$$

starting at the center after the inward traveling wave arrives there.

The totality of these surface waves are considered the source of compressional waves in a fluid surrounding the transducer. This source problem is then analyzed by applying Huygens principle to each frequency component comprising the complete surface disturbance. At selected points in the fluid, the contributions of these frequency components is summed to get pressure versus time at those points.

The coordinate system for this analysis is shown in Figure 4.

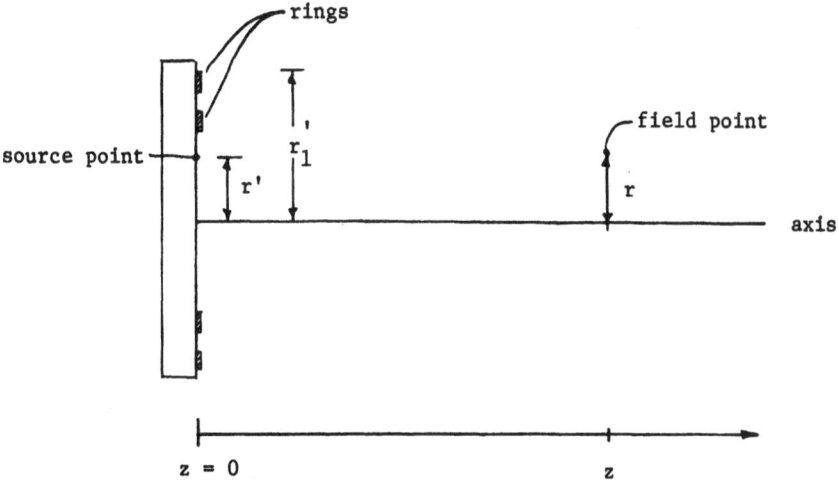

Figure 4. Cross Section of Surface and Coordinate System for Field Analysis.

Let the normal, or z component of particle velocity on the surface be $v(r',t)$ and its temporal Fourier transform be

$$V(r',\omega) = F[v(r',t)]. \tag{2}$$

The pressure, P at angular frequency ω at a point (r, z) in the field is given by the Rayleigh-Sommerfield formula [7,8]

$$P(r,z,\omega) = j \frac{\rho c k}{2\pi} \iint \frac{e^{jkR}}{R} V(r',\omega)dS \qquad (3)$$

where ρ is fluid density, c is the propagation velocity in the fluid, $k = 2\pi/\lambda$, λ = wavelength in the fluid, $R = \sqrt{(r-r')^2 + z^2}$, and dS is a surface element on the transducer. The pressure as a function of time is found from the inverse Fourier transform of equation 3.

$$p(r,z,t) = \frac{1}{2\pi} \int_{-\infty}^{\infty} e^{j\omega t} P(r,z,\omega)d\omega \qquad (4)$$

This procedure is adapted to numerical computation by discretizing the spacial variables in equation 3, the frequency variable in equation 4 and converting integrals to sums. The on-axis form of equation 3 is then

$$P(0,z,\omega) = const1 \sum_{m=1}^{200} \frac{e^{jK \sqrt{r'_q{}^2 + z^2}}}{\sqrt{r_q'{}^2 + z^2}} V(r'_q,\omega) \qquad (5)$$

where $r'_q = \frac{mr_1'}{100}$, r_1' is the outer radius of the rings, and const1 is the combination of multiplier constants. Equation 4 becomes

$$p(0,z,t) = const2 \sum_{n=-100}^{100} e^{j\omega_q t} (0,z,\omega_q) \qquad (6)$$

where $\omega_q = 2\pi n \ 10^5$.

The step sizes and limits in equations 5 and 6 are typical values used to analyzed the transducer used in the experiments. The transducer surface has approximately 15 wavelengths on it between the rings and the center and this is divided into 200 pieces. The ring spacing corresponds to a center frequency of 3 MHz and the frequency summation range covers ± 12 MHz.

While this method is not as intuitive as Farnell's [4], it is efficient for numerical computation and has uniform and adjustable precision. Furthermore, the source function can be easily modified, without effecting the rest of the computation, to account for various ring geometries and electrical driving signals.

CONSTRUCTION

Each annular device was constructed on the polished surface of a 2.54 cm diameter, 3.2 mm thick disk of PZT. Polishing was necessary to reduce attenuation of the traveling surface wave. Next, approximately 100 angstroms of chromium followed by about 5000 angstroms of copper was vacuum deposited on the surface. An additional 50 μm of copper was electroplated onto the surface so that the final band resistance of the transducer electrodes would be on the order of 0.1 ohm. The surface of the electroplated copper had to be polished to remove surface imperfections. A photomask, made from computer generated artwork, in conjunction with standard photolithographic techniques was used to create the final transducer electrode patterns.

The thickness transducer was a 12.5 mm diameter by 0.75 mm thick PZT disk. It was used in the form received: a thin film of chrome covered with a thin film of gold on each face.

Both types of device were poled in the z-direction (thickness direction).

MEASUREMENTS

Surface Wave Parameters

The laser probe tests, as described earlier and pictured in Figure 3, confirm the analytic form of the surface wave given in the theory section. This information, coupled with the measurements made on a transducer with a set of receiving rings near the center, is used to calculate the surface wave loss and center reflection loss which determine α and the ratio A_3/A_1 in the expressions for surface waves. For the transducer with inner rings, these numbers were $\alpha = 119$ (meter)$^{-1}$ and $A_3/A_1 = 0.172$.

Test Tank and Measurement Procedures

All measurements, except where noted, were made in an anechoic tank filled with water. The sides and bottom of the tank were lined with an acoustic material to minimize reflections; however, no precautions were used to minimize reflections from the surface. The transmit transducer was mounted under water at one end of the tank with driving circuity nearby above the water. Unless the transmit transducer was also being used to receive echoes, the submerged receiving transducer was mounted on an arm connected to an automated data acquisition system. This system consists of a microprocessor controlled motor-driven positioner to move the receiver to any point in the on-axis plane parallel to the water's surface. The received signal is amplified, then the largest positive voltage peak is detected. An analog-to-digital converter then converts the signal to digital format so that the peak signal information, along with corresponding position information, can be stored on computer disk.

Conventional graphical display programs were used to make graphs of the data.

Field Patterns

The waveform of a pulse received from the annular transducer as calculated and as measured by the hydrophone at 23 cm is shown in Figure 5. Both theoretical and experimental curves are from a single pair of rings as shown in Figure 1 excited by one cycle of a sine wave.

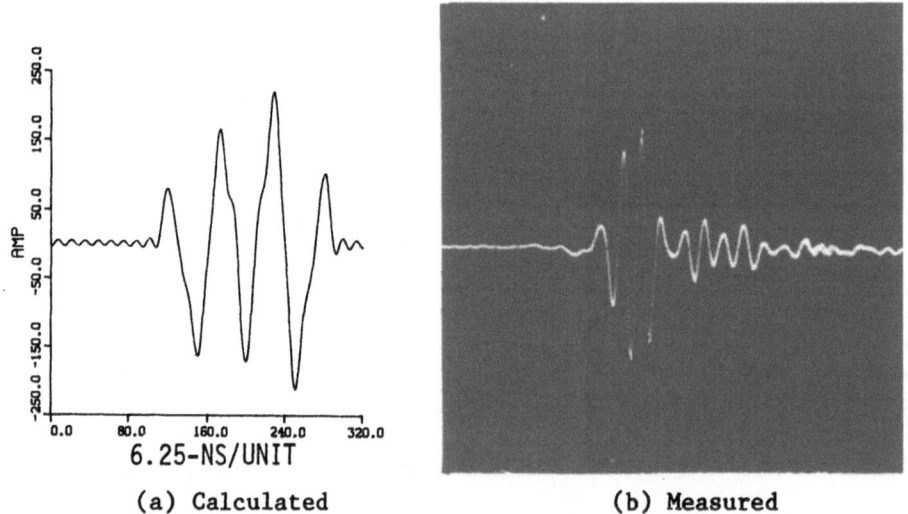

(a) Calculated   (b) Measured

Figure 5. On-axis Pulse Waveforms at 23 cm.

The peak of these waveforms is used as an indication of pulse amplitude, which is displayed in the field pattern plots. Field patterns are displayed by three dimensional plots of pulse amplitude versus z and r coordinates. Figure 6 displays the data for a back damped thickness transducer and Figure 7 for the annular transducer.

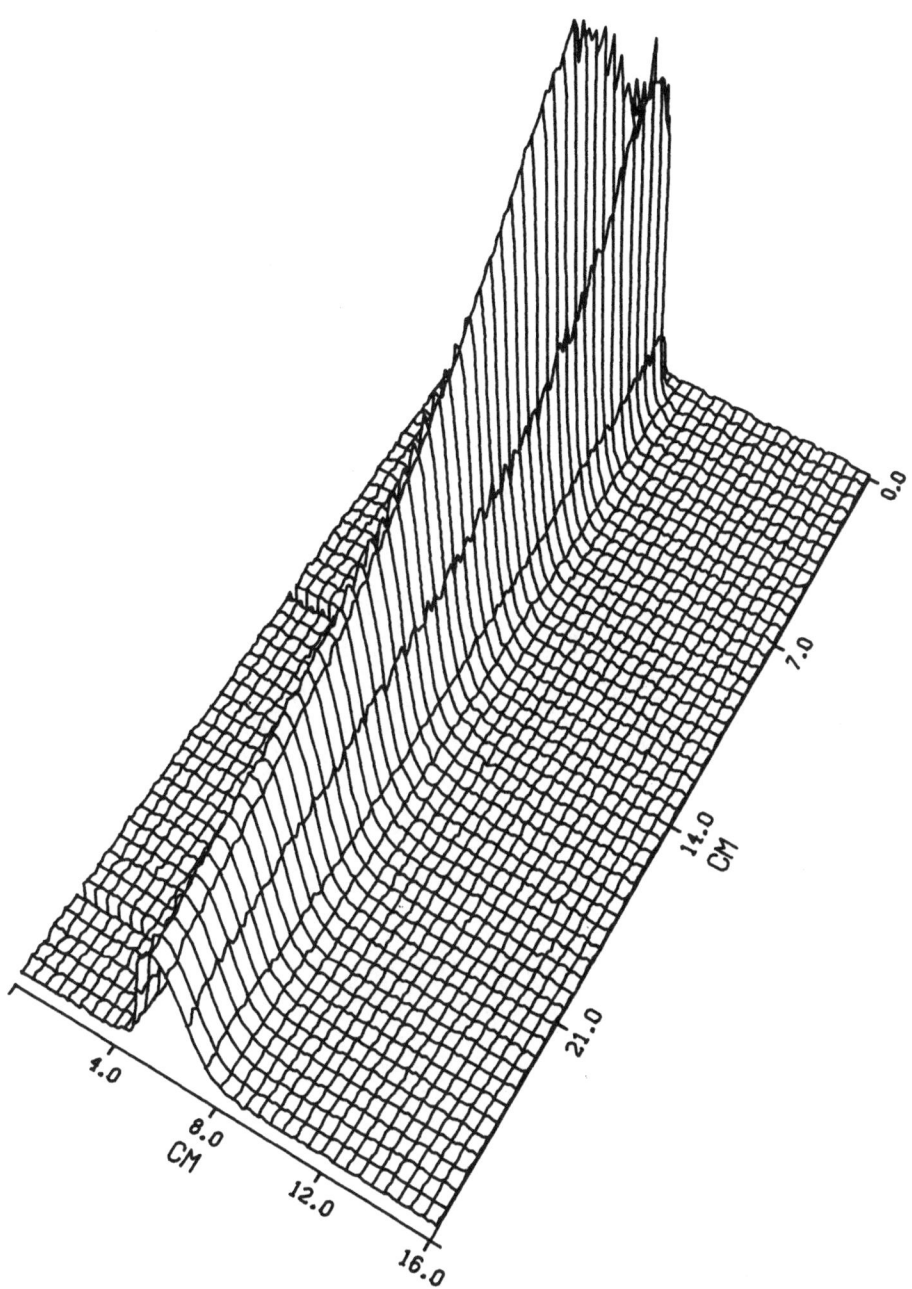

Figure 6. Field Pattern for Thickness Transducer.

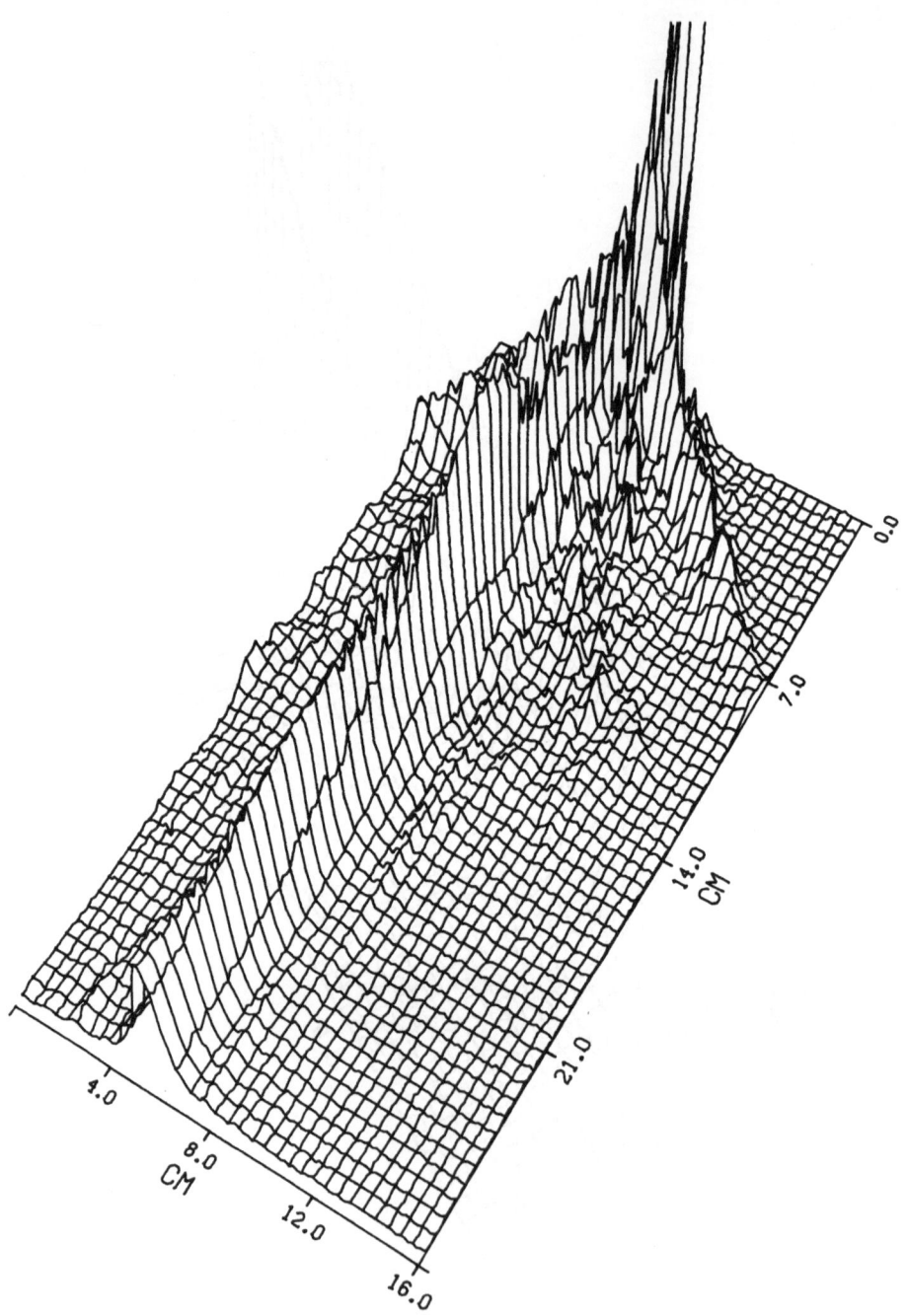

Figure 7. Field Pattern for Annular Transducer.

Figure 8 shows the calculated and measured axial pulse amplitude.

Other on-axis tests were done using thickness and annular transducers as receivers. These, along with the on-axis fields as measured by the hydrophone, are also shown in Figure 8.

Resolution Tests

Lateral resolution tests were made with two 2 mm copper wires positioned 11 cm from the annular transducer surface. The annular transducer was used as both source and receiver, and the pulse echoes were measured with an oscilloscope. When spaced at 6 mm separation, the returned pulse from an axis direction half way in between the wires was approximately one third that at the peaks. The 3dB width of the scan patterns for a single wire was 4 mm.

DISCUSSION

Comparison of the calculated and measured pulse shapes, as shown in Figure 5 show good agreement indicating that the important source contributions have been included in the analysis. The largest variation between calculation and measurement is close to the surface where the measured pulse shape changes very rapidly with transverse distance making it difficult to get accurate measurements. The measured axial pulse amplitude shown in Figure 8 shows a more nearly constant z dependence beyond the first few centimeters. This is significantly different from well known [8] thickness transducer results and different from the calculated result also shown in Figure 8. The calculated curve in Figure 8 does not include a possible contribution from the center of the disk due to nonlinear effects of the large amplitude wave at that point. This contribution should be present and the difference in the two aforementioned curves in Figure 8 indicates it is substantial.

The field plots for the annular transducer agree qualitatively with those of Farnell et. al. [3], in the region of line focus which extends from 1.5 to 2 cm along z for this experiment. Beyond the focal line the axial field amplitude decreases slowly with z out to the limits of this experiment of 29 cm. From 1 to 3 cm, peaks in the transverse pattern can be found which correspond to the inward traveling cone. Beyond that, peaks can be found corresponding to the outward traveling cone. Other lobes in the pattern result from outward traveling surface waves and diffraction effects.

Some consequences of the side lobes in the annular transducer pattern are shown in transmitter-receiver tests, in which on-axis measurements are made using the thickness or an annular transducer

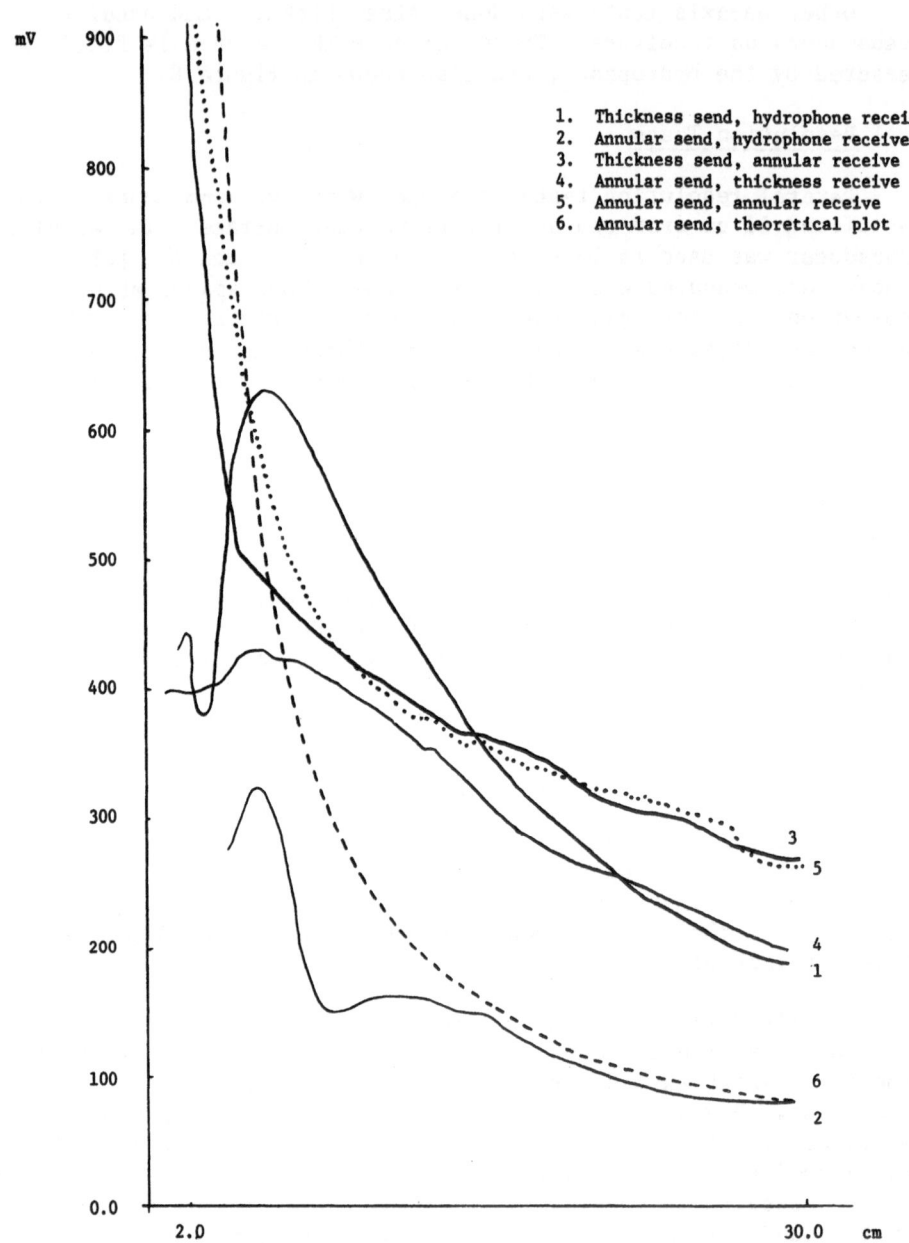

Figure 8.   On-axis Measurements.

as a field probe. If a side lobe is included in the diameter of
the receiving transducer the output is greater than contribution
of the center lobe alone. As the receiver is moved further away,
side lobes leave the area of the transducer, leading to the
irregular patterns shown in Figure 8.

Comparison of the annular with the thickness transducer is
more direct if the results for the thickness transducer is scaled
so that its radius matches that of the rings in the annular
transducer. This is done by simply scaling spatial variables for
field comparisons and scaling the input capacitance for electrical
comparisons. In what follows, these scale changes have been made.

A comparison of the near field patterns of the annular device
with the thickness transducer shows that the annular device has an
aperture equivalent to a 2 mm thickness transducer. However, the
near field pattern of a 2 mm thickness transducer at the same
frequency would extend out to only approximately 8 mm. Beyond
that the field would resemble a far field where the wave spreads
and decreases in amplitude as reciprocal distance. The annular
device, on the other hand, maintains a well defined focus of
energy out to the point where the beginning of a far field is
expected for a device of radius approximately equal to the
substrate radius.

The comparison can be made in other ways. The directivity,
D, of a thickness transducer is [8]

$$D = (Ka)^2$$

where $K = \frac{2\pi}{\lambda}$, λ = wavelength in fluid, and a = radius of
transducer. The directivity for the annular transducer can be
approximated numerically from the field curves and the formula

$$D = \frac{(\text{axial pressure})^2}{(\text{axial pressure of simple source})^2}$$

The results of these formulae are summarized in Table I. Table I
also shows the 3dB beam widths estimated from the field curves at
500 wavelengths from the transducer. The result for the thickness
transducer was scaled up to the radius of 20λ. Electrical input
impedances and power conversion ratios are the final two entries
in Table I. Power conversion ratio is evaluated from the formula

$$R = \frac{P_{ax}}{A\, P_{el}}$$

Table I

	Thickness Transducer	Annular Transducer
Radius in Wavelengths	20	20
Directivity, D at 500 Wavelengths	$5.5 \cdot 10^3$	$3.5 \cdot 10^4$
3dB Beam Width at 500 Wavelengths	4.4°	1.7°
Electrical Input Impedance	12.4 Ω	80.0 Ω
Relative Power Conversion Ratio, R at 500 Wavelengths	1.0	.15

where P_{ax} is the acoustic power incident on the hydrophone, A is the area of the hydrophone, and P_{el} is the electrical input power. The ratio R represents acoustic power per unit area in the fluid per unit electrical power applied.

Lateral resolution test results, predictable from the measured field pattern, demonstrate the possibility of high resolution instruments made with this type of transducer. Depth resolution is normally determined by pulse duration. The pulse duration of the transducers in this report is characterized by a large initial pulse of approximately 2 μsec duration and a gradually decreasing disturbance lasting approximately 10 μsec. With ordinary amplitude detection, this duration would be unacceptable; however, the pulse from this transducer has a characteristic shape so that correlation techniques [9] could be used to produce a much shorter output pulse from a returned echo.

CONCLUSIONS

The planar annular acoustic antenna, as described in this report, has promise as a medical imaging transducer. It maintains a well focused on-axis beam out to 30 centimeters, and the attenuation with distance is lower than it is for thickness transducers.

This research is sponsored by the National Institute of General Medical Sciences of the U.S. Department of Health and Human Services.

REFERENCES

1. Toda, K. and Murata, Y., "Acoustic Focusing Device with an Interdigital Transducer", J. Acoustic Soc. Am., vol. 62, pp. 1033-1036, 1977.

2. Farnell, G.W. and Jen, C.K., "Planar Acoustic Microscopy Lens Using Rayleigh to Compressional Conversion", Elec. Lett., vol. 16, pp. 541-543, 1980.

3. Farnell, G.W. and Jen, C.K., "Experiments with the Planar Acoustic Microscope Lens", Proc. IEEE Ultrasonic Symposium, pp. 547-551, 1981.

4. Farnell, G.W. and Jen, C.K., "Planar Acoustic Microscope Lens", Acoustical Imaging, vol. 12, pp. 27-36, 1982.

5. Day, C.K., and Koerber, G.G., "Annular Piezoelectric Surface Waves", IEEE Trans. on Sonics and Ultrasonics, vol. SU-19, pp. 461-466, 1972.

6. Adler, R., Korpel, A., and Desmares, P., "An Instrument for Making Surface Waves Visible", IEEE Trans. on Sonics and Ultrasonics, vol. SU-15, pp. 157-161, 1968.

7. Goodman, J.W., Introduction to Fourier Optics, McGraw Hill, New York, 1968.

8. Kinsler, L.E., Frey, A.R., Coppens, A.B., and Sanders, J.V., Fundamentals of Acoustics, Third Edition, John Wiley and Sons, New York, 1982.

9. Poo, Y.-H., El-Sherfini, A., and Chen, V.C., "ARMA Processing for Ultrasonic Reconstructive Imaging", Proc. IEEE Ultrasonic Symposium, pp. 691-695, 1982.

ACOUSTIC SPECTRAL INTERFEROMETRY: A NEW METHOD

FOR SONIC VELOCITY DETERMINATION

Michael Oravecz   and   Sidney Lees

Sonoscan, Inc.   Bioengineering Dept.
530 East Green Street   Forsyth Dental Center
Bensenville, IL 60106   140 Fenway
                                Boston, MA 02115

ABSTRACT

When sonic phase velocities are determined using the interferogram mode of a scanning laser acoustic microscope (SLAM), the field
of view is visualized directly and regional variations in velocity
can be determined with wavelength limited resolution. An early
technique described in 1975(1) has only limited, primarily biological, application. This technique is limited by an inherent 2π ambiguity in measuring the fringe shift. The new method described
here eliminates the 2π ambiguity and yields accurate and precise
velocities by combining a spectral distribution of acoustic frequencies with acoustic interferograms. This method, called spectral
interferometry, measures the relative shift of constant phase lines
in the specimen and in a reference medium (distilled water) while
the sonic frequency is changed. For demonstration, three very different materials were selected: a plastic, a ceramic and a metal.
Shear wave velocity was determined in the ceramic; otherwise, compressional wave velocity was measured. Measurements were made on
several thicknesses of the metal to show that specimen thickness is
not tightly constrained. For comparison, velocity determinations
were made using both the new spectral interferometry method and a
standard time-of-flight technique. The results of the two methods
agree within the experimental accuracy of a few percent. Additional
instrumentation will allow greater precision and automatic determination of sonic velocity. No special sample preparation is
required.

INTRODUCTION

    Acoustic phase velocity determination using acoustic interfer-
ograms obtained from a scanning laser acoustic microscope (SLAM) was
originally described by Kessler(1), and a more detailed development
of the technique was given by Goss and O'Brien(2). Applications of
the technique have been primarily in biological areas (3,4,5). The
interferogram technique permits the acoustic field to be visualized
directly and the sonic velocity to be determined at each point in
the field within the spatial resolution capability of the SLAM.

    Unfortunately there is an inherent ambiguity in the determina-
tion of sonic velocity using the interferogram technique. Velocity
determination requires the measurement of the <u>total</u> lateral displacement of a vertical field of fringes between a material of known
velocity and a material of unknown velocity. However, an interferogram only reveals the fractional part of this fringe shift. For
example, the fringe shift seen in an interferogram could represent
0.5, 1.5, 2.5, 3.5, etc. fringe spacings. Thus this technique can
rapidly reveal very small differences in sonic velocity, but the
fringe shift ambiguity makes quantification of both the overall
velocity and small differences in velocity a complicated and
laborious process (5,6).

    A newly developed method eliminates the ambiguity of the single
frequency interferogram technique and yields accurate and precise
determination of sonic velocity quickly and with little labor by
combining a spectral distribution of acoustic frequencies with
acoustic interferograms. This method is based on the movement of
the fringes as the sonic frequency is changed. The ability to visualize the acoustic field directly and to determine the sonic velocity at each resolvable point in the field are retained. The
technique is easily applied to all materials compatible with the
SLAM. It appears this process can be easily instrumented to allow
automatic determination of sonic velocity.

CONCEPTUAL DEVELOPMENT

    Figure 1 provides a geometrical construction to help visualize
the origin of the acoustic interferogram and the source of the lateral fringe shift observed at the boundary between two media having
different sonic velocities. In SLAM the image is derived from a
continuous, single-frequency, ultrasonic plane wave propagating
through the medium. The acoustic field is made visible at the surface of the medium. The lines of constant phase are shown in the
bottom of Figure 1 spaced one wavelength, λ, apart. The high frequency (10 to 500 MHz) of the sonic radiation requires the sample
be coupled through a liquid (distilled water) to the sound source.
At the water-sample interface the waves are refracted in accordance

with Snell's law:

$$\sin\theta_x/C_x = \sin\theta_o/C_o \qquad (1)$$

The refracted waves are represented by the solid lines while the waves in the liquid beside the sample are dashed lines.

When the waves strike the top of the sample, the phase fronts are equidistantly spaced by the projected wavelength, $\overline{ac}$:

$$\overline{ac} = \lambda_o/\sin\theta_o = \lambda_x/\sin\theta_x \qquad (2a);$$

which can also be expressed as

$$\overline{ac} = C_o/f \sin\theta_o = C_x/f \sin\theta_x \qquad (2b),$$

where f = frequency of the sonic radiation
C_o = sonic velocity in the liquid
C_x^o = unknown sonic velocity in the sample.

It is important to observe that the spacing between fringes is in-dependent of the propagating medium, whether the liquid or the sample.

Since lines which were continuous and in phase at the under-side of the sample reach the detection surface at different places in the sample and the liquid due to the differences in the respective propagation velocities, there is a lateral shift of one fringe pat-tern with respect to the other along the boundary between the two media. A single phase front entering the under surface arrives at point a on the upper surface of the liquid, but at point b on the upper surface of the sample. The two points define the line segment $\overline{ab}$, the total fringe displacement. Comparison of $\overline{ab}$ with $\overline{ac}$ gives the normalized total fringe shift for a single frequency, which as shown (1,2) can be expressed as

$$N = f\,T\,\sin\theta_o(\cot\theta_o - \cot\theta_x)/C_o \qquad (3)$$

Equations (1) and (3) permit the unknown sonic velocity determination

$$C_x = C_o \sin(\tan^{-1}[\cot\theta_o - (N/f)C_o/T\,\sin\theta_o]^{-1})/\sin\theta_o \qquad (4)$$

These expressions are the basis of single frequency sonic inter-ferometry.

Note that the fringe shift, N, is only partially known, since the interferogram only allows measurements of the fractional fringe shift, n:

$$n = N - I \qquad (5)$$

where I=the integral number of fringe shifts. The difficulty in

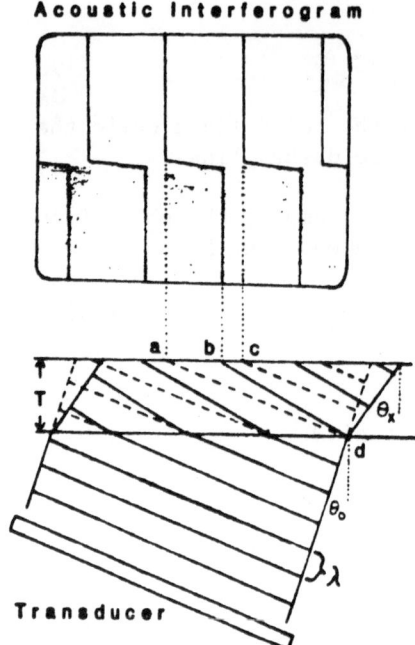

Figure 1. Geometric construction (bottom) showing phase fronts in a sample of thickness, T, and an equivalent thickness of the reference medium. The associated acoustic interferogram (top) illustrates the total lateral fringe shift calculated by Eq. (3).

Figure 2. Comparison with Fig. 1 shows how an acoustic interferogram changes when acoustic frequency is increased. The difference between the total fringe shift in Fig. 1 and the total fringe shift in Fig. 2 is calculated by Eq. (6).

determining I represents the fringe shift ambiguity which can be resolved by using many frequencies, or even a continuous range of frequencies, defined here as spectral sonic interferometry.

Figure 2 is the same as Figure 1, but the frequency has been increased. The interferogram has changed markedly. In particular the fringe displacement now exceeds the fringe spacing where before in Figure 1 it was less, i.e. $\overline{ab} < \overline{ac}$ in Figure 1 and $\overline{ab} > \overline{ac}$ in Figure 2. The change from Figure 1 to Figure 2 is associated on the acoustic monitor of SLAM with a rapid movement of the fringes while the frequency is changing. Even a small change in frequency is accompanied by a large movement of the fringes and this effect is readily detected. In addition as the frequency is changed, the fringes sweep by at different rates depending on whether they are in the liquid or the sample. Thus the fringe shift, N, is different at different frequencies. This spectral effect is calculated from Equation 3. The difference in total fringe shift between two frequencies is

$$dN=(N_2-N_1)=(f_2-f_1)T\ sin\theta_o[cot\theta_o-cot\theta_x]/C_o \qquad (6)$$

provided the sonic velocity in each medium remains constant over the frequency range. As previously, the unknown sonic velocity is given by the expression

$$C_x=C_o sin(tan^{-1}[cot\theta_o-(dN/df)C_o/T\ sin\theta_o]^{-1})/sin\theta_o \qquad (7)$$

where $df=(f_2-f_1)$.

Since

$$N_1=I_1+n_1 \quad and \quad N_2=I_2+n_2 \qquad (8)$$

the difference can also be written

$$dN=(I_2-I_1)+(n_2-n_1) \qquad (9)$$

The fractional fringe shifts, n_1 and n_2, can be found from the two interferograms at f_1 and f_2. The difference in the integer number of fringe shifts, dI, must be found by monitoring the fringe patterns as the frequency is changed from f_1 to f_2, keeping count of the total number of fringes in the sample field that go by a particular fringe in the liquid field.

In practice it is often more convenient to arrange to have $n_1=n_2=0$ and $dN=dI$. In this situation only the relative displacement of integral number of fringes between the liquid and sample fields is found for the measured change in frequency, df.

MATERIALS AND METHODS

Sample materials were selected to demonstrate the capability and flexibility of spectral interferometry. Three types of homogeneous isotropic media were tested: (1) a plastic, (2) an alumina-titanium carbide ceramic composite and (3) several thicknesses of commercial 6061-T6 aluminum machined from one piece. The ceramic was chosen to demonstrate the capability of the method for shear wave velocity determination as compressional wave velocities were measured in the other two examples. Measurements were made on several thicknesses of aluminum to show that sample thickness is not constrained by the method.

The Model 100 SONOMICROSCOPE[TM] scanning laser acoustic microscope (Figure 3) was used in all tests. Details of its operation are given elsewhere(7).

Fig. 3. Commercially available Scanning Laser Acoustic Microscope
         (SLAM).

A block diagram of the detection and display systems used for
this paper is shown in Figure 4. To make measurements, the operator
first takes advantage of the real-time ultrasonic image displayed on
the video monitor. This image helps in set-up and allows the selection of an appropriate area for measurement. Actual measurements of
dN were made by displaying the detected signal on an oscilloscope
triggered by the horizontal drive signal. This allows simultaneous
display of all 525 horizontal scans.

Figure 5 shows a typical oscilloscope display. The acoustic
image corresponding to this display consists of half water field and
half sample. The larger amplitude waveform corresponds to the water
field fringes and the smaller amplitude waveform to the fringes in
the more attenuating test sample. dN was found by watching these two
waveforms move relative to each other as the ultrasonic frequency is
changed. The change in frequency, df, where the waveforms are exactly
in phase to where they are exactly out of phase, or the reverse,
corresponds to a half-integral differential fringe shift (dN=0.5).

Ultrasonic frequencies around 100 MHz were used. The specific
frequency at which the ultrasonic transducer was driven was accurately
measured by a frequency counter. The incident angle of the sound into
the sample, θ_o, was ten degrees. Distilled water with an ultrasonic
velocity, C_o, of 1490 m/sec was used as the reference medium.

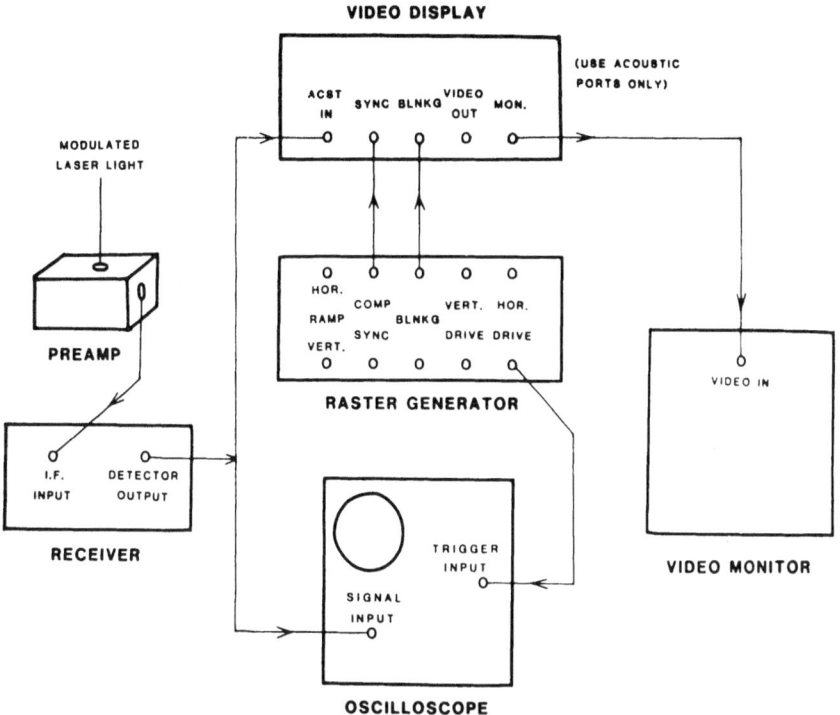

Fig. 4.   Block diagram of detection and display systems used in the measurements.

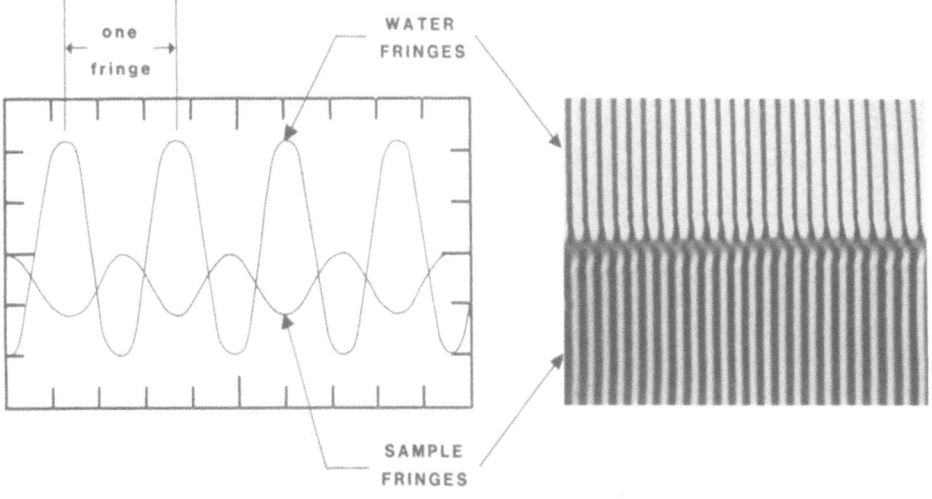

Fig. 5.   Representation of a typical oscilloscope display and a corresponding acoustic interferogram. The oscilloscope waveforms are horizontal scans across the acoustic image.

RESULTS

The data are plotted in Figures 6, 7 and 8. For Figs. 6 and 7 the abscissa is the observed sonic frequency while the ordinate is the measured differential fringe shift, dN. In Fig. 8 the abscissa is the differential frequency, df. For each set of data a least squares line was found, for which the slope is the quantity, dN/df, required in Eq. 6 to calculate the sonic phase velocity. The error reported below in the SLAM velocities only includes the uncertainty in determining the slope, dN/df. This uncertainty is expressed by the standard deviation of the slope as calculated during the least squares fit. The slope standard deviation is due to a relative precision in determining half-integral fringe shifts of 5-10%. The sonic frequency can be measured with much greater precision. In all instances the least squares line fitted to the data with a correlation of 0.999 or more and gave the smallest standard deviation of various data reduction methods.

Data for the plastic specimen, 1.702 mm thick, are shown in Fig. 6. The data can be reasonably associated with only the com-

LONGITUDINAL WAVES

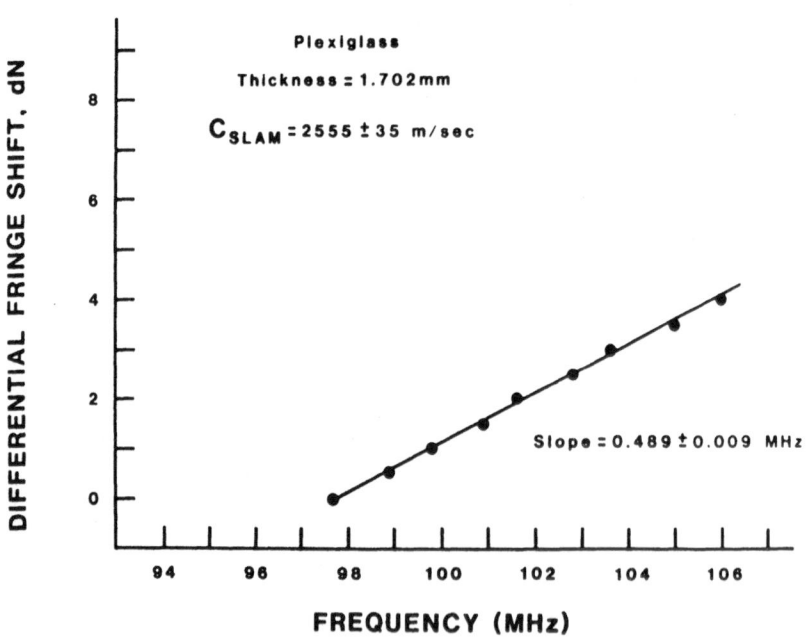

Fig. 6. SLAM data used to calculate a plastic's compressional velocity. The differential fringe shift was observed in the acoustic image of the plastic/water boundary as the frequency was changed. The pulse-echo velocity was 2500 + 75 m/sec.

pressional wave since a calculation of the relative amplitude of
compressional and shear waves, following (8), shows that the compressional wave amplitude should be at least 25 dB greater than that
for the shear waves. The compressional wave sonic phase velocity at
100 MHz was 2555 $\pm$ 35 m/sec compared with a value of 2500 $\pm$ 75 m/sec
at 15 MHz by the pulse-echo method(9). The pulse-echo error is
principally due to time-of-flight measurements with 3% relative
precision.

The second specimen, a dense alumina-titanium carbide ceramic
composite was selected to show shear velocity measurement. The
pulse-echo measured compressional sonic velocity at 15 MHz was found
to be 10.6 km/sec. The critical angle for compressional waves incident in water is 8.1 degrees. Since the angle of incidence was 10
degrees, only shear waves were transmitted. The shear wave velocity
calculated from the slope of the plot in Fig. 7 and from the measured
sample thickness was 6036 $\pm$ 110 m/sec compared with a pulse-echo
value of 6090 $\pm$185 m/sec.

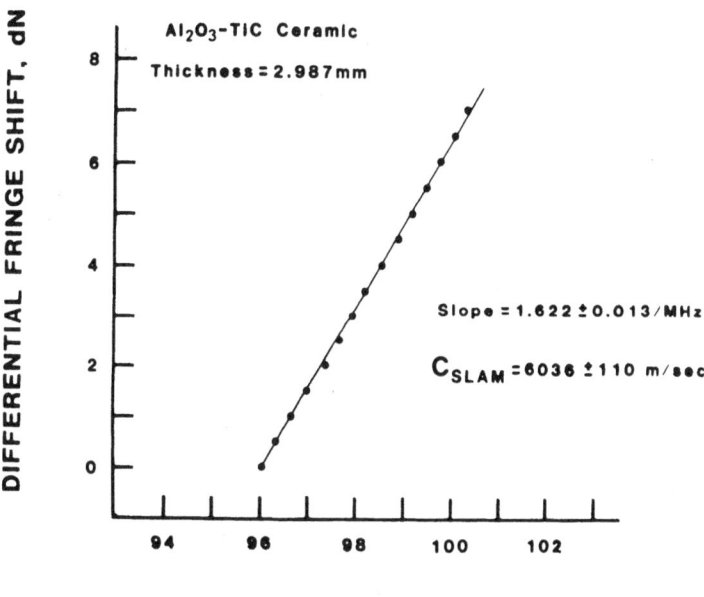

SHEAR WAVES

Fig. 7. SLAM data used to calculate a ceramic's shear velocity.
The differential fringe shift was seen in the acoustic
image of the ceramic/water boundary as the frequency was
changed. The pulse-echo velocity was 6090 $\pm$ 185 m/sec.

The compressional velocity for four different thicknesses of commercial 6061-T6 aluminum were found to show that the interferometric technique does not constrain sample thickness. The four plots shown in Fig. 8 yield the velocities shown in the table:

t(mm)	C(m/sec)
0.61	6440 +305
1.30	6410 +152
1.94	6620 +114
2.50	6290 + 61

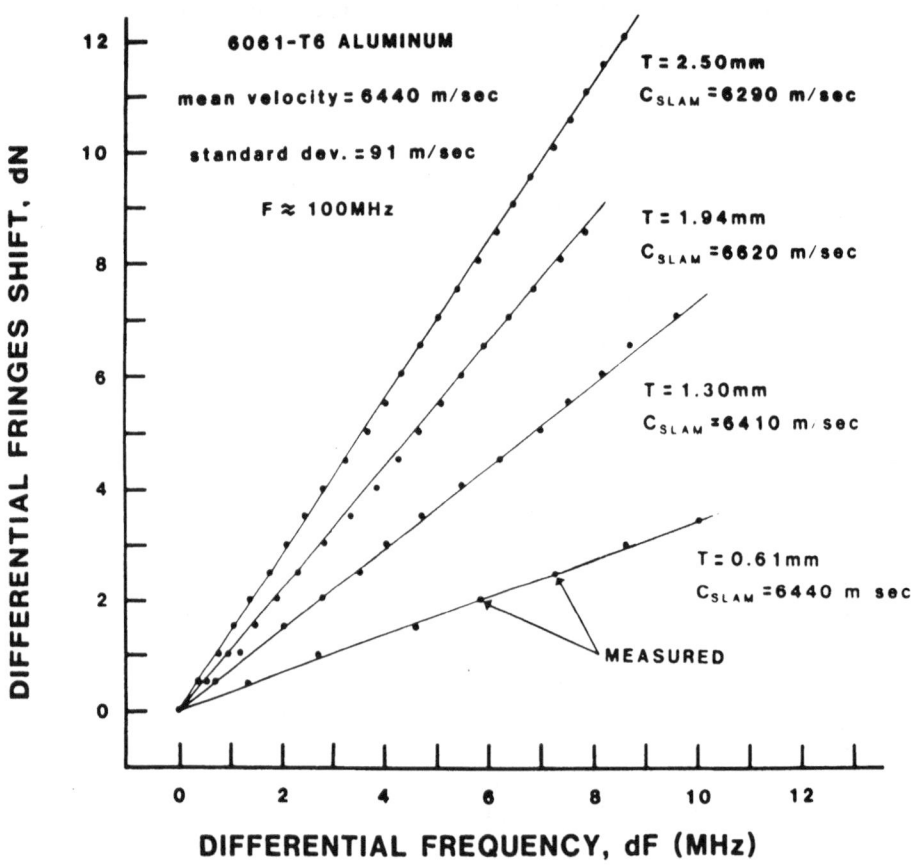

Fig. 8.   SLAM data used to calculate the velocity of several thicknesses of a metal. The differential fringe shift was seen in the acoustic images of the metal/water boundaries as the frequency was changed. The average pulse-echo measurement for the four pieces was 6460 ± 195 m/sec.

The mean value for the four thicknesses was 6440 +91 m/sec compared with the pulse-echo value of 6460 ± 195 m/sec.

The agreement between the spectral interferometry technique and the pulse-echo technique is summarized in the table:

MATERIAL	VELOCITY(m/sec)	
	SLAM	PULSE-ECHO
plastic	2555 + 35	2500 + 75
ceramic	6036 +110	6090 +185
metal	6440 + 91	6460 +195

As mentioned above, the stated error in the SLAM velocities only includes the uncertainty in determining the slope, dN/df. Improvements in measuring the differential fringe shifts, dN, could reduce this error significantly and would result in more precise velocity determinations. There are also errors associated with the values of the angle of incidence of the sound into the sample, the sample thickness and the velocity of the reference medium which must be considered. If the accuracy of the experiment is to be improved, better knowledge of these parameters is necessary. Inclusion of these errors in the present results would only increase the stated error about one or two percent, but it would explain the small variations observed in the SLAM velocities of the four aluminum samples. These variations are not adequately explained by the error resulting from the differential fringe shift uncertainty.

SUMMARY

A new method for measuring sonic phase velocity using spectral interferometry has been demonstrated. The spectral nature of this method eliminates the problem of phase ambiguity present in many interferometric techniques. The method is equally useful for compressional and shear wave modes of propagation and is independent of the instrument used to measure the sonic velocity. It was used with the scanning laser acoustic microscope because acoustic interferometry is a convenient velocity determination technique for SLAM. The limitations on accuracy have not been attained. With additional instrumentation to measure the phase more accurately, greater sonic velocity precision is possible. With SLAM it is possible to determine sonic phase velocity in a local well defined region of a specimen, i.e. within about a 100 micron diameter at 100 MHz.

ACKNOWLEDGEMENT

Sidney Lees acknowledges his partial support by National Institute on Aging Grant AG 02325.

REFERENCES

1. L. W. Kessler, "Tissue Characterization by Means of Acoustic Microscopy", Proc. of Seminar on Ultrasonic Tissue Characterization at NBS, Gaithersburg, MD, May 28-30, 1975, NBS, Spec. Publ. 453 (Issued Oct. 1976)

2. S. A. Goss and W. D. O'Brien, Jr., "Direct Ultrasonic Velocity Measurements of Mammalian Collagen Threads", J. Acoust. Soc. Am. 65(2), Feb. 1979, pp. 507-511.

3. D. E. Yuhas and L.W. Kessler, "Acoustic Microscopic Analysis of Myocardium", Ultrasonic Tissue Characterization II, M. Linzer, ed., NBS, Spec. Publ. 525 (U. S. Govt. Printing Office, Washington, D. C., 1979).

4. W. D. O'Brien, J. Olerud, K. K. Shung and J. M. Reid, "Quantitative Acoustical Assessment of Wound Maturation with Acoustic Microscopy", J. Acoust. Soc. Am. 69(2), Feb., 1981, pp. 575-579.

5. S. Lees, J. D. Heeley, J. M. Ahern and M. G. Oravecz, "Axial Phase Velocity in Rat Tail Tendon Fibers at 100 MHz by Ultrasonic Microscopy", IEEE Trans. on Sonics and Ultrasonics 30(2), March 1983, pp. 85-90.

6. M. A. McAvoy and W. D. O'Brien, Jr., "Acoustic Velocity Using the Scanning Laser Acoustic Microscope (SLAM)", Proc. of 1982 IEEE Ultrasonics Symposium, Catalog #82CH1823-4, IEEE, N. Y., pp. 634-637

7. L. W. Kessler and D. E. Yuhas, "Acoustic Microscopy - 1979", Proc. of IEEE, Vol. 67, No. 4, pp. 526-536, April 1979.

8. J. Krautkramer and H. Krautkramer, Ultrasonic Testing of Materials, 2nd Ed., Springer-Verlag, Berlin, 1977.

9. Pulse-echo time-of-flight measurements were made. A Panametrics Model 5052UA Ultrasonic Analyzer was used with a Leader LBO-518 100 MHz oscilloscope whose timebase accuracy was three percent. Panametrics V115 15 MHz compressional wave and V154 2.25 MHz shear wave contact transducers were used.

THERMOACOUSTIC IMAGING INDUCED BY DEEPLY PENETRATING RADIATION

T. Bowen, R. L. Nasoni, and A. E. Pifer

Departments of Physics and Radiology
University of Arizona, Tucson, AZ   85721

ABSTRACT

A special-purpose acoustic imaging technique is described in which heat pulses are generated within the region of interest by deeply penetrating radiation, and the resultant thermal expansions produce acoustic waves which are detected at the surface of any body sufficiently uniform for conventional ultrasonic imaging. By employing signal-averaging of successive pulses, only a fraction of a microdegree temperature rise is required in each pulse. If electromagnetic heating radiation is employed, images will sharply differentiate regions of differing electrical conductivity. Both electromagnetic and ionizing radiation show changes of thermal expansion coefficient and thermal heat capacity. In the soft-tissue regions of the body, the technique provides (a) a novel non-invasive identification of tissue characteristics which may provide a valuable supplement to the information derived from conventional imaging modalities and (b) a method to verify treatment plans in the positioning of therapeutic radiation dosage. In inanimate materials, the technique offers a new method of non-destructive evaluation (NDE) which can differentiate and locate regions of differing electrical conductivity, thermal expansion coefficient, and heat capacity not sensed by conventional NDE imaging.

INTRODUCTION

Thermoacoustic Imaging from Surface-Generated Waves

There is a considerable discussion in the scientific literature of various schemes to generate acoustic waves for imaging information by an intense beam of radiation which is rapidly absorbed near the surface of the body under examination. For example, in photoacoustic imaging,[1] a laser beam is employed to deliver a short burst of heat energy or a periodically varying heat intensity to the surface region of a material body. The mechanical strain caused by the rise in temperature then generates an acoustic wave which may be reflected, scattered, or absorbed by acoustic impedance discontinuities in other regions of the material body for the desired information. The heat pulses may also be produced by electron beam bombardment,[2] microwave rf radiation,[3] or any other physical process which generates heat rapidly varying with time at or near the surface of the material body. When the heat is generated by these means the term thermoacoustic imaging is often applied; this next paragraph distinguishes our thermoacoustic imaging method from such surface techniques.

Thermoacoustic Waves Generated by Deeply Penetrating Radiation

Pulses of heat can also be generated more or less uniformly throughout an entire material body or region of interest, rather than only near the surface. Depending upon the physical properties of the material body, such volume heating can be induced by ionizing radiation, light, rf radiation, electric currents, and ultrasound. If the body is perfectly homogeneous and the volume heating is uniform, no acoustic waves will be generated in the interior; only a gradient in the spatial heat distribution can give rise to an acoustic source. If the body is only approximately homogeneous, then gradients in the heat deposition will generate acoustic waves from the interior which may be able to reach the surface without serious degradation due to absorption or due to scattering and refraction by other inhomogeneities. Although the authors have not found discussion in the literature of imaging utilizing such volume-generated thermoacoustic wave, volume-generated waves have been studied in connection with thermoelasticity by engineers, auditory sensing of microwave radiation by biomedical researchers, and acoustic detection of charged particles by physicists. Undoubtedly volume-generated thermoacoustic radiation has been studied in many other contexts, but the above three applications will be briefly reviewed because they led one of the authors to the invention of the imaging application.[4]

Thermoelasticity. In mechanical engineering, several textbooks discuss a wide range of problems in thermoelasticity.[5,6,7] One class of solutions relates to the dynamical problem: where the thermal stress is applied suddenly, so the transit time for the propagation of elastic stresses (at the speed of sound) cannot be neglected. Heat diffusion is negligible if $c\Delta x/D \gg 1$, where Δx is the size of the smallest region to be resolved, c is the speed of sound, and D is the thermal diffusion coefficient ($D \sim 10^{-3}$ cm^2/s in soft tissue, $D \sim 1.14$ cm^2/s in copper). For $\Delta x = 10^{-2}$ cm, $c\Delta x/D$ is 1.5×10^6 in soft tissue and 4.4×10^3 in copper. Since heat diffusion can be neglected, the solution for thermoacoustic emission is fairly simple, especially if one thinks in terms of retarded potentials in analogy to those in electromagnetic theory.[8] (Reference 8 hereafter is denoted I).

Auditory Sensing of Microwave Pulses. Among biomedical researchers there has been considerable work in recent years which has established that the auditory sensing of microwave pulses can be explained by thermoacoustic waves induced by the sudden thermal stress due to absorption of energy from the microwave radiation field.[9-16] For humans, the response threshold is typically ~16 mJ/kg and for cats ~10 mJ/kg.[11] The latter figure is equivalent to 100 ergs/g-pulse = 1 rad/pulse. Since the microwave pulse lengths are typically 1-10 μsec, the acoustic frequencies extend up to ~0.1-1 MHz. As the ear responds to only a small fraction of the emitted acoustic energy spectrum, a wide-band acoustic detector optimized for maximum S/N ratio would certainly have a lower dose threshold than the auditory system.

Acoustic Detection of Charged Particles. One of the authors began calculations and participated in experimental studies on the feasibility of acoustically detecting particle cascades due to very high energy cosmic ray events.[17,18] It was clear from earlier experimental work that the acoustic signal from a cosmic ray event would be very weak, so the theoretical techniques known in thermoelasticity for calculating the thermoacoustic wave due to volume heating were combined with the signal-to-noise ratio (S/N) theory well known in sonar and radar engineering.[19] The result was a straight-forward set of formuals and calculational approximations which permit one to calculate the S/N to be expected for simple heat pulse profiles.[20] The Arizona group participated in a large collaborative effort[21] which confirmed the thermoacoustic theoretical analysis.

FEASIBILITY OF THERMOACOUSTIC IMAGING WITH VOLUME-GENERATED WAVES

In this section the following questions are discussed: Is volume-generated thermoacoustic imaging feasible at available and safe levels of heat deposition? If feasible, would it be useful?

Equations and Physical Parameters

The thermoacoustic wave is generated by differences from point to point of the source function

$$S(\underline{r}) = \beta(\underline{r})W(\underline{r})\tau/C_p(\underline{r})\rho(\underline{r}) , \tag{1}$$

where the vector $\underline{r}$ specifies the position of the point, $\beta(\underline{r})$ is the volume coefficient of expansion at $\underline{r}$, $W(\underline{r})$ is the power per unit volume deposited in the neighborhood of point $\underline{r}$, τ is the pulse length ($c\tau \lesssim \Delta x$ is assumed), $C_p(\underline{r})$ is the specific heat per unit mass at $\underline{r}$, and $\rho(\underline{r})$ is the density at $\underline{r}$. The thermoacoustic emission is generated by the gradient of the source function, $\nabla S(\underline{r})$, so no emission would occur from the interior of a uniformly irradiated homogeneous medium. Thermoacoustic imaging would give information on the function $S(\underline{r})$ relative to a constant average or background level. The radiation field which induces the heat power density W also may directly produce acoustic radiation by radiation pressure and, if it is an electric field, by electrostriction. In cases where these contributions are significant, the definition of the source function $S(\underline{r})$ can be appropriately generalized.

In order to estimate S/N, let us note that any constant background level S_{bkg} produces no signal, and let us assume that $S(r)$ differs markedly from S_{bkg} only in a small region whose profile is described by a 3-dimensional Gaussian;

$$\Delta S = S(\underline{r}) - S_{bkg} = S_0 \exp[-r^2/2\eta^2] , \tag{2}$$

where the parameter η characterizes the size of the source (the rms radius is $\sqrt{3}\eta$). Then, utilizing Eq. (26) in I for the S/N ratio for the detection of a thermoacoustic signal by a directional detector and multiplying by $N_{Hz}t$, the number of pulses signal-averaged in data collection time t, we obtain:

$$\frac{S}{N} = \frac{\pi^{1/2}A\rho c^2}{4kTz^2} \left[\frac{\beta W \tau}{\rho C_p}\right]_0^2 \eta^3 N_{Hz}t , \tag{3a}$$

$$= \frac{\pi^{1/2}A\rho c^2}{4kTz^2} S_0^2 \eta^3 N_{Hz}t , \tag{3b}$$

where A is the receiving transducer area, ρ is the average density, c is the average speed of sound, $k=1.38\times10^{-16}$ erg/$^\circ$K is the Boltzmann constant, T is the absolute temperature, z is the distance from the source to the receiving transducer, and N_{Hz} is the heat pulse repetition rate.

In order to put numbers into Eq. (3a), let us take soft tissue as an example with $\rho=1$ g/cm^3, $c=1.5\times10^5$ cm/s, T=310°K, $\beta=4\times10^{-4}$ $^\circ$C^{-1}, and $C_p=3.8\times10^7$ erg/g -$^\circ$C; these are the constants appropriate for average tissue, so $S_{bkg}=1.05\times10^{-11}(W\tau)$erg/cm^3. If we measure the peak amplitude S_0 of the signal-producing region as a fraction δ of the background level, then $S_0 = \delta S_{bkg} = \delta(\beta W\tau/\rho C_p)_{bkg}$. Lin[15] has shown that in tissue radiation, pressure and electrostriction contributions to the generation of acoustic waves may be neglected. Then, assuming the transducer is 2.5 cm dia. and is at a distance z = 10 cm, the S/N ratio becomes:

$$\frac{S}{N} = 1.3 \ N_{Hz}(W\tau)^2_{erg/cm^3} \ \delta^2 n^3_{cm} \ t_{sec} \ , \tag{4a}$$

$$= (1.3 \times 10^4) \left[N_{Hz}(W\tau/\rho)^2_{rad} \right] \delta^2 n^3_{cm} \ t_{sec} \ , \tag{4b}$$

where $W\tau/\rho$ in rads has been sustituted in Eq. (4b) (1 rad $\equiv$ 100 ergs/g). Table I lists typical values for $N_{Hz}(W\tau/\rho)^2_{rad}$, which can be regarded as a figure of merit, for several pulsed heat sources, including examples of ionizing radiation, rf electromagnetic radiation, and ultrasound. Before continuing the estimation of S/N ratios, it is worthwhile to discuss some of the considerations leading to the entries in Table I:

Ionizing radiation (electrons, x-rays). The power deposition density W and maximum pulse rate for the electron accelerators in Table I are those appropriate for Varian machines. The pulse length of the electron accelerator pulses was chosen as long as possible compatible with a resolution $n\approx 2$ mm, since the S/N ratio is proportional to τ^2. At the dose rates listed, large numbers of pulses can be applied to living tissues only for radiation cancer therapy.

RF Radiation. The pulse rate for rf radiation is limited only by the necessity to allow time for all thermoacoustic waves to propagate to the receiving transducers and to allow all reverberations to die out. At the power deposition density $W = 100 \ mW/cm^3$, pulse repetition rate $N_{Hz} = 3000$ Hz, and pulse length $\tau = 10^{-6}$ s listed, the average specific absorption rate is $0.3 \ mW/cm^3 \simeq 0.3$ W/kg. This is below the specific absorption rate of 0.4 W/kg employed as the basis for computing threshold limit values for rf or microwave exposure in the workplace.[22] If the listed power level were applied continuously, the temperature would rise 1.5°C/min, neglecting perfusion and other losses; this is comparable to the rf heating rates employed in hyperthermic cancer treatments. Of course, when the rf radiation is pulsed as indicated in Table I, the temperature rise is on the order of 0.005°C/min, which is negligible. The temperature rise per pulse is $2.6 \times 10^{-8} \ ^\circ$C.

Ultrasonic radiation. A somewhat longer time between pulses may be needed for ultrasonic heating in comparison with rf heating

Table I. A comparison of typical parameters for therapeutic
          ionizing radiation, rf and microwave electromagnetic
          radiation, and ultrasonic radiation to induce thermo-
          acoustic radiation. S/N ratio is proportional to
          $N_{Hz}(W\tau/\rho)^2$, where N_{Hz} is the number of pulses per
          second, W is the power deposition per unit volume, τ is
          the pulse width, and ρ is the density.

Radiation Source	Dose Rate or Intensity	Pulse Rate N_{Hz} [Hz]	Power W [mW/cm^3]	Pulse Length τ [μsec]	S/N Figure of Merit $N_{Hz}(W\tau/\rho)^2_{rad}$
4 MeV electrons or x-rays	178 rad/min	320	46	2	0.027
18 Mev electrons or x-rays	166 rad/min	300	46	2	0.026
30 MHz rf	1.5 W/cm^2	3,000	100	1	0.30
915 MHz microwave	0.3 W/cm^2	3,000	100	1	0.30
1 MHz ultrasonic	1 W/cm^2	1,000	100	4	1.6

because the primary wave itself requires considerable time to
reach all regions of the body and produces strong reverberations.
The ultrasonic heating level can be increased by a factor $\sim 10^3$
before observing lesions, probably due to cavitation.[23-25]
However, this advantage is offset by the disadvantages that the
ultrasonic radiation does not simultaneously heat all ponts in
the irradiated region and that the thermoacoustic emission is
accompanied by scattered, direct radiation which may be of
greater amplitude. A technique for separation is to reverse the
polarity of the received signal each pulse: the scattered
radiation from successive pulses should cancel and the
thermoacoustic radiation should coherently add.

Theoretical S/N Ratio Estimates

       Returning to the S/N estimation, let us conservatively take
$N_{Hz}(W\tau/\rho)^2_{rad} = 0.026$ from Table I for insertion into Eq. (4b) and
assume $\eta = 0.2$ cm = 2 mm:

$$\frac{S}{N} = 2.7 \ \delta^2 \ t_{sec} = 163 \ \delta^2 \ t_{minutes} \text{ [for ionizing radiation]. (5)}$$

Since a typical therapeutic dose of 200 rads would be delivered in 1.2 min. from the 18 MeV Varian (see Table I), an excellent S/N ratio could be obtained during a treatment in an electron or x-ray beam for $\delta \geq 1/2$; i.e., with a thermoacoustic source amplitude corresponding to a 50% change from the background value of $(W\tau/\rho)(\beta/C_p)$. Changes of this magnitude or greater in β/C_p are expected between tissues with high lipid content and tissues with high water content, and in $W\tau/\rho$ at the edges of the beam ($\eta \approx 2$-5 mm, depending upon field size).

If we take $N_{Hz}(W\tau/\rho)^2_{rad} = 0.3$ and $\eta_{mm} = 1$ mm, which are appropriate for the rf radiation entries in Table I,

$$\frac{S}{N} = 3.9 \ \delta^2 \ t_{sec} = 234 \ \delta^2 \ t_{minutes} \text{ [for rf radiation]. (6)}$$

Observable S/N Ratio Estimates when Corrected for Attenuation and Amplifier Noise.

The S/N ratio estimates in Eq. (5) and Eq. (6) must be revised downward to take into account attenuation of the thermoacoustic waves and noise introduced by the transducer and amplifier. Making use of Eq. (34) or Fig. 4 in I for a directional pressure detector, and assuming $\alpha c \approx 0.015$ appropriate for soft tissue, then for $z = 10$ cm and $\eta = 0.2$ cm, $\alpha c z/\eta = 0.75$, so

$$\left(\frac{S}{N}\right)_{reduction \ factor} = 0.77 \text{ for Eq. (5)}$$

and for $z = 10$ cm and $\eta = 0.1$ cm, $\alpha c z/\eta = 1.5$, so

$$\left(\frac{S}{N}\right)_{reduction \ factor} = 0.59 \text{ for Eq. (6) .}$$

Techniques are available to construct low-loss wide-bandwidth acoustic transducers[26] and low-noise amplifiers.[27] It should be possible to reduce the transducer and amplifier noise figure to ~1 dB, contributing a S/N ratio reduction by a factor 0.79 due to transducer and amplifier noise. Equations (5) and (6) for the S/N ratios in soft tissue then become:

$$\left(\frac{S}{N}\right)_{actual} = 1.6 \ \delta^2 \ t_{sec} = 98 \ \delta^2 \ t_{minutes} \quad (5')$$

[ionizing radiation, η-2 mm];

$$\left(\frac{S}{N}\right)_{actual} = 1.8 \; \delta^2 \; t_{sec} = 110 \; \delta^2 \; t_{minutes} \qquad (6')$$

$$[rf \; radiation, \; \eta = 1 \; mm];$$

S/N ratios of Volume-Generated Thermoacoustic Signals in Other Materials

In the basic S/N formula, Eq. (3b), the factors dependent upon the physical characteristics of the media are $\rho c^2 S_0^2(\eta^3)$, where the signal source peak amplitude S_0 is due to changes in $\beta W \tau / \rho C_p$ from region to region and η is a length which characterizes the size of the signal-source region. Since we can write $S_0 = \delta S_{bkg}$, where δ is the fractional change in source strength relative to the background level,

$$S/N = [E(T,A,z,N_{Hz},t)] \cdot [\rho c^2 S_{bkg}^2] \cdot [\delta^2 \eta^3], \qquad (7)$$

where $E(T,A,z,N_{Hz},t)$ depends upon the experimental arrangement, $\delta^2 \eta^3$ depends upon the size and relative amplitude of the signal-source region, and $\rho c^2 S_{bkg}^2$ depends only upon the average physical characterisitcs of the medium and the heat energy density deposited in each pulse. Separating the heat density, this factor becomes

$$\rho c^2 S_{bkg}^2 = \left[\frac{c^2 \beta^2}{\rho C_p^2}\right] \cdot [W\tau]^2 . \qquad (8)$$

The quantity $c^2 \beta^2 / \rho C_p^2$ can be regarded as a figure of merit, F, of the medium for given source size η and relative source amplitude δ:

$$F \equiv c^2 \beta^2 / \rho C_p^2 . \qquad (9)$$

Values for F are listed in Table II for various materials for comparison. It should be emphasized, however, that η and δ may be strongly dependent upon the intended application, so the figure of merit F only serves as a rough guide. In general, the thermoacoustic figure of merit does not vary much from one material to another, and soft muscle tissue whose parameters were employed for the estimates in Sections 1-3 ranks near the bottom of the scale. Hence, we conclude that the feasibility of thermoacoustic imaging in most materials will depend mainly upon the resolution η desired and the relative source contrast δ available. Of course, inanimate materials permit much greater latitude in the pulsed-heat density which can be employed safely.

In some cases, an increased energy input well above the levels suggested in Table I can eliminate the need for signal averaging of many successive pulses; i.e., a satisfactory S/N might be obtained with Nt=1 in Eq. (3a) or (3b). With the exception of the plastics, the inanimate materials listed in Table II attenuate acoustic waves much less than the amount assumed for soft tissues in Section 3.

Effect of Inhomogeneities Upon the Thermoacoustic Image

The propagation of acoustic waves in a body is affected by changes from region to region in the speed of sound c and the density ρ. These changes give rise to scattering or reflection of acoustic waves into other directions as well as attenuation and refraction of the primary wave. Since conventional ultrasonic echo imaging is affected in the same way by inhomogeneities, we can apply the extensive experience with medical and NDE echo

Table II. Thermoacoustic physical parameters for various materials. The figure of merit $F = [c(m/s)\beta(^{\circ}C^{-1})/C_p(cal/g-^{\circ}C)]^2/\rho(g/cm^3)$ in mixed units.

Material	Temp. [$^{\circ}$C]	Sound Speed c[m/s]	Vol. exp. coefficient β[$^{\circ}$C^{-1}]	Density ρ[g/cm^2]	Specific heat,C_p [cal/g-$^{\circ}$C]	Figure of merit, F= $c^2\rho^2/\beta C_p^2$
Water	37	1524	3.64×10^{-4}	0.993	0.998	0.31
Sea water	37	1560	4.0	1.019	0.93	0.44
Muscle tissue	37	1590	4.0[a]	1.06	0.75-0.91	~0.6
Fat tissue	37	1412	7.4[b]	0.93	0.55-0.62	~3.4
Olive Oil	37	1375	7.44	0.902	0.471[c]	5.2
Ice	-20	3500	1.525	0.917	0.55	1.03
Acrylic	20	2650	2.70	1.182	0.35	3.54
Polystyrene	20	1396	6.30	1.056	0.32-0.35	~6.0
Aluminum	20	6300	0.696	2.70	0.216	1.53
Silicon	20	8430	0.219	2.33	0.123	0.97
Titanium	20	6000	0.270	4.50	0.113	0.46
Iron	25	5940	0.351	7.87	0.1073	0.48
Copper	20	4760	0.498	8.93	0.092	0.74
Zinc	25	4210	0.891	6.92	0.0928	2.36
Germanium	20	4920	0.183	5.32	0.074	0.28
Lead	25	1960	0.867	11.35	0.0309	2.66

[a] β taken as β(sea water)
[b] β taken as β(olive oil)
[c] C_p at 6°C

imaging to the thermoacoustic case. In general, the image distortions would be comparable for echo and thermoacoustic imaging. Although the acoustic wave in an echo system must make a round trip to the imaged region and back, the refractions are such that the same path is followed out and back, so the distortions due to refraction are not doubled. On the other hand, the loss of signal due to attenuation is less serious for thermoacoustic imaging, since the acoustic wave only makes a one-way journey from the imaged region to the transducer.

In medical-imaging applications, refraction errors at soft tissue interfaces are small. McDicken[28] reports that typical ray deviations for a 20° angle of incidence at such interfaces range from 1° to 2°. Losses of intensity from back reflections at soft tissue interfaces are also very small. For example, McDicken[28] reports that at interfaces of fat/muscle, fat/kidney, soft-tissue/water, and muscle/blood, the respective percentages of energy back-reflected are 1.08%, 0.64%, 0.23%, and 0.07%. These deviations and losses usually result in only minor degradation of image quality as is evidenced in the wide application and acceptance of medical ultrasonic echo imaging.

It must be emphasized that the comparisons with ultrasonic echo imaging are not meant to imply that the proposed thermoacoustic imaging would be a general-purpose substitute. On the contrary, since a thermoacoustic imaging system involves considerably more apparatus than an echo system, it is envisaged as a special-purpose imaging modality capable of non-invasively supplying different information which may be of practical importance in some situations.

POTENTIAL APPLICATIONS OF THERMOACOUSTIC IMAGING INDUCED BY DEEPLY PENETRATING RADIATION

Medical Imaging in Soft Tissues

Thermoacoustic imaging offers the possibility of non-invasively obtaining information in soft tissue which is not obtainable by any other non-invasive technique. Since the heat pulses are very brief, on the order of 1 µsec, the motion of blood and perfusion have negligible effect upon the thermoacoustic signal. Figure 1 illustrates several arrangements for inducing thermoacoustic signals from soft tissues with various deeply penetrating radiations. There are several potential areas of application:

Monitoring of Therapeutic Ionizing Radiation Treatments. In connection with cancer treatment by electron and x-ray beams, thermoacoustic imaging offers the possibility of verifying the

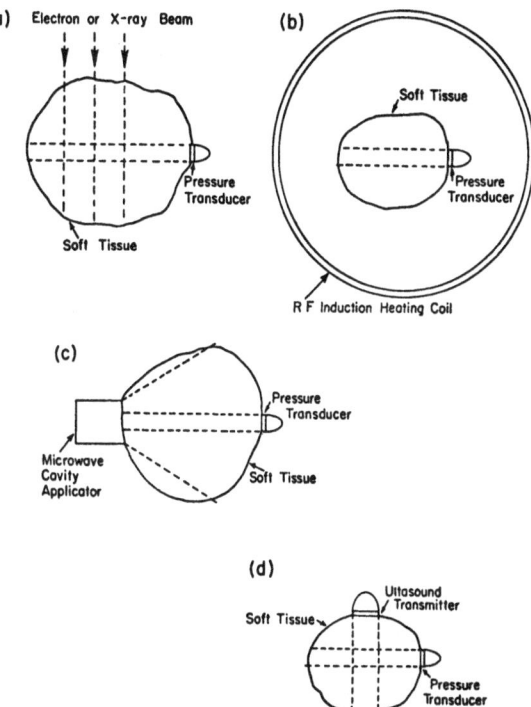

Fig. 1.   Schematic diagrams of various methods for inducing
          thermoacoustic radiation in soft tissue: (a) therapeutic
          electron or x-ray beam, (b) rf induction heating pulse,
          (c) microwave radiation pulse, (d) ultrasonic radiation
          pulse.

beam profile, the distribution of total dose, $W\tau/\rho$, and the
positioning of the dose with respect to the boundaries of organs
and   other   identifiable   tissue   structures.   When   electron
accelerators, such as the Varian machines listed in Table I, are
employed, the edges of the beam are quite sharp (2-5 mm) and are
expected to give clear signals.   The position in time of these
edge signals directly gives the edge distances from the acoustic
receiving transducer.

Between the two edge signals, one will observe signals due
to variations of the ratio of thermal expansion coefficient to
specific heat, β/C_p. This is expected to show boundaries between
organs and between differing types of tissue in relation to the
beam edges. Since these signals are proportional to the ratio of
thermal expansion coefficient to specific heat, β/C_p, the change
of this quantity from treatment to treatment might be correlated
with effects of the treatment. For example, if the water content
decreased, β/C_p would increase because water has an anomalously
low value of β/C_p.

Monitoring of Hyperthermia Treatments. Hyperthermia as a treatment to control cancer lesions is under study in many laboratories and clinics. A serious practical problem with such treatments lies in the difficulty of uniformly heating the tumor region. Thermoacoustic imaging would permit non-invasive monitoring of the heat deposition distribution. At the same time, image features due to organ boundaries and other tissue structures would appear superimposed upon the heat deposition profile. As explained in the preceding paragraph for treatments with ionizing radiation, some changes of the image with successive treatments may be correlated with effects of the treatment in a relation unique to thermoacoustic imaging.

Normally, the heating radiation is applied continuously during hyperthermia therapy. However, thermoacoustic emissions are equally well generated by turning a continuous radiation off for a brief interval (for example, 1 μsec) and repeating this interruption, say, 3000 times per second.

It is often pointed out that the most important parameter to measure is the resultant temperature, not the heat input. In response, it can be argued that it would be desireable to measure both the distribution of heat input and the resulting temperature distribution because other factors that are difficult to estimate in advance, such as perfusion, strongly affect the relationship between these factors and the outcome of the treatment. Independent acoustic methods to solve the problem of non-invasively measuring the temperature distribution have been investigated by the authors[29,30] and others.[31]

Diagnostic Imaging Induced by Radio Frequency and Microwave Radiation. Pulsed rf or microwave radiation, at safe levels of average power,[22] could generate diagnostically useful thermoacoustic images of the irradiated region. Unlike the response in any other diagnostic imaging technique, the regions of differing electrical conductivity would exhibit high contrast in this method of thermoacoustic imaging. This unique response characteristic might prove to be a valuable adjunct to density and acoustic impedance responses of conventional x-ray and ultrasonic imaging in identifying tumors and other abnormal tissue states.

Diagnostic Imaging Induced by Ultrasonic Radiation. In this case the thermoacoustic image would have a unique response to variations in the ultrasonic attenuation coefficient, since the ultrasonic heating is proportional to the rate of attenuation. This coefficient tends to be correlated with the collagen content of tissue,[32] and there are claims that attenuation is abnormal in some tumor tissues.[33] Paradoxically, although echo ultrasound signals are affected by the cumulative effects of attenuation, it is difficult to measure local variations of the attentuation with echo techniques.

Non-Destructive Evaluation of Materials

Thermoacoustic imaging with penetrating radiation or electric current pulses offers the possibility of obtaining image contrast from regions which differ only in electrical conductivity, thermal expansion coefficient, or specific heat, and which are not revealed by conventional methods. A couple of potential applications are discussed below:

NDE Imaging Induced by Electric Current Pulses in Metals. If a metal has inclusions which closely match the surrounding host in density and acoustic impedance, they will not be revealed by x-ray and ultrasonic imaging. If the inclusions differ in electrical conductivity due to differences of composition, these would be imaged in the proposed thermoacoustic system when electric current pulses are applied to the specimen. Eddy-current testing methods might also reveal such inclusions, but the thermoacoustic technique would permit resolutions on the order of 1 mm or better, even for deep-lying anomalies.

NDE Imaging of Composite Materials. If a composite material with laminations or filaments is imaged thermoacoustically with penetrating ionizing or rf radiation, the reinforcing material can act as acoustic sources for the signals received at the surface. The received signal should be very sensitive to a lack of bonding between the reinforcing laminant or filament, as well as to inhomogeneities and displacements. The method is illustrated in Fig. 2.

EXPERIMENTAL PROGRESS IN THERMOACOUSTIC IMAGING

Experiments with Soft Tissue Phantoms

The Arizona group has carried out several experiments with "phantoms," which model interfaces between different tissues to verify that thermoacoustically generated signals are observable with signal-to-noise ratios in approximate agreement with the calculations outlined in the discussion of feasibility. In one experiment,[34] electric current pulses generated thermoacoustic signals in a phantom representing alternating layers of high-lipid content and high-water-content soft tissues. In another study[35] (see following paper in this volume), an 18 MeV electron beam was used to generate a thermoacoustic signal from water-acrylic

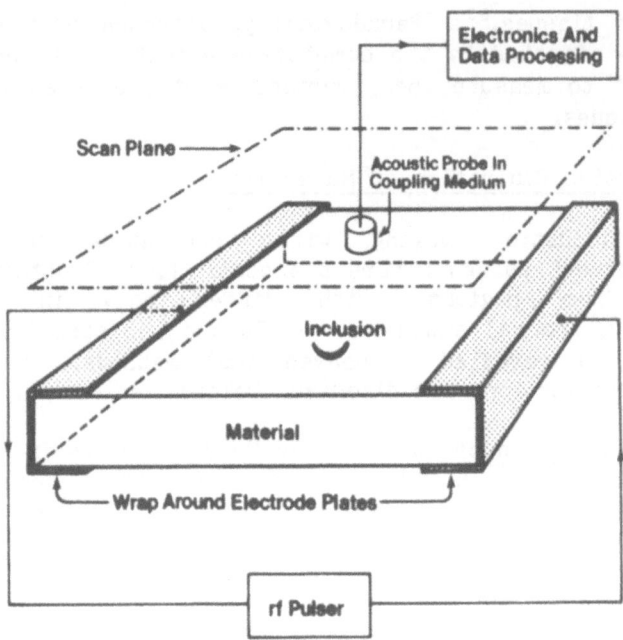

Fig. 2.   Schematic diagram of a method for inducing thermoacoustic
          radiation in graphite-filament composite materials for
          non-destructive evaluation.

interfaces.   Signal averaging was utilized in both experiments,
and the achieved S/N ratios were in reasonable agreement with
expectations.   Independently, Caspers and Conway[36] have suggested
and studied microwave-induced thermoacoustic emissions.

Experiments for NDE Applications

    In a study of potential NDE applications, the Arizona group
observed thermoacoustic emissions from a copper/stainless steel
interface induced by electric current pulses.[37]   Chamuel[38] has
reported observing thermoelastic waves induced by current pulses
in a graphite fiber of a graphite-epoxy composite.

CONCLUSIONS

    Although the concept of thermoacoustically generating sound
waves is as old as the understanding of thunder from lightning
and the thermophone,[39] the consideration of imaging applications
awaited a theoretical undersatnding of signal-to-noise ratios and
recent technological developments. Calculations show that a good
signal-to-noise ratio can be obtained in a reasonable time, but in
most cases a large number ($\sim 10^{4}$) of pulses must be signal-
averaged.   If one pulse were examined at permissible radiation

intensities, the thermoacoustic signal would be lost in the noise. Hence, the useful properties of thermoacoustic signals were not likely to be discovered "accidentally" by a tinkering experimenter, and became apparent only after the theory was developed to understand acoustic detection of high energy cosmic ray events.

Technological advances essential to thermoacoustic imaging feasibility include transducers with multiple matching layers,[26] very high gain transistors for use in the input stage of a low-noise amplifier,[27] and digital signal averagers. Averagers with the required high speed (~4 MHz) and moderate cost are now possible due to the advances in digital electronics and minicomputers of recent years. A block diagram of a thermoacoustic imaging system utilizing deeply penetrating radiation is shown in Fig. 3. At the present state of technology, the instrumentation to achieve thermoacoustic imaging would be comparable in cost to other imaging modalities. The method merits further investigations of its potential benefits for medical and NDE applications.

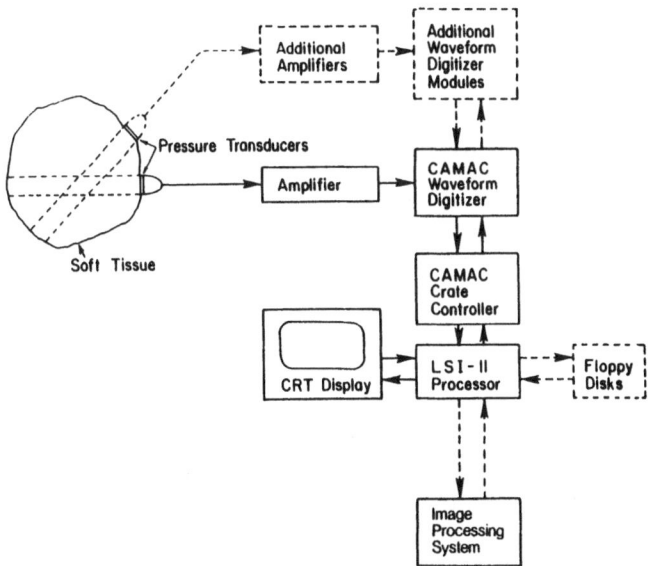

Fig. 3. Block diagram of instrumentation for a thermoacoustic imaging system.

REFERENCES

1. R. J. von Gutfeld, "Thermoacoustic generation of elastic
    waves for non-destructive testing and medical diagnostics,"
    Ultrasonics 18:175 (1980); D. A. Hutchins, R. J. Dewhurst,
    S. B. Palmer, and C. B. Scruby, "Some Applications of Laser
    Generated Ultrasound in Metals," Proceedings of 1981 IEEE
    Ultrasonics Symposium, Chicago, Oct. 14-16, 1981, p. 798.
2. A. Rosencwaig, "Thermal-Wave Microscopy of Semiconductor
    Devices," Proceedings of 1981 IEEE Ultrasonics Symposium,
    Chicago, Oct. 14-16, 1981, p. 828.
3. R. G. Olsen, Generation of Acoustical Images from the
    Absorption of Pulsed Microwave Energy, in "Acoustical
    Imaging," vol. 11, Plenum, New York (1982).
4. T. Bowen, Radiation-Induced Thermoacoustic Imaging, U.S.
    Patent No. 4,385,634, May 31, 1983.
5. W. Nowacki, "Thermoelasticity," Addison-Wesley, Reading,
    MA/Pergamon, Oxford (1962).
6. W. Nowacki, "Dynamic Problems of Thermoelasticity," Noordhoff,
    Leyden (1975).
7. H. Parkus, "Thermoelasticity," Springer-Verlag, Vienna (1976).
8. T. Bowen, "Radiation-Induced Thermoacoustic Soft Tissue
    Imaging," Proceedings of 1981 IEEE Ultrasonics Symposium,
    Chicago, October 14-16, 1981, p. 817. [Referenced as I
    in text.]
9. A. W. Guy, "Analysis of Electromagnetic Fields Induced in
    Biological Tissues by Thermographic Studies on Equivalent
    Phantom Models," IEEE Trans. Microwave Theory Tech.
    MTT-19 (2), 205 (1971).
10. K. R. Foster and E. D. Finch, Microwave Hearing: Evidence for
    Thermoacoustic Auditory Stimulation by Pulsed Microwaves,
    Science, 185:256 (1974).
11. A. W. Guy, C. K. Chou, J. C. Lin, and D. Christensen, Microwave-
    Induced Acoustic Effects in Mammalian Auditory Systems and
    Physical Materials," Annals N.Y. Acad. Sci., 247:194 (1975).
12. S. Baranski and P. Czerski, "Biological Effects of
    Microwaves," Dowden, Hutchinson, and Ross, Stroudsburg
    (1976).
13. J. C. Lin, Theoretical Analysis of Microwave-Generated
    Auditory Effects in Animals and Man, in: "Biological
    Effects of Electromagnetic Waves," Selected Papers of
    the USNDC/URSI Annual Meeting, Boulder, Colorado,
    Oct. 20-23, 1975, C. C. Johnson and M. L. Shore, eds.,
    HEW Publication (FDA) 77-8010, vol. 1, U.S. Government
    Printng Office, Washington (1976), p. 36.

14. C. K. Chou, A. W. Guy, and R. Galambos, Characteristics of
    Microwave-Induced Cochlear Microphonics," Radio. Sci.
    12 (6S):221 (1977).

15. J. C. Lin, "Microwave Auditory Effects and Applications,"
    Charles C. Thomas, Springfield (1978).

16. R. G. Olsen and W. G. Hammer, Microwave Induced Pressure
    Waves in a Model of Muscle Tissue, "Biomagnetics," 1:45
    (1980).

17. T. Bowen, "Sonic Particle Detection," Proceedings of the 1976
    DUMAND Summer Workshop, University of Hawaii, Sept. 6-19,
    1976, p. 523.

18. T. Bowen, "Sonic Particle Detection," Proceedings of the 15th
    International Cosmic Ray Conference, Plovdiv, Bulgaria,
    vol. 6, p. 277 (1977).

19. T. Bowen, "Theoretical Prediction of the Acoustic Emission
    from Particle Cascades, and the Signal-to-Noise Ratio,"
    Proceedings of the La Jolla Workshop on Acoustic Detection
    of Neutrinos, Scripps Institution of Oceanography, July
    15-29, 1977, p. 37.

20. T. Bowen, Acoustic Detection: A New Technique in 10 TeV
    Experiments, in: "Cosmic Rays and Particle Physics - 1978,"
    T. K. Gaisser, ed., American Institute of Physics, New York
    (1979), p. 72; T. Bowen, "Signal-to-Noise Ratio for the
    Acoustic Detection of High Energy Particle Cascades,"
    Proceedings of the 16th International Cosmic Ray
    Conference, Kyoto, Japan, August 1979 (Institute
    for Cosmic Ray Research, University of Tokyo, Tokyo,
    Japan), vol. 11, p. 184

21. L. Sulak, T. Armstrong, H. Baranger, M. Bregman, M. Levi,
    D. Mael, J. Strait, T. Bowen, A. E. Pifer, P. A. Polakos,
    H. Bradner, A. Parvulescu, W. V. Jones, and J. Learned,
    Experimental Studies of the Acoustic Signature of Proton
    Beams Traversing Fluid Media, Nucl. Instrum. and Methods,
    161:203 (1979).

22. "Threshold Limit Values for Chemical Substances and Physical
    Agents in the Workroom Environment with Intended Changes
    for 1981," American Conference of Government and
    Industrial Hygienists.

23. P. D. Edwards, Interactions of Ultrasound with Biological
    Structures -- a Survey of Data, "Interaction of Ultrasound
    and Biological Tissues," HEW Publication (FDA) 73-8008,
    p. 299 (1972).

24. W. D. Ulrich, Ultrasound Dosage for Non-Therapeutic Use on
    Human Beings -- Extrapolations from a Literature Survey,
    IEEE Trans. Biomed. Eng. BME-21:48 (1974).

25. P. P. Lele, "Thresholds and Mechanisms of Ultrasonic Damage to 'Organized' Animal Tissues," Proceedings of Symposium on Biological Effects and Characterizations of Ultrasound Sources, Rockwille, MD, June 2-3, 1977, p. 224.

26. J. H. Goll, The Design of Broad-Band Fluid-Loaded Ultrasonic Transducers, IEEE Trans. Sonics & Ultrasonics SU-26:385 (1979).

27. Y. Netzer, "The Design of Low-Noise Amplifiers," Proceedings of the IEEE 69:728 (1981).

28. W. N. McDicken, "Diagnostic Ultrasonics," Crosley, Lockwood, Staples, London (1976).

29. R. L. Nasoni, T. Bowen, M. Dewhirst, H. B. Roth, and R. Premovich, Speed of Sound as a Thermal Image CT Scan Parameter, in: "Acoustical Imaging," vol. 11, J. P. Powers, ed., Plenum, New York (1982), p. 563.

30. T. Bowen, Acoustic Passive Remote Temperature Sensing, in: "Acoustical Imaging," vol. 11, Plenum, New York (1982), p. 549.

31. J. F. Greenleaf, S. A. Johnson, W. F. Samayoa, and F. A. Duck, Algebraic reconstruction of spatial distributions of acoustic velocities in tissue for their time-of-flight profiles, in: "Acoustical Holography," vol. 6, N. Booth, ed., Plenum, New York (1975), p. 71.

32. W. D. O'Brien, Jr., The Role of Collagen in Determining Ultrasonic Propagation Properties in Tissue, in: "Acoustical Holography," vol. 7, L. W. Kessler, ed., Plenum, New York (1977).

33. E. K. Fry, N. T. Sanghvi, F. J. Fry, and H. S. Gallager, Frequency Dependent Attenuation of Malignant Breast Tumors Studied by the Fast Fourier Transform Technique, in: "Ultrasonic Tissue Characterization II," M. Linzer, ed., National Bureau of Standards Spec. Publ. 525, U.S. Government Printing Office, Washington (1979), p. 85; T. Kobayashi, Correlation of Ultrasonic Attenuation with Connective Tissue Content in Breast Cancers, ibid., p. 93.

34. T. Bowen, R. L. Nasoni, A. E. Pifer, and G. H. Sembroski, "Some Experimental Results on the Thermoacoustic Imaging of Tissue Equivalent Phantom Materials," Proceedings of 1981 IEEE Ultrasonics Symposium, Chicago, October 14-16, 1981, p. 823.

35. T. Bowen, W. G. Connor, R. L. Nasoni, A. E. Pifer, R. Bell, D. H. Cooper, and G. H. Sembroski, Observation of Acoustic Signals from a Phantom in an 18 MeV Electron Beam for Cancer Therapy, in: "Acoustical Imaging 83," M. Kaveh, ed., Plenum Press, New York (following paper).

36. F. Caspers and J. Conway, "Measurement of Power Density in a
     Lossy Material by Means of Electromagnetically Induced
     Acoustic Signals for Non-Invasive Determination of
     Spatial Thermal Absorption in Connection with Pulsed
     Hyperthermia," Proceedings of 12th European Microwave
     Conference, 13-17 Sept. 1982, Helsinki.

37. T. Bowen, R. L. Nasoni, and A. E. Pifer, "Thermoacoustic NDE
     Imaging Induced by Deeply Penetrating Radiation,"
     Review of Progress in Quantitative Nondestructive
     Evaluation, D. O. Thompson and D. E. Chimenti, eds.,
     Plenum Press, New York (1983), vol. 2B, p. 1029.

38. J. R. Chamuel, Stress waves induced in graphite-epoxy
     composite structures by current pulses, J. Acoust.
     Soc. Am. Suppl. 1, vol. 72, p. S27 (1982).

39. F. Braun, Ann. d. Physik 65:358 (1898); P. de Lange, On
     Thermophones, Proc. Roy. Soc. (London) 91:239 (1915);
     H. D. Arnold and I. B. Crandall, The Thermophone as a
     Precision Source of Sound, Phys. Rev. 10:22 (1917);
     E. C. Wente, The Thermophone, Phys. Rev. 19:333 (1922).

OBSERVATION OF ACOUSTIC SIGNALS FROM A PHANTOM

IN AN 18 MeV ELECTRON BEAM FOR CANCER THERAPY

T. Bowen, W.G. Connor, R.L. Nasoni, A.E. Pifer, R. Bell,
D.H. Cooper, and G.H. Sembroski

Departments of Physics and Radiology
University of Arizona, Tucson, AZ 85721

At typical dosages with radiation therapy LINAC's, a detectable acoustic signal from soft tissues is predicted[1,2] if 10^3-10^4 successive signals arising from x-ray or electron beam pulses are signal averaged. A preliminary experiment is reported to confirm the existence of this acoustic signal which is expected because of the very small, but significant, thermal stresses due to pulsed heating by the LINAC beam.

The arrangement of the phantom, electron beam, and acoustic transducer is shown in Fig. 1. The 18 MeV therapeutic electron beam from a Varian LINAC was collimated to a 2.5 cm x 2.5 cm square cross section at the entrance to the phantom, which was 65 cm from the source. The LINAC was operated at 180 rad/min measured at 100 cm distance, a 180 pulse/s repetition rate, and a 1.5 μs beam pulse width.

The phantom consisted of a 10 cm x 10 cm x 13 cm deep acrylic water-filled tank with a horizontal 1.59-mm-thick acrylic plate 6 cm above the bottom. A custom-made 19-mm-dia. 500 kHz piezoelectric transducer was located a distance z above the acrylic plate. The electron beam was directed horizontally, parallel to the acrylic plate, with the plate centered in the beam.

A block diagram of the electronics and signal-averaging system, which is similar to the arrangement described for an earlier experiment with electric current heating pulses,[3] is shown in Fig. 2. The acoustic transducer, which has an output impedance of 900 ohms at 500 kHz, was connected to a custom-built low-noise amplifier with a gain of approximately 620. The amplifier was located about

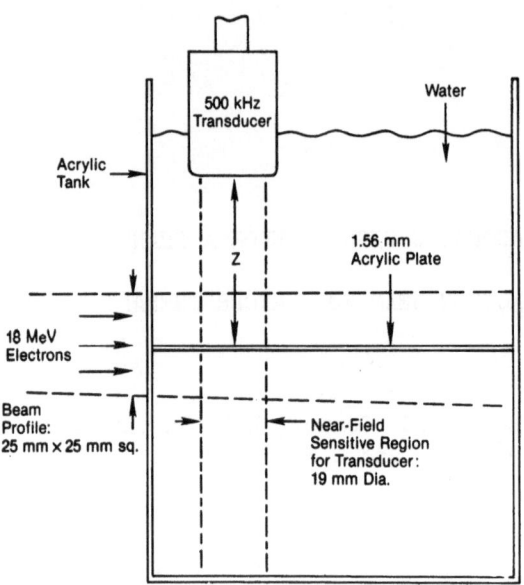

Fig. 1.   Arrangement of electron beam, phantom, and acoustic
          transducer.

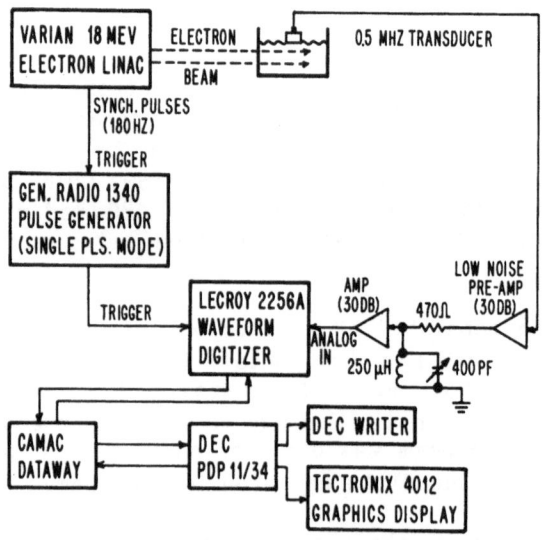

Fig. 2.   Block diagram of the electronics and signal-averaging
          system.

50 cm from the transducer, and was powered by batteries. Battery power was found to be essential to reduce electrical pickup·from the klystron power supply at the time of the beam pulse to levels such that the amplifier is not overloaded. The amplifier output entered a LeCroy 2256A 20 MHz 8-bit waveform digitizer which was set to sample at 0.1 μs intervals. The signal-averaging computations were performed by an on-line PDP11/34 computer whose output was displayed on a Tektronix 4012 graphics display terminal and also printed out. The system permitted the experimenters to write a program in FORTRAN which controlled the direct memory transfer of digitizer data via the CAMAC system, carried out the signal averaging and other numerical tasks, and reconstructed the averaged data as a waveform on the Tektronix 4012.

If $V(i,j)$ is the voltage of the ith sample (i runs from 1 to 1024) of the jth waveform (j runs from 1 to N, where a typical value of N is 1,000), then the signal averaged output $\langle V(t_i) \rangle$ at time t_i was computed as

$$\langle V(t_i) \rangle = \frac{1}{N} \sum_{j=1}^{N} V(i,j) , \tag{1}$$

where t_i = i x (0.1 μsec sampling interval). If N waveforms are averaged, the signal-to-noise amplitude ratio is increased by a factor $\sqrt{N}$. In many runs, the signals could not be discerned in single waveforms. As the signal-averaging computer code has not been optimized for rapid data acquisition, the system acquired data from approximately 5 pulses per second.

Since the time available for an experiment was limited to a few hours, due to separate time constraints on the availability of the accelerator, computer, and key personnel, the best results were obtained in the final two data-taking runs, Runs 3 and 4 listed in Table I. In this table the time delay was measured with respect to a trigger pulse from the accelerator which precedes the beam by a fixed time.

Table 1. Results based upon Runs 3 and 4 in an 18 MeV electron beam.

Run	No. of pulses sampled	Transducer distance,z	Time delay of signal	$\left(\begin{matrix}\text{Speed} \\ \text{of sound}\end{matrix}\right) \text{x} \left(\text{Time}\right)$
3	1000	45 ± 1 mm	39.5 μs	58.9 mm
4	1000	35 ± 1 mm	33.4 μs	49.8 mm
Difference		10 ± 1.5 mm	6.1 μs	9.1 mm

The signals which were obtained by signal averaging 1000 waveforms are shown in Figs. 3(a) and (b). The observed signal waveforms agree very well in shape with the expected form which is shown in Fig. 3(c). The expected signal waveform was derived from the prediction of thermoacoustic theory[1,4] that the first pressure signal is a rectangular pulse of width t_1 equal to the acoustic transit time through the acrylic plate (0.6 μs) followed a time t_2 later by an identical pressure pulse with opposite polarity. The time t_2 is equal to the width of the LINAC electron beam pulse, 1.5 μs for these runs. It was then assumed that these pressure pulses correspond to forces applied to a damped harmonic oscillator where the displacement x is given by

$$x = x_o e^{-\gamma t} \sin 2\pi f t ,$$

and where f = 0.5 MHz and $\gamma = 0.35(\mu s)^{-1}$. By inspecting the relative heights of successive peaks of the oscillatory waveforms shown in Fig. 2, it is seen that the observed signals agree well with the expected form.

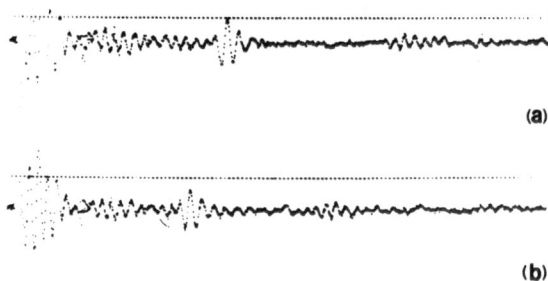

(a)

(b)

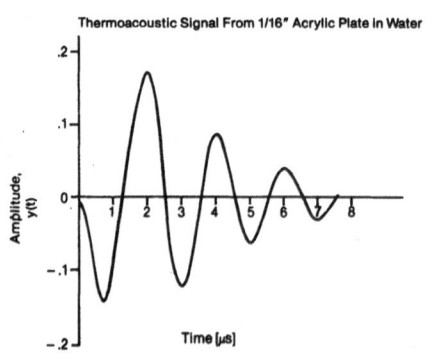

(c)

Fig. 3. Signal-averaged waveforms (a) for Run 3 and (b) for Run 4 photographed from the computer graphics display (total trace length corresponds to 102.4 μs). The expected waveform is shown in (c).

By fitting the expected signal waveform to the observed signals from Runs 3 and 4, the signal delay times shown in Table I were determined. When the time difference between Runs 3 and 4, 6.1 μs, is multiplied by the sound velocity in water, 1.49 mm/μs, the result, 9.1 mm, agrees well with the change of position of the transducer, 10±1.5 mm.

If the total time delay, for example in Run 3, is multiplied by the speed of sound, a distance of 58.9 mm is obtained which is larger than the actual distance, 45±1 mm. The additional delay of approximately 9 μs arises because the pulses which triggered the recording electronics preceded the beam.

The signal-to-noise (S/N) ratio of the observed signal v(t) was obtained from

$$S/N \cong \{2 \; \Delta f \int_{signal} [v(t)]^2 dt\}/\langle v^2 \rangle_{N \; samples} \; ,$$

where Δf is the noise bandwidth and $\langle v^2 \rangle_{N \; samples}$ is the mean-squared noise voltage obtained from portions of the waveform where there was no apparent signal. The result was $S/N \cong 850$ for the signals in Runs 3 and 4 assuming that the noise bandwidth $\Delta f = \pi \gamma = 1.1$ MHz. It is difficult to obtain a theoretical estimate of the best possible S/N in the arrangement of Fig. 1 because the plate is in the transition region between near-field and far-field of the transducer, but rough estimates give a best possible $S/N \sim 10^3$–10^4. Since the experimental system has not been optimized, the observed S/N appears reasonable.

In conclusion, we have succeeded in observing a thermoacoustic signal from a phantom in a therapeutic electron beam. With further optimization of the system, these signals offer the possibility of observing the positions of the edges of the beam in relation to structures within irradiated soft tissues.

REFERENCES

1.   T. Bowen, Radiation-Induced Soft Tissue Imaging, Proceedings of 1981 IEEE Ultrasonic Symposium, p. 817.

2.   T. Bowen, R. L. Nasoni, and A. E. Pifer, Thermoacoustic Imaging Induced by Deeply Penetrating Radiation, in: "Acoustical Imaging 83," M. Kaveh, ed., Plenum Press, New York (preceding paper).

3.   T. Bowen, R. L. Nasoni, A. E. Pifer, and G. H. Sembroski, Some Experimental Results on the Thermoacoustic Imaging of Tissue Equivalent Phantom Materials, Proceedings of 1981 IEEE Ultrasonics Symposium, p. 823.

4.   L. Sulak, T. Armstrong, H. Baranger, M. Bregman, M. Levi,
       D. Mael, J. Strait, T. Bowen, A. E. Pifer, P. A. Polakos,
       H. Bradner, A. Parvulescu, W. V. Jones, and J. Learned,
       Experimental Studies of the Acoustic Signature of
       Proton Beams Traversing Fluid Media, Nucl. Instrum.
       and Methods 161:203 (1979).

HOLOGRAPHIC IMAGING SYSTEM USING WIDEBAND CHIRPED ULTRASOUND

T. Yamamoto, S. Fujii and Y. Aoki

Department of Electrical Engineering
Faculty of Engineering, Hokkaido University
N-13,S-8, Kitaku, Sapporo 060, Japan

INTRODUCTION

The application of the ultrasound imaging techniques is spreading in many fields such as nondestructive testing, medical diagnosis or underwater imaging. Especially the B-scan imaging which visualizes a two dimensional section of a specimen is the most popular technique. In the conventional B-scan imaging system, the lateral resolution is limited by the aperture of the transducers and range resolution is limited by the pulse width of the ultrasound. As for the range resolution, the use of a short time pulse improves the resolution, however it causes a degradation of the signal to noise ratio (SNR). Another possibility to improve the resolution is to employ the holographic technique. In this case, while the lateral resolution is greatly improved by the synthetic aperture, the range resolution gets worse because the coherent illumination must be used in the holographic imaging technique. This nature had been also a problem in the field of the microwave radar until the synthetic aperture radar (S.A.R.)[1] was developed. Since the success of S.A.R. is a result of combination of a holography and the pulse compression technique, it may be possible to employ the same technique in the acoustical imaging.

In this paper, We propose a new technique of the B-scan imaging which makes it possible to obtain higher resolution and the SNR than the conventional imaging system. In the method proposed, a wideband chirped ultrasound is employed instead of the impulse or the coherent ultrasound. The range resolution is obtained by the pulse compression using chirp signal. The difference from the conventional B-scan imaging system is that the phase information remains after this processing and it is possible to improve the

435

lateral resolution by the holographic image reconstruction.   The
calculations required in an image reconstruction are the phase
rotation (complex multiplication) and the Fourier transform.   Thus
it can be calculated effectively by a digital computer.   In this
paper we briefly describe the theory of an image reconstruction and
realization of a hardware system.   Further, an experimental result
demonstrating high resolution is shown.

PRINCIPLE

    Figure 1. shows the coordinate system of the B–scan imaging
system.   In this figure, the center of an imaging area is assumed
to be $(0, z_0)$.   The receiving and transmitting transducers are
located on the straight line of z=0 and it is assumed that the
location of them is the same for the simplicity.   The transmitter
fires a burst of ultrasound at each sampling point.   The waveform
g(t) is given as follows,

$$s(t) = \begin{cases} \exp(jw_0 t + j\frac{1}{2}ut\,) & :-\frac{T}{2} < t < \frac{T}{2} \\ 0 & :\text{elsewhere}, \end{cases} \quad (1)$$

where w_0 is the angular frequency of the center of a chirp signal,
u is the chirp rate and T is the pulse duration time.   The
frequency sweep range B is given by

$$B = uT. \quad (2)$$

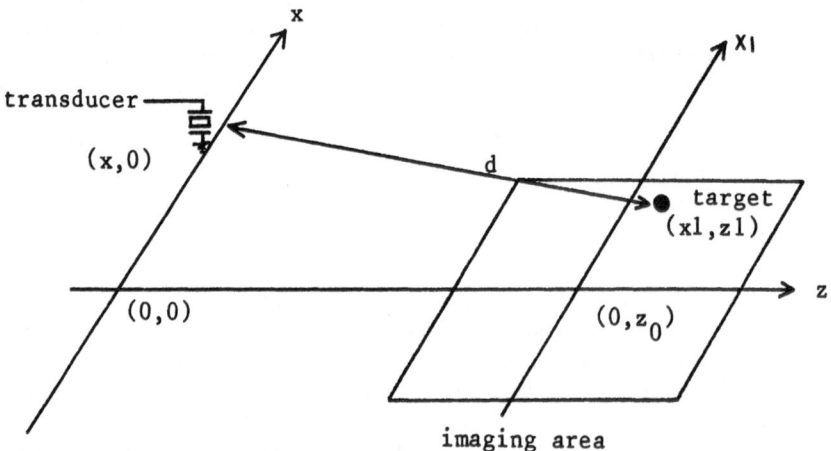

Figure 1. Coordinate system of B–scan imaging system.

The reflected wave from targets, located at the coordinate (x_1, z_1) can be written by

$$g(x,t) = R_0 \exp(jw_0(t - \frac{2d}{c}) + j\frac{1}{2}u(t - \frac{2d}{c})^2) \tag{3}$$

where

$$d = \sqrt{(x-x_1)^2 + z_1^2} . \tag{4}$$

In equation (4), x is the lateral offset of transducers and R_0 in (3) is the complex coefficient of a target reflection.   The data are collected at many points along the x axis and are stored in the memory device.   As a result, collected data become space-time signal.   Since the equation (3) has a carrier component which has no information about targets, it should be removed by the coherent detector using reference signal.   The reference signal $r_0(t)$ is given by

$$r_0(t) = \exp(-jw_0 t). \tag{5}$$

The result of detection or $g_1(t)$ is written as follows,

$$g_1(x,t) = R_0 \exp(-j\frac{4\pi d}{\lambda}) \exp(j\frac{u}{2}(t - \frac{2d}{c})^2), \tag{6}$$

where λ is the wavelength at angular frequency of w_0.   Since the first exponential function on the right side does not depend on the variable t, the information about longitudinal direction is all contained in the second exponential function.   At first, the pulse compression of the range direction must be done.   The first step of this processing is multipling a reference signal.   This reference signal $r_1(t)$ is determined by the longitudinal distance between transducers and the center of an imaging area.   In the case of figure 1, it becomes

$$r_1(t) = \exp(-j\frac{u}{2}(t - \frac{2z_0}{c})^2) . \tag{7}$$

Multipling (6) by (7), new form of $g_2(x,t)$ is obtained

$$g_2(x,t) = R_0 \exp(-j\frac{4\pi d}{\lambda}) \exp(j\frac{2u}{c^2}(z_1^2 - z_0^2)) \exp(-j\frac{2u}{c}(z_1 - z_0)t). \tag{8}$$

In (8), $g_2(x,t)$ contains the variable t only in the last exponential function on the right side and the range information $z_1 - z_0$ is

expressed by the angular frequency of the t direction.   The pulse
compression of a longitudinal direction is obtained as the result
of spectral estimation of equation (8).   The correspondence of the
z coordinate with angular frequency w can be written as follows,

$$z_1 = \frac{cw}{2u} + z_0. \qquad (9)$$

Usually, FFT algorithm is employed as the spectral estimator.
Other techniques such as the Maximum Entropy Method (M.E.M.) are
also possible to use in this compression and they may give better
resolution.   However, the M.E.M. estimates only power spectrum,
so that the phase information which is indispensable to the following lateral compression is lost.   Further, since the M.E.M.
requires more computing efforts than the FFT, it is almost impossible to construct a real time imaging system.   From the fact
described above, the use of FFT technique as the spectral estimator
is preferable.

After the longitudinal compression, the information about
targets is contained in the relevant range bins.   Next step of
image reconstruction is the lateral compression.   The information
contained in the range bin $z = z_1$ can be expressed by $g_{z1}(x)$:

$$g_{z1}(x) = R_1 \exp(-j\frac{4\pi d}{\lambda}), \qquad (10)$$

where R_1 represents the combination of a constant which is meaning-
less in the lateral compression.   Assuming that x is
sufficiently smaller than z_1, (4) can be rewritten as follows

$$d \simeq z_1 + \frac{(x - x1)^2}{2}. \qquad (11)$$

Substituting (11) into (10) we get

$$g_{z1}(x) = R_1 \exp(-j\frac{4\pi z1}{\lambda} - j\frac{2\pi(x-x1)^2}{\lambda z1}). \qquad (12)$$

The equation (12) expresses the ordinary Fresnel transformed hologram.   There are many techniques[2] which can be applied to the
image reconstruction from (12).   The most popular technique is to
transform the Fresnel transformed hologram into Fourier transformed
hologram.   This transform is given by

$$g_{z1}(x) = g_{z1}(x) \cdot P^*_{z1}(x) \qquad (13)$$

$$P_{z1}(x) = \exp(-j\frac{2\pi x^2}{z1}), \qquad (14)$$

where $P_{z1}(x)$ is the propagation function of $z=z_1$ and $*$ expresses the complex conjugate.   Equation (13) can be rewritten as follows,

$$g_{z1}(x) = R_1 \exp\left(-j\frac{4\pi z}{\lambda} - j\frac{4\pi x1^2}{\lambda z_1} \right) \exp\left(j\frac{4\pi x1}{\lambda z_1} x \right). \tag{15}$$

Since the first exponential function of equation (15) is independent from variable x, this term can be assumed as a constant value. As the result, the lateral position x_1 is obtained from the spectral estimation of (15).   The correspondence between x and the spatial angular frequency v is given by

$$x_1 = \frac{\lambda z1}{4\pi} v. \tag{16}$$

The pulse compression technique used in the longitudinal and lateral directions is derived using different principles, however, the resulted algorithms are almost same.   This nature is very useful for the design of a special purpose hardware for an image reconstruction since the required calculation becomes a simple iteration of a 1-dimensional FFT.

RESOLUTION CONSIDERATION

In this method, an image is obtained as the result of the Fourier transform.   This means that the theoretical resolution is calculated from the resolution of the spectral estimation.   The basic relationship between spectral resolution Δw and the time aperture T is given by

$$\Delta w = \frac{2\pi}{T} . \tag{17}$$

In this case, T is the pulse duration time.   Substituting (17) into (9), the following equation is obtained,

$$\Delta z = \frac{c\pi}{B} . \tag{18}$$

This equation states that the resolution is limited by the frequency sweep range of the chirp signal.

The lateral resolution is also limited by the aperture of the Fourier transform.   Since the aperture is the lateral length of a scanning area, there is no limitation on it if the beam width of transducers is wide enough.   However, another limitation arises from the pulse compression of the longitudinal direction. Because the locus of the compressed pulse is mapped on a hyperbolic curve, the compressed pulse exists over some range bins when the size of a range bin Δz becomes small.   This situation is illustrated in Figure 2.   Using the size of a range bin of equation (18), the

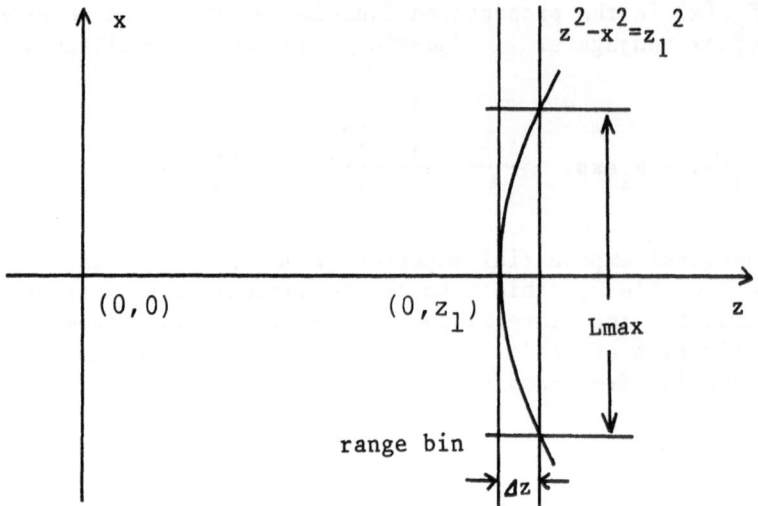

Figure 2. Available synthetic aperture length after longitudinal compression.

maximum available aperture Lmax is given as the solution of following equation

$$\Delta z = \sqrt{(Lmax/2)^2 + z_1^2} - z_1.$$ (19)

Solving equation (19) for Lmax, we obtain the maximum available aperture Lmax as

$$Lmax = 2\sqrt{\Delta z(2z_1 + \Delta z)}.$$ (20)

Using the condition of $z_1 \gg \Delta z$ and substituting (18) into (20) we obtain

$$Lmax = \sqrt{\frac{8c\pi z1}{B}}.$$ (21)

When the aperture along the x axis is limited to Lmax, the resolution in the spatial frequency domain is given as follows

$$\Delta v = \frac{2\pi}{Lmax}.$$ (22)

Substituting (21) and (22) into (16), the lateral resolution is given by

$$\Delta x = \lambda \sqrt{\frac{Bz1}{32c\pi}}.$$ (23)

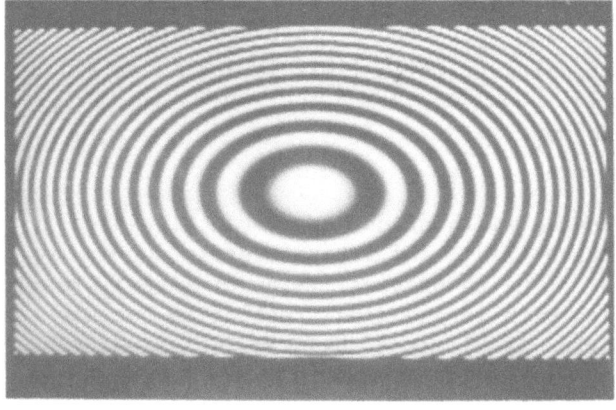

(a)

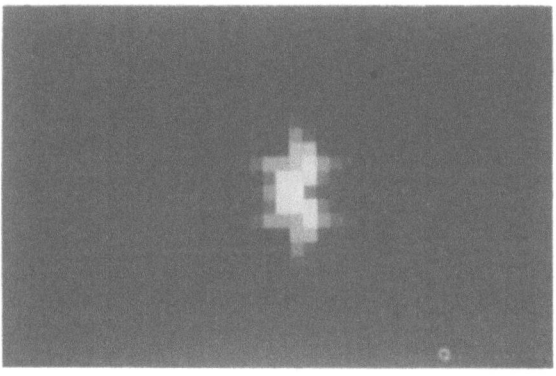

(b)

Figure 3.   (a) Simulated data from a point-like object.
            (b) Reconstructed image. The size of pixel is 0.31mm
                over z direction and   0.87mm over x direction.

Equation (23) gives the maximum available resolution, and it corresponds to the beam width of a single transducer of the diameter of
L.   However, the aperture size is sometimes restricted by other
factors such as the beam angle of the transducers or physical
limitation of a scanning width.

     The quality of an image reconstructed can be demonstrated by
the computer simulation.   Figure 3-a shows an example of the simulated data of $g_2(x,t)$ from a point-like object located at the
distance of 45Cm. The pulse duration time is 650us and the frequency sweeps from 6.6MHz down to 5.4MHz.   The form of data change
by the direction of frequency sweep.   In the case of downward sweep
as in this simulation, they become elliptic zoneplates, however

they become hyperbolic zoneplates in the case of upward sweep. Substituting these parameters into equation (18) and (21), we obtain theoretical resolution limit Δx and Δz as follows

$$\Delta z = 0.625mm, \tag{24}$$
$$\Delta x = 1.19mm. \tag{25}$$

Since the scanning width of our experimental system is greater than Lmax of equation (21), the lateral resolution is not affected by this factor. The theoretical value of lateral resolution is not good comparing with longitudinal resolution because the synthetic aperture length is limited by the longitudinal compression. However, it is almost impossible to obtain the theoretical resolution about longitudinal direction, as the result, the lateral resolution may be improved because of expansion of Lmax. The image reconstructed from the data of figure 3-a is shown in figure 3-b. This example displays the intensity of the image reconstructed and it can be regarded as the point-spread function of this imaging system. This simulation demonstrates that this method has a potential of a possible application in a high resolution ultrasound imaging system.

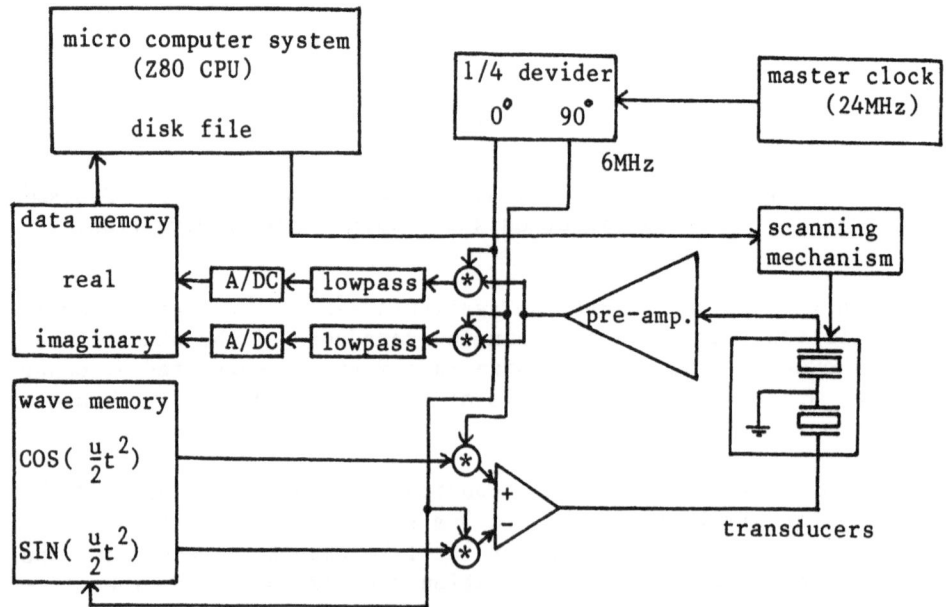

Figure 4.   The block diagram of wideband chirp signal
generator/receiver.

SYSTEM REALIZATION

To demonstrate this imaging method, we construct an experimental apparatus using 6MHz ultrasound. The block diagram is illustrated in figure 4. We designed a digital-analogue hybrid chirp signal generator in which a chirp signal of the center frequency of 6MHz is generated from the crystal controlled carrier signal and the baseband complex chirp signal stored in a digital memory. A part of a carrier signal is used as the reference signal in the coherent detector of receiver. A returned ultrasound is converted into electric signal by the transducer and the result of a coherent detection is stored in a digital memory system. The data collected are read by a computer after the completion of 1 line data collection. The imaging area about the range direction is determined by the range gate. The gate signal is generated by the timer circuit triggered at the time of firing an ultrasound. A Z80 based microcomputer system controls system timing, mechanical scanner and data storing to disk system. Furthermore, it is possible to calculate some parts or an entire image reconstruction process.

An image is reconstructed by a large computer system or micro computer system and the image resulted is displayed on a color CRT. The required calculation time to reconstruct an image depends on its size. In the case of our experimental system, The size of the data collected is 2048(t direction) X 128(x direction) and each datum contains a complex number. The sampling interval is 0.33us for t direction and 0.5mm for x direction. The required computation for an image reconstruction is almost equal to a 2-dimensional Fourier transform of the same size. Usually the image is reconstructed by a large computer system, however, recent progress of micro-computer will make it possible to build a real time imaging system based on a specialized digital signal processors.

EXPERIMENTAL RESULTS

To demonstrate the performance of this method, under water experiment were performed in a water bath. In this experiment, films of the copolymer of PVDF/TriFE were used as receiving and transmitting transducers. Figure 5 shows the object which is used as the target. This object consists of nine tin plated wires whose diameter is 2mm and the eight of them are placed along the circle of radius 3cm. The rest is placed 2cm behind this circle. This target is placed in the water bath at the distance of 45Cm from the transducers. The same parameters as in the computer simulation were used in this experiment. Figure 6-a shows an example of the data collected or $g_2(x,t)$ and figure 6-b shows the reconstructed image. The resulted image shows that this method has good resolution in both lateral and range directions.

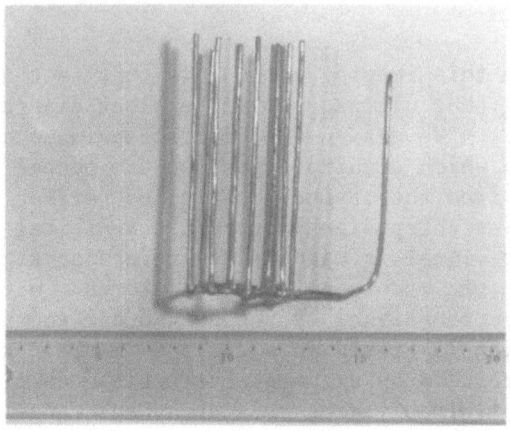

Figure 5.   Photograph of a test object.

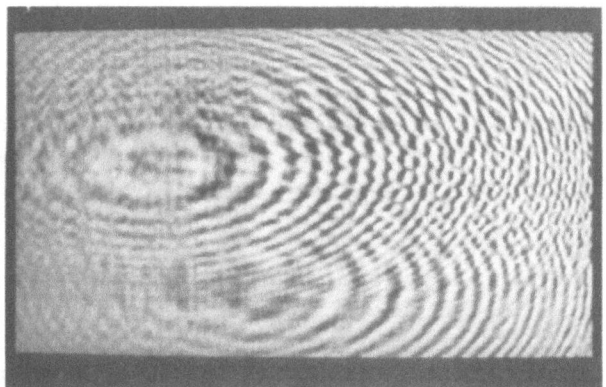

Figure 6. An example of recorded data from the test object.

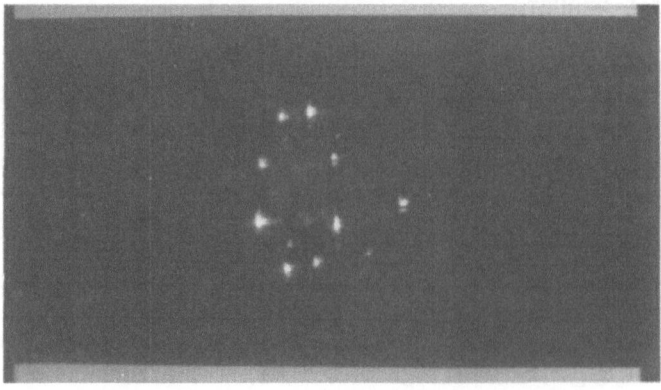

Figure 7. Reconstructed image.

CONCLUSIONS

Poor resolution is a problem in the conventional B-scan imaging especially in the lateral direction. A special feature of holography is its good lateral resolution. By combining the conventional B-scan technique with holographic imaging , it is possible to achieve a good resolution in any direction over the whole range. However some problems still remain. A problem of the method proposed is the phase incoherence of the ultrasound. In the simulation and experiment, it is assumed that the targets are stable during data collection, however, if the targets are moving objects such as alive human body, the irregular phase occurs and this causes the degradation of the reconstructed image. To avoid this effect, rapid data collection is necessary. In our experimental system it takes more than 10 minutes since the data are collected using mechanical scanner. During this period, the targets cannot move over the half of the wavelength. This restriction is quite serious especially in the application of medical diagnosis in which a human body is an object. An approach to solve this problem is the use of the array transducer system. Using an array system, whole data are collected by an ultrasound burst, so that this restriction will be greatly relaxed. Another problem is the computation time to reconstruct an image. When a Z80 based microcomputer system is used in the reconstruction, the processing time will exceed 2 hours. For the real time imaging system, sophisticated digital signal processing system must be developed.

ACKNOWLEDGMENT

The authors would like to thank Professor A. Odazima for his advice about the ultrasound transducer and Dr. K. Ohohigasi for his kind co-operation of offering ultrasound transducers. We would also thank our colleagues for helpful discussions and their technical advice.

REFERENCES

1. E.N. Lieth, "Quasi-Holographic Techniques in the Microwave Region", Proc. IEEE, Vol. 59, No. 9, 1971.
2. Y. Aoki, "Image Reconstruction by Computer in Acoustical Holography", Acoustical Holography, Vol. 5, 1967, Prenum press, New York.
3. Ermert,H., Karg, R., "Multifrequency Acoustical Holography", IEEE Trans. Sonics Ultrasonics, SU-26, No. 4, 1979, pp.279-286.
4. G. Alasaarela, K. Tervola, J.Ylitalo and J. Koivukangas, "UHB Imaging", Acoustical Imaging , Vol. 12, 1982, pp. 687-696, Prenum press, New York, 1982. New York.

REAL-TIME TWO-DIMENSIONAL DOPPLER FLOW MAPPING USING

AUTO-CORRELATION

C. Kasai, K. Namekawa, A. Koyano and R. Omoto*

Aloka Co., Ltd., 6-22-1 Mure, Mitaka, Tokyo 181, Japan

*Saitama Med. School, Moroyama, Saitama 350-04, Japan

INTRODUCTION

The tomographic imaging that employs ultrasonic echoes has achieved outstanding advances in recent years, and today, ultrasonic diagnostic equipment has become the tool that is absolutely indispensable for clinical operations.

Meanwhile, the feasibility of measuring blood flow in the heart and vessels by the use of Doppler effect in ultrasonic waves is a well known fact. With respect to the method of blood flow measurement, there are two kinds which employ continuous wave and pulse wave so called pulsed Doppler. Since the pulsed Doppler is capable of providing blood-flow information at any depth on the sound beam axis simultaneously with 2-D and M-mode images, it is steadily increasing in popularity at the present time.

The pulsed Doppler system, however, has a disadvantage in that information only within a narrow range of a sampling site on a beam axis is obtained. In order to acquire the whole flow profile on the axis, a Multi-Channel method that employs an increased number of pulsed Doppler sampling gates [1, 2, 3] and the MTI (Moving Target Indication) method [4, 5, 6] has been devised.

Advancing the step further, systems are reported in which the ultrasonic beam is scanned in a certain cross section for the two-dimensional mapping of blood flow [7, 8, 9]. However, the images obtained by the system are still images, and no real-time observation of actual flow dynamism has been possible.

We have previously developed an equipment that in real-time,

447

displays blood-flow movements on a cross-section in the heart or blood vessels by the use of an auto-correlation technique, and we have already disclosed its outline [10]. Here in this paper, we describe the details of its principle and the instrumentation. Clinical evaluation data in a hospital are also presented.

PRINCIPLE

In the well-known B-mode instruments, only the amplitudes of echoes reflected at tissues are imaged. However in the blood flow visualization, the frequency change or phase shift as well as amplitude of echoes returning from corpuscles must be detected.

Important aspects that provide flow information effective for diagnosis may be considered to be the following three factors:

(1)   Flow Direction

In the pulse echo instrument employing one ultrasonic transducer, the flow component onto the sound beam axis is measured. By detecting the polarity of Doppler frequency of echoes with respect to that transmitted, the flow direction (i.e. toward flow or away flow) is discriminated.

(2)   Mean Flow Velocity

Mean blood flow velocity is estimated from the frequency spectra of echoes. Denoting the power spectra of echoes with $P(\omega)$, the mean angular frequency $\overline{\omega}$ is expressed as a primary moment of frequency with respect to the point of origin of frequency as follows:

$$\overline{\omega} = \frac{\int_{-\infty}^{+\infty} \omega P(\omega) \, d\omega}{\int_{-\infty}^{+\infty} P(\omega) \, d\omega} \tag{1}$$

(3)   Turbulence

The extent of turbulence in blood flow may be inferred from the spectrum variance. Now, denoting the standard deviation of spectra with σ, the variance σ^2 may be represented as a secondary moment with regard to the point of origin of ω by the following equation:

$$\sigma^2 = \frac{\int_{-\infty}^{+\infty} (\omega - \overline{\omega})^2 P(\omega) \, d\omega}{\int_{-\infty}^{+\infty} P(\omega) \, d\omega} = \overline{\omega^2} - (\overline{\omega})^2 \tag{2}$$

Next, by denoting with $R(\tau)$ the auto-correlation function of echo signals, the following relationship will pertain between $R(\tau)$ and $P(\omega)$ due to the Wiener-Khinchine's theorem:

$$R(\tau) = \int_{-\infty}^{+\infty} P(\omega)\, e^{j\omega\tau} d\omega \tag{3}$$

Expressing with $\dot{R}(\tau)$ and $\ddot{R}(\tau)$ the primary and secondary differentials, respectively, of $R(\tau)$ by τ, the following equations are derived from Eqs. (1), (2), and (3):

$$j\overline{\omega} = \frac{\dot{R}(0)}{R(0)} \tag{4}$$

$$\sigma^2 = \left\{ \frac{\dot{R}(0)}{R(0)} \right\}^2 - \frac{\ddot{R}(0)}{R(0)} \tag{5}$$

Because direct computation of Eqs. (4) and (5) is extremely time consuming, we will determine a simplified method. First, we assign:

$$R(\tau) = |R(\tau)|\, e^{j\phi(\tau)} \tag{6}$$

and by considering $|R(\tau)|$ as an even function and $\phi(\tau)$ as an odd function, we obtain the following equations:

$$\overline{\omega} = \frac{\phi(T) - \phi(0)}{T} = \frac{\phi(T)}{T} \tag{7}$$

$$\sigma^2 = \frac{2}{T^2} \left\{ 1 - \frac{|R(T)|}{R(0)} \right\} \tag{8}$$

where, T denotes the emission interval of ultrasonic pulses. (See Appendix.)

Equations (7) and (8) demonstrate the feasibility of obtaining the mean angular frequency and variance from auto-correlation values and phases at $\tau = 0$ and $\tau = T$.

Figure 1 shows the circuit of a complex auto-correlator for the real-time computation of $R(T)$ and phase $\phi(T)$.

A pair of complex Doppler signals obtained by a quadrature detector,

$$f_1(t) = f_{1R}(t) + j f_{1I}(t) \tag{9}$$

are each branched into two, and they are fed to a complex multiplier where each of the two signals are input directly and the other two are supplied via a pair of delay lines having the delay time T. The complex multiplier executes the following computation:

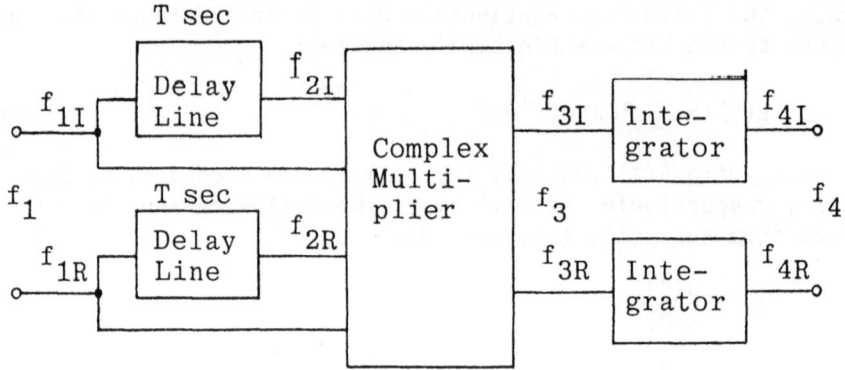

Fig. 1   Complex Auto-correlator for calculating R(T).

$$f_3(t) = f_1(t) \times f_2^*(t) \tag{10}$$

where,

$$f_2^*(t) = f_{1R}(t - T) - jf_{1I}(t - T) \tag{11}$$

is the conjugate complex of function $f_1(t)$ that has been delayed by time duration T. The outcome of integration of $f_3(t)$ carried out by integrators is $f_4(t)$:

$$f_4(t) = f_{4R} + jf_{4I}(t) \tag{12}$$

The correlated output, R(T), is obtained as follows:

$$|R(T)| = |f_4(t)| = \sqrt{f_{4R}^2(t) + f_{4I}^2(t)} \tag{13}$$

The phase, $\phi(T)$, is obtained as the slope of imaginary to real parts of the complex function $f_4(t)$ by the following equation:

$$\phi(T) = \tan^{-1} \frac{f_{4I}}{f_{4R}} \tag{14}$$

In addition, R(0) is obtained by integrating $f_1(t)$ as follows:

$$R(0) = \int (f_{IR}^2(t) + f_{1I}^2(t)) \, dt \tag{15}$$

SYSTEM

    Described below is the two-dimensional blood-flow mapping system that employs the subject auto-correlator.

    Figure 2 shows a block diagram of the system which is equipped

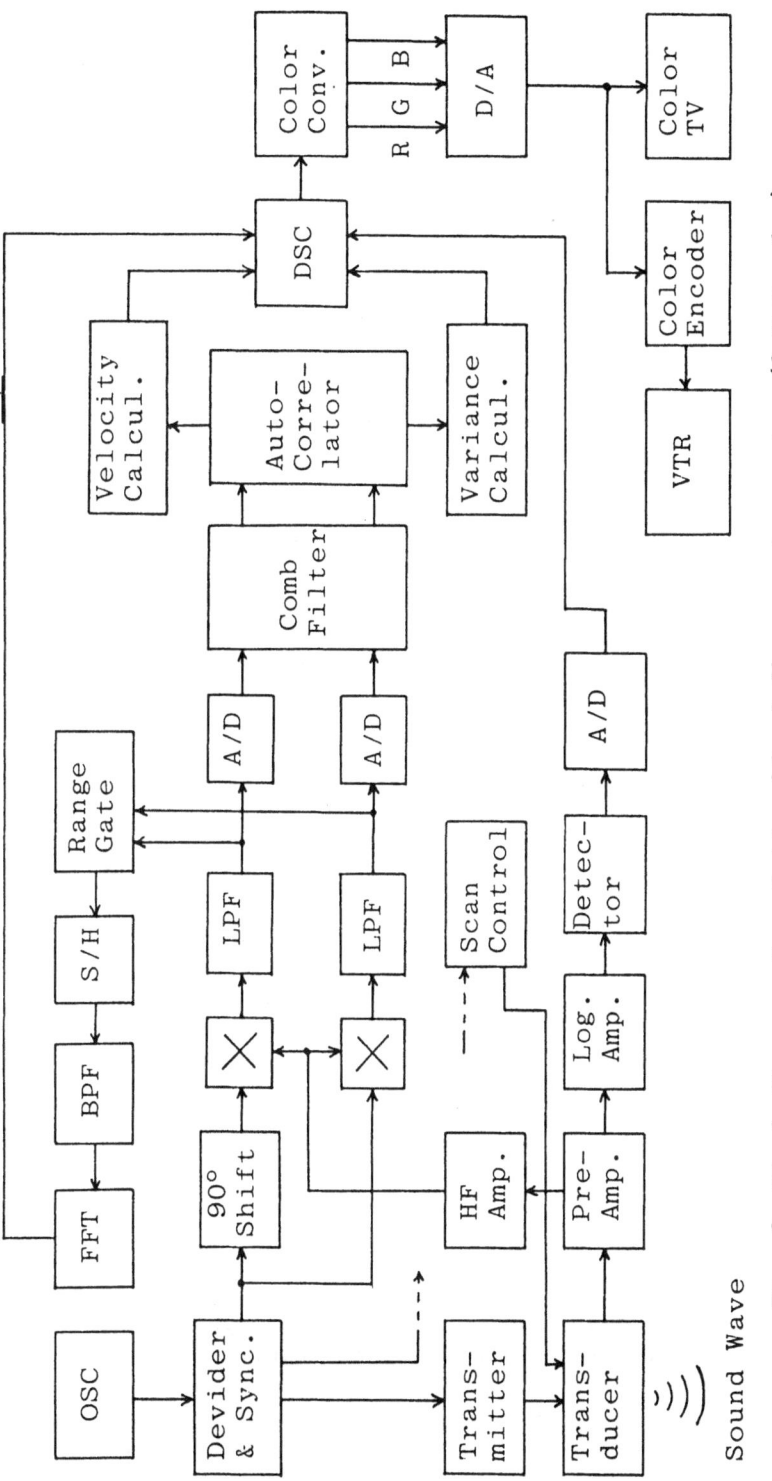

Fig. 2   Block diagram of real-time blood-flow mapping system (2-D Doppler).

with a conventional 2-D imaging unit and a pulsed Doppler unit provided with FFT. Images obtained by these units and flow mapping images are displayed overlapped together, simultaneously.

OSC is a high frequency oscillator whose output is divided to provide the clock pulses that trigger various units and to provide the continuous wave that is required for the demodulation of Doppler signals.

Signals received through the transducer are first amplified by pre- and HF-amplifiers, and then conveyed to a pair of quadrature detectors where the phases of the mixing reference frequency differ by 90°. Since this reference frequency is made the same as that of the transmitting burst wave, the outputs from the low pass filter (LPF) become the Doppler frequencies that have been shifted by Doppler effects, and the pair of outputs also become complex signals with phases that differ by 90°. The pair of signals, after conversion to digital signals by A/D converters, are passed through comb filters which suppress clutter signals, and then are supplied to the complex auto-correlator that has been described in Fig. 1. Its output is conveyed to the velocity calculator and the variance calculator that respectively calculate the mean value and the variance of Doppler signals. The outcome is recorded into a digital scan converter (DSC). The DSC additionally records the 2-D or M-mode images that have been obtained with conventional equipment, and FFT analyzed spectra of the blood flow at any sampling point specified.

The color converter serves the purpose of converting the data stored in the DSC to chrominance signals.

Firstly, with regard to the phase $\phi(T)$ that has been obtained with the velocity calculator, if it is in the first and second quadrants (i.e. $0° < \phi < 180°$), the color converter gives a red color. This signifies that the Doppler frequency shift is positive, and therefore the blood flows toward the transducer. If $\phi(T)$ is in the third and fourth quadrants (i.e. $180° < \phi < 360°$) it gives a blue color. This signifies that the Doppler frequency shift is negative and the blood flows away from the transducer. The faster the blood flow, the brighter the color will become.

Secondly, with regard to the variance, the larger its value, the greater the blend ratio of a green color will become. Since variance represents the flow turbulence, the color hue changes according to its extent. That is, red color approaches a yellow color and blue color approaches a cyan color.

On the other hand, 2-D echo images, M-mode and FFT analyzed blood flow data are all converted to black/white as in the conventional way.

The output from the color converter is transformed to analog signals by a D/A converter, and is displayed on a color TV screen in real-time.

The scanning of the ultrasonic beam may be carried out in either way by using a mechanical or a phased array transducer, whichever is preferred.

EXPERIMENTS

Experiments were conducted to examine the performance capabilities of the equipment.

First, a thread was stretched aslant across two pulleys placed in a water tank, and the inclination was set such that the angle between the thread and the ultrasonic beam was about 60°. A motor was coupled to one of the pulleys to enable the thread to move at any speed desired. The transmitting frequency of the transducer was 3 MHz.

Figure 3 shows the relationship between the speed of the thread movements and the output from the velocity calculator. The graph indicates that the output voltage from the auto-correlator increases almost linearly over the entire range examined of the moving speed of the thread up to 130 cm/s. This demonstrates that the auto-correlator is functioning properly. Although the graph represents the case when the thread was moved to an upward slanting direction, the output voltage will necessarily be negative when it is moved downward.

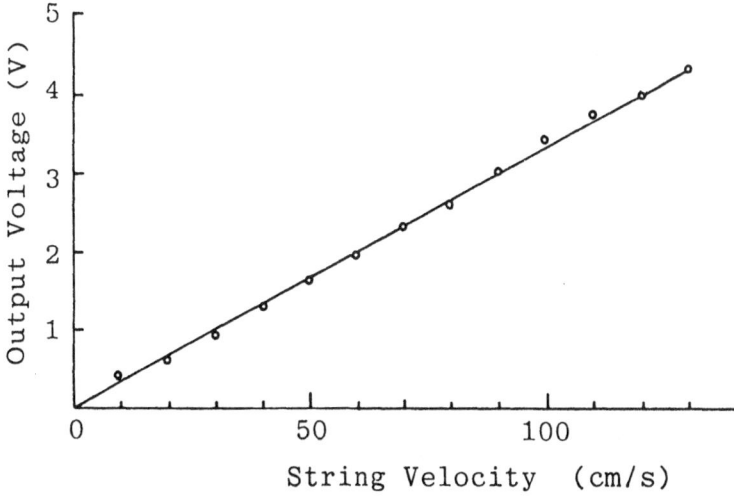

Fig. 3 Output voltage of velocity calculator versus
        string velocity.

Next, to study the relationship between moving objects and the colors on the TV monitor, a vinyl tube was laid horizontally inside the tank as shown in Fig. 4, and water was made to flow through it. The ultrasonic beam was scanned in a two-dimensional sector form.

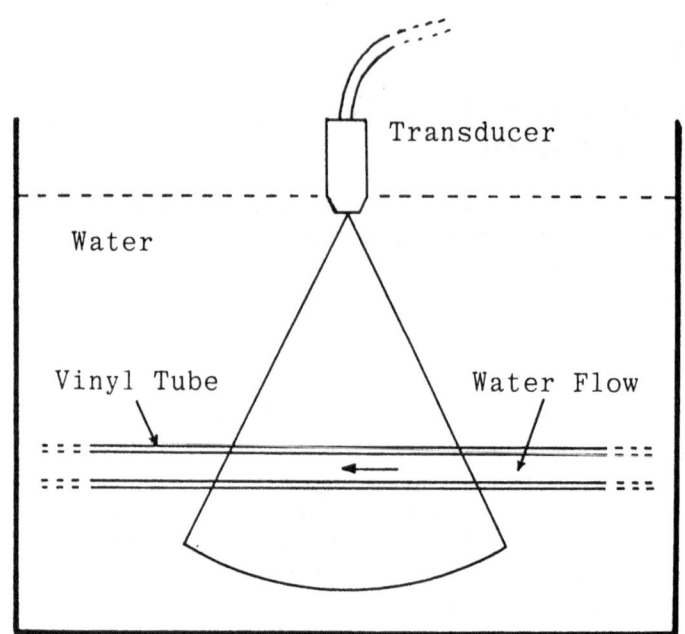

Fig. 4 Experimental setup for the color mapping of flow.

Figure 5 shows the 2-D color image that was displayed on the TV screen. The image shows no coloring in the central part of the vinyl tube where ultrasonic waves are incident almost at right angles with the water flow and therefore creating no Doppler frequency shifts. However in the other portions of the tube, colors are displayed. Since water is flowing from

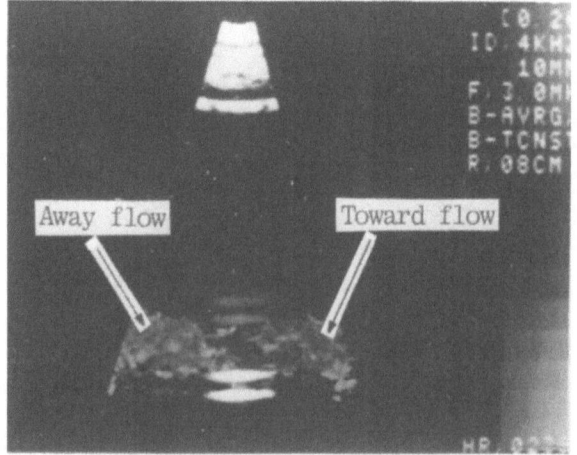

Fig. 5 2-D color-coded velocity mapping of water flow in a vinyl tube.

right to left, the color at the right side is red and that at the
left side is blue. Owing to the slight turbulence, some green
color is mixed. It is also recognized that the color brightness
increases according to the distance from the central part due to
the increase of Doppler frequency.

CLINICAL TESTS

Presented next are the clinical data that have been obtained
by applying the equipment to the mapping of blood-flow within the
heart.

Figure 6 is an example with a normal person that shows the
blood flow mapping for a long axis cross-section of the heart. (a)
is a 2-D image in diastole where the inflow from the left atrium
to the left ventricle is displayed in red. (b) is an image in
systole where the outflow from the left ventricle to the aorta is
displayed in blue.

Figure 7 is an M-mode Doppler display for the same person obtained with the ultrasonic beam fixed in the direction of the
mitral valve. The difference in blood-flow directions in diastole
and systole may readily be interpreted from the variations in
color.

Figure 8 is an example with a patient suffering from mitral
regurgitation and stenosis together, where (a) is a diastolic
image and (b) a systolic image. Due to the stricture of the valve,
the inflow in diastole shows a fast turbulent flow. In systole,
on one hand, alongside the outflow into the aorta, a regurgitant
flow returning from the left ventricle to the atrium is also observed at the same time. A comparison of colors of the outflow
with those of the regurgitant flow reveals the regurgitant flow to
be more turbulent. This situation may also be understood from the
M-mode Doppler display in Fig. 9.

Figure 10 is a case of tricuspid regurgitation and (a) gives
a diastolic image while (b) gives a systolic image. In systole,
the regurgitant flow returning from the right ventricle to the
right atrium is clearly displayed. Figure 11 shows the M-mode
Doppler display.

X-ray angiographic photographs have also been taken of the
above two cases, and their coincidence with the Doppler angio-
graphy using the equipment under investigation has been verified
in both cases.

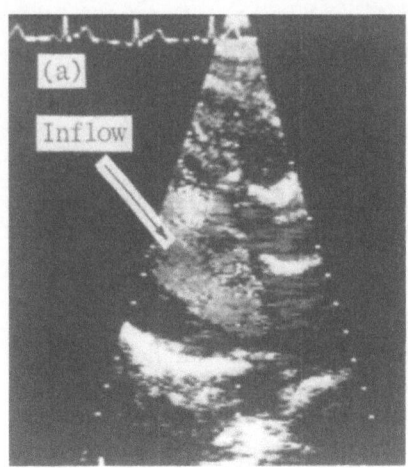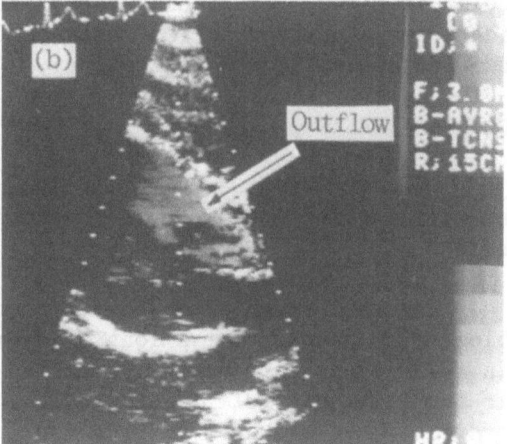

Fig. 6   Two-dimensional blood-flow images in the heart of a normal
person.

(a)   In diastole: Inflow is shown in warm (red) colors.

(b)   In systole :   Outflow is shown in cool (blue) colors.

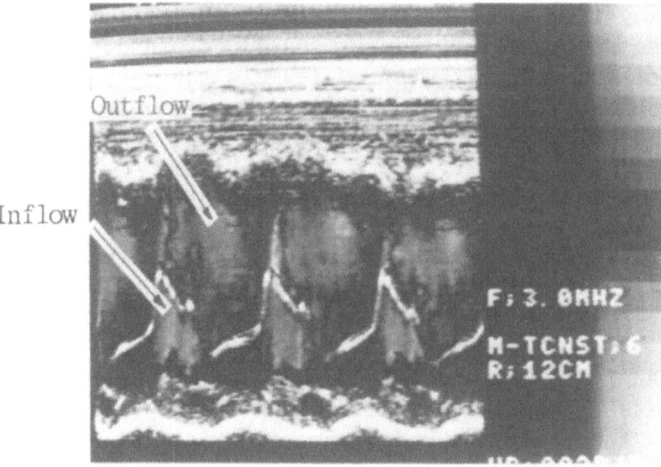

Fig. 7   M-mode Doppler display of the same
person as in Fig. 6.

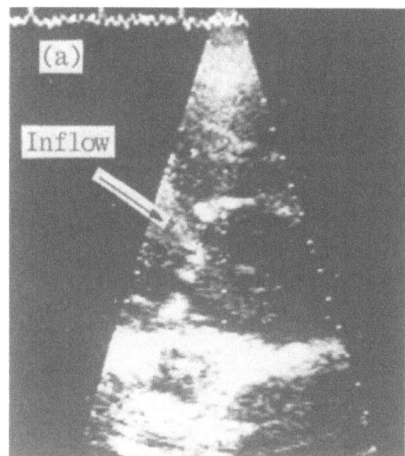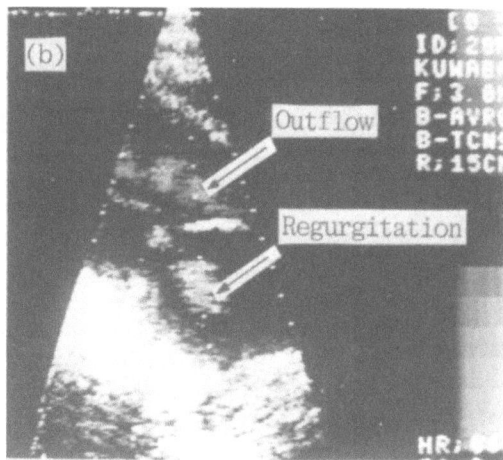

Fig. 8 2-D blood flow images in the heart with a valvular disease
        of mitral regargitation and stenosis.

        (a) In diastole: High speed turbulent inflow is shown
                           in mixed colors.
        (b) In systole : Regurgitation with turbulence as
                           well as outflow are shown.

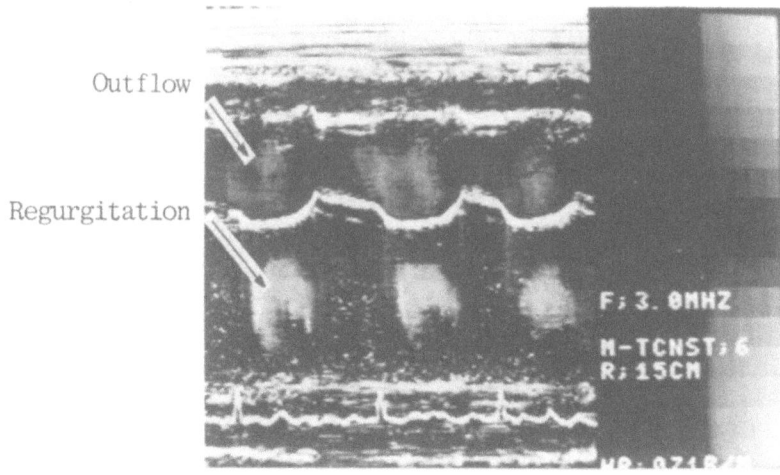

Fig. 9 M-mode Doppler display of the same
        patient as in Fig. 8.

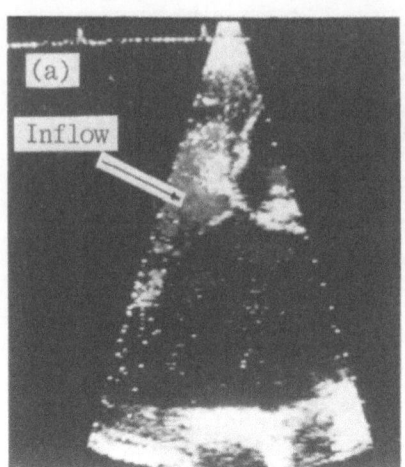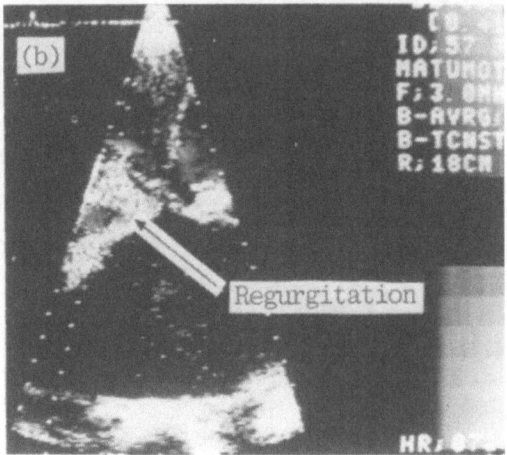

Fig. 10   2-D blood-flow images in the heart with a tricuspid regurgitation disease.

(a)   In diastole: Inflow from the right atrium to
ventricle is shown.

(b)   In systole :   Highly turbulent regurgitation
is shown in mixed colors.

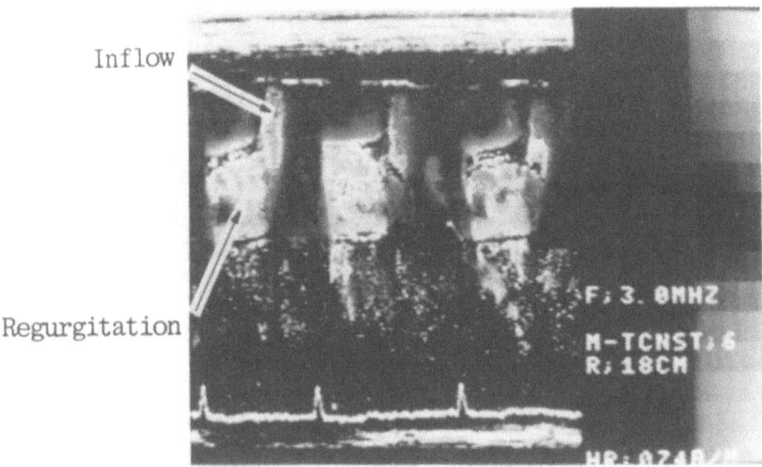

Fig. 11   M-mode Doppler display of the same
patient as in Fig. 10.

CONCLUSION

The principle and operation of equipment that is capable of observing in real-time the two-dimensional blood flow in heart and blood vessels using auto-correlation processing of Doppler signals, have been described.

Experiments were conducted by the use of phantoms, and the auto-correlator of the invention has been verified as operating correctly and properly.

The equipment has been applied to patients in a hospital suffering from heart diseases, and the two-dimensional mapping of blood flows in the heart has been performed. The results have verified the feasibility of the equipment in a clinical field as valvular diseases, septum defects and other diseases are diagnosed quite easily and non-invasively from the real-time two-dimensional information of blood flow.

APPENDIX

Defining

$$R(\tau) = |R(\tau)| e^{j\phi(\tau)} = A(\tau)e^{j\phi(\tau)} \tag{A1}$$

and considering that $A(\tau)$ is an even function and $e^{j\phi(\tau)}$ an odd function, we find:

$$\dot{R}(\tau) = (\dot{A}(\tau) + jA(\tau)\ \dot{\phi}(\tau))e^{j\phi} \tag{A2}$$

$$\therefore\ \dot{R}(0) = jA(0)\ \dot{\phi}(0) \tag{A3}$$

$$R(0) = A(0) \tag{A4}$$

From Eqs. (4), (A3) and (A4), we find:

$$\overline{\omega} = \dot{\phi}(0) \cong \phi(T)/T \tag{A5}$$

where, T is the emission interval of ultrasonic pulses, during which we assume approximately constant velocity of movements of the reflective bodies. By differentiating Eq. (A2) with τ, and inserting $\tau = 0$, we obtain:

$$\ddot{R}(0) = -(\dot{\phi}(0))^2 A(0) + \ddot{A}(0) \tag{A6}$$

From Eqs. (5), (A3), (A4) and (A6), we know:

$$\sigma^2 = -\ddot{A}(0)/A(0) \tag{A7}$$

By spreading $A(\tau)$ out in a series relative to τ, and considering that $A(\tau)$ is an even function, we find:

$$A(\tau) = A(0) + \frac{\tau^2}{2} \ddot{A}(0) + \cdots\cdots \qquad (A8)$$

Neglecting the third and later terms of Eq. (A8) as small enough in value, and using Eqs. (A6) and (A7) with insertion of $\tau = T$, we obtain:

$$\sigma^2 \cong \frac{2}{T^2} \{ 1 - \frac{A(\tau)}{A(0)} \} = \frac{2}{T^2} \{ 1 - \frac{|R(T)|}{R(0)} \} \qquad (A9)$$

REFERENCES

[1]   D. W. Baker, "Pulsed Doppler Blood Flow Sensing", IEEE Trans. SU-17, No. 3, 170-185 (1970).

[2]   P. J. Fish, "Multichannel Direction Resolving Doppler Angio-graphy", Abstract of 2nd European Cong. on Ultras. in Med., 72 (1975).

[3]   M. Brandestini, "Topflow—A Digital Full Range Doppler Veloci-ty Meter", IEEE Trans. SU-25, No. 5, 287-293 (1978).

[4]   P. A. Grandchamp, "A Novel Pulsed Directional Doppler Veloci-meter, the Phase-detection Profilometer", Proc. of 2nd Europ. Cong. on Ultras. in Med., 122-132 (1975).

[5]   M. Brandestini, "Application of the Phase Detection Principle in Transcutaneous Velocity Profile Meter", ibid 144 (1975).

[6]   A. Nowicki, J. M. Reid, "An Infinite Gate Pulsed Doppler", Ultras. in Med. & Biol. 7, 41-50 (1981).

[7]   G. B. Curry, D. N. White, "Color Coded Ultrasonic Differential Velocity Arterial Scanner (Echo Flow)", Ultras. in Med. & Biol., 4, 27 (1978).

[8]   B. A. Coghlan, M. G. Taylor, "A Carotid Imaging System Utiliz-ing Continuous-Wave Doppler-Shift Ultrasound and Real-Time Spectrum Analysis", Med. & Biol. Eng. & Compt., 16, 739-744 (1978).

[9]   M. Asao et al., "Flow Mapping of Intracardiac Blood Flow by Computer-Based Ultrasonic Multigated Pulsed Doppler Flow-meter", Proc. of 20th Conf., Japan Soc. of Med. Elect. & Biol. Eng. (in Japanese), 44 (1981).

[10] K. Namekawa, C. Kasai et al., "Imaging of Blood Flow Using Auto-Correlation", Ultras. in Med. & Biol., Vol. 8, Supple-ment 1, 413 (1982).

IMAGING OF THE ULTRASONIC VELOCITY AND REFLECTION PROPERTIES OF OBJECTS BY A C-SCAN METHOD

K. Tervola,[+] J. Koivukangas, E. Alasaarela, and
J. Ylitalo
University of Oulu, SF 90570, Oulu 57, Finland
[+]Present address: Bioacoustics Research Laboratory
University of Illinois, Urbana, IL 61801

A method and imaging system for the measurement of the ultrasonic velocity and reflection properties of objects are described. Some preliminary images of test cylinders and brain specimen are shown and the various imaging difficulties are discussed. The measured velocity values are in agreement with those found in the literature and in the images a brain tumor is to be seen.

The method is similar to the C-scan method. In every imaging point an ultrasonic pulse of 4 MHz is sent to the object and the reflected echoes are digitized and stored in the memory of a computer. The echoes, which are reflected from the surface and bottom of the object, are sought. The velocity is calculated from the data of the imaging geometry and from the time which elapsed when the ultrasound traverses from surface to bottom and back again. Also, two normal C-scan images from given depths and their difference are calculated. The images are shown on a monitor in 16 grey levels and are recorded by a videorecorder.

INTRODUCTION

Ultrasonic imaging is used as a diagnostic tool in medicine because it is a noninvasive and harmless method. Images are normally made from the gradient of the acoustic impedance by the B-scan method. But the acoustic impedance is only one property of the object. Therefore, in the study of tissue characterization many research groups are examining other acoustical properties of tissues and the various mechanisms by which ultrasound interacts with tissues. The results can be used to compensate imaging errors and for pattern recognition.

The method and imaging system of the ultrasonic velocity and reflection properties of objects will be described. Some preliminary images of physical objects and brain specimen will be shown and the difficulties of such imaging will also be discussed.

2. THE METHOD AND IMAGING SYSTEM

The structure of the imaging system and the flowcharts of the process are shown in Fig. 2.1 - 2.4. First, the imaging parameters are read, see Fig. 2.5 and 2.6. The parameters are listed in Table 2.1. For every imaging point a burst is sent to the object and the reflected echo is digitized after a given time

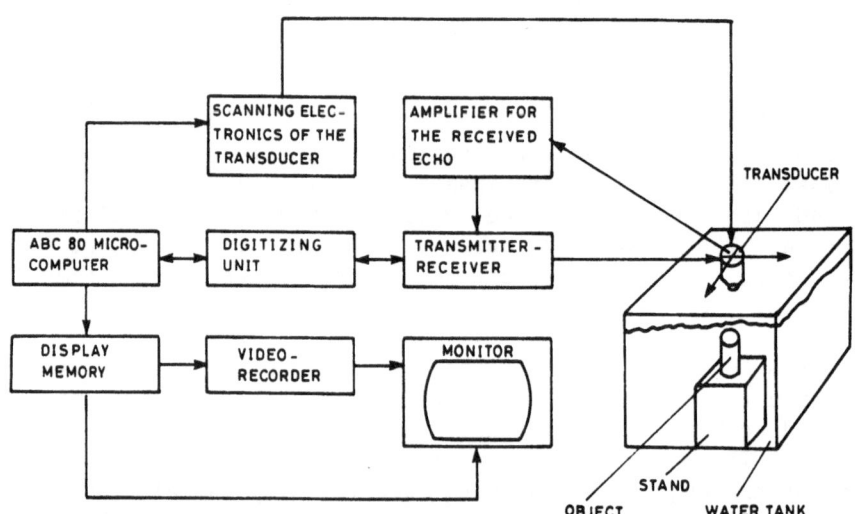

Figure 2.1.   The structure of the imaging system.

delay with a sampling frequency of 16 MHz and 8 bit accuracy to 1024 samples [1-3]. The ultrasonic frequency used is 4 MHz, so four samples are taken per wavelength. From the samples are calculated 256 intensity values, which are stored in an array. The method for calculating the velocity is illustrated in Fig. 2.6 and the flowchart of the velocity subroutine is shown in Fig. 2.3. First, the positions of the surface and bottom are determined from the intensity values.

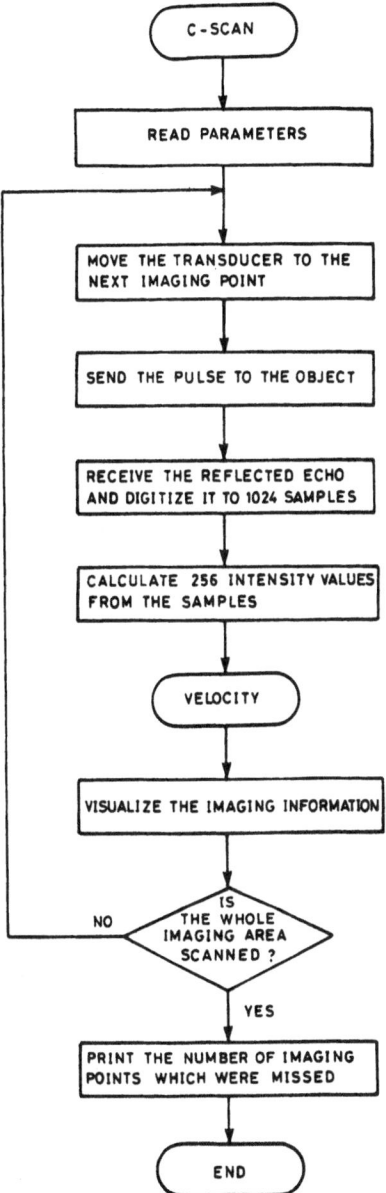

Figure 2.2. The flowchart of the imaging.

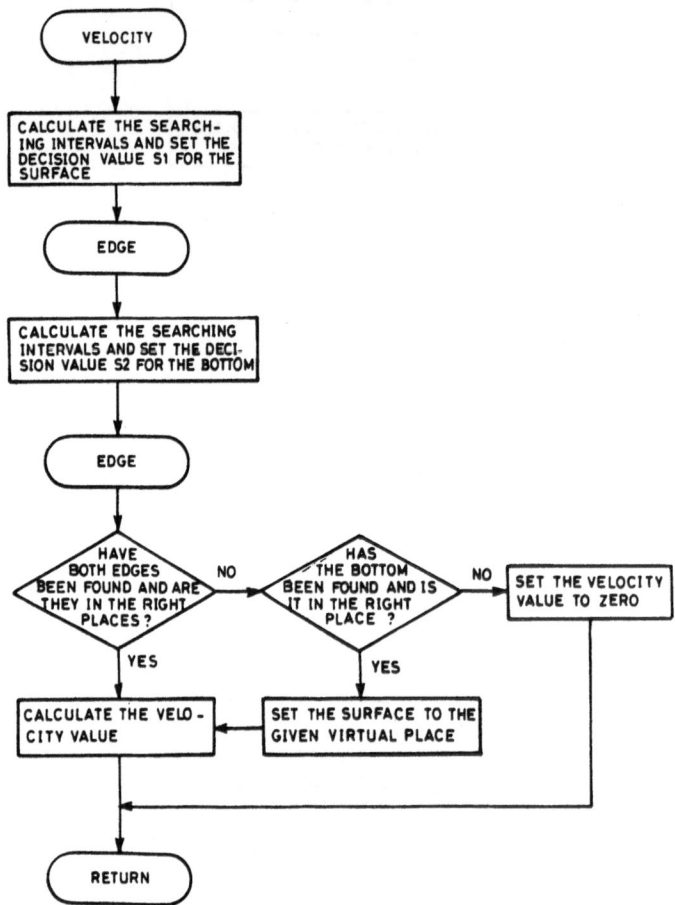

Figure 2.3   The flowchart of the velocity subroutine.

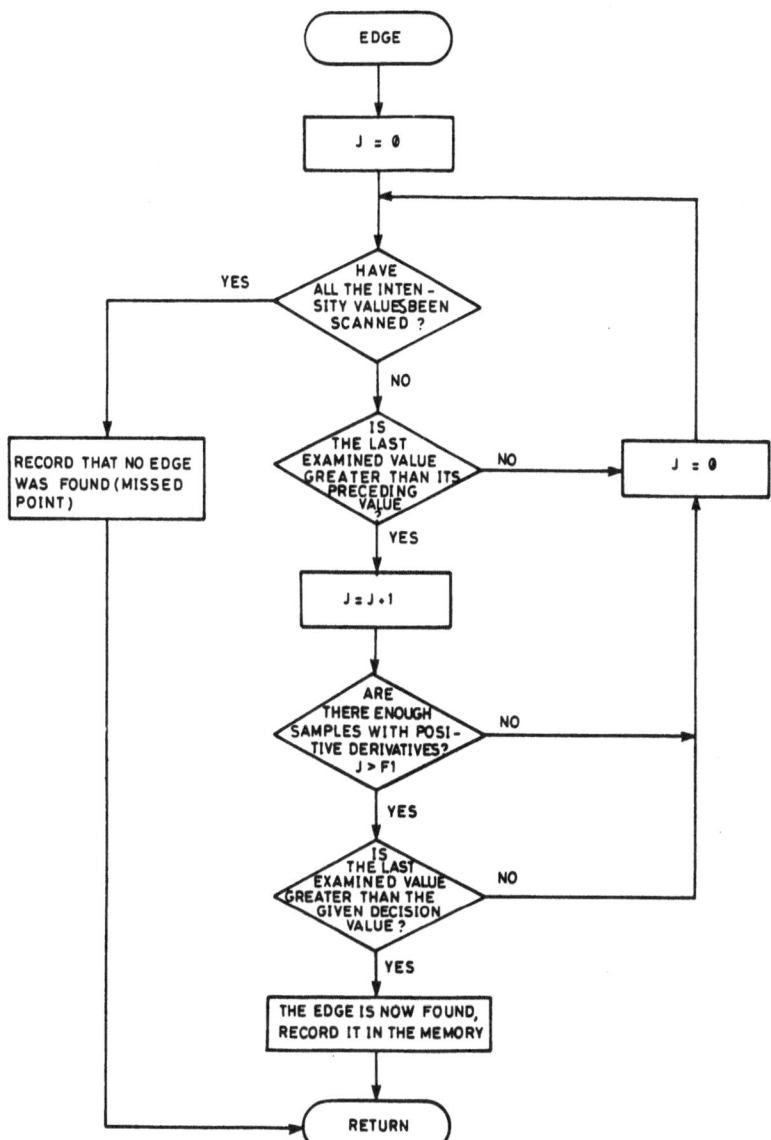

Figure 2.4.   The flowchart of the edge subroutine.

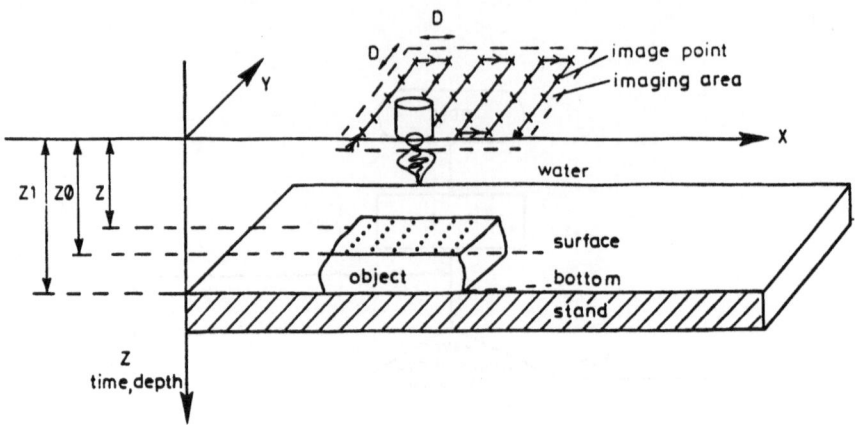

Figure 2.5.   The imaging geometry.

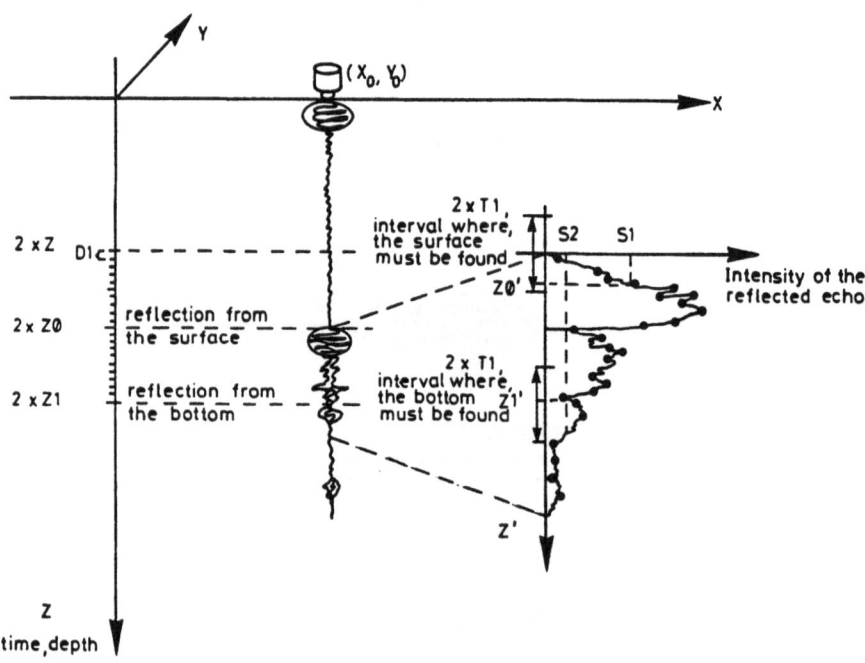

Figure 2.6.   An illustration of the method for calculating the
              velocity value for every imaging point.

Table 2.1.   The Imaging Parameters

M1	is the number of sampling points in the Y-direction
M2	is the number of sampling points in the X-direction
D	is the distance between sampling points in units of 0.094 millimeters in the X-Y plane
D1	is the distance between sampling points in water in the z-direction
Z1	is the distance between the transducer and the stand
Z0	is the distance between the transducer and the surface of the object
Z	is the depth, which corresponds to the given time gate value
S1	is the decision value, when searching for the position of the surface Z0 from the echo
S2	is the decision value, when searching for the position of the bottom Z1´ from the echo
F1	is the number of intensity values with positive derivate which must be found when searching for the surface or bottom
2 x T1	is interval, where the surface and bottom must be found
2 x V5	is the examined number of intensity samples, which are examined when searching for the surface or the bottom
E5	is the virtual position of the surface, when no real surface is found
M9, C8	are ordinals of the intensity values, which are visualized as normal C-scan images
C1, A7	are the minimum velocity value, Vmin and the dynamic range v, see Fig. 2.7.

The flowchart of the edge subroutine is shown in Fig.   2.4.
An edge is recorded when enough intensity values with positive
derivative have been found and the last examined intensity value
is greater than the given decision value. Then, the positions of
the edges are checked in terms of the given parameters of the
imaging geometry.   If the surface has not been found, then the
virtual value is used instead of the real one.   This is because
some parts of the imaging area contain only water and then no
surface is found. The velocity value V is calculated as follows:

$$V = 2(Z1 - Z0')/N \cdot \Delta T, \text{ where} \tag{2.1}$$

Z1   is the distance between the transducer and the stand
Z0'   is the distance between the transducer and the surface
N   is the number of intensity values which are taken
      between the surface and bottom
ΔT is the time interval elapsed between the sampled
      intensity values, 250 ns

The velocity value is coded to 16 gray levels using the curve
shown in Fig. 2.7. If no bottom has been found, the velocity
value is set to zero. When the measurement has run to its end,
the number of missed points is printed out.

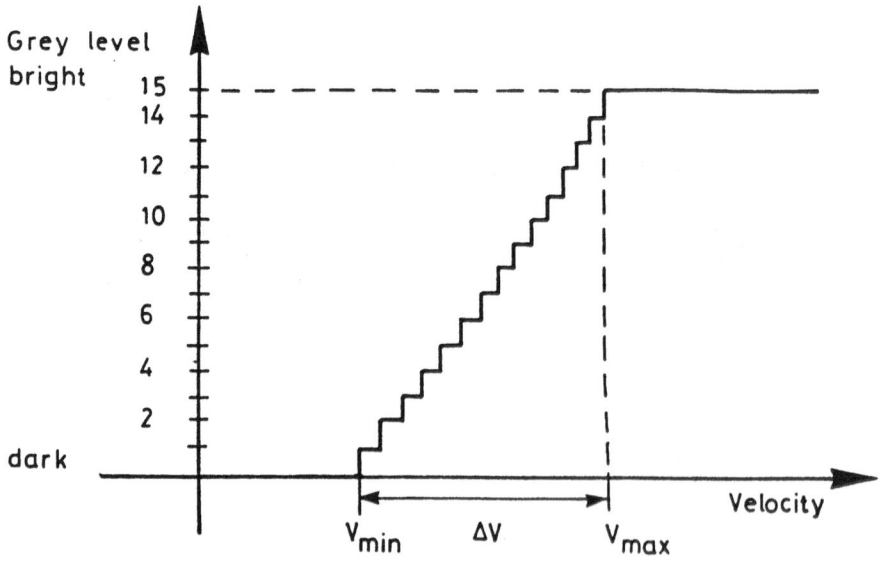

Figure 2.7.   The curve used for coding the velocity values to 16
             grey levels.

In addition, the intensities of two echoes from given depths and their difference are visualized. This gives information about the reflection and absorption properties of the object.

The objects of the present study were four cylinders and formalin-fixed human brain specimen. The cylinders, shown in Fig. 2.8, were 52 mm. long made of aluminum, steel, brass and plastic. The brain specimen are illustrated in Fig. 2.9 and 2.10.

Before imaging, the objects were immersed in the water tank for many hours. The temperature of the water was 20°C. 10% formalin solution had been used for fixing the brain specimen. The brain specimen was placed on a plastic stand and the test cylinders on a styrofoam plate. The imaging time was 16 hours.

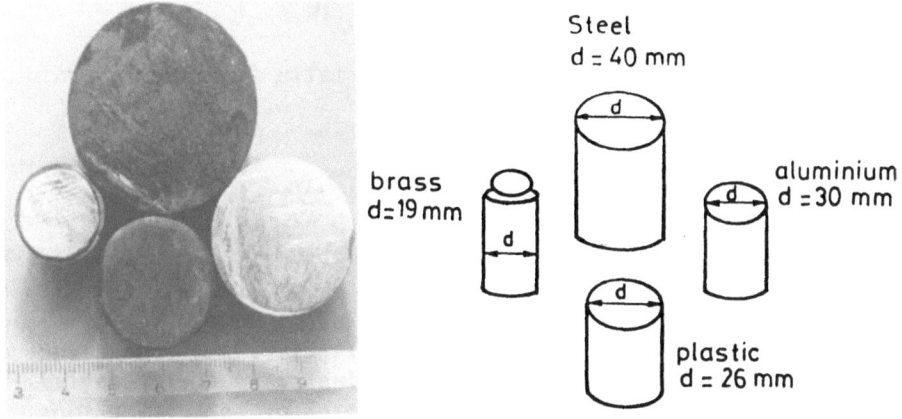

Figure 2.8.   The test cylinders.

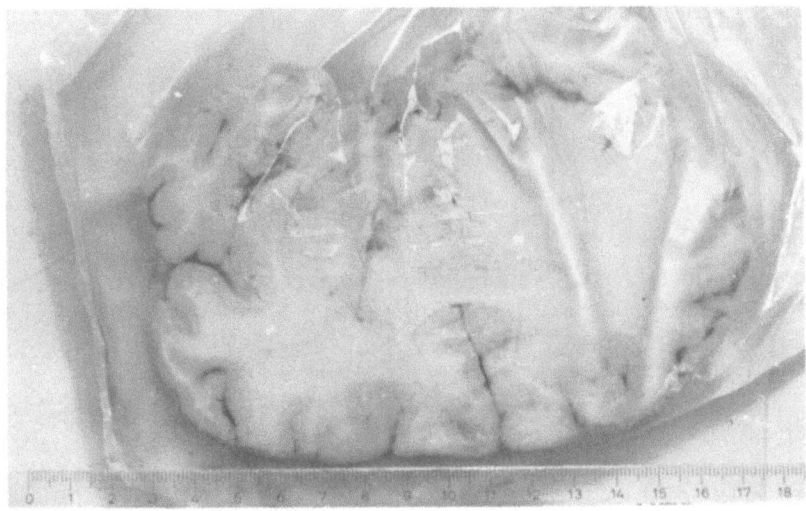

Figure 2.9.   The formalin fixed human brain specimen.

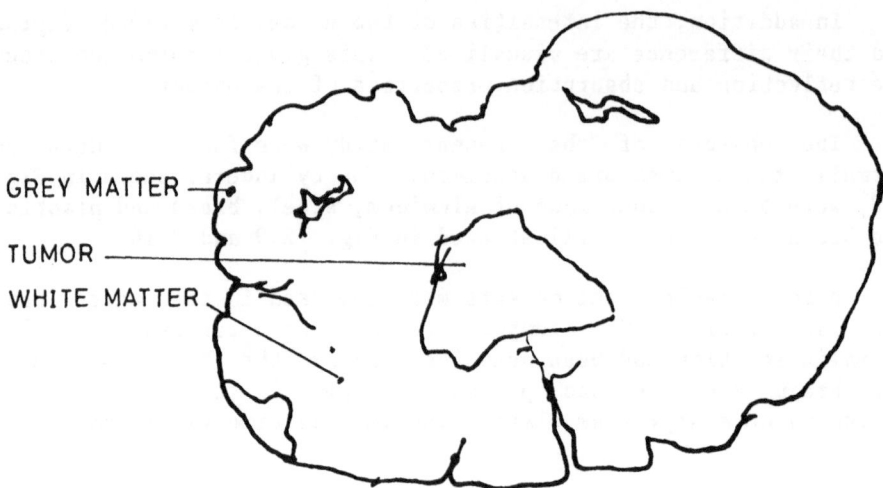

GREY MATTER

TUMOR

WHITE MATTER

Figure 2.10.   The structure of the formalin fixed human brain
             specimen.

3.   RESULTS

    Images of the test cylinders are shown in Fig.   3.1.   The
imaging parameters are listed in Table 3.1.   The average velocity
values for aluminum, steel, brass and plastic were 6.0,   5.4,   3.0
and 2.9 kilometers per second, respectively.   The C-scan image on
the right side of the velocity image illustrates the intensity of
the   surface   reflection.   The intensities of the reflections, in
rising order, were those of steel, brass,   aluminum and plastic.
Because the velocity of ultrasound in the different cylinders
varied so widely, no C-scan images of the deeper echoes were
visualized.

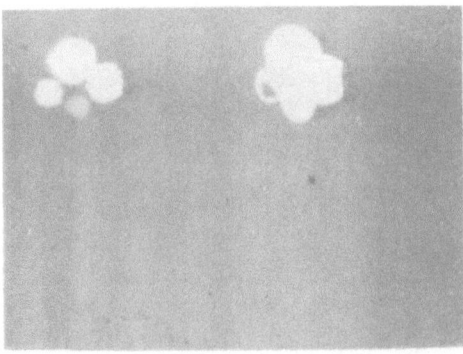

Figure 3.1.   An image of the test cylinders shown in Fig. 2.8. The
             imaging parameters are listed in Table 3.1.

Table 3.1. Imaging Parameters

M1		64	128
M2		64	128
D	mm	1.6	0.4
D1	mm	0.4	0.4
Z1	mm	92	68.0
Z0	mm	37	52.0
Z	mm	32	44.0
S1		80	12
S2		80	80
F1		2	2
2 x T1	mm	10	10
2 x V5		44	24
E5	mm	37	52
M9		30	40
C8		–	50
C1	km/s	2.000	1450
A7	km/s	4.800	0.2
Figure		3.1	3.2

Figure 3.2 shows an image of the brain specimen (Fig.   2.9).
The imaging parameters are listed in Table 3.1. It was observed
that the ultrasonic velocity in the white matter is greater than
that of the grey matter and that the value of the tumor was
intermediate. The C-scan images in the upper and lower right
corners were taken from different depths and the imaging planes
were 2 mm from each other. The tumor is shown as a "butterfly" in
the images. The difference of these two images was also a
"butterfly" pattern, which means that the specimen was nearly
isotropic in the Z-direction, which satisfies the condition for
meaningful velocity distribution imaging.

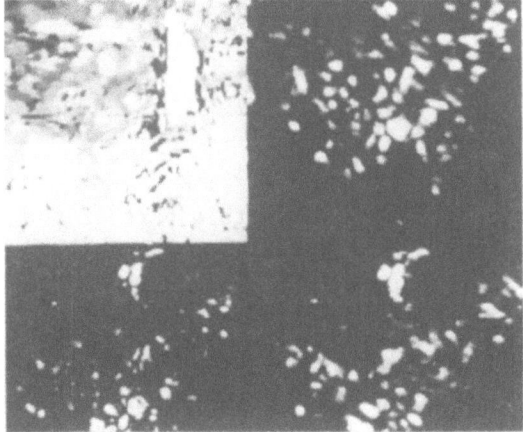

Figure 3.2.   An image of the brain specimen shown in Fig. 2.9.
The imaging parameters are listed in Table 3.1.

4. DISCUSSION

Figure 4.1 illustrates the generation of surface and bottom echoes when imaging the ultrasonic velocities in the object. τ_1, τ_2, α_1, and α_2 are reflection and transmission coefficients. Z1, Z2 and Z3 are the acoustical impedances of water, object and stand, respectively. First, the transmitted pulse reaches the surface of the object and one portion of the ultrasonic pressure amplitude $P_{surface}$, is reflected back to the transducer.

$$P_{surface} = \tau_1 \cdot P \text{ , where} \tag{4.1}$$

$$\tau_1 = (Z2 - Z1)/(Z1 + Z2) \quad \text{and} \tag{4.2}$$

P is the pressure amplitude of the transmitted pulse.

Then, the pulse passes through the object and is reflected from the bottom back towards the transducer, but at the surface a third reflection occurs and only part of the upulse passes through the surface to the transducer. The pressure amplitude reflected from the bottom, P_{bottom}, can be determined as follows, if we assume the system to be lossless.

$$P_{bottom} = (\alpha_1 \cdot P) \cdot \tau_2 \cdot \alpha_1 \tag{4.3}$$

$$= \alpha_1^2 \cdot \tau_2 \cdot P \text{ ,} \tag{4.4}$$

where

$$\alpha_1 = (2Z2)/(Z1 + Z2) \quad \text{and} \tag{4.5}$$

$$\tau_2 = (Z3 - Z2)/(Z3 + Z2) \tag{4.6}$$

There are several difficulties in this part of the imaging. The object surface must be smooth. The transmitted pulse must come perpendicularly to the surface and bottom. Ray bending and scattering cause various propagation paths. The object must be isotropic in the z-direction. The size of the focal point is finite. Finally, there may be difficulties in finding the surface and bottom echoes from the noisy signal. The given results, however, are in agreement with those found in the literature [4-6].

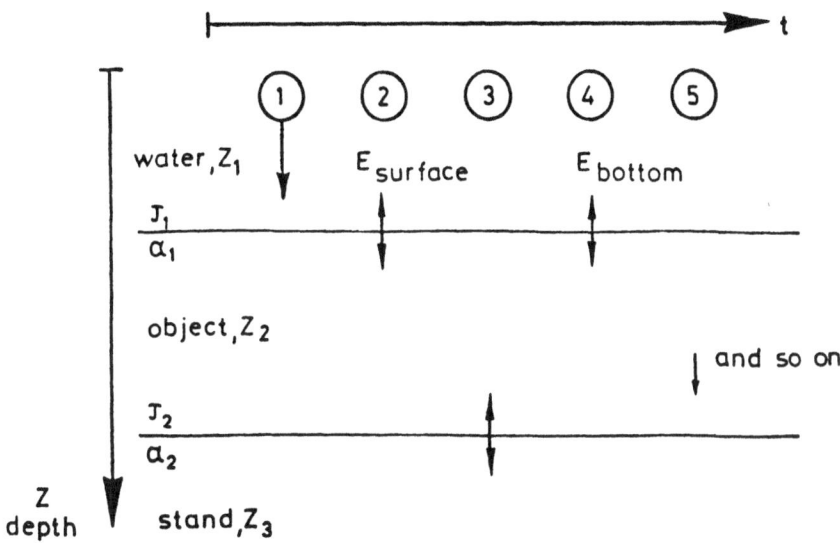

Figure 4.1 The generation of the surface and bottom echoes

The accuracy of the imaging system is strongly related to the thickness of the object. The sampling and ultrasonic frequencies were 16 MHz and 4 MHz, respectively. Thus, in water one sample is taken after every 0.375 mm because one intensity sample is calculated from 4 samples. Figure 4.2 shows a graph where the number of the samples is plotted as the function of the thickness of the object. The ultrasonic velocity in the object is considered to be the same as in water.

Figure 4.3 shows a graph where the difference in the number of samples is plotted as a function of the velocity difference. The thickness of this object is 10 mm and the reference velocity is that of water. If the ultrasonic velocity in the object is twice that in water, then the numbers of samples are 50 and 25, respectively. If the difference in ultrasonic velocity values is 20%, then N is only 8.

Another strategy was also used to search for the bottom echo. A part of the tissue specimen was placed on a water window in the stand and the bottom was considered to be found when there were no more echoes in the reflected signal. This method gave faulty results because the bottom of the tissue specimen was physically nonplanar.

This algorithm and the one described previously take into account the curvature of the surface of the specimen.

The imaging system will be provided with a PDP 11 computer. Then, the imaging time will be a fraction of that now needed.

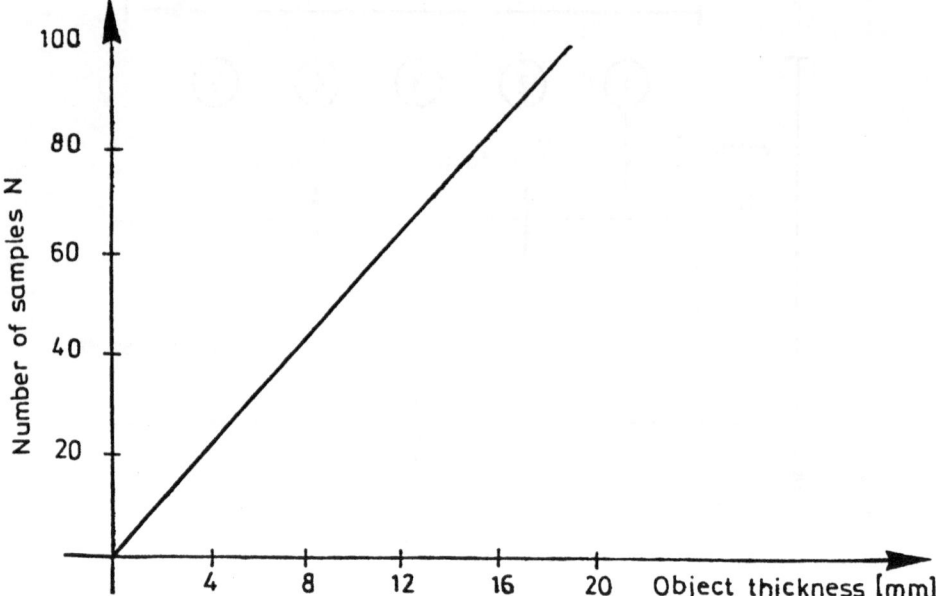

Figure 4.2.   The number of samples as a function of the depth of the
               object. The ultrasonic velocity used is that of water.

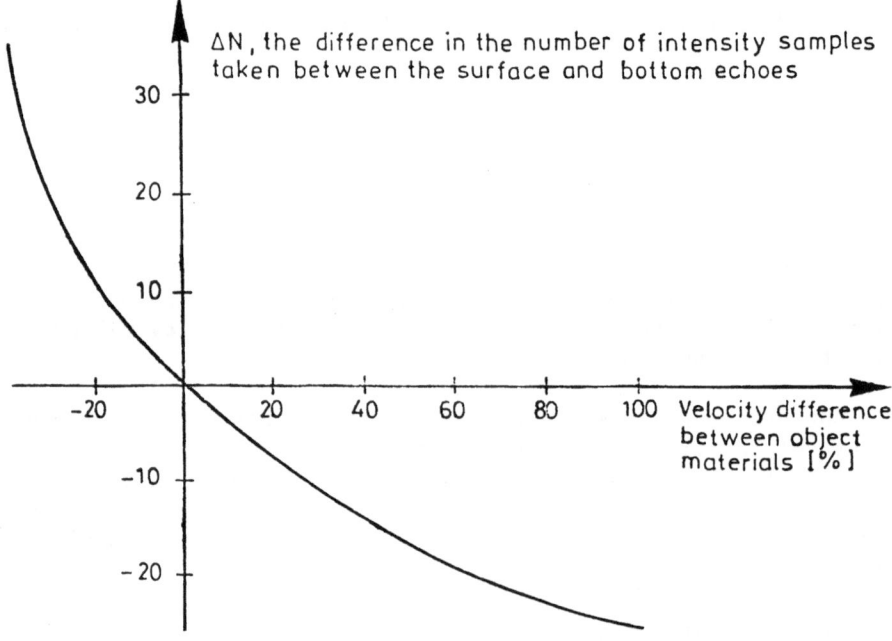

Figure 4.3.   The difference in the number of intensity samples
               taken between the surface and bottom echoes as a
               function of the velocity difference betwee object
               materials.   The thickness of the object is 10 mm.

CONCLUSIONS

A method and imaging system for the determination of the ultrasonic velocity and reflection properties of objects is described. The preliminary images confirm the functioning of the system, and the obtained ultrasonic velocity values are in agreement with those of the literature. The difficulties of the imaging are discussed in terms of improvements for better results.

ACKNOWLEDGEMENTS

We thank Professors Stig Nyström and Antti Tauriainen for their continuous support. We thank also engineer Päivikki Honkanen, who always helped us when there was something wrong in the imaging system.

The financial support of the Academy of Finland, OMP-Yhtyma Oy, Orion-Yhtymä Oy, Finnish Culture Fund, Emil Aaltosen Saatio, Tauno Tönningin Säätiö and Oulun yliopiston Tukisäätiö are gratefully acknowledged.

REFERENCES

1. J. Ylitalo, A measurement and communication system for ultrasonic holographic B-scan imaging apparatus. Diploma thesis. Oulu, 1981 University of Oulu, Department of Electrical Engineering. (In Finnish).

2. Alasaarela, E. Tranceiver unit for a measurement system of ultrasonic holograms. Licentiate thesis. University of Oulu, Faculty of Technology, Oulu 1977 (In Finnish).

3. M. Rikola, A high speed data acquisition unit for holographic B-scan imaging. Diploma thesis. Oulu, 1981. University of Oulu, Department of Electrical Engineering. (In Swedish).

4. Blitz, J., Ultrasonics: Methods and Applications. Butterworths, London, 1971.

5. Handbook of Chemistry and Physics, 48 Edition, the Chemical Rubber Co., Cleveland, 1967.

6. W. D. O'Brien, Jr. The role of collagen in determining ultrasonic propagation properties in tissue. Acoustical Holography, Vol. 7, pp. 37 -50, 1977.

IMPROVEMENT AND QUANTITATIVE ASSESSMENT OF B-MODE IMAGES

PRODUCED BY ANNULAR ARRAY/CONE HYBRIDS

M. S. Patterson and F. S. Foster

Ontario Cancer Institute
500 Sherbourne St.
Toronto, Canada

ABSTRACT

We have developed a hybrid ultrasound imaging system consisting of a spherically focused annular array transmitter and a conical receiver. The high resolution (0.3 mm at 4 MHz) of this system has provided superior images of the female breast. Recently, we have applied two signal processing techniques that offer further improvement in image quality. In the first, phase insensitive sector addition (PISA), the cone is divided into eight sectors and images are obtained with each sector. The final image, made by adding these independent images, has much smoother speckle at the expense of some loss of resolution. In the second method, multiplicative processing (MP), the B-scan is formed from the produce of the eight rf signals. This technique, which also provides smoother speckle with some loss of resolution, reduces the sensitivity of the hybrid to scatterers more than one millimetre off axis and thus suppresses some artifacts characteristic of axicon focusing. To quantitatively assess these signal processing methods, and indeed any B-mode imaging system, we have devised a measure that we call the contrast to speckle ratio (CSR). Images are made of a scattering phantom containing anechoic cylindrical holes. The CSR is a measure of the contrast of these objects in the image relative to the contrast fluctuations caused by speckle. Plots of CSR vs. hole size show the superiority of PISA and MP over conventional coherent processing. This result correlates with subjective impressions of images of the test object and freshly excised breast tissue. It appears that some of the resolution of the hybrid system can be "traded off" for smoother and less distracting speckle to achieve even better breast images.

IMAGE INTERPOLATION FOR REAL-TIME COMPOUND DUAL SECTOR SCANS

Etsuro Machida, Junji Miyazaki*, Norimasa Sugaya*,
Keiichi Murakami, Takaki Shimura and Hajime Hayashi

Fujitsu Laboratories Ltd., *Fujitsu Limited
Kamikodanaka, Kawasaki 211, Japan

1. INTRODUCTION

Many technical advantages and clinically valuable evaluations have been reported since simultaneous multifrequency ultrasonography was introduced[1-2]. This newly devised technique enables dual ultrasonic images to be simultaneously obtained. One application of this technique--real-time compound dual sector scanning--offers more valuable information than conventional ultrasonography.

Two major problems must be solved, however, before digital scan converters (DSCs) can be introduced in real-time compound dual sector scanning. First, static images remaining each time the relative position of the two probes is changed must be eliminated. Second, Moire patterns must be removed because the number of display pixels is restricted and straight scan lines become jagged. To meet the first requirement, we developed a new removal algorithm.[3] This paper discusses how we met the second requirement.

Real-time compound dual sector scanning is reviewed for better understanding of the following sections, DSC concepts and specification requirements are introduced, and two methods for eliminating Moire patterns are proposed and evaluated.

2. REAL-TIME COMPOUND DUAL SECTOR SCANNING

2.1 Concepts

Although current ultrasonic diagnostic equipment provides

good-quality images, defects remain. The most common are coarse
scan-line density and the lack of echoes at certain incident
angles.

Conventional contact compound scanning solves these problems
but cannot be applied in real time to imaging a moving heart, for
example.

Completely simultaneous multiple functions such as imaging
both axial views and simultaneous real-time combinations of
Doppler and B mode have yet to be achieved. The only approach
devised for multiple functions--i.e., time sharing--reduces
scan-line density, frame rate, or maximum Doppler-detectable
velocity.

To eliminate these problems, we propose that multiple
frequency ultrasonic pulses be transmitted and received,
discriminating among pulses with filters[4]. We call this
approach simultaneous multifrequency ultrasonography. Real-time
compound dual sector scanning is one application of this technique.

2.2   Real-time Compound Dual Sector Scanning

Two ultrasonic probes with different frequencies are
mechanically linked by two arms and three nodes. The relative
position of the two probes is calculated by a microprocessor using
angle data provided by the three nodes (Figure 1). Each echo
signal is discriminated using a filter which reduces cross talk.

The filtered signal is written into frame memory. When data
is written, the memory locations of the two sectors are determined
using the angle data previously obtained. Frame memory is read in
a TV-raster direction and superimposed on a single NTSC TV signal.

Real-time compound dual sector scanning has the following
advantages:

1.   A wider view is obtained, even in sector apexes.
2.   A perfect image is obtained by restoring echo shadows.

A wider area covering, for example, a whole human heart, can
be obtained by overlapping sector images (Figure 2-a). Moreover,
dropout or echo shadows caused when a single probe fails to reveal
organ boundaries lying parallel to ultrasonic beams can be
restored by using a second probe from a different direction
(Figure 2-b).

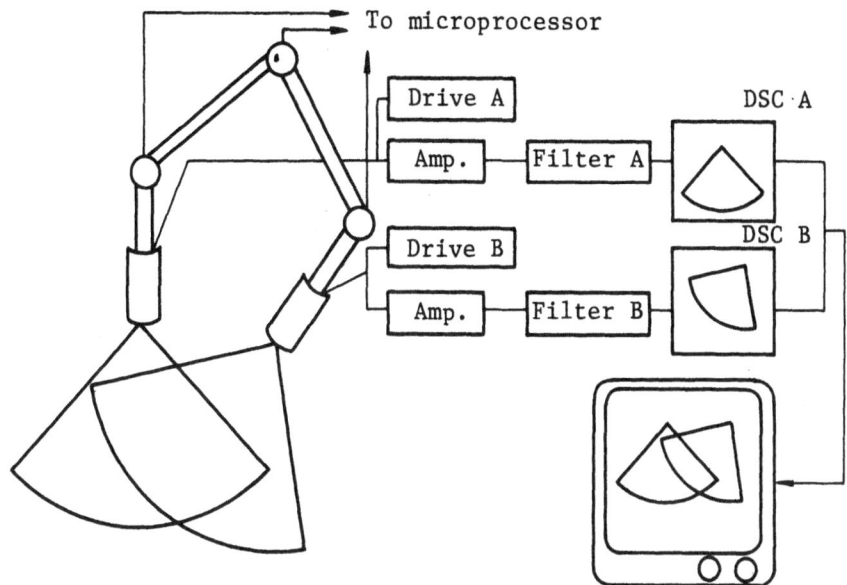

Figure 1   Real-time compound dual sector scanning

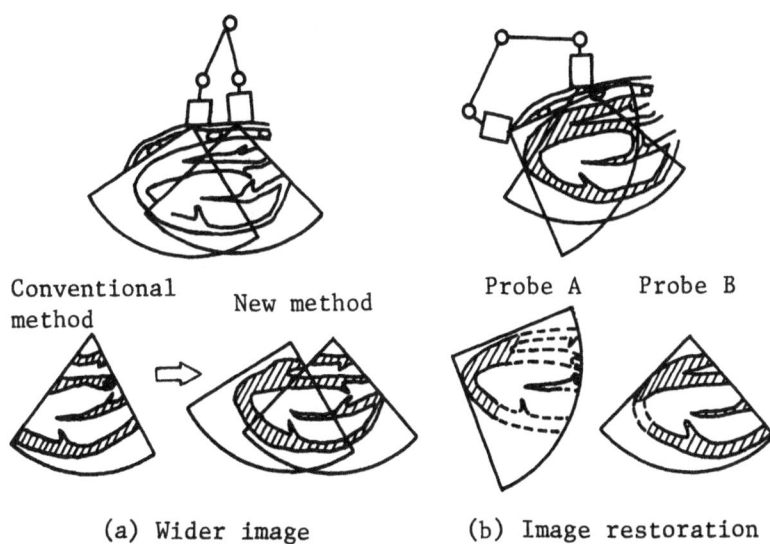

(a) Wider image   (b) Image restoration

Figure 2   Advantages of real-time compound dual sector
           scanning

3.   DIGITAL SCAN CONVERTERS

Received data is written into frame memory along with ultrasonic beam scan lines, then read out in a TV-raster direction by the DSC. Writing and reading are generally time-shared. Write addresses for received data are generated by the "nearest-neighbor" method[5] (Figure 3).

Assuming that the scan line is represented in line P_0R_0, points Q_1, Q_2, ..., Q_n are sampling points for received data. Q_n points must be mapped to pixels in frame memory. Candidates for point Q_1 are $P_0(x_0,y_0)$, $p_1(x_0+1,y_0)$, $P_2(x_0+1,y_0+1)$ $p_3(x_0,y_0+1)$. Their discriminants are

$$dx_0 \leq 1/2 \text{ and } dy_0 \leq 1/2 \longrightarrow \text{ mapped pixel } = P_0$$
$$dx_0 > 1/2 \text{ and } dy_0 < 1/2 \longrightarrow \text{ mapped pixel } = p_1 \qquad (1)$$
$$dx_0 > 1/2 \text{ and } dy_0 > 1/2 \longrightarrow \text{ mapped pixel } = p_2$$
$$dx_0 \leq 1/2 \text{ and } dy_0 > 1/2 \longrightarrow \text{ mapped pixel } = p_3$$

where

$$dx_n = Q_n - [Q_n]|x$$
$$dy_n = Q_n - [Q_n]|y$$

$[Q_n]$ is Q_n's maximum integer value, not more than Q_n.

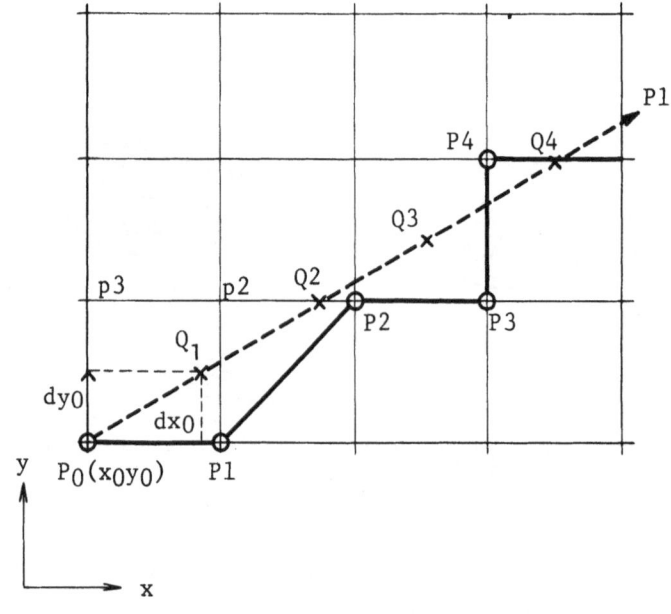

Figure 3   Nearest-neighbor method
- - - - - -   Original scan line
_____   Mapped line
       x   Sampling point

When this method is introduced into a sector scanning
ultrasonic diagnostic system, Moire patterns occur because the
angle between adjacent scan lines is small and constant (Photo 1).

The easiest way to reduce Moire occurrence is to increase the
number of pixels in frame memory because the mapping error (dx,dy)
becomes smaller. However, this makes the TV band width wider, and
recording cannot be done directly to VTRs without degrading image
quality.   This necessitates a Moire-removal technique which does
not increase the number of pixels.

If a sector is displayed in a fixed position, the proper
Moire removal is realized by interpolating in a TV-raster
direction when reading from frame memory. This can be done by
adding a small amount of hardware to read circuits.

Applying this to real-time compound dual sector scanning
often degrades images in a range direction (Figure 4), because the
sector display position moves in accord with the relative position
of the two probes. Images completely free of Moire patterns must
be obtainable in whatever position sectors are displayed.

To summarize, DSCs must have three capabilities:

1.   All Moire patterns must be completely removed in real time in
     whatever position sectors are displayed.
2.   Image degradation must not occur.
3.   The number of pixels in frame memory must not be increased.

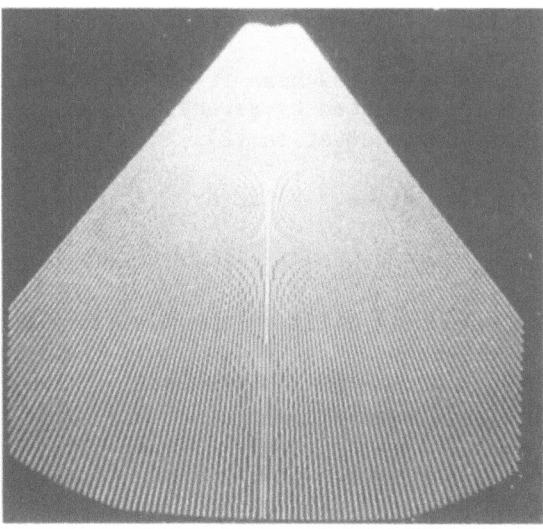

Photo 1 Nearest neighbor method

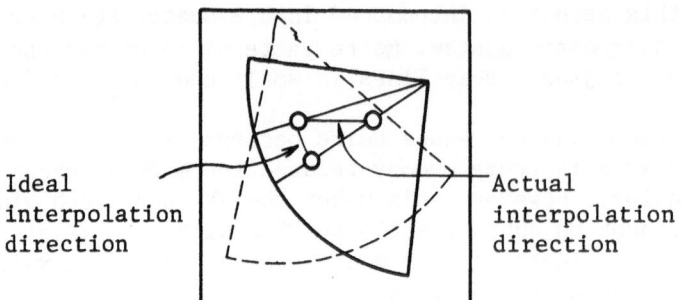

Figure 4   Interpolation for tilted sectors

4.   MOIRE REMOVAL

We developed two types of Moire removal. One is to leave
original scan lines using what we called the "smoothed-line
method", and the other is to fill in the whole area within a
sector in what we called the "fill method". The smoothed-line
method is presented in Section 4.1. The fill method is introduced
in Section 4.2. The two methods are compared in terms of image
quality and hardware implementation in Section 4.3.

4.1 Smoothed-line Method

Many methods described for smoothing jagged or staircased
lines[6-9] cannot be applied to real-time operation because of the
excessive amount of computation they require. We developed a
simplified method and confirmed its effectiveness.

Lines are smoothed by low-pass filtering, in which one item
of sampled data is represented by several pixels. Data for each
pixel is calculated using equation (2).

$$Y(i,j) = \iint X(x,y) \, H(i-x,j-y) \, dx \, dy \qquad\qquad (2)$$

where

$X(x,y)$:   Sampled data at $(x,y) = (u \Delta r \cos(v \Delta \theta), u \Delta r \sin(\Delta \theta))$
u: Sampling number on a scan line (integer)
v: Scan line number (integer)
Δr :   Sampling interval
$\Delta \theta$:   Angle between adjacent scan lines
$H(x,y)$:   Two-dimensional smoothing filter
$Y(i,j)$:   Pixel data at (i,j)

Equation (2) shows one-pixel data is calculated by using less than 4 items of sampled data if a conical smoothing filter (Figure 5) is used. Because this equation is difficult to calculate in real time, however, we simplified equation (2) to make equation (3):

$$Y(i,j) = X(x,y) \; H(i-x \;, j-y) \qquad\qquad (3)$$

Equation (3) produces 1-to-4 mapping (Figure 6). Because the same pixel is often assigned different data, we used the simplest method of giving priority to newly written data by overwriting.

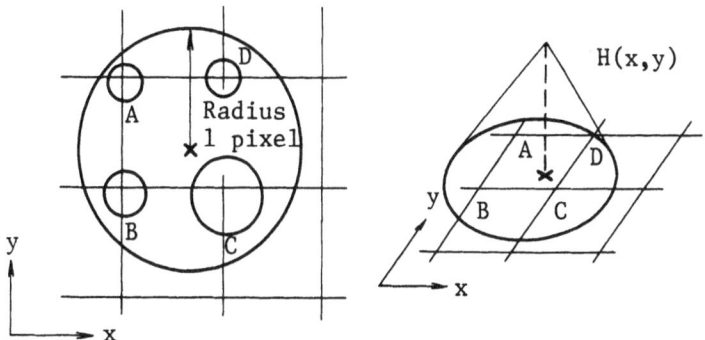

Figure 5   Conical smoothing filter

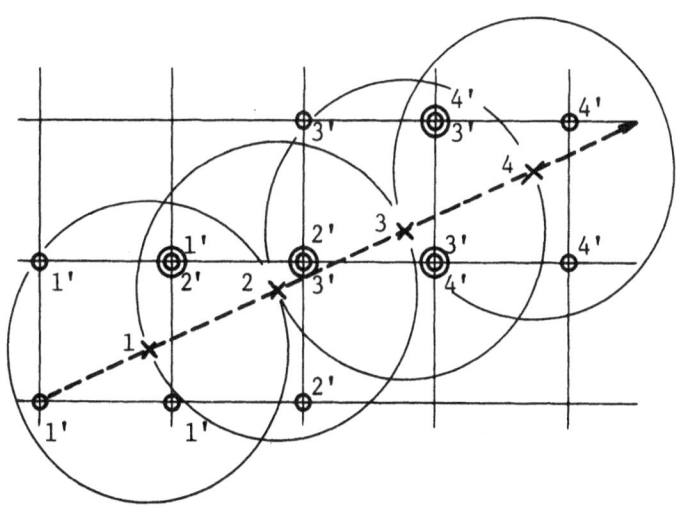

Figure 6   Smoothed line
           (using the smoothing filter in Figure 5)
           x: Sampling point
           o: Mapped pixel

In the next step, we used a more simplified method because of
limitations on the speed of writing to frame memory. As mentioned
previously, on item of sampled data is represented by one to four
pixels. This means the speed of writing must be four times as
fast as the nearest-neighbor method to maintain the same writing
speed for sampled data. Hardware implementation is difficult. We
must decrease the number of written pixels, decreasing the number
of pixels. We used a new smoothing filter (Figure 7) which is one
pixel wide along the scan line and two pixels wide at right angles
to the scan line.

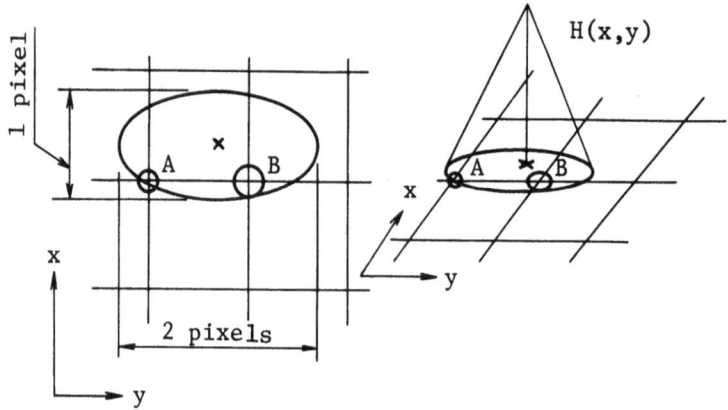

Figure 7   Improved 2-D smoothing filter

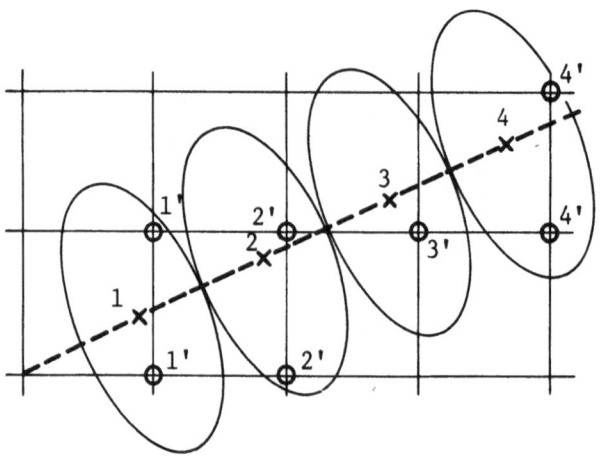

Figure 8   Smoothed line
           (using the smoothing filter in Figure 7)
           x: Sampling point
           o: Mapped pixel

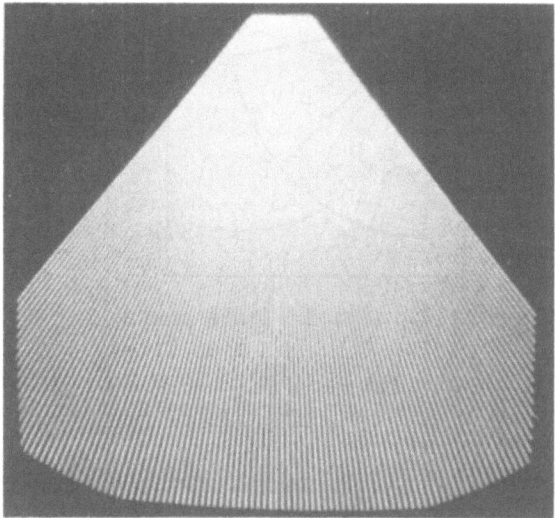

Photo 2   Smoothed-line method

    The same scan line as in Figure 6 is mapped in Figure 8.   The
number of overwritten pixels is almost the same.   One item of
sampled data is expressed using one or two pixels.   The required
speed for writing is twice as fast as that for the
nearest-neighbor method, which is achievable in real time.
Simulated data is shown in Photo 2.

4.2 Fill Method

    In the fill method, we interpolated empty areas between scan
lines.   Ideally, data of empty areas is interpolated by using only
the four nearest sampling points as indicated in equation (4).

$$Y(i,j) = \sum_{k=0}^{1} \sum_{l=0}^{1} a_{kl} \, X((u+k) \, \Delta r \cos \Delta \sigma (v+1),$$

$$(u+k) \, \Delta r \sin \Delta \theta (v+1)) \tag{4}$$

where

$$(u \, \Delta r \cos (v \Delta \theta), u \, \Delta r \sin (v \Delta \theta)) \leq (i,j) \leq ((u+1) \, \Delta r \cos((v+1) \Delta \theta),$$

$$(u+1) \, \Delta r \sin ((v+1) \Delta \theta)))$$

and a_{kl} is the weighting value.

    This interpolation must be done independent of a sector's
tilt angle and display position.

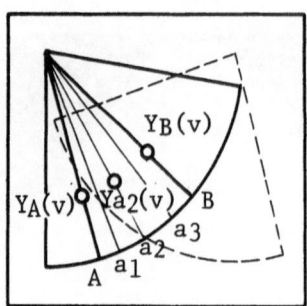

Figure 9   Fill method (Write phase)
A,B: Original scan lines
a_1, a_2, ..., a_i: Subscan line
$$Y_{ak}(v) = \frac{(i-k)\ Y_A(v)+kY_B(v)}{i}$$

We developed an algorithm--consisting of two phases, write
and read--which approximates equation (4).

Write phase.   In writing to frame memory, subscan lines are
inserted interpolating laterally.   This narrows spacing between
scan lines.   Range degradation is minimized because the length of
interpolation becomes shorter even though interpolation is done in
a TV-raster direction.

As shown in Figure 9, i subscan lines a_1, a_2, ...,a_i
are inserted between adjacent scan lines A and B.   Equation (5)
gives data for each subscan line,

$$Y_{ak}(v) = \frac{(i-k)\ Y_A(v)+k\ Y_B(v)}{i} \qquad (5)$$

where $Y_A(v)$, $Y_B(v)$, $Y_{ak}(v)$, which is the vth item of sampled
data for scan line A, B, a_k(k=1, 2, ..., i).

Read phase.   Reading from frame memory consists of two types
of interpolation--horizontal and vertical.

In horizontal interpolation, first-order interpolation is
done in a TV-raster direction (Figure 10).   If the distance
between two items of interpolation data exceeds 1, interpolation
is not done.   Because scan density is (i+1) times, there is little
image degradation, but breaking off interpolation causes "holes"
when the sector tilts.

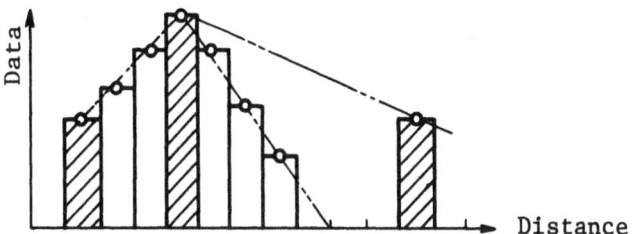

Figure 10   Horizontal interpolation

▨ Original data

▭ Interpolated data

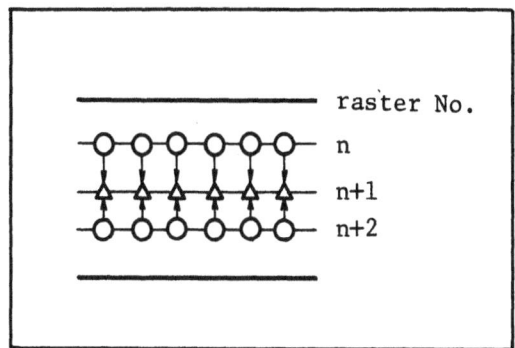

Figure 11   Vertical interpolation

In vertical interpolation, frame memory is read in standard TV format and (n+1)st raster data is generated by (n)th and (n+2)nd data (Figure 11).

In defining parameters, the number of subscan lines, i, and the maximum length of the interpolation in a TV-raster direction, l, are calculated assuming that a whole sector must be filled. The results are

$i = 3, l = 3$

4.3 Comparison

Comparison of image quality for the smoothed-line and fill
methods becomes unavoidably subjective because no good index
applicable to echocardiograms currently exists.

First, we investigated simulated data for the smoothed-line
method (Photo 2). Moire patterns could not be completely removed
using the smoothed-line method, but we could not determine from
simulated data the extent to which Moire patterns influence
diagnosis.

Second, we constructed experimental systems to compare
real-time image quality (Photo 3). When the smoothed-line method
was used, Moire patterns remained conspicuous in images of a
moving heart. The fill method produced satisfactory, Moire-free
images without image blur.

From the viewpoint of hardware implementation, the
smoothed-line method requires a writing speed twice as fast and
the fill method one four times as fast that for the nearest
neighbor method. The fill method required more complicated
hardware than the smoothed-line method because address calculation
for the smoothed-line method is made by simple offset, but
requires start/end address calculation for every three subscan
lines.

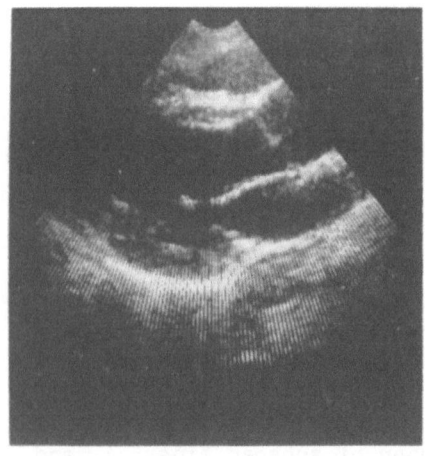
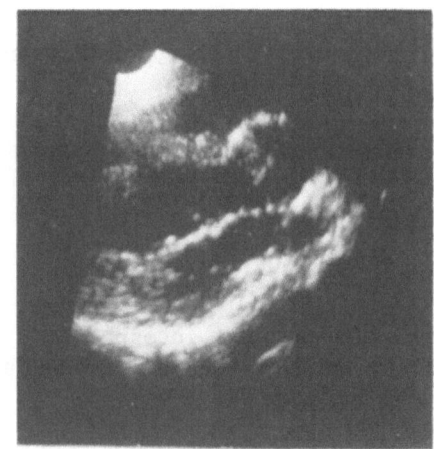

    (a) Smoothed-line   (b) Fill

Photo 3 Echocardiogram

Because image quality was our top priority, however, we used
the fill method for our practical system.

5.   HARDWARE IMPLEMENTATION

Development of a fill method using a practical memory meant
that the following requirements had to be met:

1.   Speed of writing to frame memory had to be increased to four
     times that for the nearest neighbor method.
2.   Scan-line addresses required four times as much calculation
     as that for the nearest neighbor method

Because meeting the first requirement was very difficult, we
introduced a simplified method in which subscan lines started at
the middle of a sector, decreasing overwriting.   We also
introduced a new frame-memory configuration which enabled writing
two pixels simultaneously.

To meet the second requirement, we further introduced a new
high-speed, single-chip digital signal processor (DSP) for address
calculation to lessen the amount of hardware and increase image
display format versatility.

The echocardiogram obtained by using our practical system is
shown in Photo 4.

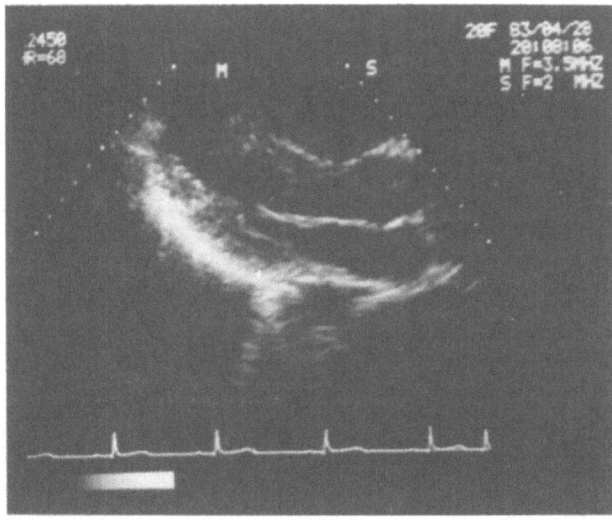

Photo 4   Compound image

6.   CONCLUSIONS

We proposed and compared two Moire-removal techniques applicable to real-time compound dual sector scanning.

The comparison showed that the fill method produced images superior to those yielded by the smoothed-line method. Satisfactory compound images free of Moire patterns were then obtained--regardless of a sector's tilt angle or display position--by implementing the fill method using a new digital signal processor and a new frame memory configuration which enabled writing at twice speeds for the nearest neighbor method.

REFERENCES

1.   T. Shimazu, M. Matsumoto, M. Fukushima, K. Yasui, T. Maeda, M. Inoue, H. Abe, K. Murakami, T. Shimura, H. Hayashi, H. Miwa, Clinical experience with simultaneous multifrequency 2-D echocardiogram, Proc. 27th Ann. AIUM Conv., 116 (1982).

2.   H. Miwa, H. Hayashi, T. Shimura, K. Murakami, Accurate dimensional measurement of the heart by simultaneous dual sector scans: Simultaneous multifrequency ultrasonography, Proc. 27th Ann. AIUM Conv., 178 (1982).

3.   K. Murakami, J. Miyazaki, T. Shimura, H. Hayashi, H. Miwa Real time digital scan converter for compounding dual sector scans, Proc. IEEE Ultrason. Symp., 705-708 (1982).

4.   H. Miwa, H. Hayashi, T. Shimura, and K. Murakami, Simultaneous multifrequency ultrasonography, Proc. IEEE Ultrason. Symp., 655-659 (1981).

5.   J. Ophir and N. F. Maklad, Digital scan converters in diagnostic ultrasound imaging, Proc. IEEE, 67:4 (1979).

6.   J. E. Bresenham, Algorithm for computer control of a digital plotter, IBM Sys. Journ., 4:1 (1965).

7.   S. Gupta and R. F. Sproull, Filtering edges for gray scale displays, Comput. Graph., 15:3, 1-5 (1981).

8.   F. C. Crow, The use of gray scale for improved raster display of vectors and characters, Comput. Graph., 12, 1-5 (1978).

9.   J. Barros and H. Fuchs, Generating smooth 2-D monocolor line drawings on video displays, Comput. Graph., 13:2, 260-269 (1979).

A DIGITAL PULSE-ECHO SYSTEM BASED ON ACOUSTIC INVERSION

J. Ridder, A. J. Berkhout and J. A. Van Woerden

Institute of Applied Physics TNO-TH
Dept. of Applied Physics, University of Technology
P.O. Box 155, 2600 GA Delft, The Netherlands

ABSTRACT

    At the 10th and 12th Acoustical Imaging Conferences we pre-
sented the theory of a multi-dimensional acoustic inversion technique. This theory is based on a quantitative description of the
wave propagation through inhomogeneous media. During the last two
years we have developed and built an imaging system for medical
applications which utilizes our inversion technique for the re-
consruction of images of 2D cross-sections. The system can be
divided in four main parts: 1) data-acquisition system, 2) reconstruction system, 3) display system, 4) control and initialization
system.

    The data is acquired over a large aperture area (typical 100 mm)
by means of a lineair array transducer. The nearly omnidirectional
elements of the transducer are sequentially used for both the transmission of ultrasound pulses and the reception of the echos. So the
medium is "illuminated" by very wide angle beams. The acquired r.f.
data is digitized and stored in a semi-conductor bulk memory. The
system can acquire in 100 msec a maximum of 1M samples. This data
is the input for the reconstruction algorithm. For the reconstruction an array-processor is used. The reconstructed image, i.e. a
high quality still-frame, is stored in a video-scanned memory and
displayed on a t.v. screen. The lateral resolution in the plane of
scanning obtained with this system is better than 0.4 mm (-6dB) and
1.0 mm (-20dB) over a depth range of 10 mm upto 150 mm. A typical
reconstruction time for a dataset of 1 Msample is 1 minute. During
the presentation examples will be shown.

A DUAL B-SCAN FOR ATTENUATION

ACOUSTICAL IMAGING

B. Ho, A. Jayasumana and C.G. Fang

Electrical Engineering & Systems Science Dept.
Michigan State University
East Lansing, MI 48824

INTRODUCTION

Extensive work has been done on medical ultrasonic imaging. Echocardiography is very useful in detecting intracardiac thrombus. However, the pulse-echo technique suffers from a rather high incidence of false positive diagnoses due to the backscatter and attenuation. The reflected ultrasound signal from a target by the conventional ultrasound pulse-echo technique, such as the B-scan, is determined by the reflection coefficient at the interface as well as the attenuation of the substance along the path. It is therefore impossible to retrieve these two types of information separately by the knowledge of a single reflected signal. Additional information is required to separate the two types of information. The work reported here is the development of a method to obtain the necessary information by a second reflected signal from the opposite side of the object.

METHOD

Consider a layered structure of N + 1 layers with N interfaces as shown in Figure 1. The first and last layers are assumed to be the same material, which normally is the case because the object is immersed in water. The reflected signal $y(t)$ is related to the incident signal $x(t)$ on either side of the object by

$$y(t) = h(t)*x(t) \qquad (1)$$

where $h(t)$ is the impulse response of the object. The impulse response has the general form of

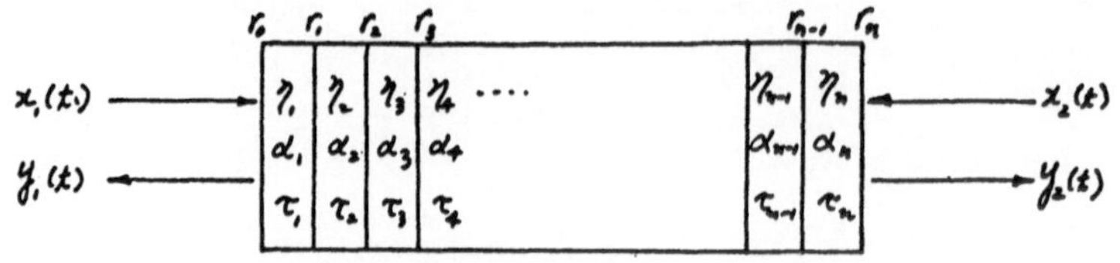

n = acoustic impedance
α = attenuation coefficient
r = reflection coefficient
τ = propagation time

Figure 1. A layered structure.

$$h(t) = \sum_i a_i \delta(t - t_i) \tag{2}$$

where a_i is the echo amplitude at $t = t_i$. This amplitude a_i can be expressed in terms of attenuation constant α propagation time τ and reflection coefficient r as

$$a_i = \exp(-2\alpha_0\tau_0)r_i \prod_{k=1}^{i-1} (1 - r_k^2)\exp(-2\alpha_k\tau_k) \tag{3}$$

where the reflection coefficient is

$$r_i = \frac{n_i - n_{i-1}}{n_i + n_{i+1}} \tag{4}$$

The amplitude of the successive peaks is

$$\frac{a_i}{a_{i+1}} = \frac{r_i}{r_{i+1}(1 - r_i^2)\exp(-2\alpha_i\tau_i)} \tag{5}$$

Similarly, the amplitude ratio of the echo response from the opposite side is

$$\frac{b_i}{b_{i+1}} = \frac{\exp(-2\alpha_i\tau_i)r_i(1 - r_{i+1}^2)}{r_{i+1}} \tag{6}$$

The product of the last two ratio expressions is then

$$\frac{a_i b_i}{a_{i+1} b_{i+1}} = \frac{r_i^2 (1 - r_{i+1}^2)}{(1 - r_i^2) r_{i+1}^2} \qquad (7)$$

If we define a parameter R_i as

$$R_i = \frac{r_i^2}{1 - r_i^2} \qquad (8)$$

then the expression in Eq. (7) can be reduced to

$$\frac{a_i b_i}{a_{i+1} b_{i+1}} = \frac{R_i}{R_{i+1}} \qquad (9)$$

This relates the successive reflection coefficients to the amplitudes of impulse responses.

From the amplitude ratios one can also obtain the following relationship

$$\frac{a_i b_{i+1}}{a_{i+1} b_i} = \frac{r_i r_{i+1}}{r_{i+1}(1 - r_i^2)\exp(-2\alpha_i \tau_i) r_i (1 - r_{i+1}^2)\exp(-2\alpha_i \tau_i)} \qquad (10)$$

Hence the attenuation constant of the ith layer can be expressed in terms of the impulse response amplitudes and the reflection coefficients as

$$\alpha_i = \frac{1}{4\tau_i} \ln \left[\frac{a_i b_{i+1}}{a_{i+1} b_i} (1 - r_i^2)(1 - r_{i+1}^2) \right] \qquad (11)$$

EXPERIMENTAL PROCEDURE AND RESULTS

The echo responses were manually sampled using the delayed sweep feature of the Tektronix 465 oscilloscope. A sampling rate of 10 MHz was chosen, as the significant frequency components of the incident .signal were below 5 MHz. Both the incident and reflected wave forms are then put through the Fourier transforms to obtain $X(f)$ and $Y(f)$ respectively. Deconvolution technique is then used to recover the impulse response $h(t)$. A model consists of five layers - three layers of plexi glass, a layer of oil and a layer of milk - was used as the test object. The impulse response functions from both side of the test object are shown in Figure 2. The results are summarized in Table 1. The property of material is shown in Table 2.

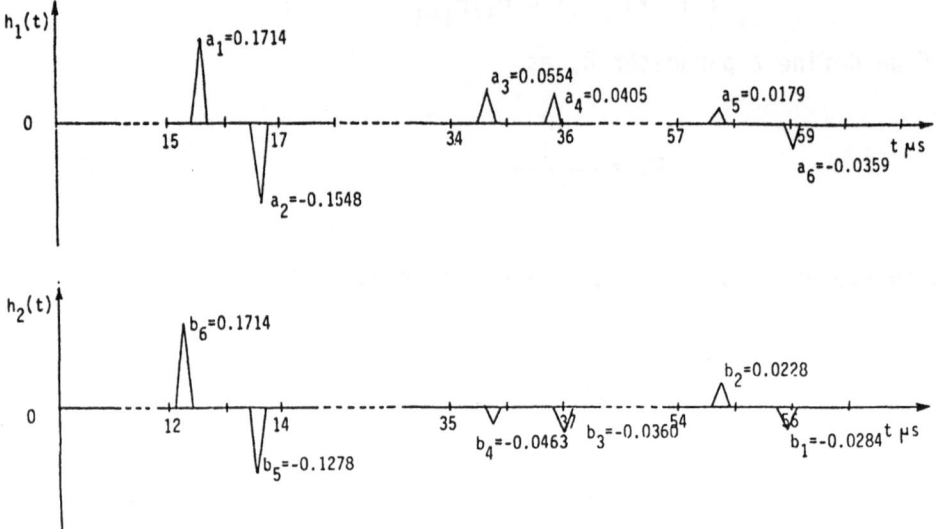

Figure 2.   Impulse response functions.

Table 1.   Experimental results.

Layer	Normalized impedance	Loss factor	Attenuation coefficient (Np/s)
Water	1.00	1.0000	0.0000
Acrylic cast 3	1.41	1.0000	0.0000
Oil	1.14	0.4042	0.0754
Acrylic cast 2	0.94	0.7611	0.2220
Milk	0.77	0.4826	0.0304
Acrylic cast 1	1.01	1.0000	0.0000
Water	0.74	1.0000	0.0000

Table 2. Material parameters from direct measurement.

Material	Normalized impedance*	Attenuation coefficient (Np/s)
Acrylic cast 1,2,3	1.50	0.1300 - 0.1800
Oil	1.03	0.0350 - 0.0440
Milk	0.98	0.0147 - 0.0217

* Normalized with respect to impedance of water.

CONCLUSION

The conventional ultrasound imaging systems make use of the amplitude information of the reflected or transmitted signals. Basically, the image is derived from the acoustic impedance variation on the boundaries. However, there are some situations where the attenuation coefficient would be a better imaging parameter than the impedance.

When an ultrasonic pulse gets reflected from an internal discontinuity of an object, the resulting echo is determined by the reflection coefficient at the interface as well as the attenuation of the substances in between the transducer and the reflecting point. The reflection coefficient is a function of the change in impedance. These two types of information- attenuation and impedance variation along the path of propagation- contained in the reflected signal cannot be separated by the knowledge of reflected signal alone. The method outlined here can extract both the variation of the attenuation coefficient and the variation of impedance along the path of propagation. The expressions developed for a layered structure were put to test by using a model composed of oil, milk and plexi glass. The experimental results are within 30% of the known values of the material under test. It should be noted that for imaging purpose, the relative attenuation coefficient is needed rather than the absolute attenuation coefficient.

IN-SITU SILT DENSITY MEASUREMENTS USING A BROADBAND PARAMETRIC

SOUND SOURCE

L.F. van der Wal*, D.Ph. Schmidt* and A.J. Berkhout**

* Institute of Applied Physics TNO-TH,
  P.O. Box 155, 2600 AD Delft, The Netherlands
**University of Technology
  Laboratory of Acoustics and Seismics
  P.O. Box 5046, 2600 GA Delft, The Netherlands

ABSTRACT

In many harbours, which are located at the mouth of a river, the accumulation of silt or fluid mud forms a serious problem with respect to accessibility. Since dredging silt is a very inefficient operation, accurate measurement of silt density profiles is very important in these situations. In a previous paper [1], the authors discussed the requirements of an acoustic measurement system to obtain continuous and accurate recordings of sub-bottom density profiles. They also suggested the use of a parametric sound source to meet these requirements.
In this paper we will discuss a number of in-situ recordings obtained with the parametric system, covering both ample and poor silt conditions. Main emphasis will be given to the way data processing is influenced by signal quality and variation in bottom topography.
It will be shown that many problems of the system are related to the poor signal to noise ratio of the parametric signals. Finally, the calibration of the acoustic density curves using spot measurements, obtained with a nuclear density gauge, will be discussed in some detail as well.

1. INTRODUCTION

For a number of years now, the Netherlands Rijkswaterstaat (RWS) carries out weekly hydrographic surveys to measure silt density profiles at a large number of pre-specified locations in the Dutch

harbours of Rotterdam and Europoort, using nuclear density gauges
[2].
Based on these measurements combined echo-depth/density charts are
produced, from which the nautical depth in the area is derived.
Since the measuring locations form a rather coarse grid on the
region of interest, intermediate density values have to be obtained
by linear interpolation [3].

In 1974 the use of an acoustic system was suggested by Kirby and
Parker [4] to obtain continuous recordings of sub-bottom density
profiles. A suggestion which is based on the assumption that the
sub-bottom density profile can be calculated from its acoustic
reflectivity function [5]. In addition a well designed acoustic
system should need only few nuclear measurements for calibration,
thus minimizing both time and costs of the hydrographic surveys.

In a previous paper [6] it was shown that - to achieve the best
possible estimate of the sub-bottom reflectivity - an acoustic
system should meet the following requirements:

a. A broad frequency spectrum, for adequate depth-resolution.
b. Sufficient energy at low frequencies for pre-specified depth-
   penetration.
c. A narrow beamwidth to minimize the influence of scattering
   noise.

Since these requirements are difficult to meet within a single
conventional transducer assembly, it was considered to use a
parametric sound source. Following successful tests with a
prototype system in 1977, the acoustic system was based on a
parametric sub-bottom profiler, initially developed at the
Birmingham University and further developped and manufactured by
Ulvertech Limited (Ulverston, Cumbria, U.K.). The complete
transducer assembly is mounted on a stabilized platform to minimize
the influences of ship-movement.

In this paper the performance of the combined nuclear/acoustic
measuring system will be discussed on the basis of a number of
in-situ recordings. But first we will briefly review the demands
which are made on acoustic data processing.

2. DATA PROCESSING CRITERIA

Data processing aims at an accurate determination of the sub-bottom
reflectivity function. The assumption that the sub-bottom density
profile can be calculated from its reflectivity is based on a
theoretical model in which the silt layer - at least locally - is
regarded as a two-dimensional horizontally stratified medium [5].
(Note the requirement of a narrow beamwidth to minimize the

influence of scattering noise, which was stated in the previous
section.)
Hence data processing of the sub-bottom reflections may be interfered with by the ships vertical motion (heave-movements), silt
layers showing rather steep tilt angles and most of all bottom
topography in the case of poor silt conditions.
Best system performance is therefore to be expected under ample
silt conditions, when a distinct stratified silt layer is present
and firm-bottom reflections can be easily recognized.

In the course of the research project several measurement campaigns
were carried out in the Europoort area, covering various silt
conditions. During each campaign nuclear spot measurements were
performed for calibration purposes as well. Acoustic sub-bottom
reflections were recorded aboard ship on an analog recorder and
processed off-line on a mini-computer.
Since the recorded data were mainly used to optimize data
processing the data sets were rather limited in size.
Off-line data processing is described by the authors in a previous
paper [1] and consists of various processing steps:

    1. A/D-conversion
    2. Bandpass filtering
    3. Heave-correction
    4. Spectral equalisation
    5. In-phase addition
    6. Envelope detection
    7. Integration
    8. Calibration

Due to the rather low efficiency of the parametric effect the
spectral bandwidth of the generated difference frequencies is
limited to 5 kHz. Although the generation of secondary signals with
a higher bandwidth is possible, the resulting signal to noise ratio
is unacceptable.
To improve the depth-resolution of the profiler, therefore three
band-limited difference frequencies are generated sequentially,
i.e. 7.5 - 12.5 kHz, 12.5 - 17.5 kHz* and 17.5 - 22.5 kHz. Thus the
desired broad-band system response is obtained by in-phase addition
of the processed recordings (step 5).

---

* Actually the system generates a 2.5 - 7.5 kHz frequency band.
  Under sailing conditions however, this 5 kHz signal showed a very
  low signal to noise ratio. During data processing therefore, the
  15 kHz signal, which is generated as a first uneven harmonic, is
  used instead.

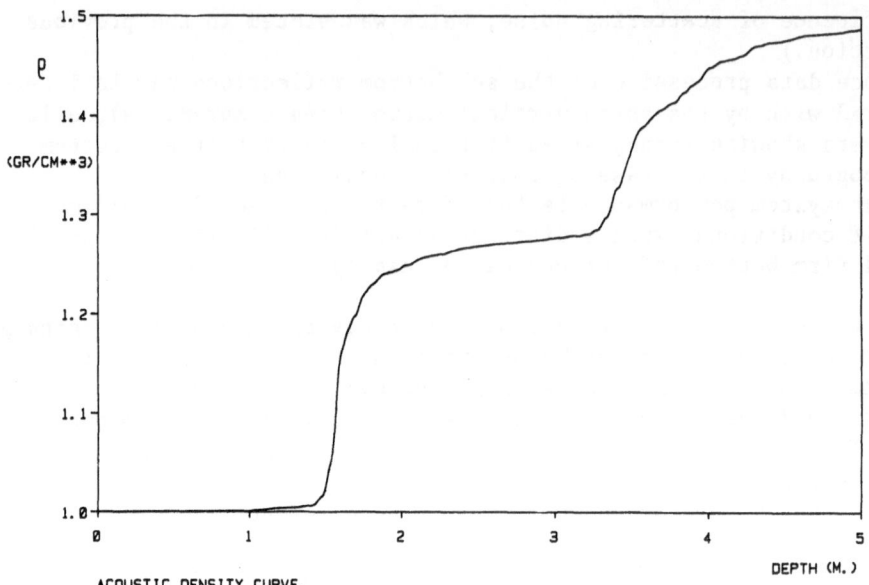

ACOUSTIC DENSITY CURVE.

Figure 1: Acoustic density curve after data processing and
          calibration on the basis of a nuclear gauge measurement,
          shown in figure 2.

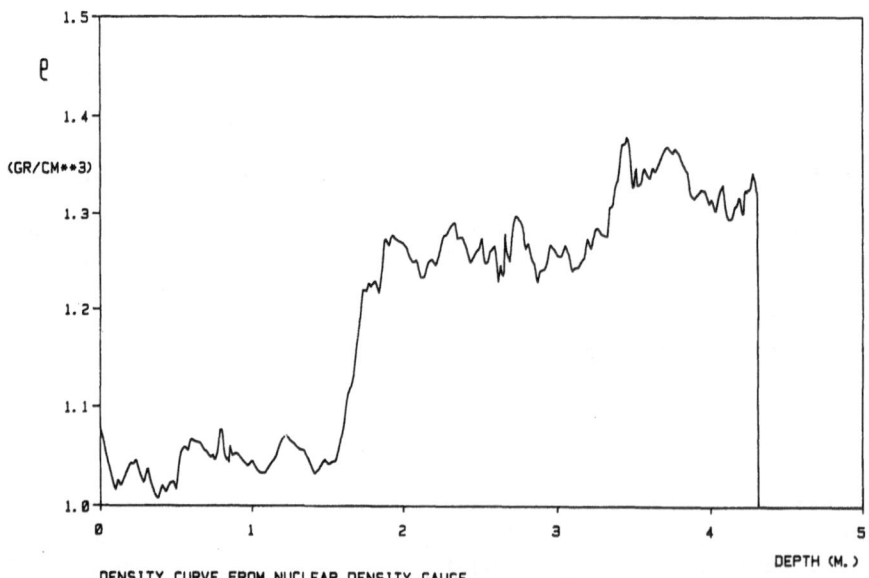

DENSITY CURVE FROM NUCLEAR DENSITY GAUGE.

Figure 2: Nuclear gauge density curve, taken at a reference
          location and used for calibration of acoustic data.

Since the source strength of the secondary signals depends on (frequency)^2 sequential recordings show a rather large variation in signal amplitude. The amplitude ratio equals ca. 0.5 : 0.2 : 1.0 for the 10, 15 and 20 kHz signals respectively.

The process of spectral equalisation (step 4) is discussed in detail in a related paper [7]. It is necessarily applied here to remove all frequency-dependent effects from the recorded data, e.g. the differences in source-strength and geometrical spreading and the relative differences with respect to sub-bottom attenuation. After spectral equalisation, the broad-band sub-bottom reflectivity function is obtained by in-phase addition.

Finally the last processing steps, i.e. envelope detection and integration, result in an estimate of the density variations within the silt layer, the so-called "acoustic density curves".

Calibration parameters are derived from the correlation of acoustic density curves and nuclear gauge measurements taken at a given number of reference points.

Given ample silt conditions good agreement is found between the nuclear and acoustic density curves in most situations (see section 3). When silt conditions become poor, the calibration process becomes very delicate, due to a decrease in reliability of both the nuclear and acoustic density curves (see section 4).

3. IN-SITU PERFORMANCE UNDER AMPLE SILT CONDITIONS

As mentioned in the previous section best system performance is to be expected under ample silt conditions.

Figure 1 shows an acoustic density curve after data processing. A distinct first silt impact can be seen, followed by a 1.75 m silt layer showing a very slow increase in density, until the firm-bottom reflection is reached.

Calibration of the acoustic density curve is performed on the basis of a nuclear gauge measurement, taken at the same reference location, and shown in figure 2.

Note the nuclear density curve shows a rather low reliability. This may be explained from the fact that the γ-radiation sources, used in the nuclear gauges, show large, high-frequent variations in source strength.

In addition the downward path of the gauge may be interfered with by vertical ship-movements, causing a drop in measurement accuracy as well.

Accurate evaluation of nuclear gauge measurements therefore requires quite an amount of pre-processing.

Nevertheless the nuclear density curve in figure 2 shows a distinct trend in silt density, which is very similar to the trend in the

acoustic density curve, shown in figure 1.
Hence absolute density values can be ascribed to the acoustic
measurements, and the average density of the silt layer is
calculated to be 1.25 gr/cm^3.

In summary:

1. Given ample silt conditions, both nuclear and acoustic
   density curves show a high degree of similarity, which allows
   for adequate calibration.
2. Nuclear gauge measurements show rather low reliability due to
   the stochastic nature of the radiation source. Measurement by
   accuracy may be interfered with by vertical ship-movement as
   well.
3. Due to a low signal to noise ratio (acoustic data) and low
   reliability (nuclear data), acoustic sub-bottom density
   profiling involves a large amount of data processing.

4. IN-SITU PERFORMANCE UNDER POOR SILT CONDITIONS

In July 1982 a large measurement campaign was carried out as a
final test to system performance. Covering a square area of
150 x 150 m^2, sub-bottom reflections were recorded along six
stretches, as shown in figure 3. At the four corners of the square,
nuclear density measurements were taken for calibration. Due to
recent dredging and constant fair weather, silt conditions were
known to be poor.

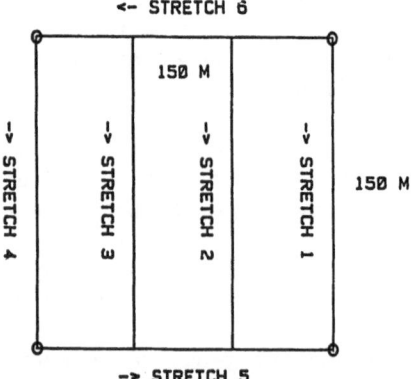

Figure 3: Schematic illustration of the final test area, covering a
square of 150 x 150 m^2, Sub-bottom reflections were
recorded along six stretches (——▶indicates sailing
direction).

Figure 4 represents the raw acoustic data (20 kHz only), obtained
at stretch 5 with a sailing speed of ca. 2.5 m/s. The vertical and
horizontal dimensions amount to 4.5 m and 150 m respectively. Note
the high noise level, the strong variations in bottom topography
and the absence of a distinct silt layer.

Following acoustic data acquisition the RWS-crew carried out
another measurement campaign, in which the same stretches were
characterized by means of a large number of nuclear gauge
measurements (5 or 6 spot measurements on each stretch). The
nuclear data were processed independently of the acoustic
measurements at RWS research facilities. Mutual comparison of both
processed nuclear and acoustic data will be discussed from figures
5, 6 and 7.

Figure 5a shows the raw acoustic data obtained at stretch 2 (20 kHz
only).
Data processing steps 1 - 7 (see section 2) result in the acoustic
density profiles shown in figure 5b. Based on nuclear gauge
measurements the acoustic density curves are calibrated and the
location of a given number of distinct density-transitions (i.e.
1.02, 1.10, 1.15, 1.20, 1.25 and 1.30 gr/cm^3) is calculated (figure
5c). These calibrated acoustic density profiles are then compared
with the nuclear profiles, as supplied by RWS (figure 5d).

A similar process was carried out on the data obtained at the other
stretches. Results of stretch 5 and stretch 6 are shown in figures
6 and 7 respectively.

Regarding figures 5a, 6a and 7a, in which the raw 20 kHz
registrations are shown, the low signal to noise ratio of the
acoustic data can be clearly seen. Given the lower source strength
of the 10 and 15 kHz registrations, here the S/N-ratio is even
worse.

Due to low S/N ratio, large variations in bottom topography and
vertical ship-movements, the in-phase addition of the narrow-band
10, 15 and 20 kHz registrations becomes a very delicate process.
Even so, the acoustic density profiles, shown in figures 5b, 6b and
7b, clearly illustrate bottom topography and sub-bottom density
variations.

The correlation between the calibrated acoustic density curves
(figures 5c, 6c and 7c) and the nuclear profiles from RWS (figures
5c, 6d and 7d) is disputable.
Regarding the first three density-transitions (i.e. 1.02, 1.10 and
1.15 gr/cm^3) a rather high similarity is found. With respect to
subsequent transitions however the nuclear profiles definitely show
a more gradual increase in density, wich is most clearly shown in
figures 5c and 5d.

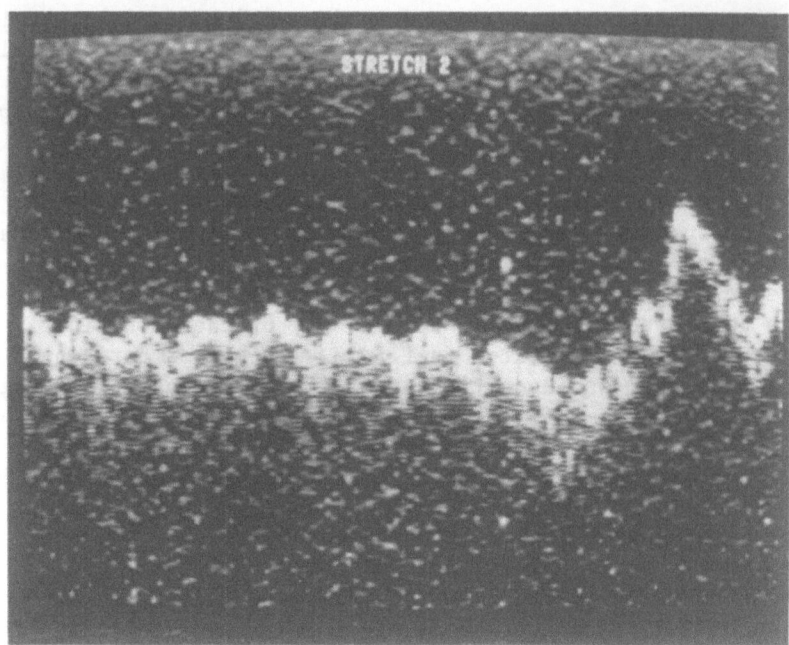

Figure 4: Unprocessed 20 kHz registrations obtained at stretch 5.

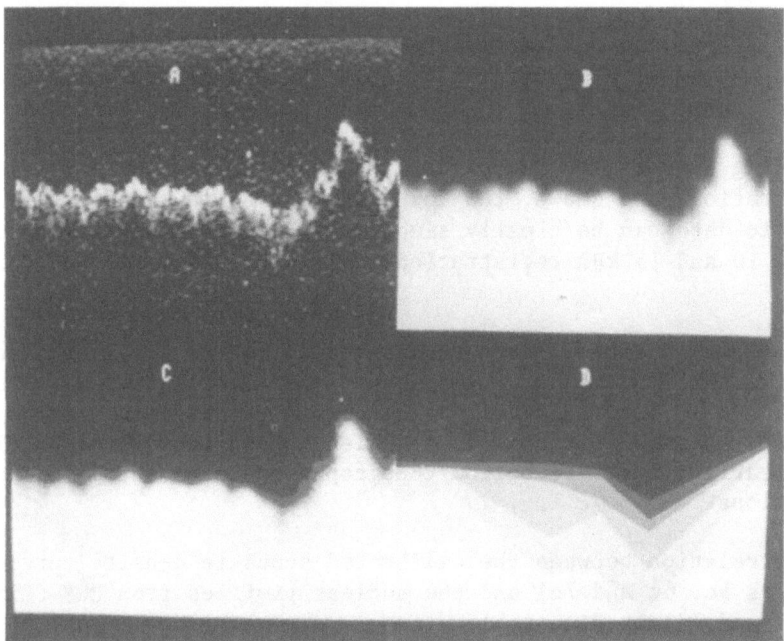

Figure 5: Acoustic and nuclear data obtained at stretch 2.
Unprocessed 20 kHz registrations (a); acoustic density
profiles before (b) and after calibration (c); nuclear
density profiles (6 spot measurements) supplied by
RWS (d).

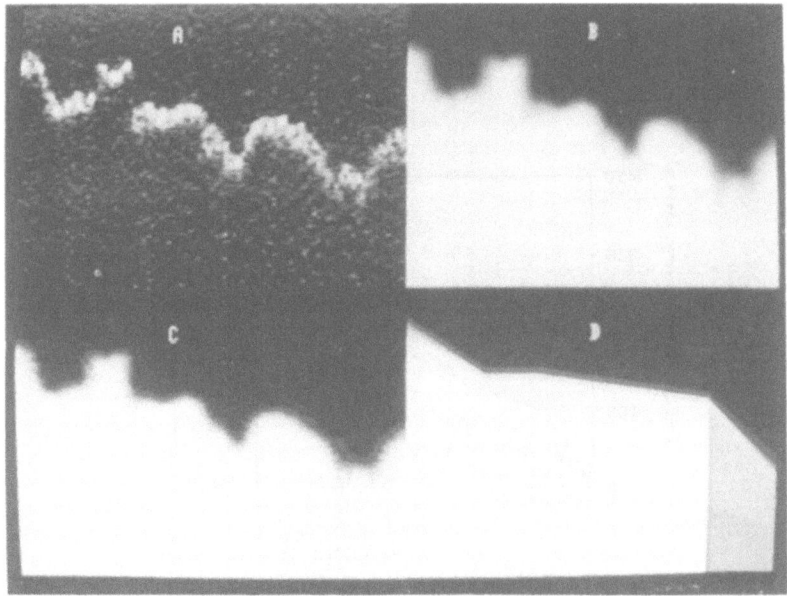

Figure 6: Acoustic and nuclear data obtained at stretch 5.
Unprocessed 20 kHz registrations (a); acoustic density
profiles before (b) and after calibration (c); nuclear
density profiles (5 spot measurements) supplied by
RWS (d).

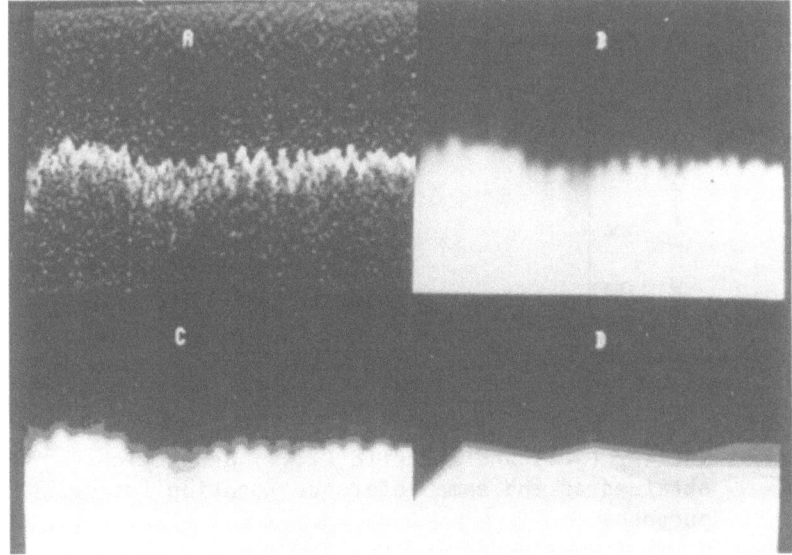

Figure 7: Acoustic and nuclear data obtained at stretch 6.
Unprocessed 20 kHz registrations (a); acoustic density
profiles before (b) and after calibration (c); nuclear
density profiles (6 spot measurements) supplied by
RWS (d).

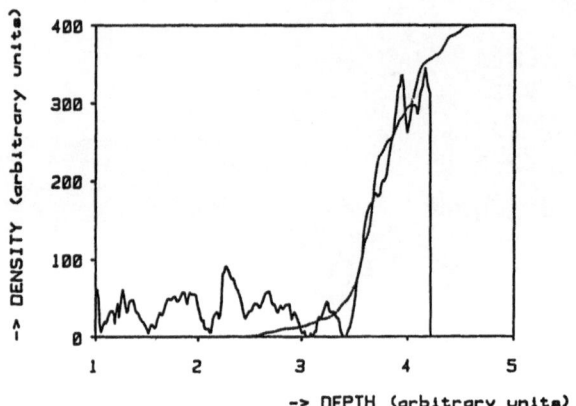

Figure 8: Nuclear (——) and acoustic (....) density curve,
obtained at the same reference location for calibration
purposes.

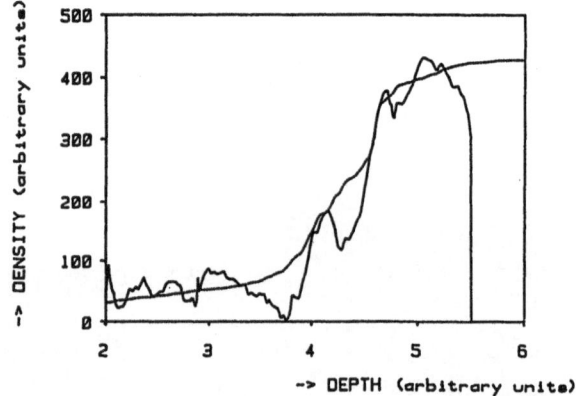

Figure 9: Nuclear (——) and acoustic (....) density curve,
obtained at the same reference location for calibration
purposes.

Although the absolute accuracy of nuclear gauge measurements may be questioned, these disappointing results are best explained by the low signal to noise ratio of the acoustic data and the strong variations in bottom topography. Both factors interfere with the process of in-phase addition and therefore will limit both depth-resolution and sub-bottom penetration.

To conclude this section we would like to illustrate the delicate process of calibration under poor silt conditions, on the basis of two examples.
Figure 8 shows both a nuclear and an acoustic density curve, taken at the same reference location. In this case accurate calibration is greatly interfered with by the strong fluctuations, shown by the nuclear curve in front of the firm-bottom location.
Figure 9 shows a similar situation, in which the nuclear curve seems to indicate a negative density-transition. Although negative density-transitions may be caused by local gas-formation, their significance is quite often questioned.
Figures 8 and 9 clearly indicate that correlation between nuclear and acoustic density profiles becomes very difficult under poor silt conditions.

CONCLUSIONS AND FUTURE DEVELOPMENTS

In this paper the performance of an acoustic sub-bottom density profiler is discussed on the basis of in-situ recordings.
Both data acquisition and data processing are based on a theoretical model in which the sub-bottom is locally regarded as a horizontally stratified medium. Good system performance is there-fore obtained under ample silt conditions and smooth bottom topography. Under these circumstances the continuous acoustic registrations accurately reproduce sub-bottom density variations and show detailed bottom topography. Good correlation is found with nuclear gauge spot measurements in most situations.
When silt conditions become poor, data processing is interfered with by low acoustic signal to noise ratio and large variations in bottom topography. Worse system performance is therefore to be expected.

In the near future the system will be modified to improve the acoustic signal to noise ratio:

- henceforth the 15 kHz difference frequency will be generated directly, and not as a first uneven harmonic.

- the use of sweep-signals ('chirps'), instead of narrow-band signals, will be evaluated.

Presently work is in progress to develop an on-line processing unit, which will enable the generation of acoustic density curves aboard ship in real-time. Following data reduction and storage, the acoustic density curves will be correlated with nuclear spot measurements in a processing centre ashore, resulting in continuous, calibrated recordings of sub-bottom density profiles.

ACKNOWLEDGEMENTS

The authors would like to thank their colleagues of the Netherlands Rijkswaterstaat for their indispensable assistance during the measurements in the Europoort area.
This work is supported by the Hydrographic Survey Division, Lower Rhine Directorate of the Netherlands Rijkswaterstaat.

REFERENCES

1. Wal, L.F van der, Schmidt, D.Ph, and Berkhout, A.J.: "Acoustic Determination of Sub-Bottom Density Profiles using a Parametric Sound Source", Acoustical Imaging, Vol. 12, pp. 721 - 732, Plenum Press, New York - London, 1982.

2. Parker, W.R., Sills, G.C. and Paske, R.E.A., "In situ Nuclear Density Measurements in Dredging Practice and Control", Symposium on Dredging Technology, 17 - 19 Sept. 175, published by BHRA Fluid Engineering, Cranfield, Bedford, England.

3. Bakker, D.J., "Density Measurements in conjunction with Echo-Sounding", The Hydrographic Journal, no. 14, pp. 21 - 27, April 1979.

4. Kirby, R and Parker, W.R., "Seabed Density Measurements related to Echo-Sounder Records", The Dock and Harbour Authority, pp. 423 - 424, March 1974.

5. Schmidt, D.Ph and Janssen, H.C. "Acoustic Silt-Density Measurements, a Feasibility Study", Institute of Applied Physics, Techn. rep. no. 910-281, July 1980 (in Dutch).

6. Schmidt, D.Ph, Janssen, H.C. and Berkhout, A.J., "Prediction of Sediment Density Profiles by means of Echo-Sounding", ULTRASONICS INTERNATIONAL 81, pp. 165 - 170, published by IPC Science and Technology Press Limited, Guildford, Surrey, England 1981.

7. Wal, L.F. van der, Dijkhuizen, A.C. and Peels, G.L.: "Spectral Equalisation: A Signal Processing Technique to Increase Pulse-Echo Signal Bandwidth", Acoustical Imaging, Vol. 13, Plenum Press, New York - London, 1983.

COMPUTER-AIDED IMAGING WITH OPTICALLY-SCANNED ACOUSTIC

TRANSDUCERS

M. Salahi and C.W. Turner

Department of Electronic & Electrical Engineering
King's College, London, England

ABSTRACT

Continuing development of an optically-scanned acoustic transducer previously described has led to a novel approach to image
correction. Digital correction techniques implemented on
inexpensive microprocessors are shown to offer a simple and
convenient means of eliminating the effects of system non-
uniformities. The application of this technique to transducer
evaluation and near-field imaging is discussed in the light of the
experimental results presented.

INTRODUCTION

In the past, we have reported[1,2,3] on the development of an
optically-scanned acoustic imaging transducer, as shown in Fig. 1,
capable of producing high-resolution, two-dimensional amplitude and
phase images of incident acoustic fields in real-time. Initial
operation at around 3 MHz has been extended to 10 MHz [4].

In parallel with the development of this transducer for operation
over a wide range of frequencies, digital processing is being
introduced in order to build a comprehensive system capable of imaging
the more complex field distributions obtained in practical situations
e.g. radiation patterns from composite transducers, arrays etc. In
this paper we will describe the initial application of this digital
processing capability and provide preliminary results on a novel form
of digital/analog correction technique that could be very useful in
such applications as transducer evaluation.

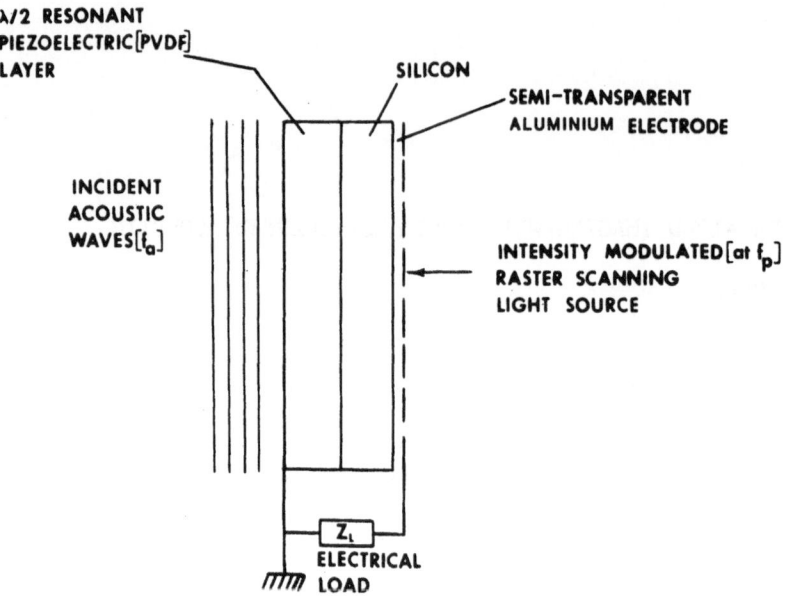

Fig. 1.   The Optically Scanned Transducer

THE DIGITAL SYSTEM

     In keeping with the simplicity and low cost of this system, it is
decided to choose the new hardware accordingly. A general-purpose,
relatively slow 8-bits A/D converter (ADC 0804) is used for digitizing
the conditioned image signal. Data handling, storage and most of the
processing is done on an Apple II microcomputer. It is decided to
divide the image frame into 64 lines and take 128 samples per line,
this being sufficient to satisfy the Nyquist-Shannon sampling criteria
for the aperture used under 2 cm diameter) at 3 MHz for diffraction-
limited imaging in water. Provision is made in the software to
increase the resolution to 256 x 128 samples if necessary. For
128 x 64 resolution, the frame rate for the raster scan is just over
1 Hz. This limitation on speed is not significant in the case of
transducer evaluation. Much faster rates can be obtained using a
faster A/D converter.

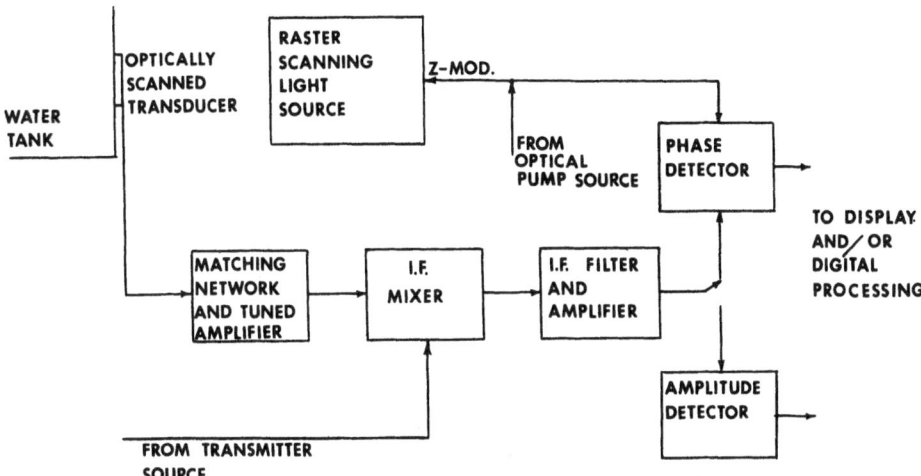

Fig. 2.   System block diagram

For the above 128 x 64 pixels chosen, each image occupies about 8.2k bytes of memory. For mass storage floppy disks with 125k bytes capacity are used. The display can take the form of colour-coded or line-scan images. The former are displayed using a B.B.C. Model B microcomputer because of its higher resolution colour graphics capability. The data transfer between the two computers is carried out through the serial interfaces, RS 232C or RS 423. The resulting system is shown in the block diagram of Fig. 2.

DIGITAL PROCESSING

The above system enables the application of a wide range of relatively simple processing algorithms to improve image presentation. These include background separation by thresholding, noise suppression by time averaging (adding multiple images), edge detection and enhancement, variable perspective, pseudo-colour, line scans with hidden-line elimination etc. For more sophisticated algorithms, the data can be transferred into more powerful computers, the processed data being returned for display in the usual manner. For example, Fig. 3 shows the acoustic (amplitude) image of a 1 cm high letter K cut in rubber, after it has been Fourier analyzed on a LSI 11/23 processor to study its spatial frequency content.

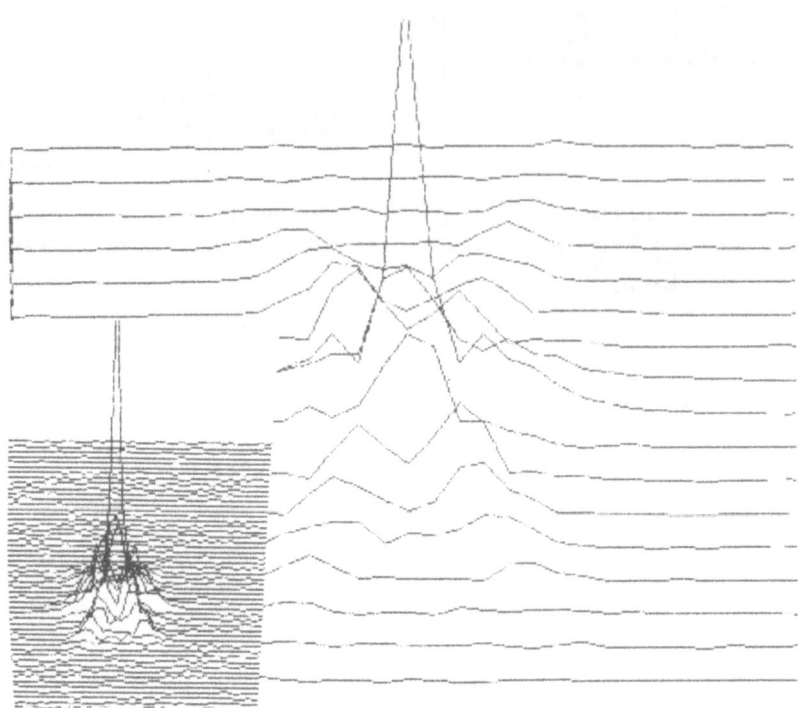

Fig. 3. Fourier spectrum (original printout and the magnified
         version) of the acoustic image of a 1 cm high letter K.

For the specific case of transducer evaluation, the main area
where digital processing can help is in eliminating the effects of
spatial non-uniformities in the receiving tranducer response,
therefore ensuring that the image obtained is a true indication of the
radiation pattern of the transducer under investigation. In our
particular case, this would mean eliminating the effects of defects
and inhomogeneities in both the piezoelectric and the photoconductive
layers as well as non-uniformity in the coupling between them.

The obvious way of achieving this objective is to obtain a
"correction frame", containing information required to eliminate the
effect of the mentioned non-uniformities, and to apply this correction
to each successive image once it is digitized. The correction frame
and the necessary information for it can be derived from two sources :

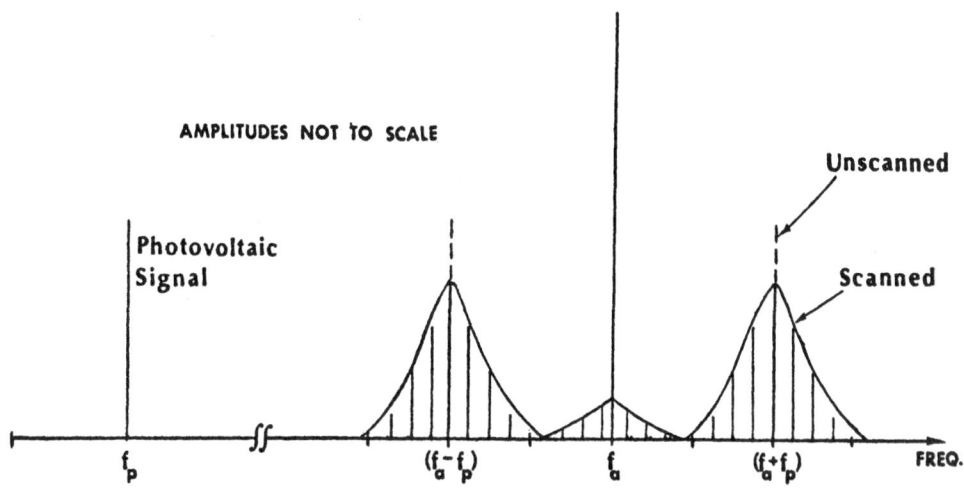

Fig. 4. Frequency content of output signal from a scanned
         transducer.

(1) Photovoltaic Correction.

     Referring back to Fig. 1, it should be noted that a Schottky
barrier is formed at the Silicon/Aluminium junction. The non-linearity
thus introduced is not important as far as the acoustic imaging is
concerned, the silicon bulk resistance modulation by photocarrier
generation and injection being the dominant acoustic signal generation
mechanism[5]. Nevertheless, a photovoltaic signal, at the optical pump
frequency, is present at the output of the composite transducer, in
addition to the acoustically generated signals, as shown in Fig. 4.
In the usual mode of operation, this "extra" signal is filtered out
and only the acoustic "sideband" is recovered. If, however, the
photovoltage map is obtained by locking onto this signal, information
about defects and inhomogeneities in the silicon as well as non-
uniformities in the thickness of the semi-transparent aluminium
electrode and the illumination by the flying spot scanner can be
derived. These factors would obviously affect the acoustic sideband
generation mechanism by varying the optical switching ratio for each
pixel.

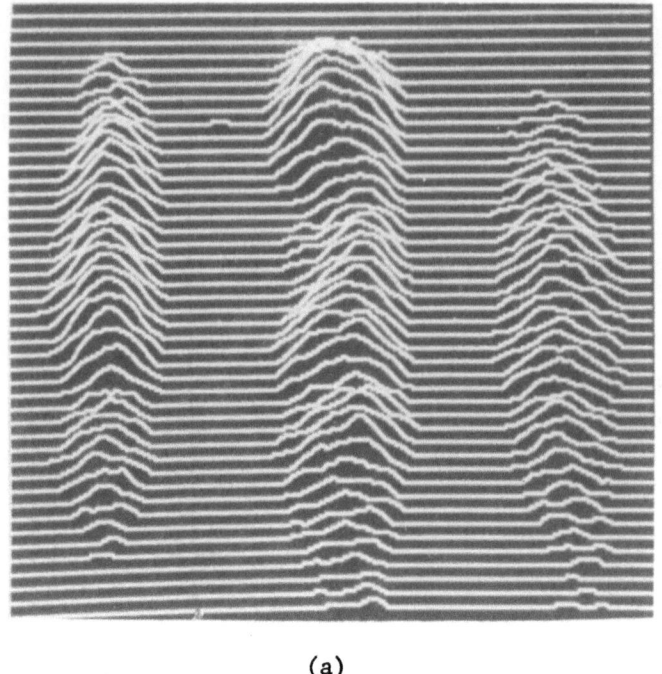

(a)

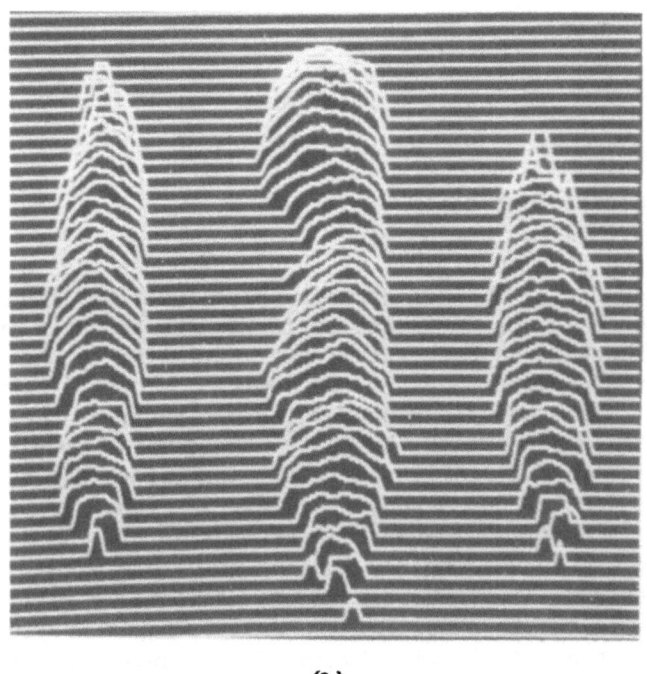

(b)

Fig. 5. Acoustic images of 3mm wide slits cut in rubber before,
(a), and after, (b), the "photovoltaic correction".

The way in which this information is utilized to develop a
correction algorithm can take several forms. One method, which is
valid for small deviations, is the following :

Let $[I]$ and $[P]$ be 128 x 64 matrices representing the digitized
acoustic images and the photovoltage maps respectively. Also, let
$[M]$ represent another 128 x 64 matrix whose elements are all 255s ;
i.e. the maximum possible value for each sample in an 8-bit system.
$[I]+[[M]-[P]]$ will therefore produce a "corrected" image, $[I_c]$, from
which features introduced by any spatial non-uniformities have largely
been removed. Elements of $[I_c]$ are rescaled to ensure that no
saturation will result from this process. Figs. 5a and 5b show
acoustic images of 3mm wide slits cut in rubber, before and after the
photovoltaic correction. In practice in many cases the non-uniform
response of the silicon photoconductor is less significant than that
arising from acoustic factors.

(2) Acoustic Correction.

A more effective correction algorithm can be derived by utilizing
the information provided by analyzing the response of the receiving
transducer to a plane wave from a "standard" transmitting transducer,
such as a point source in the far field. Replacing the photovoltage
map matrix, $|P|$, with the acoustic image matrix, $|A|$, of the plane
wave, subsequent images can be corrected as :

$$[I_c] = [I] + \left[[M] - [A] \right]$$

This method will substantially reduce the effects on image
quality of non-uniformities and defects in both the photoconductive
and the piezoelectric layers, as well as the coupling between them.
Therefore, algorithms based on this correction matrix are used when
evaluating transducers. The photovoltaic correction, however, is very
useful since it can be used to isolate the response of the piezoelectric layer of the composite transducer, hence provide a "self-test"
facility. Figs. 6a and 6b show the acoustic amplitude image of a 1 cm
high letter K cut in rubber, before and after the application of the
acoustic correction algorithm.

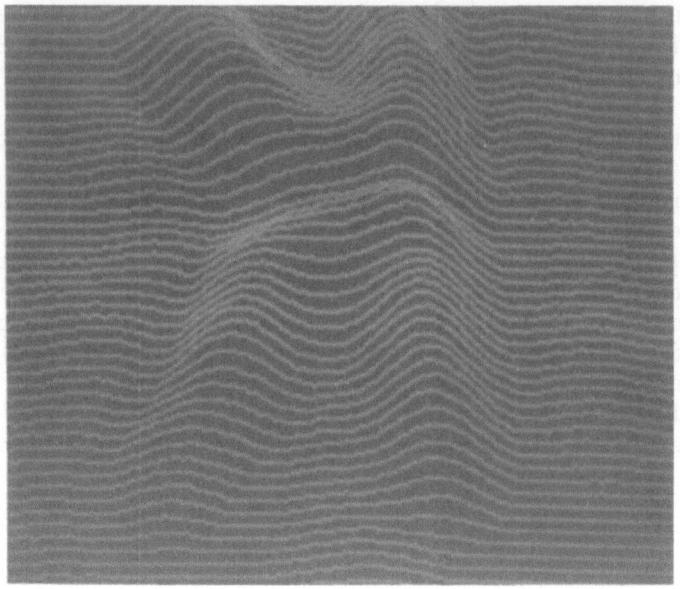

(a)

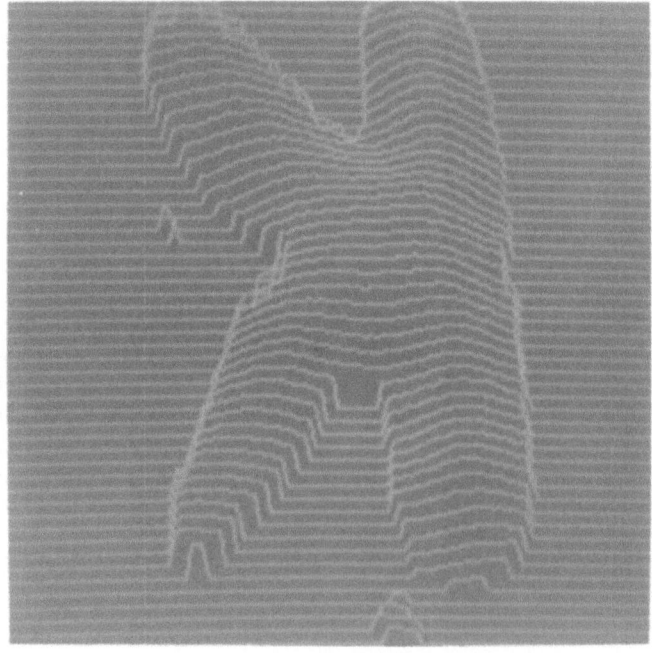

(b)

Fig. 6. Acoustic amplitude images of a 1cm high letter K, cut in
         rubber, before, (a), and after, (b), the "acoustic
         correction".

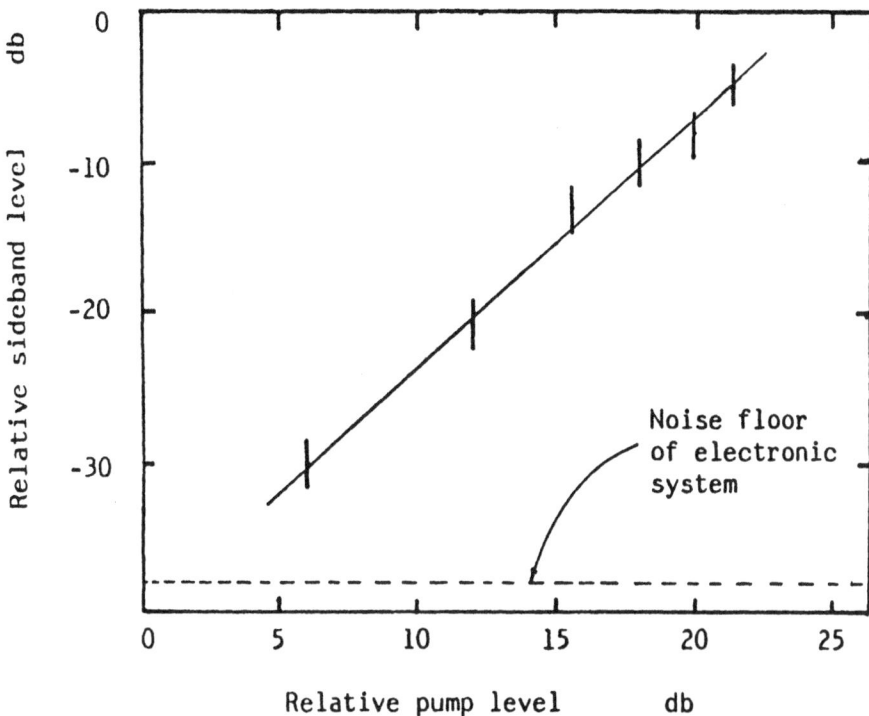

Fig. 7. Acoustic sideband level versus optical pump
amplitude.

OPTICAL CORRECTION

The corrections described above can be applied in a manner which
makes full use of the particular properties of the optically-scanned
transducer. Once a correction matrix, based either on the photovoltage maps or the acoustic response to a plane wave, is obtained,
the data contained in it can be used to modulate the intensity of the
scanning light beam. By synchronizing the raster scan of the beam
with the rate at which data is sent out of the correction matrix, it
is possible to ensure that the correction applied at each pixel is
properly registered. The relationship between the acoustic sideband
level, from a given pixel, against the optical pump level is shown on
the graph of Fig. 7. It is effectively this relationship which allows
the use of the optical correction method.

It should be noted that the optical beam is already intensity
modulated sinusoidally at 80 kHz. Therefore, the much lower frequency
correction signal is used to modulate the amplitude of this 80 kHz
sinusoid. The resulting composite signal, together with the line and
frame blanking pulses are applied to the cathode of the scanner tube.
Alternatively, the amplitude of the 80 kHz optical pump can be kept
constant while the d.c. bias applied to the grid is varied by the
correction signal, once again modulating the intensity as required.

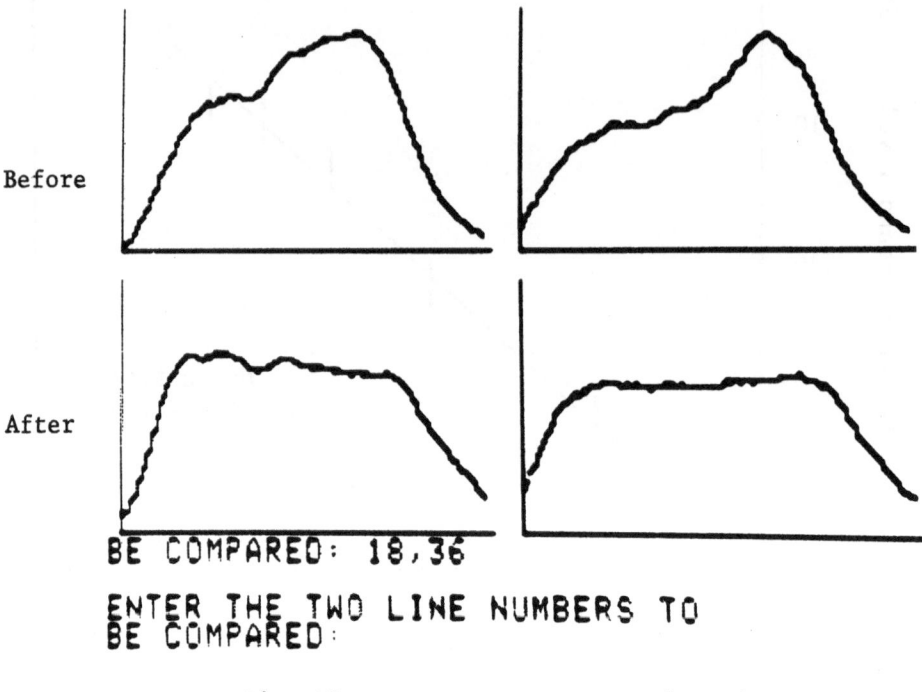

Before

After

BE COMPARED: 18,36

ENTER THE TWO LINE NUMBERS TO
BE COMPARED:

Line 18   Line 36

Fig. 8.   Line scans from the photovoltage map before and
           after optical correction.

The main advantage of this method is that once the correction
matrix is computed, every subsequent image will be "automatically"
corrected before digitization whereas in the previous methods each
image is digitized and the correction algorithm is then applied to it.
Hence, great savings in processing time as well as a reduction in
software complexity can be achieved. A second Apple II is used to
store the correction matrix and to send its contents out, through a
D/A converter, at the correct rate under the control of a machine code
program. An alternative arrangement would be to replace the second
Apple with a dedicated microprocessor-controlled circuit, with
E.P.R.O.M.s being used for image storage. A different set of
E.P.R.O.M.s will then be plugged into the circuit for each different
composite transducer being used.

Fig. 8 shows line scans (lines 18 and 36) from the photovoltage
map of the same silicon specimen, before and after the application of
optical corrections, clearly demonstrating the improvement in the
uniformity of the output.

CONCLUSIONS

We have described the developments in our system, resulting from
the introduction of a digital processing capability. We have pointed
out two sources that can provide information about deficiencies in the
spatial uniformity of the transducer response and have shown that even
the simplest correction algorithms derived using this information can
be effective. Furthermore, we have developed a technique whereby such
corrections can be applied through optical scanning rather than by
computation. This demonstrates that optically-scanned acoustic
imaging is much easier to "adapt" than, say, a phased-array system.
Hence, a very simple but effective approach to high fidelity amplitude
and phase mapping of acoustic fields is obtained that can successfully
be applied in areas such as transducer evaluation and near-field
imaging for non-destructive testing.

ACKNOWLEDGEMENTS

Grateful acknowledgement is made of the support of the U.K. Science
and Engineering Research Council. PVF_2 films were kindly supplied
by Thorn-E.M.I. Central Research Laboratories. The F.F.T. algorithm
used was written by A.I. Mitchell.

REFERENCES

1. C.W. Turner, S.O. Ishrak and D.R. Fox, "Optically-Scanned
       Amplitude and Phase Probing of Acoustic Fields", paper
       presented at the 1979 Ultrasonics Symposium.

2. C.W. Turner and S.O. Ishrak, "Comparisons of Different
       Piezoelectric Transducer Material for Optically-Scanned
       Acoustical Imaging, Vol. 10, P. Alais and A. Metherell,
       Eds., Plenum Press, New York, pp. 761-778, 1980.

3. C.W. Turner and S.O. Ishrak, "Two-dimensional Imaging with a
       High Resolution PVF_2/Si Optically-scanned Receiving Trans-
       ducer," Acoustical Imaging, Vol. 11, J.P. Powers, Ed.,
       Plenum Press, New York, 1981.

4. C.W. Turner, A. Ayoola, S.O. Ishrak and M. Salahi, "Physical
       Limitations of Optically-scanned Acoustic Imaging Transducers
       at Ultrasonic Frequencies Above 10 MHz", Acoustical Imaging,
       Vol 12, E.A. Ash and C.R. Hill, Eds., Plenum Press, New York,
       1982.

5. S.O. Ishrak, "Optically-scanned Piezoelectric/Semiconductor
       Transducers for Acoustic Imaging", Ph.D. Thesis, pp. 264-267-
       University of London, 1981.

SIMULATION OF IMAGING SYSTEMS FOR UNDERWATER VIEWING

Kjell Dalland and Magne Vestrheim

Norwegian Underwater Technology Center (NUTEC)
5034 YTRE LAKSEVÅG
Bergen, Norway

INTRODUCTION

The exploration and development of offshore oil and gas resources at increasing water depths challenges present-day technology. In order to make deep-water operations feasible and cost efficient, the use of unmanned, remotely controlled operations and sensing becomes increasingly attractive. Because of the better transmission and penetration strength compared to optic waves, acoustic waves offer interesting capabilities for the underwater remote sensing and transmission of information.

To study and evaluate the potentials of acoustic imaging systems for underwater observation and inspection purposes, work is in progress at NUTEC for the development and use of a versatile computer-based simulation tool. The simulation system uses input from computer-calculated sound fields for simulated acoustic objects to evaluate the imaging capabilities of acoustic imaging methods. The system is intended to be used together with experimental data for designing acoustic imaging systems for underwater applications.

Some of the work on the simulation system will be briefly discussed here, and some early results will be presented. The system is intended to be further developed with respect to choice of simulated objects, imaging methods and flexibility in imaging hardware configuration. The software for the system is described in detail in /1/.

PARAMETERS USED

Our interest has mainly been concentrated around ranges from 5 to 50 meters. This has led to a choice of a monochromatic system with frequency 500 khz. The frequency has to be kept low to assume no attenuation in the water.

Square arrays with equidistantly-spred receiving elements have been simulated. To give an impression of the parameters involved, it can be mentioned that the standard test example used in most comparisons has been a quadratic array with one meter in each direction and with 32x32 elements. The number of elements have been kept relative low in the simulations, both to keep the computation time down and to easier detect weaknesses with the algorithms. The input to each element has contained both phase and amplitude information.

THEORIES FOR RECONSTRUCTION

In our reconstruction we assume there is a pressure disturbance $p(r_0)$ at a plane a distance zo from the array. Assuming a linear theory and no other objects, the Rayleigh integral gives the pressure field $p(r)$ at the array:

$$p(\underline{r}) = \iint p(\underline{r}_0) \cdot \exp\{ikR\}/2\pi R \; dS \qquad (1)$$

where
$$R = [\,(x-x_0)^2 + (y-y_0)^2 + z_0^2\,]^{\frac{1}{2}}$$

x_0, y_0 and z_0 denote the coordinates of the object and x,y and $z=0$ are the array coordinates. The time factor $\exp(i\omega t)$ is neglected in the equations since it is a common factor in all terms.

One of the main prospects of this work has been to compare two methods of getting an expression for an image from the input $p(r)$ at the array . This is not necessarily $p(r_0)$, but a function proposional to the sound intensity is approximated. The two approaches are:

1. Fresnel Integral Method (FIM) /2/

By using the Fresnel approximation:

$$R= \begin{cases} z+[(x-x_0)^2+(y-y_0)^2]/2z & \text{in the exponent} \\ z & \text{in the denominator} \end{cases}$$

eq.(1) can be written as a Fourier Transform

$$p(x,y) = C(z_0)\exp\{ik(x^2+y^2)/2z_0\}\cdot FT_{xy}[p(x_0,y_0)e^{ik(x_0^2+y_0^2)/2z_0}]$$

where $C(z_0)$ is a variabel only dependent on z_0 and FT_{xy} denotes the Fourier Transform. Inverting according to the rules for Fourier Transforms gives

$$p(x_0,y_0) = C^1(z_0)e^{-ik(x_0^2+y_0^2)/2z_0}\cdot IFT_{xy}[p(x,y)e^{-ik(x^2+y^2)/2z_0}]$$

For presentation the relative squared amplitude is taken and constants neglected

$$I(x_0,y_0)\alpha|p(x_0,y_0)|^2\alpha|IFT[\exp\{-ik(x^2+y^2)/2z_0\}\cdot p(x,y,0)]|^2$$

2. Spatial Frequency Method (SFM)/3/

Writing the Rayleigh integral

$$p(x,y,0) = \iint p(x_0,y_0,z_0)\cdot G(x-x_0,y-y_0,z_0)\,dx_0\,dy_0$$

this represents a convolution with a Transfer function G

$$G(x,y,z) = \exp\{ik(x^2+y^2+z^2)^{\frac{1}{2}}\}/\{2\pi(x^2+y^2+z^2)^{\frac{1}{2}}\}$$

and is evaluated by the means of Fourier Transforms

FT(Image) = FT(Array)/FT(Transfer func.)

Problems arise when the high spatial frequencies $(k_x^2+k_y^2>k^2)$ force the Fourier Transform of the transfer function to an exponential decay. The denominator goes to zero for these frequencies and noise and errors are strongly amplified. My implementation is limited in resolution since FFT is used and only perserves frequencies up to $kx=2\pi/dx$. And as long as $dx >$ lamda even frequencies below the critical point for the transfer function are neglected.

Differences between the Methods

While the SFM is exact, the FIM uses the Fresnel Approximation. A criteria for using this is

$$(z_0)^3 >> (\pi/4\lambda)\cdot[(x-x_0)^2+(y-y_0)^2]_{max}^2$$

which gives a maximum phase change contribution, by the
next-higher order term in the exponent less than one
radian. For the test example of a square array with 32
elements and 1.0 meter in each direction, zo theoretically
has to be much greater than 9.6 meters as the criteria
gives. But as many authors refer this is not a necessary
requirement since other effects disturb the picture.
For this example the Fresnel Approximation gives good
results for zo down to 3.0 meters. Then the noise level
in the image starts to rise. So FIM works for the ranges
in question.

The main practical difference between the methods
is the resolution. While the FIM has a constant angle
between the reconstructed points, the SFM has the sampling
distance between the image points for all ranges. The
image from the FIM method is always the largest image
possible without getting aliasing. The SFM gives aliasing
at short ranges and lose information at long.

SIMULATION PROGRAM

The simulation program was implemented on a VAX
11/750 computer to test some features of the reconstruction algorithms and later evaluate imaging systems designs.
The program generated the input data, reconstructed them
and displayed the resulting image as a three-dimentional
plot of intensity.

In most of the input generating routines, it was
assumed that an object in the object plane (z=zo) was
radiating sound at the given frequency. This object can
either be a collection of point sources, a circular piston
or a rectangular piston. Known theories /4,5,6/ were used
in generating the pressure field from the objects. For
the point source routine, special features as some of the
points are out of focus, phase difference between the
points and directivity of the receiving elements could
be taken into account. In the last the elements are
assumed to be small circular pistons with the radius
given as input. Then the far-field Bessel directivity
corresponding to this radius is calculated and taken into
account in each point source/element configuration.

One choice is to get input as scattering from a
cylinder. This routine assumes a point source at the edge
of the array and a cylinder in the object plane. Since
the ka values of interest are in the order of 1000, a
geometric optic solution /7/ is used.

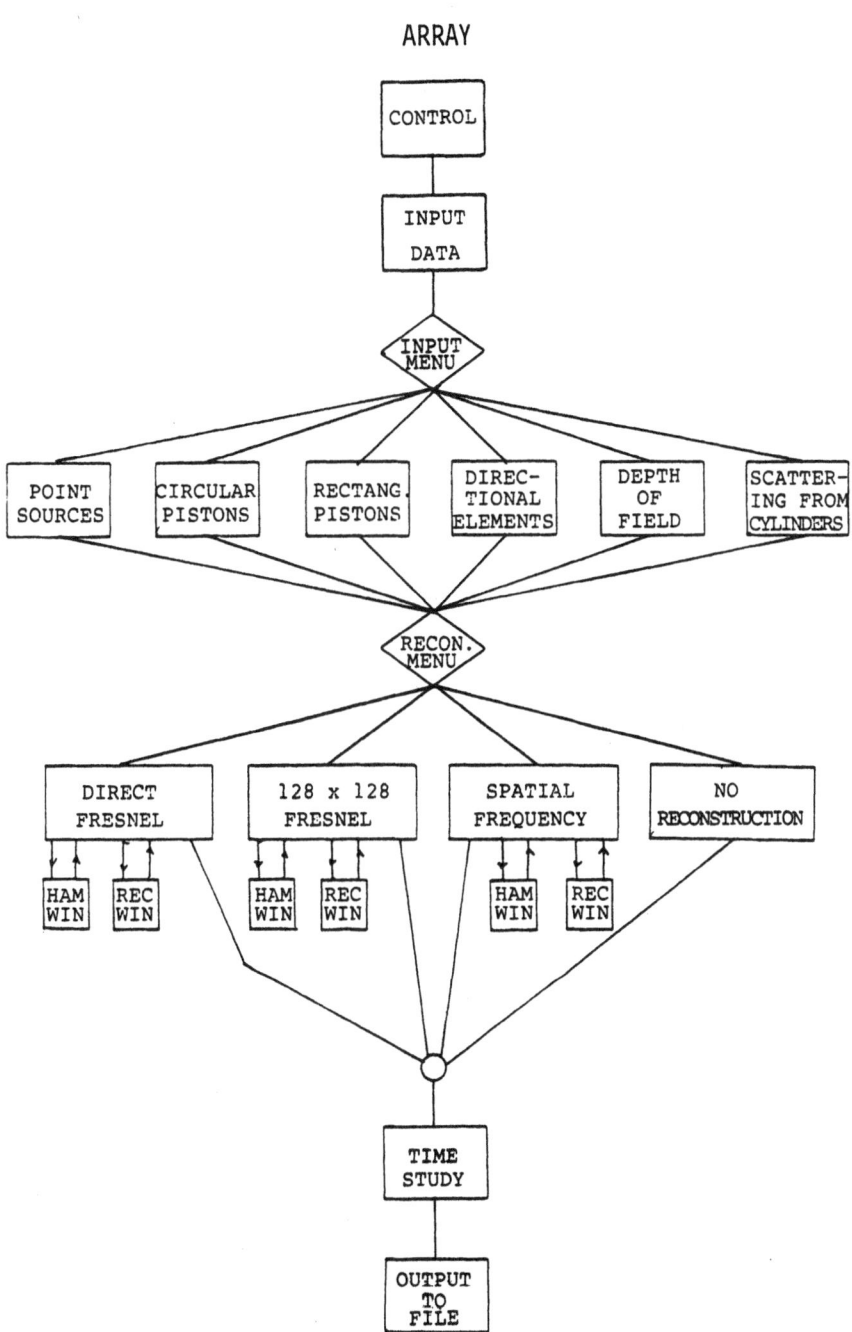

Fig. (1) Program elements of the Simulation Program.

In the reconstruction part both the Fresnel Integral
and the Spatial Frequency methods were implemented in
different variants with use of a two-dimentional FFT.
The use of Hamming window is optional with both methods.
Some of the best results are achieved with an interpolation in the image, done by filling the input matrix
with zeros outside the part with real data and using a
larger FFT.

Efficiency

With an Analogic AP-400 array processor reconstruction can work in real time. On the array processor the
memory is limited. Accordingly the time studies have been
run on the implementations at the VAX 11/750. The time
is taken on the necessary time between each image, from
when the data has come into the computer to it is ready
to be taken out. It includes FFTs,multiplication with
precalculated constants and calculating the squared
magnitude. Input is from 32x32 elements.

Method 1:   Direct Fresnel
            (focus factors and 32x32 FFT) 1 sec.

Method 2:   128x128 Fresnel
            (focus factors and a 32x32 input
            in a 128x128 FFT) 40 sec.

Method 3:   Spatial Frequency
            (128x128 convolution by use of FFT) 80 sec.

DEPTH OF FIELD

For the most practical applications the assumption
that the object is in a plane parallel to the array may
seem strong, and it is necessary to see how critical an
unfocusing of a part or the whole object is.

Using a continuous variant of the Fresnel Integral
method and a point source input one can achieve an
analytical expression for the image as a function of the
defocusing.

$$p(x,y;\mu) = C(z_1) \cdot \{ [K\{\mu(D_1+\beta)\}-K\{\mu\beta\}] \cdot [K\{\mu(D_2+\gamma)\}-K\{\mu\gamma\}]/\mu^2 \}$$

where
$$\mu = \{2(z_2-z_1)/z_2 \cdot z_1\}$$
$$\beta = 2(x-x_0)/(\lambda \cdot z_1 \cdot \mu^2)$$

$$\gamma = 2(y-y_0)/(\lambda \cdot z_1 \cdot \mu^2)$$

$$K(x) = \int_0^x \exp(-i\pi T^2/2)\, dT = C(x) + iS(x) \quad \text{Fresnel Integrals}$$

z_1 is the distance to the point source
z_2 is the distance used in the reconstruction

When z2=z1 then μ=0 and p reduces to a product of
two sinc functions. As μ increases by reconstruction at
different z2s the energy in the image spreads out. A
measure for image focusing has been the range where the
top has fallen less than 3dB from the value when μ=0.
This region can be given as

$$\left| \frac{1}{z_2} - \frac{1}{z_1} \right| < \frac{(2.22)^2}{D_1 \cdot D_2} \cdot \frac{X}{4}$$

Fig. (2) shows a reconstruction at 40.0 meters of
six point sources. Three points at 20.0 meters and three
points at 40.0 meters, in both cases at coordinates
corresponding to the lower left corner, the center and
the upper right corner of the array. The center points
add and enforce since they are in phase at the array.
The outer points at 20.0 meters move out to the side and
spread out. In this case the loss in intensity due to
further spherical spread of the 40.0 meters points about
balance the defocusing of the points at 20.0 meters.

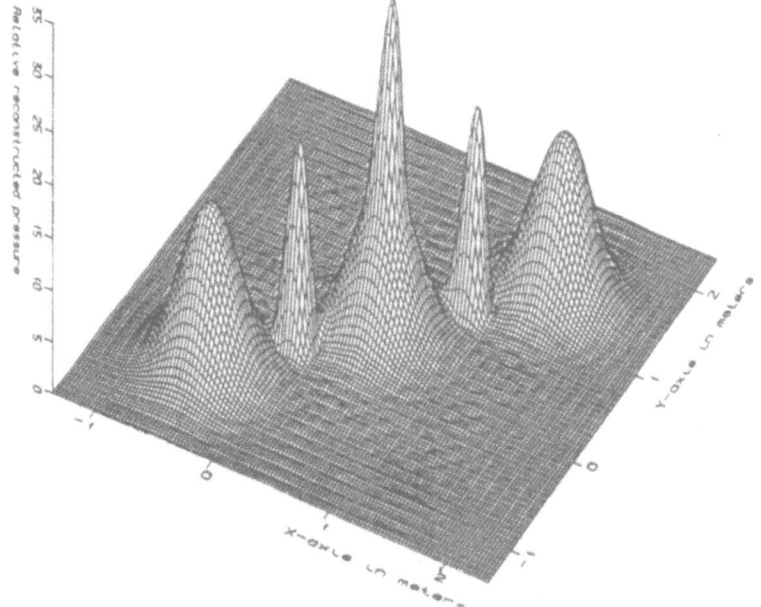

Fig. (2) Effects of defocusing(see text).

CIRCULAR PISTONS

    An important feature of the system has been to have
an input not only from point sources, but from extended
objects of actual size. While a point source is undirectional, a circular piston with uniform velocity distribution has a directional sound field. Neither the
undirectional point source field or the exstreme
directional piston field are realistic models for a
subsea input, but an imaging system's reaction these
exstreme cases can help us evaluate it. With a point
source input the sound field amplitude is almost uniform
over the array and the information lies mostly in the
phase. But as seen from Fig. (3) the field from a piston
with radi .3 meters at 20.0 meters has a very defined
main lobe. If this falls outside the array, or even on
the edge, reconstruction is impossible, only some contours
of the piston remain in the reconstruction. At this point
it is necessary to say that the expression used for the
field works best near the axis. The expression is an
integral formula developed by Naze Tjøtta and Tjøtta /4/
by a parabolic high ka approximation of the Helmholtz
equation.

$$p(\sigma,\xi) = \exp\{[i(ka)^2/2]\sigma\}$$
$$\times\{1-e^{i/\sigma}+2\cdot e^{i/\sigma}\cdot\int_0^\tau e^{i\sigma(kas)^2}J_1(2kas)\ ds\}$$

where σ and ξ are cylindrical coordinates

$\sigma = 2z/ka^2$

$\xi = \sqrt{(x-x_0)^2+(y-y_0)^2}/a$

$\tau = \xi/ka\sigma$

Fig. (3)
Field from a
piston (see
text).

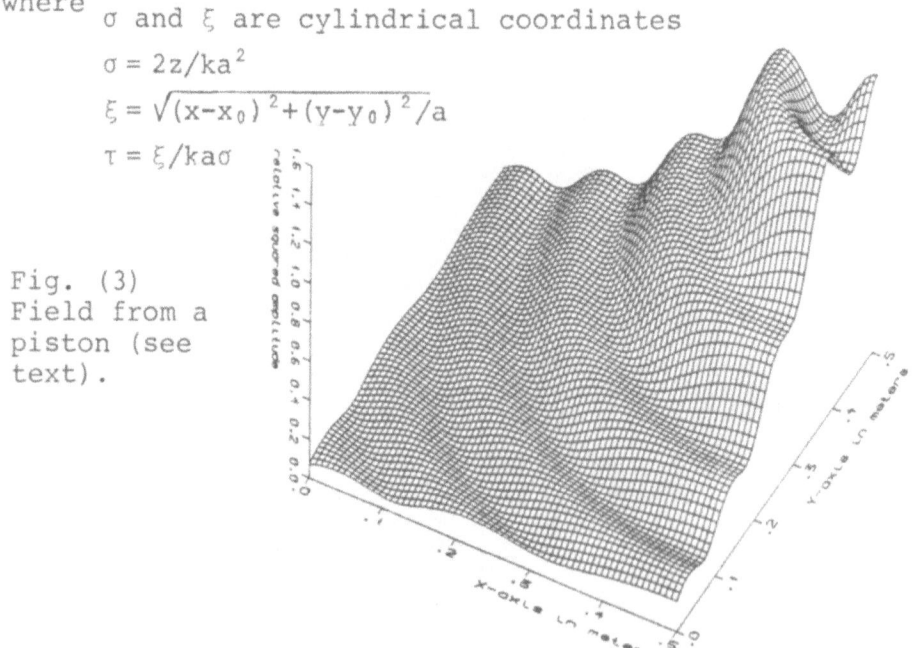

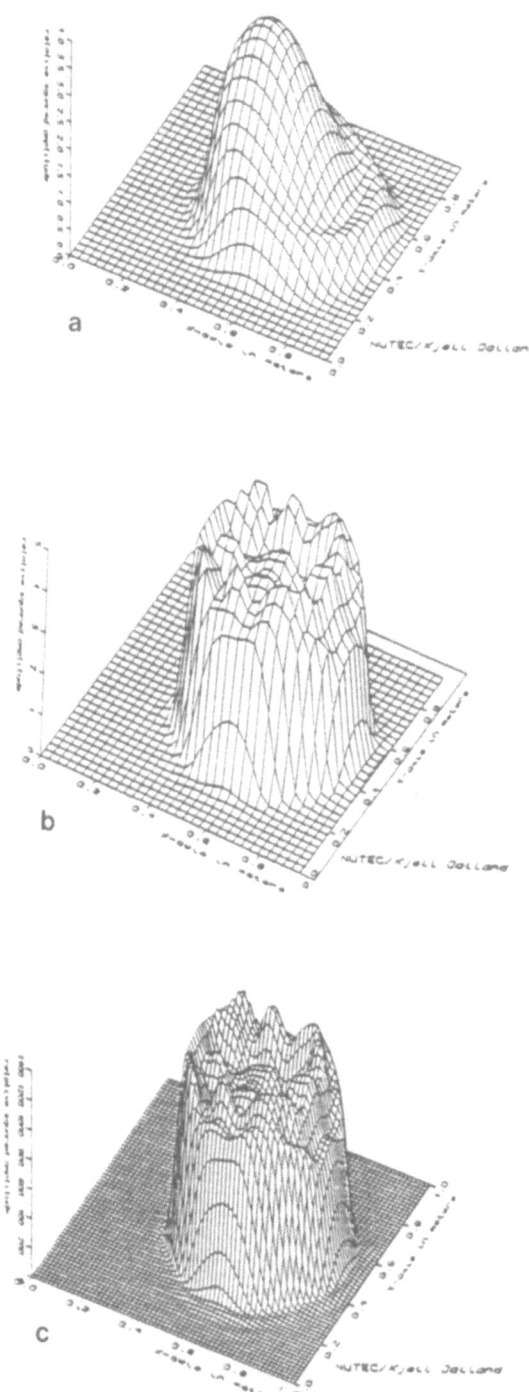

Fig.(4) Reconstruction of an input from a piston
source, with a)FIM with Hamming window, b) FIM with
rectangular window and c) SFM with rectangular window.

To see some weakness with use of Hamming window,
which is successful with point source input, an example
with the piston with a 30 cm radius. (The field shown in
Fig. (3).) This was taken as input to a 64x64 array
(1x1 meters) and where the central axis of the piston
hits the array at (.6,.5), a little off the center.
The image is reconstructed with Hamming or rect. window
FIM and the SFM with rectangular window (Fig. 4 a,b and c).
One can see that with a Hamming window not only has the
expected cylinder shape lost its edges, but the effect
of getting nearer the edge of the array is much stronger.

SCATTERING FROM CYLINDERS

Few results are obtained from these simulations up
till now. The only conclusion one can draw is that with
high frequencies and cylinders of the interesting sizes
mirror effects will be a great problem. Fig. (5) shows a
contour plot of a reconstruction of a cylinder, infinite
in y-direction and 1.0 meter in radius. The distance is
20.0 meters.

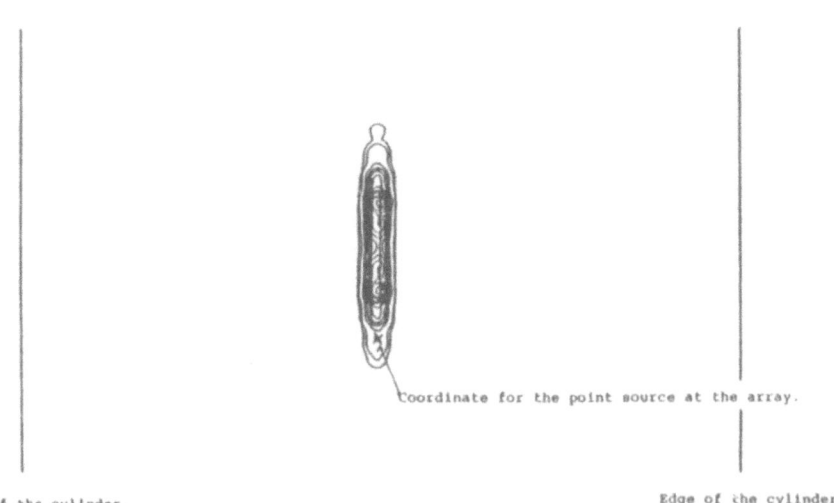

Coordinate for the point source at the array.

Edge of the cylinder.   Edge of the cylinder.

Fig. (5)   Reconstruction of an infinite cylinder with rad-
           ius 1.0 meter Lowest contour line - 13 dB

CONCLUSION

The simulation system developed is a powerful tool
in studying acoustic imaging characteristics. The work
is still in an early phase. Further development of near-
field models for extended objects, reconstruction methods,
and flexibility in system configuration is needed. By
using the simulation system together with experimental
data, efficient system designs for specific underwater
imaging applications can be evaluated.

ACKNOWLEDGEMENT

The authors wish to thank their colleague Erik
Jacobsen and Professors Sigve Tjøtta and Jaqueline Naze
Tjøtta, Department of Mathematics, University of Bergen,
for helpful discussions and comments.

This work was supported by the Royal Norwegian
Council for Scientific and Industrial Research.

REFERENCES

1. K. Dalland, "Simulation of an Underwater Acoustic
    Imaging System," NUTEC Report No. 47-83,
    Bergen (1983).
2. J. L. Sutton, "Underwater Acoustic Imaging,"
    Proc. IEEE 67, pp. 554-566 (1979).
3. B. Hosten and J. J. Roux, "Théoreme de Green ou
    Spectre Angulaire en holographic acoustique
    plane," C. R. Acad. Sc. Paris, 285, pp. 317-319
    (1977).
4. J. Naze Tjøtta and S. Tjøtta, "Analytical model
    for the near-field of a baffled piston trans-
    ducer," J. Acoust. Soc. Am. 68, pp. 334-339
    (1980).
5. P. Morse and U. Ingard, "Theoretical Acoustics,"
    Chapter 7, McGraw-Hill, New York (1968).
6. A. Freedman, "Sound Field of a Rectangular
    Piston," J. Acoust. Soc. Am. 32, pp. 197-209
    (1960).
7. J. J. Bowman, et al. (ed), "Electromagnetic and
    Acoustic Scattering by simple shapes,"
    Chapter 2, Wiley, New York (1969).

IMAGING OF CLUTTERED TARGETS IN COLOURED ABSORPTIVE MEDIA

Roderick C. Bryant and Robert E. Bogner

Department of Electrical and Electronic Engineering
The University of Adelaide
Adelaide, South Australia, 5000

INTRODUCTION

In this paper we address the problem of imaging a solid surface using ultrasonics through a medium that exhibits highly coloured absorption properties and random scattering, while ambient noise arises from acoustical and electrical sources.

The immediate application is the imaging of sheepskin through the wool for the control of an automated shearing system. Two different modes of control may be envisaged. In one mode a substantial area is imaged and then shorn. In the other, a strip is imaged just ahead of the cutter and transverse to the direction of cutter motion.

The first mode imposes delays during the shearing operation but the second mode entails severe restrictions on the size of the focussing aperture because of mechanical constraints. In both modes, the technique adopted is to range points on the skin using a-priori knowledge of the surface characteristics to aid in false target discrimination.

We examine suitable and novel transduction and focussing systems, image reconstruction algorithms, and signal optimization.

TRANSDUCERS

Because the attenuation increases rapidly with frequency, we are forced to use a frequency band below about 150KHz. However the 1mm range resolution required demands a system bandwidth of around 80KHz.

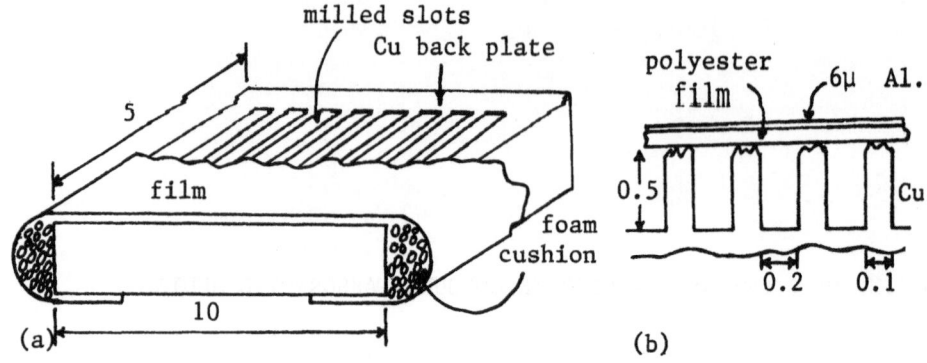

Fig. 1.   Transducer Element. (Dimensions in mm.)

Solid transducers operating in air are incapable of such wide relative bandwidths because the gross impedance mismatch between the transducer material and the medium prohibits rapid response.

Figure 1 illustrates the construction of a typical electro-static transduction element. A 6 micron polyester film is stretched across the surface of a grooved and roughened metal plate. The signal voltage, plus a bias, is applied between this plate and an aluminium coating on the film on the side away from the plate. We have found that the use of a grooved backplate provides much enhanced sensitivity over other surface profiles. Although similar technology is quite widely used there does not appear to be a satisfactory theoretical model except for the case of a uniform airgap. Thus the film tension, the grooves, and the roughening of the plate are the result of an extensive empirical optimising procedure.

As a result of this experimentation we have formulated the following hypotheses which form the basis of our design technique. They are applicable for frequency ranges up to about 150KHz.

1.   The rails between the grooves provide the excitation while the air in the grooves effectively controls the compliance of the diaphragm above the rails and the grooves.
2.   The rails need to be sufficiently rough and narrow to allow good air communication between the grooves and the air pockets above the rails.
3.   The grooves should be narrow compared to a wavelength at high frequencies to prevent unwanted resonance effects.
4.   The groove depth controls the effective diaphragm compliance which resonates with the diaphragm mass to achieve the required frequency response.

The price paid for wideband performance is that peak power is limited by the puncturing stress of the diaphragm. As a result, it

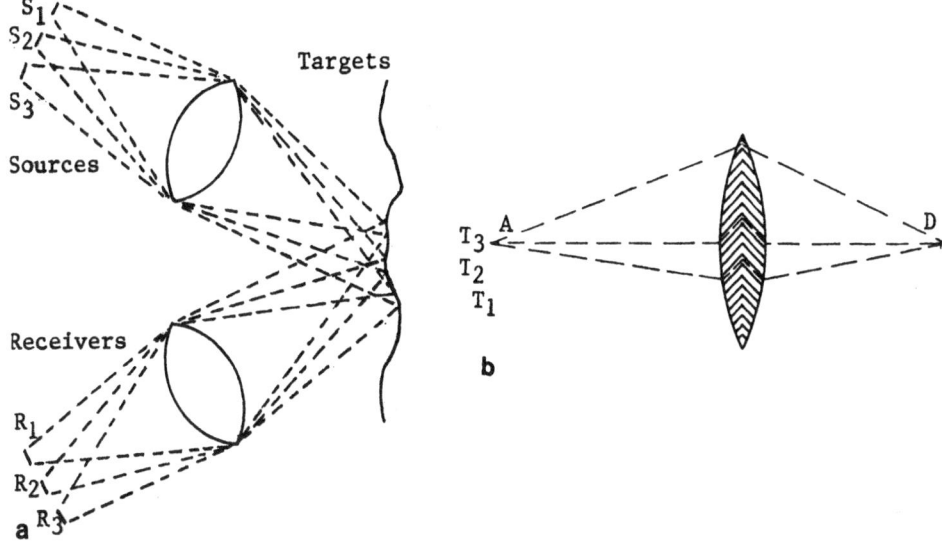

Figure 2   a) Lenses and Retinas.
           b) Path Length Lens.

is necessary to transmit constant-envelope signals of extended
duration which are compressed at the receiver by the optimal receive filter.

FOCUSSING

     The elemental transducers of Figure 1 may be used in trans-
ducer arrays with electronic beam formation. They may be placed
on a concave surface to form a focussing aperture, without electrical delays being necessary. Alternatively several may be used
to form a retina in conjunction with a lens.

     Figure 2A illustrates the lens and retina approach. 2 lenses
and 2 retinas appear to be necessary in our application because the
very large attenuation (up to 100dB) and relatively small round
trip delay renders transmit-receive switching impracticable. The
reason is that very small residual vibrations of the diaphragm for
several milliseconds after pulse transmission interfere with the
received signal.

     Figure 2B shows the novel path-length lens which we have dev-
eloped for airborne focussing applications. The device is constructed of bent vanes which are assembled and then machined to
form a spherical lens. Acoustic delays in the lens are achieved
by varying the path length rather then the velocity. The latter
approach is precluded by the severe impedance mismatches between
air and any self supporting lens material.

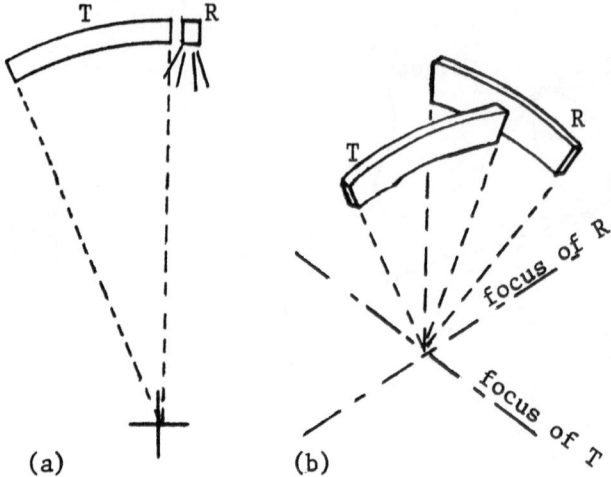

Fig. 3. Focussing with 'Bananas'.

The gaps between the vanes need to be small in order to avoid multimode propagation at higher frequencies with attendant dispersion. However, narrow gaps result in attenuation which increases with lens size.   The final design is a compromise between these factors.

Figure 3 illustrates the use of singly-curved transducers or 'bananas'. Two bananas oriented at right angles provide effective focussing by focussing the transmitter and microphone along lines which cross at right angles. Multi-point focussing is envisaged by stacking bananas. Results presented later in this paper were achieved using such bananas.

In the design of any focussing system for this application a compromise must be reached between transverse resolution which improves with aperture size and depth of focus which decreases with aperature size. One way to alleviate this problem is to use a beam-formed array with dynamic focussing for reception but the electronic complexity of this approach is less attractive. Dynamic focussing of the transmitting system is not practicable because of the duration of the transmitted signal and because it is necessary to illuminate targets at various ranges simultaneously.

Figure 4 demonstrates how the performance of a focussing system depends on the signal spectrum.

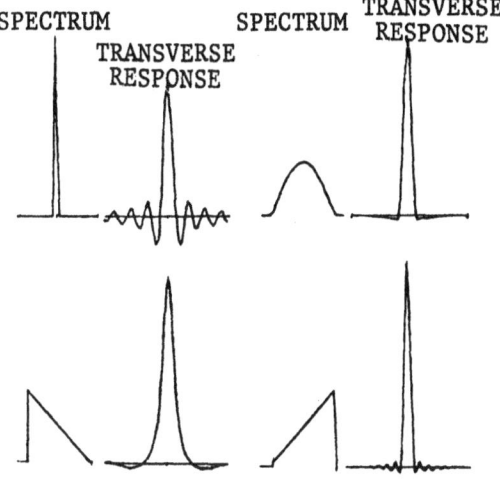

SPECTRUM

TRANSVERSE
RESPONSE

SPECTRUM

TRANSVERSE
RESPONSE

Fig. 4.   Spectra and Focussing.

TRANSMISSION PATH MODEL

Our model for the transmission channel is shown in Figure 5.
One path, via the deterministic filter with known impulse response,
$h_T(t)$, represents transmission of the target echo. The path via
stochastic impulse response, $h_C(t)$, models the effects on the received waveform of scattering within the medium. Both paths include focussing effects since the model target is a square patch
on the surface of dimension equal to the required transverse resolution. Reflections from the surface outside this patch add to the
clutter and not to the target signal. As we have seen it is important that our signal design take focussing effects into account.

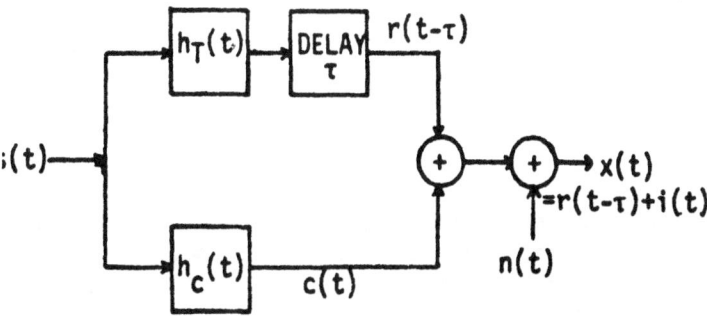

Fig. 5.   Transmission Path Model.

M.A.P. ESTIMATION

Now, using Bayes' rule it is easily shown that:

$$p(\tau|x(t),\hat{\underset{\sim}{\tau}}_p) \propto p(x(t)|\tau,\hat{\underset{\sim}{\tau}}_p) \; p(\tau|\hat{\underset{\sim}{\tau}}_p)$$

where $\hat{\underset{\sim}{\tau}}_p$ is a vector of estimates for nearby points on the surface while $\underset{\sim}{x}$ and τ are defined in Figure 5.

It is clear, also, that for a slowly moving target where Doppler shift is negligible, $x(t)$ does not depend on $\hat{\underset{\sim}{\tau}}_p$ if the focussing is perfect. Hence, we obtain the approximate relation:

$$p(\tau|x(t),\hat{\underset{\sim}{\tau}}_p) \propto p(x(t)|\tau) \; p(\tau|\hat{\underset{\sim}{\tau}}_p)$$

It is convenient to generate in our estimator a function, $F(\tau)$, which is monotonic with this conditional posterior density, where:

$$F(\tau) = \ln[p(x(t)|\tau)] + \ln[p(\tau|\hat{\underset{\sim}{\tau}}_p)]$$

We will assume that the second term is the log of a normal density, whence:

$$\ln[p(\tau|\hat{\underset{\sim}{\tau}}_p)] = \frac{(\tau - \bar{\tau})^2}{2\sigma_\tau^2} + K$$

where K is an uninformative constant which may be omitted, $\bar{\tau}$ is predicted target delay based on $\hat{\underset{\sim}{\tau}}_p$ and σ_τ^2 is the mean squared prediction error.

If the statistics of the target range function are stationary then σ_τ^2 may be estimated apriori and $\bar{\tau}$ may be the result of a Wiener filtering of the vector, $\hat{\underset{\sim}{\tau}}_p$, since such a process yields a minimum mean squared prediction. The coefficients of this filter will depend on the statistical model used for the surface.

More sophisticated prediction algorithms may be used if a more accurate apriori model of the surface characteristics is available.

In the current application a single estimation pass appears to be adequate but iterative algorithms based on this approach can be envisaged.

Other algorithms have been investigated which use the complete log likelihood functions $\left(LL(\tau|x(t)) = \ln[p(x(t)|\tau)]\right)$ for some nearby points rather than the corresponding delay estimates. These algorithms are based on approximating the function, $\ln[p(\tau_n/\hat{\underset{\sim}{\tau}}_p, x_n(t), x_m(t))]$ where the subscripts n and m denote different points on the skin surface. The results of Figure 6 for the 1-dimensional case

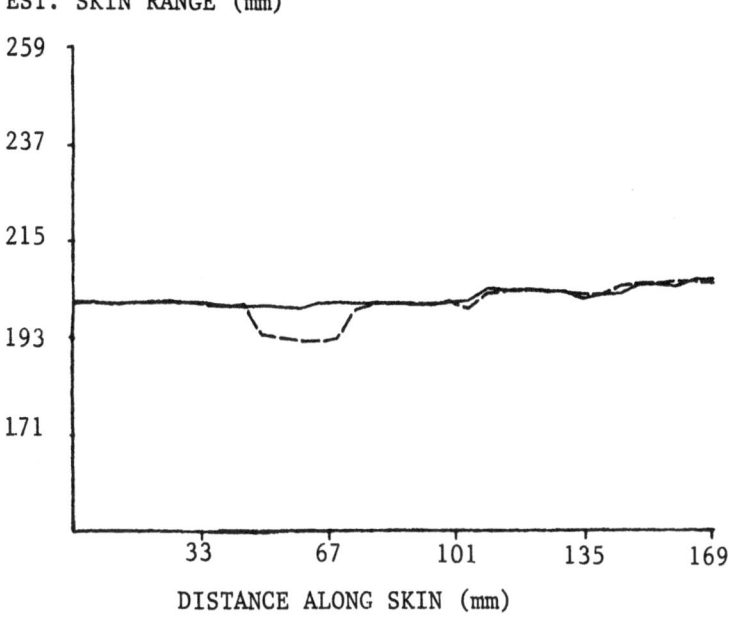

EST. SKIN RANGE (mm)

DISTANCE ALONG SKIN (mm)

Fig. 6.   1 D. Range Track.

were obtained using such a 'look-ahead' algorithm.

In that figure, the dotted line graph represents results obtained after inserting an object in the wool with its top surface
several millimetres above the skin.   The particular virtue of our
approach over range-gating is that rapid range variations can be
accommodated whilst effective discrimination against false targets
is possible.   Such discrimination would not be possible with a range
gate wide enough to allow the detection of the object.

THE LOG-LIKELIHOOD FUNCTION

The first term in $F(\tau)$ is known as the log-likelihood function
of delay, τ, given the waveform, $x(\tau)$, $LL(\tau|x(t))$. It is well
known [1] that:

$$LL(\tau|x(t)) = \left| \int_{-\infty}^{\infty} x(t)q*(t,\tau)dt \right| \tag{1}$$

where

$$r(t_1-\tau) = \int_{-\infty}^{\infty} K_i(t_1,t_2)q(t_2,\tau)dt_2 \tag{2}$$

and x, r and i are complex analytic functions defined in Figure 5
while $K_i(t_1,t_2)$ is the autocovariance function of the interference, i.

For stationary interference with zero mean, $K_i(t_1,t_2) = \Phi_{ii}(t_1,t_2)$, an autocorrelation function. (2) is a convolution so that:

$$q(t,\tau) = q(t-\tau)$$

and is given by:

$$Q(f) = \frac{R(f)}{S_{ii}(f)}$$

where Q is the Fourier transform of the optimal reference, q, R is the transform of r and S_{ii} is the transform of Φ_{ii} and hence the interference power spectrum.

This result is also obtained, for an appropriate definition of Φ_{ii}, when a local stationarity condition is met. Inspection of (2) leads to the requirement that:

$$K_i(t_1,t_2) \approx \Phi_{ii}(t_1-t_2) \nabla t_1, t_2 : \overline{\tau}-T<t_1,t_2<\overline{\tau}+T \qquad (3)$$

where $\overline{\tau}$ is expected target range and T is the effective signal duration.

However, highly coloured absorption characteristics cause echoes from greater range to suffer more attenuation and colouration than those from smaller range. Under these conditions, (3) may not be met because the high attenuation forces us to use signals of small dynamic range and long duration, T.

Such signals, having sufficient effective bandwidth, B, however, may be compressed in time by the factor BT by means of a matched filter inserted at the receiver. This enables us to satisfy (3) and so obtain Q(f) for our application.

SIGNAL DESIGN

The signal design is based on maximum likelihood estimation (MLE) of delay, with noise and clutter considered to be independent additive interferences. The received signal is coloured by the path attenuation characteristics, and so the overall performance (range variance) achieved with the MLE receiver depends on the transmitted signal. The actual transmitted signal has been designed to minimise the resultant range variance, and the design depends on the models that have been derived for the clutter, the signal reflection, the noise, and the transducers. The principle of the estimator is illustrated in Figure 7.

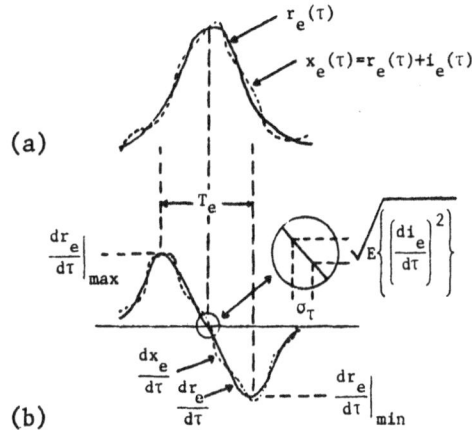

(a)

(b)

Fig. 7. Estimate Variance Formulation.

The full line waveform of Figure 7a depicts the envelope of the output of a suitably matched filter. The true range corresponds to the peak of this output, but the additive interference perturbs the position of the peak. Figure 7b shows the derivative of the perturbed envelope, and the corresponding perturbation of the zero crossing is readily related to the nominal slope and the mean square value of the output interference derivative. The nominal slope is estimated as the peak-peak signal output, U divided by the effective pulse duration T_e,

$$dv/dt = U/T_e$$

so that the mean square delay error is

$$\sigma_\tau^2 = \left(\frac{T_e}{U}\right)^2 \times (\text{mean square derivative of interference}) \qquad (4)$$

The resultant optimal transmitted signal is defined by its spectrum, and the signal has to be implemented as a constant envelope signal with this spectrum and the appropriate duration or spread to ensure that sufficient energy is transmitted. The duration is restricted by the need to avoid confusion with succeeding pulses, and a practical compromise is a duration of lms. We are using nonlinear chirps, tailored in spectrum by causing the instantaneous frequency to reside near each value for a time proportional to the spectral density required there.

The expression for delay estimate variance may be minimised analytically under certain conditions of practical interest. In particular, consider the condition when the interference power spectrum is dominated by the clutter:

$$S_{cc}(f) >> S_{nn}(f) \forall f.$$

If we can assume that this condition applies when the spectrum is optimal, then we can design our estimator accordingly. This simplifies the optimization problem considerably.

Under the additional simplifying assumptions,

$$r_0(t) >> |i_0(t)| \text{ near the peak of } r_0$$

$$\beta_r << \beta_n$$

where r_0 and i_0 are post-correlation signal and interference and β_r and β_n are effective bandwidths of the post-correlation signal and noise, the problem reduces, approximately, to minimising

$$I_2 = \int_0^\infty \frac{(f-f_0)^2 \, S_{nn}(f) \, |H_T(f)|^2}{S_{ss}(f) \, |H_c(f)|^4} \, df$$

subject to

$$E_T = \int_0^\infty S_{ss}(f) \, df = \text{constant}$$

where f_0 is the single-sided centroid of the function,

$$\frac{|H_T(f)|^2}{|H_C(f)|^2}$$

$S_{nn}(f)$ is the ambient noise power spectrum, $H_T(f)$ and $H_C(f)$ are the model target and clutter filter transfer functions and $S_{ss}(f)$ is the transmitted power spectrum.

This problem is solved exactly via an Euler-Lagrange variational procedure to give:

$$S_{ss}(f)_{optimal} \propto \frac{|f-f_0| \, |H_T(f)| \, (S_{nn}(f))^{\frac{1}{2}}}{|H_C(f)|^2} \tag{5}$$

This expression agrees well with that obtained by maximizing the detection index [2,3] under conditions ensuring that the assumptions of both derivations are satisfied. The only difference is the

$|f-f_0|$ factor which arises in the derivation as a result of the differentiation of the likelihood function in order to find the peak.

This result can be interpreted as providing an optimal compromise between the various factors affecting the range variance. The $|H_T(f)|$ factor attempts to tune the spectrum to the target channel. The $1/|H_c(f)|^2$ factor attempts to keep the spectrum out of the clutter channel while the $(S_{nn}(f))^{\frac{1}{2}}$ factor attempts to maintain a high signal to ambient noise ratio over the whole band. The $|f-f_0|$ factor has the effect of concentrating the envelope noise at low frequencies while maintaining the high frequency components in the target signal and thus enhancing the locatability of the envelope peak.

CONCLUSIONS

To obtain a high resolution image of a cluttered surface through a coloured absorbing medium using airborne ultrasonics it is necessary to pay close attention to every aspect of the design.

It is possible to use a-priori knowledge of the surface characteristics in reconstructing the image without sacrificing the ability to detect irregularities.

Using electrostatic transducers it is possible to obtain wide system relative bandwidths. It then becomes necessary to use that bandwidth most effectively when peak power is limited. The optimal spectrum for the transmitted signal has been found in closed form, as in equ (5). The optimality of this signal depends upon careful modelling of the transmission and cluttering characteristics. In particular, the target must be defined with reference to the transverse resolution requirement so that the signal design takes account of the dependence of the focussing performance on the signal spectrum.

ACKNOWLEDGEMENTS

The authors are grateful to the Australian Wool Corporation and the Australian Research Grants Scheme for financial support and to Mr. H. Nissink for his contributions in the critical early phases of the project.

REFERENCES

1. H.L. Van Trees, "Detection, Estimation and Modulation Theory", Vol. 1, p. 301, Wiley, (1969).

2. T. Kooij, "Optimum Signals in Noise and Reverberation", NATO Advanced Study Institute, Enschede, Netherlands, (1968).

3.   P.H. Moose, "Signal Processing in Reverberant Environ-
      ments", Signal Processing - a NATO Advanced Study In-
      stitute, Academic Press, (1973).

SEISMIC TOMOGRAPHY APPLIED TO SURFACE WAVE ANALYSIS

Nigel J. Wattrus

Department of Geology and Geophysics
University of Minnesota
Minneapolis, MN

ABSTRACT

A method to tomographically image the spatial variation of group velocity across a region of geologic interest has been developed. The tomographic algorithm used represents a development of the Simultaneous Iterative Reconstruction Techniques (SIRT). It includes a ray tracing capability. The principal advantage of this method over previously described methods is that it permits the dispersion function to vary continuously over the region. Computer simulations and the results from some field tests in Texas demonstrate the utility of this method.

INTRODUCTION

Since the advent of the digital computer, computerized tomography has been extensively applied to a wide range of scientific problems. Perhaps its most frequent use occurs in medical imaging. Its application to seismic imaging situations has up to this point been rather limited (Mason, 1981; Lytle and Dines, 1979; Dines and Lytle, 1980).

A tomographic technique is one which reconstructs an N-dimensional object from a collection of its N-1 dimensional projections. In a seismic situation we try to reconstruct a velocity field from a collection of travel times across the field.

SEISMIC TOMOGRAPHY

Several different tomographic methods have been described, of which only the iterative series expansion methods are suitable for application to seismic imaging problems. This is primarily because in most cases it is unreasonable to expect that a seismic ray will follow a straight path across a velocity field between the source and detectors. Also, it is usually impossible to collect travel time information from sources and receivers using fixed geometries since in many cases the source location is not within our control (eg. earthquake seismology) or access to source and detector sites is restricted.

All the iterative series expansion methods operate under the same principle, that the original function may be approximated by a finite, digital image which is iteratively adjusted until the model represents an acceptable approximation to the true field. The iterative adjustments to the model are calculated from the difference between the observed projections and those calculated across the image.

Many different forms of the iterative series expansion method have been described, but most fall into one of the following two types. The first are known as Algebraic Reconstruction Techniques (ART). They were first described by Gordon, Bender and Herman (1970). In ART algorithms the image is adjusted after each projection has been processed. The second group of iterative series expansion methods are known as the Simultaneous Iterative Reconstruction Techniques (SIRT). They were first described by Gilbert (1972). In a SIRT algorithm, a new image is calculated only after the entire set of projections has been processed.

SIRT methods have been used in this study because they are, in general, much less susceptible to interference from noise than are the ART methods. In the absence of noise however, convergence should occur faster with an ART algorithm (Gilbert, 1972).

The SIRT algorithm developed for this study operates in the following manner. The model is divided into a collection of overlapping cells. Each cell is assigned a velocity which represents the average velocity over the area of the cell. Seismic rays are traced across the trial model, which may initially be set as a constant field, between the sources and detectors. The path length of each ray in each cell is calculated and this is summed for all the rays crossing the cell. In addition the ratio of the total ray length to the i'th ray segment is used to determine what fraction of the difference between the observed and the calculated travel times should be assigned to the i'th cell. Thus the difference is distributed throughout the grid for each ray traced. These are summed for each cell as the rays are traced. Once all the rays have been traced a new model may be calculated. The new velocity at the i'th cell is given by

    1/VNEW = 1/VOLD + RF * (ΣERROR)/(ΣSEGMENT LENGTH)

where RF is a relaxation factor, an appropriate choice of which can help speed convergence.

    This process is repeated until the reconstructed image reaches an acceptable approximation to the true field. The root-mean-square of the difference, RMSD, between the observed and the calculated travel times is used to measure the accuracy of the reconstructed image. Iteration is terminated once the RMSD has stabilized, or if it becomes smaller than the sampling period of the waveforms.

SURFACE WAVE ANALYSIS

    Surface waves are commonly observed in both exploration and earthquake seismology. Their propagation velocity is related to the average velocity over the depth of penetration. Since this is frequency dependent so too is the propagation velocity and hence surface waves are usually dispersive.

    The energy in a dispersive wave travels at a different velocity from the phase velocity. This velocity is referred to as the group velocity and it may be faster or slower than the corresponding phase velocity. Since a large amplitude arrival in the wave represents a concentration of energy, most observations are interpreted in terms of group velocity , whereas most theory is based upon phase velocity. Group velocity, U(w), can be obtained from the phase velocity, V(w), by using the equation

    U = V + k*dV/dk

where k is the wavenumber.

Over the last three decades extensive investigations of the fundamental and higher modes of surface waves, and observations of the Earths free oscillations have provided much information about the global structure of the Earth's crust and mantle. By comparison, not much has been published on variations in surface wave dispersion due to regional lateral variations in the Earth's structure. The group velocity dispersion curves for purely oceanic and purely continental crust are shown in Figure 1 where it can be seen that for periods shorter than 75 seconds the two curves differ substantially. Therefore, unless the surface wave has travelled across a structure without any lateral variation of its dispersive character, the observed dispersion curve represents the average effect of all the different geologic structures between the source and the observation point.

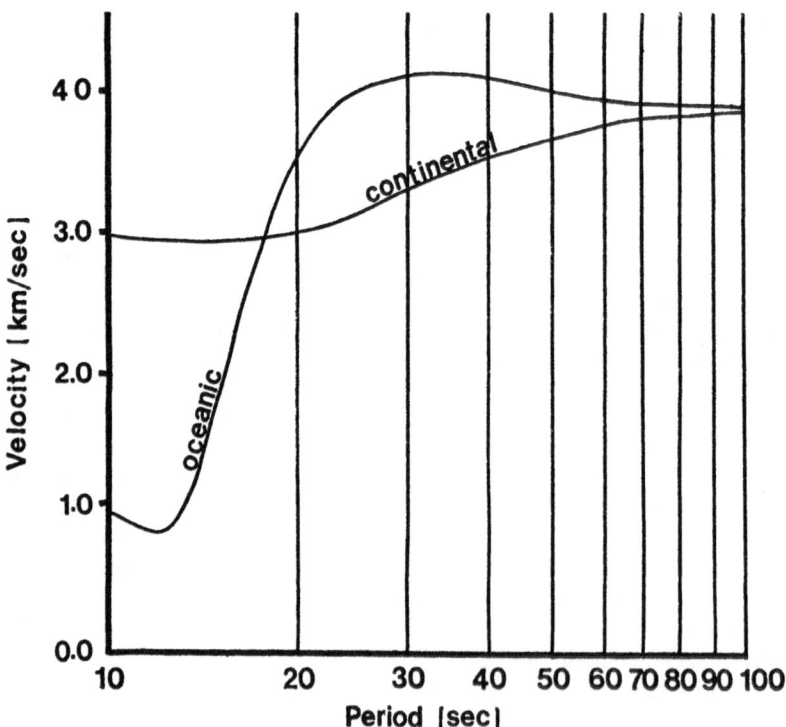

Fig.1. Group velocity dispersion curves for purely oceanic and continental ray paths (after Tarr, 1969).

If a surface wave crosses several different geologic terranes, the composite group velocity dispersion curve may be calculated if curves are available for each individual terrane. The important point to remember is that the total travel time must equal the sum of the individual travel times across each block. If a raypath crosses two blocks, A and B, with ray segments of length xa and xb, then the travel time for a frequency with a group velocity of U_a in block A and U_b in block B is

$$T_a = xa/U_a$$

in block A, and

$$T_b = xb/U_b$$

in block B. The total travel time for the complete ray path is

$$T_c = T_a + T_b$$

and the composite group velocity for the complete path is

$$U_c = (xa + xb)/T_c$$

This method has been used to remove averaging effects by several authors (Ewing and Press, 1950; Shurbet, 1960, 1961, Sheridan, 1972).

In an extensive series of investigations, Santo (1960, 1961, 1965, 1966) produced a global map of regional variation in surface wave dispersion (Figure 2). Santo used a technique known as the "crossing raypath" technique to produce his map. The crossing raypath technique assumes that the area of interest is covered by many criss-crossing raypaths. The area is divided into a collection of blocks of unknown extent, each of which has a characteristic "standard dispersion curve". The boundaries of the areas are chosen so as to minimize the difference between the observed and the calculated travel time for each waveform.

A related method to the crossing raypath technique has been described by Tarr (1969). This has been called the "raypath network" technique. As in the crossing raypath technique, the region of interest is covered by a series of criss-crossing raypaths. In this method

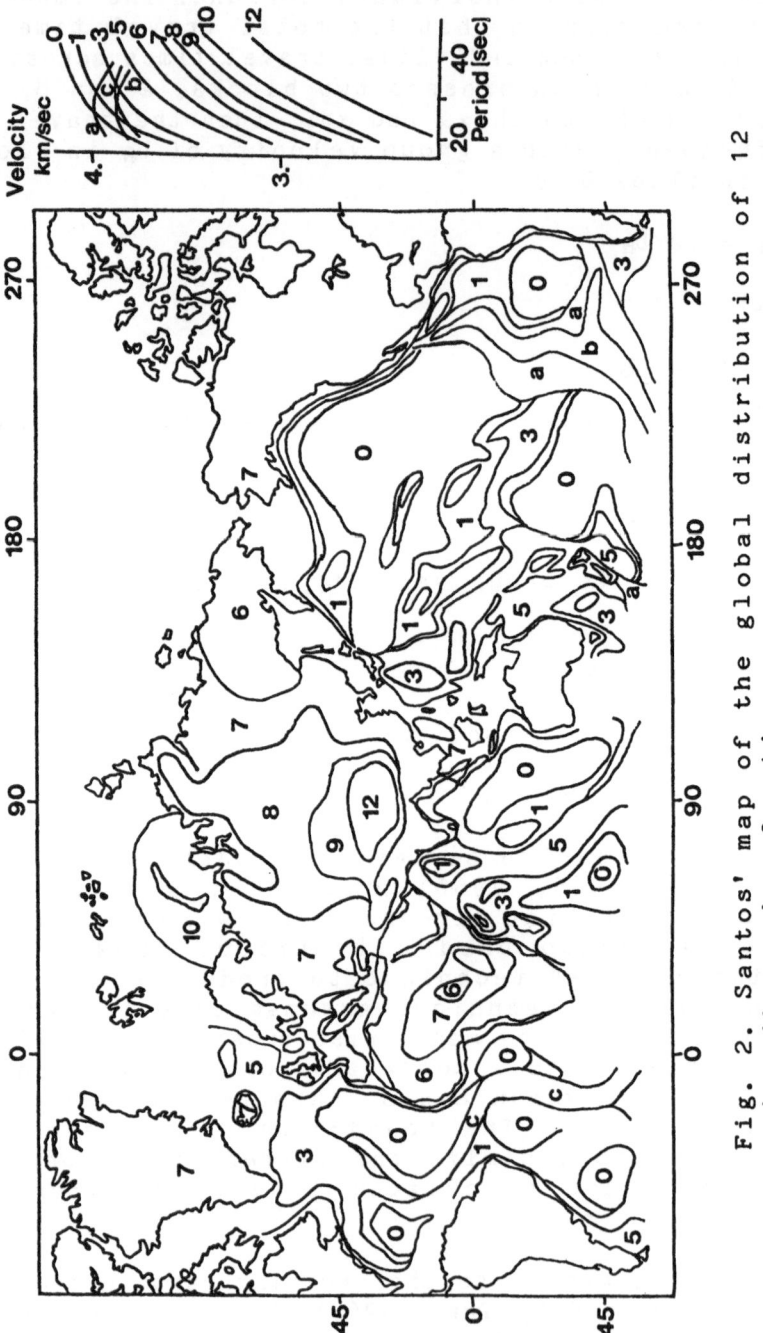

Fig. 2. Santos' map of the global distribution of 12 major dispersion functions.

however, the area is divided into cells of known
dimension and an unknown dispersive character. If m
surface waves cross a region which has been divided into
n cells, (m > n), a system of linear equations may be
set up to solve for the unknown dispersion
characteristics.

A principal difficulty with both of these
techniques is that neither of them permit a continuous
variation across the region of interest. Consequently a
transition from purely continental crust to purely
oceanic crust must be represented by a model of two (or
more) crustal blocks separated by discontinuities,
where each block has its own characteristic dispersion
function. A much better model of the situation might
show the junction as a more gradual transition. Such a
model is not feasible with any of the previously
described techniques. Seismic tomography offers a way of
obtaining such a model.

COMPUTER SIMULATIONS

Computer simulations were carried out to test
seismic tomography on some synthetic surface waves.
Since the algorithm would later be applied to surface
wave data collected on a test site 1210 feet square,

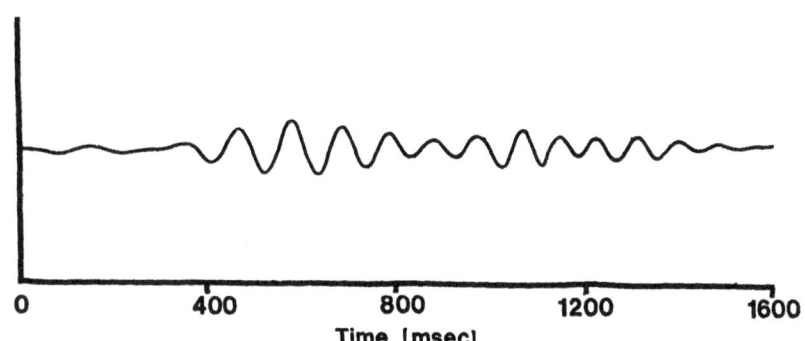

Fig.3. An example of the synthetic surface waves used in
the computer tests.

similar dimensions and source/receiver geometries were
used in the simulations. Abo-Zenas' (1979) matrix
method was used to calculate the phase velocity
dispersion function for all the synthetic surface wave
simulations. The group velocity dispersion function was

obtained by using the standard relationship between
group and phase velocities previously mentioned. From
this information, a simple dispersed synthetic surface
wave may be calculated. Figure 3 shows an example of
the type of synthetic waveform generated.

Two situations were considered during the
modelling. In the first no lateral variation in the
dispersion function was permitted. The second model
showed radial variation of the dispersion function about
a point in the image region.

Synthetic Rayleigh surface waves were calculated
using only the first or fundamental mode of the
dispersion function. The object of the simulations was
to accurately reconstruct the variation of group

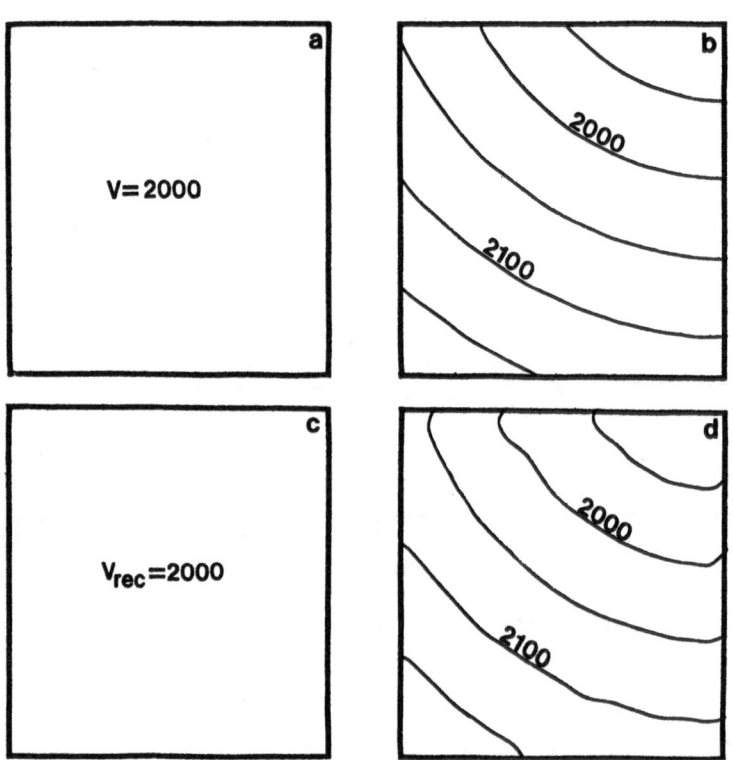

Fig.4. Examples of the computer tests. Velocity models
are shown in a and b, and the corresponding
reconstructions in c and d.

velocity within the region at three different
frequencies. The arrival time information for each
frequency component was obtained by applying Multiple
Filter Analysis (MFA) techniques to the data set. The
MFA technique applied is based upon one described by
Dziewonski, Bloch, and Landisman (1969). The MFA applies
a sharply restricted bandpass filter to each waveform,
centered at the desired frequency. The complex envelope
of the filtered signal is calculated and its peak is
picked as the frequency group arrival time. Figures 4
a,b illustrate the variation of the group velocity at 10
Hz in the original fields. Figures 4 c,d shows the
reconstructed 10 Hz group velocity fields. Clearly the
technique works satisfactorily.

FIELD TESTS

The new technique was applied to a set of surface
waves collected at a test site in N.E. Texas. Nearby
boreholes showed that the near surface geology consists
of a sedimentary sequence of claystones and marls which
are thinly bedded near the surface and became more
massive deeper in the section. An eight layer velocity
model of the test site is given in Table 1. This was
produced from information obtained from the boreholes
and some seismic refraction profiles. Group velocity
curves were calculated for the model up to and including
the fifth mode. These curves are shown in Figure 5.

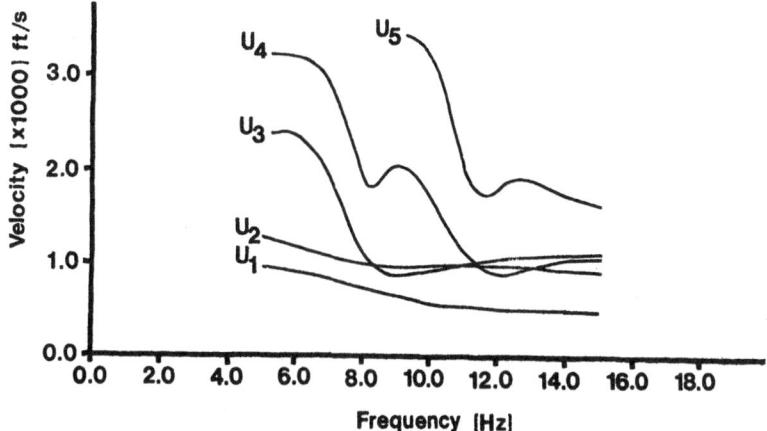

Fig.5. Group velocity curves for the first five modes of
the Rayleigh dispersion function, calculated for the
eight layer model of the test site.

Table 1. Parameters for the eight layer model of the test site.

Depth [feet]	Vp [ft/sec]	Vs [ft/sec]
11	1200	600
23	2520	800
44	3620	1000
90	6240	1400
150	6960	1500
210	6055	1900
1060	7300	3480
1980+	8000	4000

The surface waves were digitally recorded at a sampling rate of 2 ms using standard exploration geophones and small explosive charges as a source. The shot/receiver locations are shown in Figure 6. Each shot was recorded by 12 geophones positioned on the opposite side of the square test site. An example of the type of waveform recorded is shown in Figure 7a. Its amplitude spectrum is shown in Figure 7b. Most of the energy within the waveform is contained within a small frequency band centered about 10 Hz. On the basis of this information, the group velocity field at 10 Hz was the first one imaged.

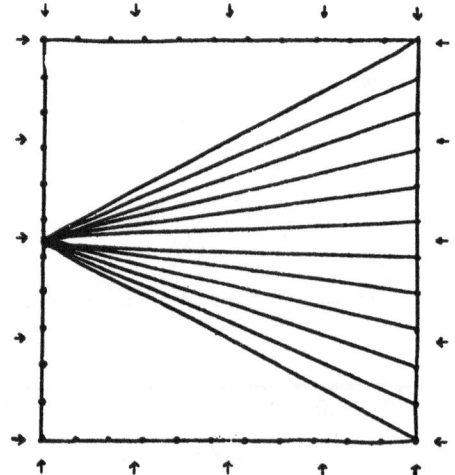

Fig.6. Shot-receiver locations used for the surface wave experiment.

The 10 Hz group velocity image is shown in Figure
8. It can be seen that over the field, the average
velocity appears to be around 1850 feet/s. From the
calculated group velocity curves (Figure 5) we may
conclude that the 10 Hz component of the waveform
represents the fourth mode of the dispersion function.
Dominant features of the imaged field are a pronounced
high velocity ridge which crosses the center of the
image approximately in a WNW–ESE direction, and a
velocity minimum on the northern boundary. The group
velocity field was also imaged at 7.5 Hz and 15 Hz. The
resultant fields are shown in Figure 9 a,b. Both of

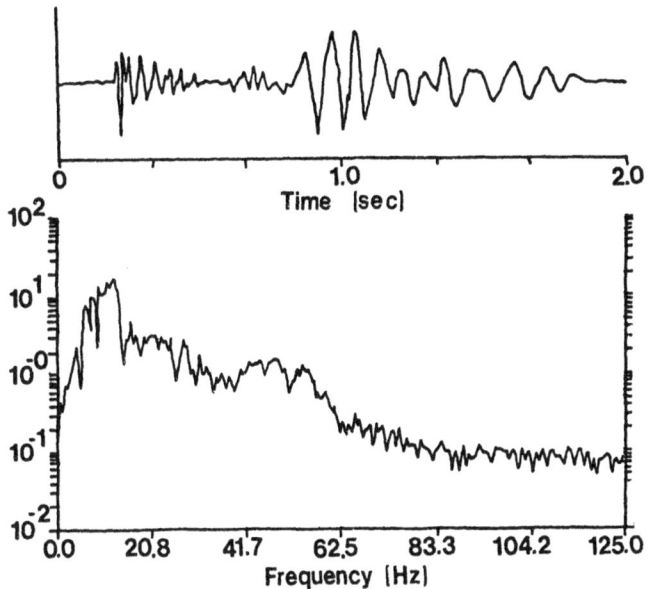

Fig.7. An example of the type of signal recorded
together with its amplitude spectrum.

these reconstructions exhibit gross features which
resemble those seen in the 10 Hz reconstruction, however
their average group velocities, both around 1950 feet/s,
do not appear to represent the fourth mode. Rather the
7.5 Hz reconstruction appears to represent the third
mode, and the 15 Hz reconstruction could represent the
fifth mode. Therefore the surface waves are not as
simple as they might at first appear to be. It seems
that the data are extremely overmoded, so that several
different modes arrive at the geophones at roughly the
same moment.

CONCLUSION

It has been shown that seismic tomography is a realistic possibility, which may be a useful technique for the analysis of regional variations in the dispersive character of the Earth's crust and upper mantle. Its use, however, is not restricted to problems of a regional scale. Provided that there is sufficent data available, it has been demonstrated that seismic

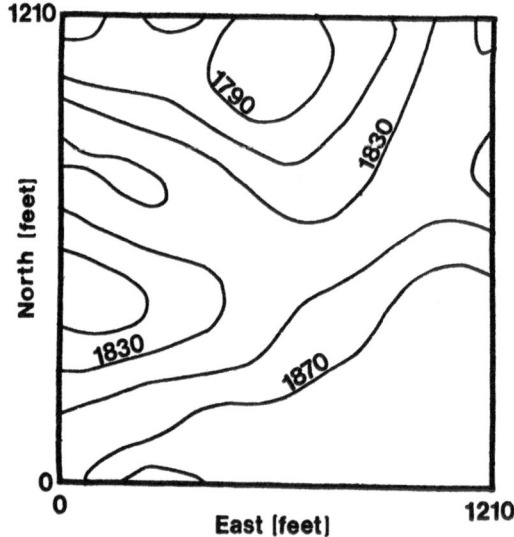

Fig.8. The velocity field imaged at 10 Hz.

tomography may have some promise at a local level.

Surface wave analysis is not the only area where tomography is of use to seismologists. Indeed, it has already been applied to cross-borehole experiments, (Lytle and Dines, 1979; Dines and Lytle, 1980), and its future application to exploration seismology appears to be inevitable.

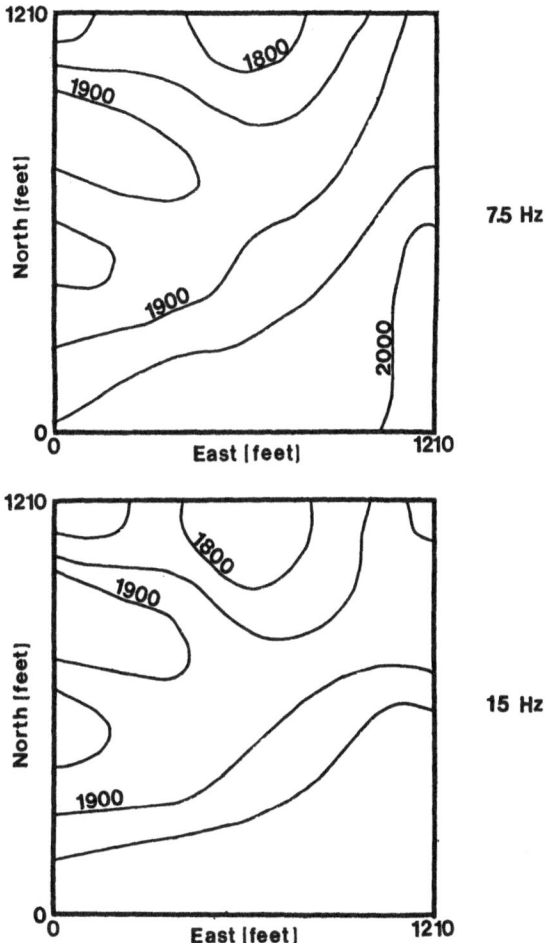

Fig.9. The velocity field imaged at 7.5 Hz and 15 Hz.

REFERENCES

Abo-Zena, A., 1979, Dispersion function computation for unlimited frequency values: Geoph. J., v.58, p91.

Dines, K.A., and Lytle, R.J., 1979, Computerized geophysical tomography: Proc. IEEE, v.67, p1065.

Dziewonski, A., Bloch, S., and Landisman, M., 1969, A technique for the analysis of transient seismic signals: Bull. Seis. Soc. Am., v.59, p427.

Ewing, M., and Press, F., 1950, Crustal structure and surface wave dispersion: Bull. Seis. Soc. Am., v.40, p271.

Gilbert, P., 1972, Iterative methods for the three-dimensional reconstruction of an object from projections: J. Theor. Biol., v.36, p105.

Gordon, R., Bender, R., and Herman, G.T., 1970, Algebraic Reconstruction Techniques (ART) for three-dimensional electron microscopy and x-ray photography: J. Theor. Biol, v.29, p471.

Lytle, R.J., and Dines, K.A., 1980, Iterative ray tracing between boreholes for underground image reconstruction: IEEE Trans. Geosci. and Remote Sensing, GE-18

Mason, I.M., 1981, Algebraic reconstruction of a two-dimensional velocity inhomogeneity in the High Hazles seam of Thoresby colliery: Geophysics, v.46

Santo, T.A., 1960, Rayliegh wave dispersion across the oceanic basin around Japan, 2: Bull. Earthquake Res. Inst. Tokyo Univ., v.38, p385.

Santo, T.A., 1961, Division of the south-western Pacific area into several regions in which Rayliegh waves have the same dispersion charadters: Bull. Earthquake Res. Inst. Tokyo Univ., v.39, p603.

Santo, T.A.,1965, Lateral variation of Rayliegh wave dispersion character, 2, Eurasia: Pure Appl. Geophys., v.62, p67

Santo, T.A.,1966, Lateral variation of Rayliegh wave dispersion character, 3, Atlantic Ocean, Africa, and Indian Ocean: Pure Appl. Geophys., v.63, p40.

Sheridan, R., 1972, Crustal structure of the Bahama platform from Rayliegh wave dispersion: JGR, v.77, p2139.

Shurbet, D.H., 1960, The effect of the Gulf of Mexico on Rayliegh wave dispersion: JGR, v.65,p40.

Shurbet, D.H., 1961,Determination of sedimentary thickness in the Mexican geosyncline by Rayliegh wave dispersion: JGR, v.66, p899.

Tarr, A.C., 1969, Rayliegh wave dispersion in the North Atlantic Ocean, Caribbean Sea, and Gulf of Mexico: JGR, v.74, p159.

APPLICATION OF ACOUSTICAL HOLOGRAPHY TO NOISE SUPPRESSION

Woon S. Gan

Acoustical Services
29 Telok Ayer Street
Singapore 0104

INTRODUCTION

The principle of Active Noise Absorption (ANA) states that we can cancel a noise source completely if we can produce another noise source of exactly same intensity but opposite phase. We call the former the primary noise source and the latter, the cancelling noise source, the secondary source. The principle of ANA is obtained from Huygens' principle[1] which states that the sound field due to the unwanted primary sources can be reproduced exactly by suppressing the primary sources and replacing it with an array of secondary sources. To do this we must convert the

primary noise sources into Huygen sources (K^H) first.

Then the absorbing secondary sources (K^A) will simply

be $K^A = - K^H$. We have to stress the importance that Huygens' principle is basically the principle of information. It states that information is invariant. So we must be extremely careful that information is totally preserved when replacing the primary noise sources by secondary noise sources (that is, to produce an identical noise field in the secondary sources as in the primary sources). A sound field identical with the primary sound field requires an infinite number of infinitely small secondary sources[2]. Mangiante[1] has shown in a computer simulation that if directional secondary sources of sufficient number and appropriate

565

spacing are employed, it is theoretically possible to achieve significantly large attenuation everywhere in the far field. The degree of noise attenuation will increase with the increasing number of small secondary sources used. Nowadays in practice, one could only obtain a reduction of about 20 dBS in Active Noise Absorption works although in theory, one should have 100% noise reduction. The reason is that information is not completely preserved in the secondary noise sources. Hence we propose the use of acoustical holography in ANA works because in holography we introduce the reference wave during the recording of the primary noise sources and this ensures that information is totally preserved in the secondary sources.

DERIVATION OF THE ABSORBING SOURCES IN THE PRESENCE OF THE REFERENCE WAVE

We start with the simple case of linear acoustics. We use the continuity equation, the Euler's equation and the equation of the conservation of the moment of momentum:

$$\frac{\partial \rho}{\partial t} + \text{div} \ (\rho_E \vec{v} + \rho \vec{v}_E) = q = L_1 \ (\rho , \vec{v}) \tag{1}$$

$$\frac{\partial}{\partial t} \ (\rho_E \vec{v} + \rho \vec{v}_E) + \text{div} \ (\vec{\sigma} + \rho \vec{v}_E \otimes \vec{v}_E + \rho_E \vec{v} \otimes \vec{v}_E$$

$$+ \rho_E \vec{v}_E \otimes \vec{v}) = F = L_2 \ (\rho , \vec{v}, \sigma) \tag{2}$$

$$\nabla \wedge \vec{v} = T = L_3 \ (\rho , \vec{v}, \sigma) \tag{3}$$

where $\rightarrow$ denotes vector, $\otimes$ symbolizes a tensor product, ρ = acoustic density, v = particle velocity, σ = acoustic strain, q = monopole density, F = dipole or impulse source density, $\vec{T}$ = quadrupole or vector source density, q, F, $\vec{T}$ = primary noise sources and quantities with subscript E refer to a basic flow.

Let us introduce a common perturbation function s' so that the field functions ρ, $\vec{v}$ and $\vec{\sigma}$ become

$$\rho' = s'\rho, \ \vec{v} = s'\vec{v} \text{ and } \vec{\sigma}' = s'\vec{\sigma} \tag{4}$$

The primary source functions q, F and $\vec{T}$ then become

$$q' = s'q + q^A, \ F' = s'F + F^A, \ \vec{\mathbb{T}} = s'\vec{\mathbb{T}} + \vec{\mathbb{T}}^A \qquad (5)$$

The perturbation function s' (x,y,z) is chosen so that s' = 0 in the part of space where one needs silence.

The necessary absorbing sources q^A, F^A and $\vec{\mathbb{T}}^A$ will be defined by

$$q^A = L_1(\ s'\rho, \ s'\vec{v} \ ) - s'q \qquad (6)$$

$$F^A = L_2(\ s'\rho, \ s'\vec{v}, \ s'\vec{\sigma} \ ) - s'F \qquad (7)$$

$$\vec{\mathbb{T}}^A = L_3(\ s'\rho, \ s'\vec{v}, \ s'\vec{\sigma} \ ) - s'\vec{\mathbb{T}} \qquad (8)$$

We now introduce the reference waves R_q, R_F and R_T for the recording of the primary sources q, F and T respectively. Then

$$(\ q + R_q \ )' = s' \ (\ q + R_q \ ) + q_H^A \qquad (9)$$

$$(\ F + R_F \ )' = s' \ (\ F + R_F \ ) + F_H^A \qquad (10)$$

$$(\ \vec{\mathbb{T}} + R_T \ )' = \quad (\ \vec{\mathbb{T}} + R_T \ ) + \vec{\mathbb{T}}_H^A \qquad (11)$$

where q_H^A, F_H^A, T_H^A = absorbing secondary sources in the presence of the reference wave.

From (9) to (11)

$$q_H^A = (\ q + R_q \ )' - s'q - s'R_q$$

$$= L_1(\ s'\rho, \ s'\vec{v}, \ R_q \ ) - s'q - s'R_q$$

$$= \frac{\partial s'\rho}{\partial t} + \mathrm{div}(s'\rho_E s'\vec{v} + s'\rho s'\vec{v}_E) + R_q - s'q -$$

$$s'R_q$$

$$= L_1 (s'\rho, s'v) + R_q - s'q - s'R_q$$

$$= q^A + R_q - s'R_q$$

$$= q^A + (1 - s')R_q$$

$$= q^A + sR_q \text{ where } 1 - s' = s \tag{12}$$

Similarly $F_H^A = F^A + (1 - s')R_F$ \hfill (13)

$$\vec{T}_H^A = \vec{T}^A + (1 - s')R_T \tag{14}$$

It was previously shown by Jessel and Mangiante[3] that

$$q^A = (\rho_E\vec{v} + \rho\vec{v}_E) \cdot \overrightarrow{\text{grad}} \ s' \tag{15}$$

$$F^A = (\vec{\sigma} + \rho\vec{v}_E \otimes \vec{v}_E + \rho_E\vec{v} \otimes \vec{v}_E + \rho_E\vec{v}_E \otimes \vec{v}) \cdot \overrightarrow{\text{grad}} \ s' \tag{16}$$

$$\vec{T}^A = (\rho_E\vec{v} + \rho\vec{v}_E) \wedge \overrightarrow{\text{grad}} \ s' \tag{17}$$

INTEGRATION OF THE HOLOGRAPHIC ABSORBING SOURCES OVER THE THREE-DIMENSIONAL SPACE

We shall integrate the holographic absorbing sources over three dimensions using Cartesian coordinates for the general case of three dimensional radiation of noise. For the perturbation function s', we choose s' (x,y,z) = Bx + Dy + Gz to account for the three dimension nature.

For the Holographic Monopole Absorbing Source

Density q_H^A

We choose the reference wave as

$$R_q = \rho_E \frac{\partial^2 \phi}{\partial x^2} + \rho_E \frac{\partial^2 \phi}{\partial y^2} + \rho_E \frac{\partial^2 \phi}{\partial z^2} \tag{18}$$

Then $\overline{q_H^A} = \int_0^d \int_0^d \int_0^d \left\{ (\rho_E v_x \frac{\partial s'}{\partial x} + \rho v_{Ex} \frac{\partial s'}{\partial x}) + (\rho_E v_y \right.$

$\frac{\partial s'}{\partial y} + \rho v_{Ey} \frac{\partial s'}{\partial y}) + (\rho_E v_z \frac{\partial s'}{\partial z} + \rho v_{Ez} \frac{\partial s'}{\partial z}) + (1 - Bx$

$\left. - Dy - Gz)(\rho_E \frac{\partial^2 \phi}{\partial x^2} + \rho_E \frac{\partial^2 \phi}{\partial y^2} + \rho_E \frac{\partial^2 \phi}{\partial z^2}) dxdydz \right.$

$$\tag{19}$$

$= (\rho_E v_x B + v_{Ex} B)d^3 + (\rho_E v_y D + \rho v_{Ey} D)d^3 + \rho_E v_z G$

$+ \rho v_{Ez} G)d^3 + d^2 \left\{ (k \rho_E A_x - B \rho_E A_x kd) \sin(\omega t - kd) \right.$

$\left. + 2B \rho_E A_x \sin(\omega t - kd/2) \sin(kd/2) - k \rho_E A_x \sin \omega t \right\}$

$+ k \rho_E A_x d^3 D \cos(\omega t - kd/2) \sin(kd/2) + d^3 k \rho_E A_x G$

$\cos(\omega t - kd/2) \sin(kd/2) + d^2 \left\{ (k \rho_E A_y - D \rho_E A_y kd) \right.$

$\left. \sin(\omega t - kd) + 2D \rho_E A_y \sin(\omega t - kd/2) - k \rho_E A_y \sin \omega t \right\}$

$- Gd^3 \rho_E \left\{ - kA_y \cos(\omega t - kd/2) \sin(kd/2) \right\} + \left\{ (k \rho_E A_z \right.$

$- G \rho_E A_z kd) \sin(\omega t - kd) + 2G \rho_E A_z \sin(\omega t - kd/2)$

$$\sin(kd/2) - k\rho_E A_z \sin\omega t\} + Dd^3 \rho_E kA_z \cos(\omega t - kd/2)$$

$$+ B\rho_E d^3 kA_y \cos(\omega t - kd/2)\sin(kd/2) - (1/2)B^2\rho_E^2 d^6$$

$$kA_z \cos (\omega t - kd/2)\sin (kd/2) \qquad\qquad (20)$$

by keeping only the real terms. The limit of integration d is chosen to be of the same dimension as the testing laboratory.

For the Holographic Dipole Absorbing Source Density F_H^A

We choose the reference wave as

$$R_F = 2j\rho_E k^3 \{\vec{A}_x^2 e^{j(\omega t - kx)} + \vec{A}_y^2 e^{j(\omega t - ky)} + \vec{A}_z^2$$

$$e^{j(\omega t - kz)}\} \qquad\qquad (21)$$

because it is of the same dimension as F.

Then $\overline{F_H^A} = \int_0^d \int_0^d \int_0^d (\vec{\sigma} + \rho\vec{v}_E \otimes \vec{v}_E + \rho_E \vec{v} \otimes \vec{v}_E + \rho_E$

$\vec{v}_E \otimes \vec{v}) \cdot \overrightarrow{\text{grad}} \, s' \, dxdydz + (1 - s')\{2\, j\rho_E A_x^2 k^3$

$e^{2j(\omega t - kx)} + 2j\rho_E A_y^2 k^3 e^{2j(\omega t - ky)} + 2j\rho_E A_z^2 k^3$

$e^{2j(\omega t - kz)}{}_{dxdydz}$

$$= d^3\{B(\sigma_x + \rho v_{Ex}^2 + 2\rho_E v_x v_{Ex}) + D(\sigma_y + \rho v_{Ey}^2 + 2\rho_E$$

$v_y v_{Ey}) + G(\sigma_z + \rho v_{Ez}^2 + 2\rho_E v_z v_{Ez})\} + \rho_E k^2 d^2$

$\{2 \cos(2\omega t - kd)\sin kd\} \{- (1 - Dy - Gz)\vec{A}_x^2 - (1 - $

$$Bx - Gz)\vec{A}_y^2 - (1 - Bx - Dy)\vec{A}_z^2] + 2\rho_E k^3 d^2 \{(d/2k)$$

$$\cos 2(\omega t - kd) + (1/4k^2)\sin 2(\omega t - kd) - (1/4k^2)$$

$$\sin 2\omega t\} (BA_x^2 + DA_y^2 + GA_z^2) \tag{22}$$

by keeping only the real terms. Again the limit of integration d is chosen to be of the same order of magnitude as for the case of q^A_H.

<u>For the Holographic Quadrupole Absorbing Source</u>
<u>Density</u> T^A_H.

We choose the reference wave as

$$R_T = \rho_E \frac{\partial^2 \phi}{\partial x^2} + \rho_E \frac{\partial^2 \phi}{\partial y^2} + \rho_E \frac{\partial^2 \phi}{\partial z^2} \tag{23}$$

which is of the same dimension as T.

Then $\overline{T^A_H} = \int_0^d \int_0^d \int_0^d \{(\rho_E \vec{v} + \rho \vec{v}_E) \wedge \overrightarrow{\text{grad }} s'\} dx dy dz$

$$+ \int_0^d \int_0^d \int_0^d (1 - s')(\rho_E \frac{\partial^2 \phi}{\partial x^2} + \rho_E \frac{\partial^2 \phi}{\partial y^2} + \rho_E$$

$$\frac{\partial^2 \phi}{\partial z^2}) dx dy dz \tag{24}$$

$$\int_0^d \int_0^d \int_0^d \{(\rho_E \vec{v} + \rho \vec{v}_E) \wedge \overrightarrow{\text{grad }} s'\} dx dy dz \tag{25}$$

$$= d^3 \{i(G\rho_E v_y - D\rho_E v_z + G\rho v_{Ey} - D\rho v_{Ez}) + \vec{j}(-G\rho_E v_x$$

$$+ B\rho_E v_z - G\rho v_{Ex} + B\rho v_{Ez}) + \vec{k}(D\rho_E v_x - B\rho_E v_y + D\rho v_{Ex}$$

$$- B\rho v_{Ey})$$ (26)

$$\int_0^d \int_0^d \int_0^d (1 - s')\left(\rho_E \frac{\partial^2 \phi}{\partial x^2} + \rho_E \frac{\partial^2 \phi}{\partial y^2} + \rho_E \frac{\partial^2 \phi}{\partial z^2}\right)$$

dxdydz is in scalar form and we have to vectorize it for uniformity in the expression for T_H^A. To do this, we choose $\vec{1} = a\vec{i} + b\vec{j} + c\vec{k}$ where $\sqrt{a^2 + b^2 + c^2} = 1$.

$$\int_0^d \int_0^d \int_0^d (1 - s')\left\{\rho_E \frac{\partial^2 \phi}{\partial x^2} + \rho_E \frac{\partial^2 \phi}{\partial y^2} + \rho_E \frac{\partial^2 \phi}{\partial z^2}\right.$$

dxdydz becomes

$$\int_0^d \int_0^d \int_0^d (a\vec{i} + b\vec{j} + c\vec{k} - B\vec{i} - D\vec{j} - G\vec{k}) \wedge (\vec{i}\left(\rho_E\right.$$

$$\frac{\partial^2 \phi}{\partial x^2}x + \vec{j}\,\rho_E \frac{\partial^2 \phi}{\partial y^2}y + \vec{k}\left(\rho_E \frac{\partial^2 \phi}{\partial z^2}z\right) \text{ dxdydz}$$

$$= \vec{i}\left[\left[k(b - 1)\rho_E d^2 A_z\left\{\sin(\omega t - kd) - \sin\omega t\right\}\right] - \right.$$

$$\left[k(c - G)\rho_E d^2 A_y\left\{\sin(\omega t - kd) - \sin\omega t\right\}\right]\right] + \vec{j}$$

$$\left[\left[k(c - G) A_x\rho_E d^2\left\{\sin(\omega t - kd) - \sin\omega t\right\} - \left[k(a - B)\right.\right.\right.$$

$$A_z\rho_E d^2\left\{\sin(\omega t - kd) - \sin\omega t\right\}\right] + \vec{k}\left[\left[k(a - B)\rho_E A_y\right.\right.$$

$$\left\{\sin(\omega t - kd) - \sin\omega t\right\}\right] - \left[k(b - D)\rho_E d^2 A_x\left\{\sin \omega t\right.\right.$$

$$\left.\left.\left. - kd) - \sin \omega t\right\}\right]\right]$$ (27)

$$\overline{T_H^A} = \vec{i}\left[d^3(G\rho_E v_y - D\rho_E v_z + G\rho v_{Ey} - D\rho v_{Ez}) + \right.$$

$$\left[k\rho_E d^2\{\sin(\omega t - kd) - \sin\omega t\}\right]\{A_z(b - D) - A_y$$

$$\left.(c - G)\}\right] + \vec{j}\left[d^3(- G\rho_E v_x + B\rho_E v_z - G\rho v_{Ex} + B\rho v_{Ez})\right.$$

$$+ k(c - G)A_x \rho_E d^2\{\sin(\omega t - kd) - \sin\omega t\} - k(a - B)$$

$$\left. A_z \rho_E d^2\{\sin(\omega t - kd) - \sin\omega t\}\right] + \vec{k}\left[d^3(D\rho_E v_x - B\rho_E\right.$$

$$v_y + D\rho v_{Ex} - B\rho v_{Ey}) + k(a - B)\rho_E A_y \sin(\omega t - kd)$$

$$\left. - \sin\omega t\} - k(b - D)\rho_E d^2 A_x \sin(\omega t - kd) - \sin\omega t\right]$$

$$(28)$$

by keeping only the real terms. Again the limit of integration d is chosen to be of the same order of magnitude as for the case of q_H^A.

VERIFICATION OF OUR HYPOTHESIS

So far we have derived the theories. The next step would be to test them. We shall use computer simulation and digital method and the whole test will be performed by a digital microcomputer. To start with, we have to design the algorithm for this test. To perform the computer simulation, we need to incorporate an Adaptive Sound Controller(ASC) in our scheme.

We will have the following operation schemes: (i) K on, m on, d off and q off, then H2 = H21, h2 becomes h21, ω_k becomes ω_{kl}, h3 becomes h31, u_k becomes u_{kl}, ur_k becomes ur_{kl}, y_k becomes y_{kl}, ϵ_k becomes ϵ_{kl}, x_k

becomes x_{kl}, r_k becomes r_{kl}, v_k becomes v_{kl} and h4

becomes h41 where K = primary noise source, m = monopole secondary noise source, d = dipole secondary
noise source, q = quadrupole secondary noise source,
H2 = transfer function for monopole, h3, h4 = other
transfer functions which will be evident from our
schematic diagram, ω_k = coefficient of the adaptive

sound controller system, u_k = primary signal, ur_k =

residual signal after active noise absorption, y_k =

output of the adaptive sound controller system, x_k =

input of the adaptive sound controller system, r_k =

reference wave signal for the application of holography,
ϵ_k = input signal for the adaptive filter coefficient

update system and v_k = output sugnal from H4; (ii) K

on, m off, d on and q off, then H2 = H22, h2 becomes
h22, ω_k becomes ω_{k2}, h3 becomes h32, u_k becomes

u_{k2}, ur_k becomes ur_{k2}, y_k becomes y_{k2}, x_k becomes x_{k2},

r_k becomes r_{k2}, ϵ_k becomes ϵ_{k2}, v_k becomes v_{k2} and h4

becomes h42 where the symbols have the same meanings
as (i), (iii) K on, m off, d off and q on, then
H2 = H23, h2 becomes h23, ω_k becomes ω_{k3}, h3 becomes

h33, u_k becomes u_{k3}, ur_k becomes ur_{k3}, y_k becomes y_{k3},

x_k becomes x_{k3}, r_k becomes r_{k3}, ϵ_k becomes ϵ_{k3}, v_k

becomes v_{k3} and h4 becomes h43 where the symbols have

the same meanings as (i), (iv) K on, m on, d on and q
on, then H2 = H21 + H22 + H23, h2 becomes h21 + h22 +
h23, ω_k becomes $\omega_{k1} + \omega_{k2} + \omega_{k3}$, h3 becomes h31 + h32

+ h33, u_k becomes $u_{k1} + u_{k2} + u_{k3}$, ur_k becomes ur_{k1} +

$ur_{k2} + ur_{k3}$, y_k becomes $y_{k1} + y_{k2} + y_{k3}$, x_k becomes

$x_{k1} + x_{k2} + x_{k3}$, r_k becomes $r_{k1} + r_{k2} + r_{k3}$, ϵ_k

becomes $\mathcal{E}_{k1} + \mathcal{E}_{k2} + \mathcal{E}_{k3}$, v_k becomes $v_{k1} + v_{k2} + v_{k3}$

and h4 becomes h41 + h42 + h43 where the symbols have the same meanings as (i).

We shall show below the schematic for computer simulation for the operation scheme (iv). For operation schemes (i) to (iii), the diagram will be exactly the same except that the symbols H2, h2, ω_k, h3, u_k,

ur_k, y_k, x_k, r_k, $\mathcal{E}_k$, v_k and h4 will be replaced by

symbols as stated in that particular scheme.

For the derivation of the expressions for the algorithm for the computer simulation, we shall present here only the case when K on, m on, d on and q on. For the other three operation schemes, the procedure is exactly the same except with the symbols H2, h2, ω_k,

h3, u_k, ur_k, y_k, x_k, r_k, $\mathcal{E}_k$, v_k and h4 replaced by

symbols as stated in that particular scheme.

With K on, m on, d on and q on:

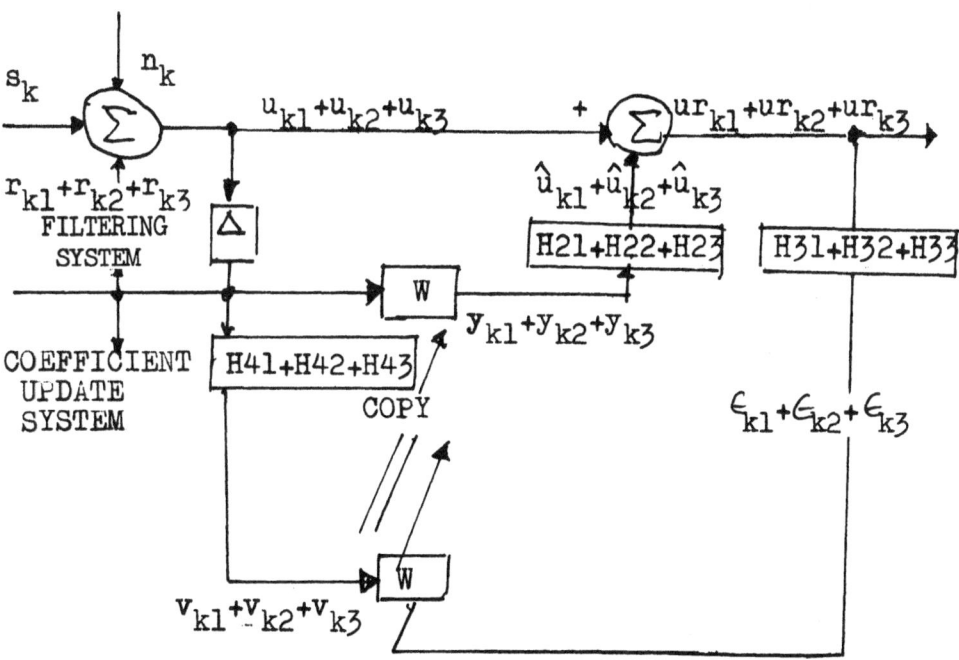

Fig.   1.   Schematic for computer simulation for the
              operation scheme of K on, m on, d and q on.

$$y_{k1}^{H} + y_{k2}^{H} + y_{k3}^{H}$$

$$= \sum_{\ell=0}^{L-1} (\omega_{k1,\ell} x_{k1-\ell}^{H} + \omega_{k2,\ell} x_{k2-\ell}^{H} + \omega_{k3,\ell} x_{k3-\ell}^{H})$$

$$= \sum_{\ell=0}^{L-1} \omega_{k1,\ell} (x_{k1-\ell} + r_{k1-\ell}) + \omega_{k2,\ell} (x_{k2-\ell} + r_{k2-\ell})$$

$$+ \omega_{k3,\ell} (x_{k3-\ell} + r_{k3-\ell}) \tag{29}$$

$$x_{k}^{H} = u_{k-\Delta}^{H} \tag{30}$$

$$\hat{u}_{k1}^{H} + \hat{u}_{k2}^{H} + \hat{u}_{k3}^{H}$$

$$= h21 * y_{k1}^{H} + h22 * y_{k2}^{H} + h23 * y_{k3}^{H} \tag{31}$$

$$(u_{k}^{H} + \hat{u}_{k1}^{H}) + (u_{k}^{H} + \hat{u}_{k2}^{H}) + (u_{k}^{H} + \hat{u}_{k3}^{H})$$

$$= ur_{k1}^{H} + ur_{k2}^{H} + ur_{k3}^{H} \tag{32}$$

$$\varepsilon_{k1} + \varepsilon_{k2} + \varepsilon_{k3}$$

$$= h31 * ur_{k1} + h32 * ur_{k2} + h33 * ur_{k3} \tag{33}$$

Combining (29), (30), (31), (32)

$$ur_{k1}^{H} + ur_{k2}^{H} + ur_{k3}^{H}$$

$$= (u_{k}^{H} + \hat{u}_{k1}^{H}) + (u_{k}^{H} + \hat{u}_{k2}^{H}) + (u_{k}^{H} + \hat{u}_{k3}^{H})$$

$$= \left\{ (u_k + r_{k1}) + h21 * y_{k1}^H \right\} + \left\{ (u_k + r_{k2}) + h22 * \right.$$

$$\left. y_{k2}^H \right\} + \left\{ (u_k + r_{k3}) + h23 * y_{k3}^H \right\}$$

$$= \left\{ (u_k + r_{k1}) + h21 * \omega_{k1} * (x_k + r_{k1}) \right\} + \left\{ (u_k + \right.$$

$$\left. r_{k2}) + h22 * \omega_{k2} * (x_k + r_{k2}) \right\} + \left\{ (u_k + r_{k3}) + h23 \right.$$

$$\left. * \omega_{k3} * (x_k + r_{k3}) \right\}$$

$$= \left\{ (u_k + r_{k1}) + h21 * \omega_{k1} * (u_{k-\Delta} + r_{k1}) \right\} + \left\{ (u_k \right.$$

$$\left. + r_{k2}) + h22 * \omega_{k2} * (u_{k-\Delta} + r_{k2}) \right\} + \left\{ (u_k + r_{k3}) \right.$$

$$\left. + h23 * \omega_{k3} * (u_{k-\Delta} + r_{k3}) \right\} \tag{34}$$

Using the z transform, equation (34) becomes

$$UR_{k1}(z) + UR_{k2}(z) + UR_{k3}(z)$$

$$= \left\{ U1(z) + R1(z) \right\} + H21(z) \left\{ U1(z) + R1(z) \right\} W_{k1}(z)$$

$$z^{-\Delta} \right\} + \left[\left\{ U2(z) + R2(z) \right\} + H22(z) \left\{ U2(z) + R2(z) \right\} \right.$$

$$W_{k2}(z) z^{-\Delta} \right] + \left[\left\{ U3(z) + R3(z) \right\} + H23(z) \left\{ U3(z) + \right. \right.$$

$$R3(z) \right\} W_{k3}(z) z^{-\Delta} \right] \tag{35}$$

using the following useful auxiliary transfer functions:

$$\widetilde{W}_{k1}(z) = \hat{U}_{k1}(z) / \left\{ U1(z) + R1(z) \right\} = H21(z) W_{k1}(z) z^{-\Delta} \tag{36}$$

$$\tilde{H}_{kl}(z) = UR_{kl}(z)/\left\{Ul(z) + Rl(z)\right\} = 1 + \tilde{W}_{kl}(z) \tag{37}$$

etc. and W = nonrecursive (FIR) filter and the superscript H of the symbols stand for holographic.

For the adaptive filter coefficient update system,
the signal we have available is ϵ_k, not ur_k. The
method of steepest descent using the gradient of the
present values of ϵ^2_{kl}, ϵ^2_{k2}, ϵ^2_{k3} give

$$W_{kl+1} + W_{k2+1} + W_{k3+1}$$

$$= W_{kl} - \mu' \nabla(\epsilon^H_{kl})^2 + W_{k2} - \mu' \nabla(\epsilon^H_{k2})^2 + W_{k3} - \mu'$$

$$(\epsilon^H_{k3})^2 \tag{38}$$

Using (30), (33), (34) together with

$$\nabla(\epsilon^H_{kl})^2 + \nabla(\epsilon^H_{k2})^2 + \nabla(\epsilon^H_{k3})^2$$

$$= 2\epsilon^H_{kl}\nabla\epsilon^H_{kl} + 2\epsilon^H_{k2}\nabla\epsilon^H_{k2} + 2\epsilon^H_{k3}\nabla\epsilon^H_{k3}, \text{ we find}$$

$$W_{kl+1} + W_{k2+1} + W_{k3+1}$$

$$= (W_{kl} - 2\mu'\epsilon^H_{kl}V^H_{kl}) + (W_{k2} - 2\mu'\epsilon^H_{k2}V^H_{k2}) + (W_{k3} -$$

$$2\mu'\epsilon^H_{k3}V^H_{k3}) \tag{39a}$$

$$V^H_{kl} + V^H_{k2} + V^H_{k3} = h41 * x^H_k + h42 * x^H_k + h43 * x^H_k \tag{39b}$$

h41 + h42 + h43

= h21 * h31 + h22 * h32 + h23 * h33   (39c)

Equations (39) represent the key theoretical results in this paper.

CONCLUSIONS

We have derived the algorithm for our computer simulation. The next stage would be to carry out the computer digital processing works. We expect the noise suppression could be near 100% but not exactly 100% due to unavoidable instrumental errors such as electronic noise etc.

REFERENCES

1.   G. A. Mangiante, Active sound absorption, J.A.S.A., 61:1516 (1977).
2.   O. L. Angevine, Active acoustic attenuation of electric transformer noise, Inter-noise 81 Proceedings, 303 (1981).
3.   M. J. M. Jessel and G. A. Mangiante, Active sound absorbers in an air-duct, Journal of Sound and Vibration, 23:383 (1972).

PROBABILISTIC IMAGE RECONSTRUCTION USING EXPERIMENTAL ULTRASONIC DATA

K.A. Marsh, R.C. Addison, and J.M. Richardson

Rockwell International Science Center
Thousand Oaks, CA 91360

ABSTRACT

Ultrasonic pulse-echo measurements at a center frequency of 2.25 MHz have been made of a multiwire target in water (consisting of 3 thin parallel wires separated by 0.8 mm), and a prolate spheroidal void in titanium, whose major axes were 800 x 1600 microns. The data consisted of digitized r.f. waveforms corresponding to 4 incident directions, obtained using a transducer array. Calibration of the acoustical system response functions for the two experiments discussed above were based on similar measurements of a thin single wire, and of a spherical void in titanium, respectively.

A probabilistic image reconstruction technique has been applied to these data, using a measurement model based on the Born approximation or a model of the same mathematical structure. The technique is non-parametric, each pixel being treated as an independent random process, with various assumed statistical distributions at each location. Three different a priori distributions were tried, namely: (1) a gaussian distribution (2) a gaussian distribution with a positivity (or negativity) constraint, (3) a distribution involving two possible values (corresponding, globally, to an inclusion of known material but unknown boundary). In addition, a fourth image reconstruction technique was tried, which made use of the spatial deconvolution algorithm known as CLEAN.

In the case of the void, the reconstructed image is degraded by the low signal-to-noise ratio of the measurements, and also shows evidence of the deficiencies in the measurement model resul-

ting from the use of the Born approximation. The results for both
scatterers illustrate the effect of a priori information on the
fidelity of the reconstructed image. The highest image fidelity
was obtained using the CLEAN algorithm, which is probably due to
its inherent sharpness bias. Such a bias is quite appropriate to
the present situation in which the scene consists of a small number of localized objects.

INTRODUCTION

Imaging of the spatial distribution of acoustic impedance
deviations within a medium using an ultrasonic array is usually
accomplished by the phased-array technique in which the region to
be imaged is scanned by a single narrow beam whose direction is
determined by the time delay between the firing of successive
transducers. The image is produced by adding the received waveforms with a similar time delay, video detecting, and displaying
in appropriate coordinates. Such an image is often referred to as
a sector scan, and the technique is under wide usage both in medical imaging and NDE. In order to achieve high spatial resolution, however, the technique requires a large number of transducer
elements, since these elements must be spaced at approximately one
half wavelength in order to avoid grating lobes. An alternative
to this "filled aperture" approach is to use only a small number
of widely spaced transducers, and to fill in the missing spatial
frequencies using techniques based on a priori knowledge of the
statistical nature of the scatterer. One approach to this problem
has been described by Richardson and Marsh (1983), whereby each
pixel of the image is represented as an independent random process, with various assumed statistical distributions of acoustic
impedance values, namely:

(1) gaussian distribution

(2) gaussian distribution with positivity (or negativity)
     constraint

(3) a distribution involving only 2 possible values of
     acoustic impedance (corresponding to an inclusion of
     known material but with an unknown boundary), the
     corresponding algorithm being referred to as the
     inclusion algorithm.

This probabilistic technique uses pulse-echo measurements
together with a measurement model based on the Born approximation.
In the time domain this measurement model takes the form:

$$f(t,\underset{\sim}{e}) = \alpha \sum_{\underset{\sim}{r}} p''\left(t - \frac{2\underset{\sim}{e} \cdot \underset{\sim}{r}}{c}\right)\zeta(\underset{\sim}{r})\ \delta\underset{\sim}{r} + \nu(t,\underset{\sim}{e}) \tag{1}$$

where e is a unit vector in the incident direction, $p(t)$ represents the reference pulse, δr is a volume increment, α is a factor dependent on the material properties, $\zeta(r)$ is the spatial distribution of acoustic impedance, and $\nu(t,e)$ is the noise, assumed to be gaussian.

It should be noted that although this measurement model is based on the Born approximation, the inversion technique is equally applicable to a measurement model of the same mathematical structure even though the underlying scattering process may be different. An example is our triple-wire target to be discussed in the following sections. Provided the axes are aligned in the x-direction, we can express the measurement model as:

$$f(t,\underset{\sim}{e}_n) = \sum_{y,z} g(t - 2|\underset{\sim}{e}_y y + \underset{\sim}{e}_z z - \underset{\sim}{r}_n|/c)\Gamma(y,z) + \nu(t,\underset{\sim}{e}_n) \qquad (2)$$

where y,z are points on a suitably defined grid in the yz-plane, r_n is the position of the nth transducer whose incident direction is $\underset{\sim}{e}_n$, and g is the measured system response function for a single wire. $\Gamma(y,z)$ is a function having values 0 or 1 at each point on the y,z grid depending on whether a wire is present. Equation (2) represents an accurate scattering model, the only approximations being the neglect of multiple reflections between wires, and the assumption that the fractional variation of the range $|\underset{\sim}{e}_y y + \underset{\sim}{e}_z z - \underset{\sim}{r}_n|$ is small.

In the frequency domain Eq. (1) takes the form:

$$f(\omega,\underset{\sim}{e}) = \alpha\omega^2 p(\omega) \sum_{\underset{\sim}{r}} \zeta(\underset{\sim}{r}) \, e^{i \frac{2\omega}{c} \underset{\sim}{e} \cdot \underset{\sim}{r}} \, \delta\underset{\sim}{r} + \nu(\omega,\underset{\sim}{e})$$

It is apparent from this expression that $f(\omega,e)/[\alpha\omega^2 p(\omega)]$ corresponds to the spatial frequency function $\bar{\zeta}(q)$ at spatial frequency $\underset{\sim}{q} = (2\omega/c)\underset{\sim}{e}$, which is related to $\zeta(\underset{\sim}{r})$ by:

$$\bar{\zeta}(\underset{\sim}{q}) = \sum_{\underset{\sim}{r}} \zeta(\underset{\sim}{r}) \, e^{i\underset{\sim}{q} \cdot \underset{\sim}{r}} \, \delta\underset{\sim}{r}$$

The image $\zeta(\underset{\sim}{r})$ can thus be obtained from $\zeta(q)$ by simple Fourier inversion, provided $\zeta(q)$ is known at a sufficient number of points. If we wish to sample the image on a rectangular grid of dimensions L_x, L_y, L_z at intervals of Δx, Δy, Δz, then we must know $\zeta(q)$ on a corresponding grid in q space, of dimensions $2\pi/\Delta x$, $2\pi/\Delta y$, $2\pi/\Delta z$ and intervals of $2\pi/L_x$, $2\pi/L_y$, $2\pi/L_z$. The image reconstruction technique can be thought of as performing an interpolation (and extrapolation) of the measured spatial frequencies onto the unmeasured portion of this grid. In the case of pure gaussian prior statistics, the net result of this procedure

is not too different from the conventional imaging approach in
which grid cells lying near measured spatial frequencies are set
at the mean value within some averaging window, and unsampled
cells are set at zero. If the averaging window used is a simple
box, the interpolation technique is often referred to as box convolution. One noteworthy property of the box convolution technique (which is also shared by the probabilistic technique based
on gaussian statistics) is that the resulting image $D(\underset{\sim}{r})$ is
related to the true image $\zeta(\underset{\sim}{r})$ via a linear operator, representing
convolution with the theoretical response to a point scatterer,
$B(\underset{\sim}{r})$. $D(\underset{\sim}{r})$ and $B(\underset{\sim}{r})$ are often referred to as dirty image and synthesized beam, respectively. We can then write:

$$D(\underset{\sim}{r}) = \sum_{\underset{\sim}{r}'} B(\underset{\sim}{r} - \underset{\sim}{r}') \, \zeta(\underset{\sim}{r}') + \nu(\underset{\sim}{r}) \tag{3}$$

where $\nu(\underset{\sim}{r})$ represents the noise resulting from measurement error.

One technique inverting this equation is the CLEAN algorithm,
developed for radio astronomy (Hogbom 1974). For the purpose of
this technique we consider the true image as a series of delta
functions, which when convolved with the synthesized beam and
linearly superposed, constitute the dirty image. The CLEAN algorithm attempts to determine the amplitude and location of each of
these delta functions. The steps involved in the ith iteration of
the algorithm are as follows:

(1) Locate the peak in $D(\underset{\sim}{r})$. Let its location be $\underset{\sim}{r}_i$ and its
amplitude be A_i.

(2) Scale $B(\underset{\sim}{r})$ by $c_i = \varepsilon A_i$ where ε is a parameter between 0
and 1 called the "loop gain".

(3) Subtract $c_i \cdot B(\underset{\sim}{r} - \underset{\sim}{r}_i)$ from $D(\underset{\sim}{r})$, and store c_i, $\underset{\sim}{r}_i$.

(4) Repeat the process from step (1) until the peak of the
remaining image is below some preset threshold. Denote
this residual image $R(\underset{\sim}{r})$. We can then express the
original dirty image $D(\underset{\sim}{r})$ as:

$$D(\underset{\sim}{r}) = \sum_i c_i \cdot B(\underset{\sim}{r} - \underset{\sim}{r}_i) + R(\underset{\sim}{r}) \quad .$$

If we define:

$$I(\underset{\sim}{r}) = \sum_i c_i \delta(\underset{\sim}{r} - \underset{\sim}{r}_i)$$

then $D(\underset{\sim}{r})$ represents the convolution of $I(\underset{\sim}{r})$ with $B(\underset{\sim}{r})$,
plus a noise term $R(\underset{\sim}{r})$. $I(\underset{\sim}{r})$ thus represents a possible
solution for the true image $\zeta(\underset{\sim}{r})$ on the basis of Eq. (3).

(5) Form the final "clean" image by convolving I(r) with a
simple function such as a gaussian whose width is chosen
to correspond to the desired spatial resolution, and add
this smoothed image to the residual image, R(r).

The statistical basis of CLEAN has been investigated for the
1-dimensional case by Schwarz (1978), and found to select the
sharpest image consistent with the data in a least-squares sense.
It might thus be expected to give a faithful reconstruction of the
scatterer in the case where most of the structure is in the form
of clumps. Since this is typically the case in NDE, we decided to
include CLEAN in with the set of image reconstruction techniques
to be used with our ultrasonic array data.

MEASUREMENTS

An ultrasonic array was used to obtain data sets consisting
of pulse-echo waveforms in 4 incident directions, from various
samples in water. The instrument has been described by Addison et
al. (1982). It consists of 2 subarrays of 32 transducers, each of
center frequency 2.25 MHz with a frequency range of approximately
3:1. The spacing between the two subarrays may be varied. The
data consist of digitized r.f. waveforms in the time domain. Individual pulse-echo waveforms were obtained by summing groups of 4
elements, chosen to give a favorable coverage of incident angles.

Our objective was to investigate the possibility of using the
array, together with the image reconstruction techniques discussed
above, to make high resolution images of two different scatterers,
namely:

(1) a target consisting of 3 thin parallel wires in water,
the wire separation being 0.8 mm.

(2) a prolate spheroidal void (major axes 800 x 1600 microns)
in a block of titanium.

The amplitude and phase calibration for each of these two
experiments was to be accomplished by making similar measurements
of (respectively):

(1) a single wire in water

(2) a spherical void of diameter 1200 microns in a similar
block of titanium.

The measurement configurations for the two experiments are
shown in Fig. 1. In order to use the measured waveforms for image
reconstruction we need to know the transducer response p(t) for
each transducer, together with the incident angles, referred to
some spatial origin.

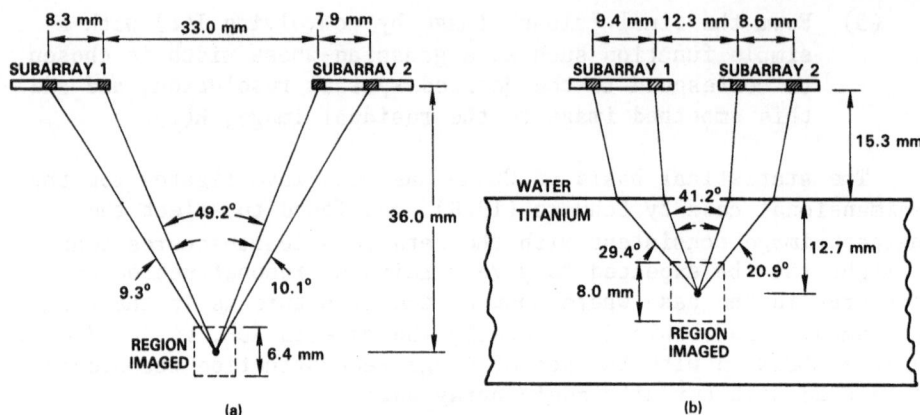

Fig. 1. The measurement configurations for: (a) the triple-wire
target in water, and (b) the prolate spheroidal void in
titanium.

In the case of the wire images, the incident angles were calculated from a knowledge of the transducer locations, deduced from
simple geometric considerations based on observed travel times.
The single-wire waveforms were used to deduce p(t) in each case.
By using the set of p(t) obtained in this way, the single-wire
measurements not only serve as an amplitude calibration for our
image of the triple-wire target, but also serve as a phase calibration, with the origin of phase corresponding to the location of
the single wire.

In the case of the voids in titanium, a different set of
spacings between element groups was used, their exact values being
determined in the same way as above. In order to calculate the
incident angles, however, it was of course necessary to take refraction into account. The set of transducer response functions
p(t) was obtained from measurements of the 1200 micron spherical
void, taking into account the theoretical impulse response of such
a scatterer. The centroid of the spherical void then defined the
spatial origin for the image of the prolate void.

One problem which arose in the case of measurements made in
titanium was that the flaw signal was swamped by the strong signal
due to the ring-down caused by of the front face echo from the
titanium block. To suppress this ringing, an additional set of
waveforms was taken, with the array displaced from the flaw so
that the only signal present was the front-face echo. These
waveforms were then subtracted from those containing the void
signal.

RESULTS

Figure 2 shows the images of the triple-wire target obtained with the 4 image reconstruction techniques discussed earlier. The field of view in each case is 6.4 mm × 6.4 mm, and the orientation is the same as for Fig. 1. It can be seen that the CLEAN algorithm (Fig. 2d) gave a good reconstruction, in which the 3 wires can be clearly distinguished, while the other images contain various spurious effects. These spurious effects are most serious for the pure gaussian case (Fig. 2a), which is of course not surprising in view of the lack of measurement data together with the weakness of the statistical model. Similar spurious effects were obtained in an image of a single-wire target using gaussian statistics by Addison et al. (1982). The situation is somewhat improved by the application of the positivity constraint (Fig. 2b), although the wires still appear as streaks rather than points. In the case of the inclusion algorithm (Fig. 2c), only the central wire has been located. The superior performance of CLEAN is

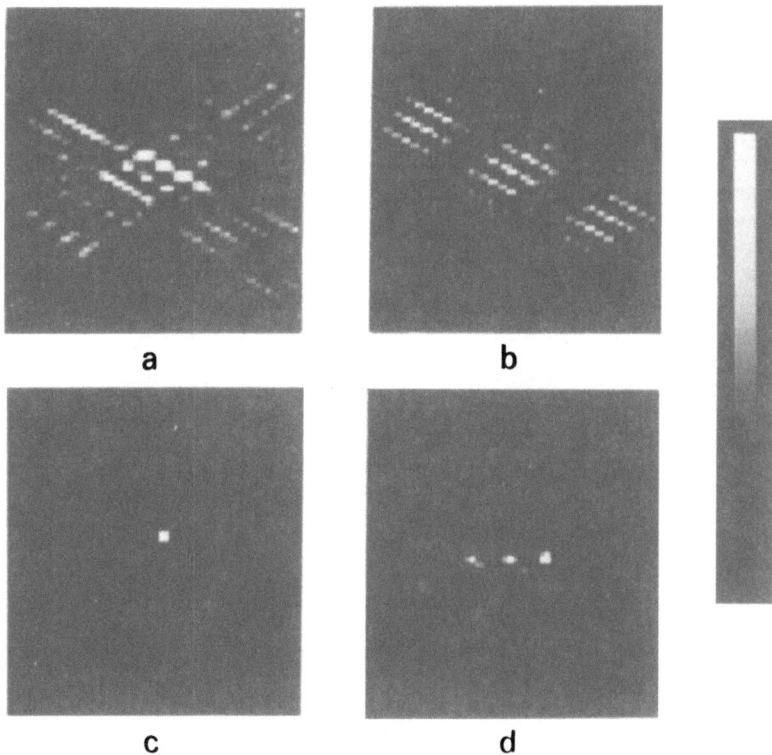

Fig. 2.   Reconstructed images of the triple-wire target using
(a) gaussian prior statistics, (b) gaussian prior
statistics with positivity constraint, (c) inclusion
algorithm, and (d) CLEAN algorithm.

probably due to the appropriateness of its inherent sharpness
bias, as applied to a scene with a strong degree of spatial
clumping.

We note that the separation of adjacent wires as inferred
from Fig. 2(d) is 0.8 ± 0.05 mm, which compares favorably with the
true value of 0.79 mm.

In the case of the prolate spheroidal void in titanium, one
problem which arises is that the incident directions were all in
the same plane. This did not matter for the wires since the scatterers could be considered 2-dimensional. Since this was not the
case for the voids, a full reconstruction was not possible, and
for this reason the inclusion algorithm could not be used. The
other 3 algorithms were, however, applicable provided the images
were interpreted as 2-dimensional projections of the scatterer.
The results are shown in Fig. 3, with a field of view of 8 mm ×
8 mm. Again we see that CLEAN (Fig. 3c) gave superior performance, as might be expected from the highly compact nature of the

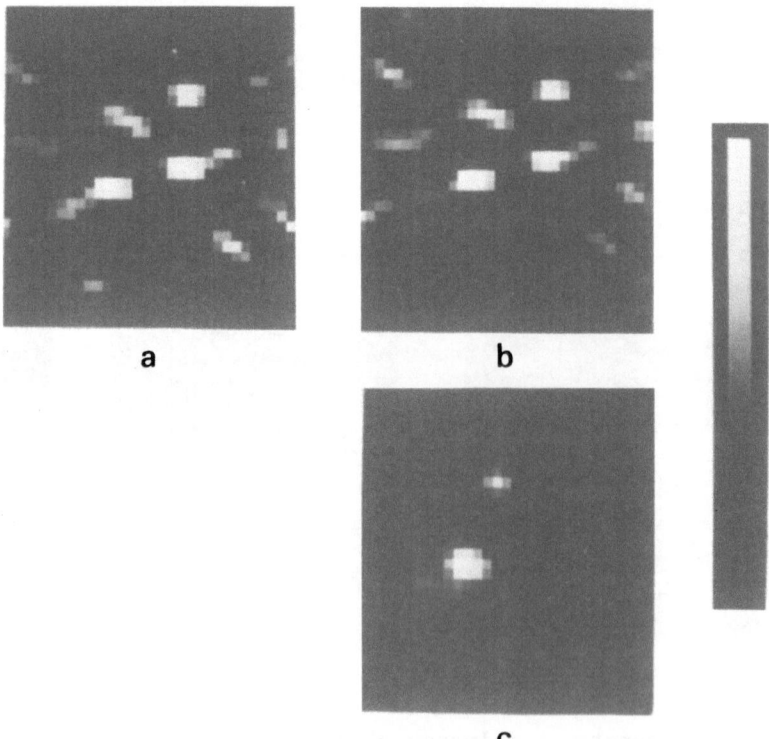

Fig. 3. Reconstructed images of the prolate spheroidal void
         in titanium using (a) gaussian prior statistics,
         (b) gaussian prior statistics with positivity constraint
         (c) CLEAN algorithm.

scatterer. Unfortunately, no indication of the prolateness of the
scatterer is apparent. This may be due to the lack of a back face
echo in the impulse response of a void, coupled with an insuffici-
ent diversity of incident directions necessary to make adequate
use of the front face echo. This of course reflects the inadequacy of the Born approximation in the case of a strong scatterer.
Other spurious effects in Fig. 3(c) are probably due to the low
signal-to-noise ratio of the waveforms, resulting from the inadequacy of the front-face ringing subtraction. The resulting image
gives a reasonable estimate of the size and location of the void,
but not its shape.

CONCLUSIONS

    We conclude that probabilistic image reconstruction tech-
niques can produce high resolution images with very few measurements provided that sufficiently strong a priori information is
supplied to the algorithm. It appears that with sparse data and
high measurement noise, treating each pixel as an independent
random process does not result in sufficiently strong constraints
being imposed on the image. The image quality can, however, be
considerably improved using stronger criteria which allow for the
presence of spatial correlations between pixels, an example being
the sharpness bias imposed by the CLEAN algorithm. This algorithm
appears to be quite appropriate for images which are known
a priori to contain a large degree of clumping, and it did in fact
perform quite satisfactorily in the two experiments described
here.

ACKNOWLEDGEMENT

    This work was supported by Rockwell International Independent
Research and Development funding.

REFERENCES

Addison, R.C., Marsh, K.A., Richardson, J.M., and Ruokangas, C.C.
    1982, "NDE Imaging with Multielement Arrays", in Acoustic
    Imaging, 12, E.R. Ash and C.R. Hill ed., (Plenum, New York).
Hogbom, J.A. 1974, Astron. Astrophys. Suppl 15, 417.
Richardson, J.M. and Marsh, K.A. 1983, "Probabilistic Approach to
    the Inverse Problem in the Scattering of Elastic and
    Electromagnetic Waves", in Proceedings of S.P.I.E. Vol. 413
    "Inverse Optics," p. 79.
Schwarz, U.J. 1978, Astron. Astrophys 65, 345.

ERRATUM:"GENERALIZATIONS OF GABOR'S THEORY-------

THE THEORY OF MULTI-BEAM HOLOGRAPHIC INTERFERENCE"

ACOUSTICAL HOLOGRAPHY, 3:363(1971)

Woon S. Gan

Acoustical Services
29 Telok Ayer Street
Singapore 0104

An error was found in that paper, that is, all the words electromagnetic and optical should be replaced by acoustical. Also clarifications need to be made on: (i) the type of holographic interference discussed in that paper was acoustical holographic interference, not optical holographic interference. As can be seen from the starting equation we used was Helmholtz equation which is an important equation in acoustics. We subsequently made use of the electrostatic potential and extended it to the acoustical case, not electromagnetic (em) case, because for em case, we need to incorporate the E- vector and H- vector which described the vector nature of em wave. We next made use of Schwinger's paper for the diffraction of wave by screen which again was meant for acoustical wave and not em wave which we have to clarify, (ii) all the plane wave in that paper meant sound wave, (iii) we have to point out that paper worked out theoretically the effect of diffraction on image formation in acoustical holography, (iv) the frequency range of the acoustical holographic interference used in that paper was audio frequency range, applicable to the analysis of the membrane of microphones, and (v) that paper presents a rigorous theory of acoustical holographic interference.

PARTICIPANTS

R. C. Addison
Rockwell International
1049 Camino Dos Rios
PO Box 1085
Thousand Oaks, CA 91360
USA

P. Alais
Paris University
2 Pl. de la Gare de Ceinture
Saint Cyr 78210
France

C. Aloysius
Mayo Clinic
200 1st Street S.W.
Rochester, MN 55905
USA

E. K. Balcer-Kubiczek
University of Maryland
600 W. Redwood
Baltimore, MD 21201
USA

M. Berggren
University of Utah
Dept. of Bioengineering
Merrill Engr. Bldg.
Salt Lake City, UT 84103
USA

G. Beylkin
Schlumberger-Doll Research
PO Box 307
Ridgefield, CT 06877
USA

F. Breimesser
Siemens Ag.
Paul Gossen Str. 100
D-8520 Erlangen
West Germany

Y. Brun
Etudes et Productions
BP202 - 26 Rue de la Cavee
92142 Clamart Cedex
France

R. C. Bryant
University of Adelaide
North Terrace
Adelaide S. Aust. 5000
Australia

C. Burckhardt
Diagnosic Ultrasound
Hoffman-LaRoche
Grenzacherstrasse
Basel, Switzerland

D. Bylski
University of Michigan
1405 E. Ann Street
Ann Arbor, MI 48109
USA

D. Carpenter
Ultrasonics Institute
44 Bellambi St.
Northbridge N.S.W.
Australia 2000

M. Chubachi
Tohoku University
Aramaki
Sendai, Japan

K. Dalland
Norwegian Underwater Tech. Ctr.
Gravdalsveien 255
N-5034 Ytre Laksevag
Norway

S. Datta
Mayo Clinic
200 1st Street S.W.
Rochester, MN 55905
USA

D. D. Day
5613 Trego
Tuecolony, TX   75056
USA

J. Dybedal
Bentech R/S
%Elab - N-7034
Trondheim-nth
Norway

P. M. Embree
Universit of Illinois
1406 W. Green Street
Urbana, IL 61801
USA

E. Furgason
Purdue University
Electrical Engineering Dept.
West Lafayette, IN 47906
USA

W. S. Gan
Acoustical Services
29 Telok Ayer Street
Singapore 0104

M. Gebel
Med. Hochschule
Konstanty Gutschow Str. 7
3 Hannover
West Germany

R. A. Giblin
Stanford University
620 St. Claire Dr.
Palo Alto, CA 94306
USA

J. F. Greenleaf
Mayo CLinic
200 1st Street S.W.
Rochester, MN 55905
USA

D. J. Guyomar
Naval Postgraduate School
200 Park Ave #5
Monterey, CA 93340
USA

M. J. Haney
University of Illinois
1406 W. Green St.
Urbana, IL 61801
USA

L. E. Hargrove
Office of Naval Research
P.D.C. 412
800 N. Quincy St.
Arlington, VA 22217
USA

G. H. Harrison
University of Maryland
660 W. Redwood St.
Baltimore, MD 21201
USA

B. Ho
Michigan State University
East Lansing, MI 48824
USA

S. A. Johnson
University of Utah
Dept. of Bioengineering
Merrill Engr. Bldg.
Salt Lake City, UT 84103
USA

T. K. Johnson
University of Minnesota
Dept. of Radiology
Box 292
Minneapolis, MN 55455
USA

H. Jones
Tech. Univ. of Nova Scotia
PO Box 1000
Halifax, Nova Scotia
Canada

J. P. Jones
University of California
Dept. of Radiological Sciences
Irvine, CA 92717
USA

H. T. Kaarmann
Siemens Corp.
Henkes M127
8520 Erlangen
West Germany

M.-L. Kao
Racal-Milgo
8600 N.W. 41st Street
Miami, FL 33166
USA

C. Kasai
Aloka Co. Ltd.
6-22-1 Mure
Mitaka, Tokyo 181
Japan

H. Kaur
3M
270-4N-09 3M Center
St. Paul, MN 55144
USA

M. Kaveh
University of Minnesota
123 Church Street S.E.
Minneapolis, MN 55455
USA

J. Kushibiki
Tohoko University
Sendai 980
Japan

J. D. Larson
Hewlett-Packard
1501 Page Mill Road
Bldg. 28C
Palo Alto, CA 94304
USA

B. B. Lee
Purdue University
902 North 9th Street
Lafayette, IN 47904
USA

S. Leeman
University of California
Dept. of Radiological Sc.
Irvine, CA 92717
USA

S. Lees
Bioengineering Dept.
Forsyth Dental Center
140 Fenway
Boston, MA 02115
USA

R. Lerch
Siemens Research Center
Paul Gossen Strasse 100
D-8520 Erlangen
West Germany

Z.-C. Lin
University of California
Dept. of ECE
Santa Barbara, CA 93106
USA

E. Machida
Fujitsu Laboratories Ltd.
1015 Kamikodanaka
Nakshara-Ku
Kawasaki 211
Japan

K. Marsh
Rockwell International
1049 Camino dos Rios
Thousand Oaks, CA 91360
USA

B. Mayo
Philips Ultrasound
2722 S. Fairview St.
Santa Ana, CA 92704
USA

J. F. McDonald
Rensselaer Polytechnic Inst.
Troy, NY 12181
USA

P. D. Mesdag
Delft University of Technology
V. Leeuwenhueksimgel 32
Delft 2611AB
The Netherlands

R. K. Mueller
University of Minnesota
123 Church Street S.E.
Minneapolis, MN 55455
USA

R. Nasoni
University of Arizona
Physics Dept.
Tucson, AZ 85721
USA

J. Nicholson
Mayo Clinic
200 1st Street S.W.
Rochester, MN 55905
USA

O. Tretiak
Drexel University
2nd and Chestnut
Philadelphia, PA 15104
USA

M. Oravecz
Sonoscan Inc.
530 E. Green St.
Bensenville, IL 60106

M. Pappalardo
Ist. Acustica CNR
Via Cassia 1215 Rome
Italy

S. B. Park
Korean Adv. Inst.
of Sci. & Tech.
PO Box 150
Chongyang, Seoul
Korea

M. S. Patterson
Ontario Cancer Institute
500 Sherbourne St.
Toronto, M4X 1K9
Canada

P. C. Pedersen
Drexel University
Biomedical Eng. & Sci. Inst.
Philadelphia, PA 19104
USA

G. M. Pollari
University of Iowa
2312 Princeton Road
Iowa city, IA 52240
USA

J. P. Powers
Naval Postgraduate School
E. E. Dept.
Monterey, CA 39343
USA

J. M. Reid
Drexel University
Philadelphia, PA 19104
USA

M. Restori
Moorfields Eye Hospital
City Road
London, GC1 2PD
England

J. Ridder
TPD TNO-TH
PO Box 5046
Delft
The Netherlands

B. Robinson
Mayo Clinic
200 1st Street S.W.
Rochester, MN 55905
USA

R. L.Rylander
3M Company
3M Center
Bldg. 201-3E-03
St. Paul, MN 55144
USA

M. Salahi
Kings College - London
Dept. of Elect. Eng.
Strand, London
England WC2 R2LS

C. Schueler
Santa Barbara Research Ctr.
75 Coromar Drive
Goleta, CA 93117
USA

H. Schwetlick
Technische Universitat
Einsteinufer 19
Elec. Sekr. E3
1000 Berlin 10
West Germany

R. Y. Liu
ADR Ultrasound
734 W. Alameda
Tempe, AZ 85282
USA

D. A. Seggie
Queen Elizabeth College
Campden Hill Road
London, England

R. Seznec
Metravib
64 Chem. des Mouilles
Ecully 69130
France

R. E. Soldner
Siemens Corp.
Henkes M127
8520 Erlangen
West Germany

M. Somekh
Oxford University
Dept. of Metallurgy
Parks Road
Oxford, England

B. Soumekh
University of Minnesota
EE Dept.
123 Church Street
Minneapolis, MN 55455
USA

M. Soumekh
Worcester Polytechnic Inst.
Electrical Engineering Dept.
Worcester, MA 01609
USA

K. Tervola
Bioacoustics Dept.
1406 West Green
Urbana, IL 61801
USA

P. Thomas
Mayo Clinic
200 1st Street S.W.
Rochester, MN 55905
USA

M. Ueda
Tokyo Inst. of Tech.
4259 Nagatsuta Midori-ku
Yokohama 227
Japan

L. F. van der Wal
T.N.O.
Sjielsjesweg 1
Delft
The Netherlands

J. A. van Woerden
Inst. of Applied Physics
Stieltjesweg 1
Delft
The Netherlands

R. Vogel
University of Iowa
4408 E. B.
Iowa City, IA 52242
USA

R. C. Waag
University of Rochester
PO Box 648
Rochester, NY 14642
USA

G. Wade
University of California
E.C.E. Dept.
Santa Barbara, CA 93106
USA

N. J. Wattrus
University of Minnesota
Geology Dept.
106 Pillsbury Hall
Pillsbury Drive S.E.
Minneapolis, MN 55455
USA

Ake V. Wernersson
National Def. Res. Ins.
PO Box 1165
S-581 11 Linkoping
Sweden

K. D. Wyatt
Phillips Petroleum Co.
265-GB PRC
Bartlesville, OK 74004
USA

T. Yamamoto
Hokkaido University
N-13 S-8 Kitaku
Sapporo 060
Japan

J. Ylitalo
Acadamy of Finland
Linnanmaa 90570
Oulu 57
Finland

AUTHOR INDEX

SUBJECT INDEX